D1527155

*Advances in*

# ECOLOGICAL RESEARCH

VOLUME 44

*Advances in Ecological Research*

Series Editor: GUY WOODWARD
*School of Biological and Chemical Sciences*
*Queen Mary University of London*
*London, UK*

*Advances in*

# ECOLOGICAL RESEARCH

VOLUME 44

*Edited by*

GUY WOODWARD

School of Biological and Chemical Sciences,
Queen Mary University of London,
London E1 4NS, UK

AMSTERDAM • BOSTON • HEIDELBERG • LONDON
NEW YORK • OXFORD • PARIS • SAN DIEGO
SAN FRANCISCO • SINGAPORE • SYDNEY • TOKYO
Academic Press is an imprint of Elsevier

Academic Press is an imprint of Elsevier

32 Jamestown Road, London NW1 7BY, UK
Linacre House, Jordan Hill, Oxford OX2 8DP, UK
Radarweg 29, PO Box 211, 1000 AE Amsterdam, The Netherlands
225 Wyman Street, Waltham, MA 02451, USA
525 B Street, Suite 1900, San Diego, CA 92101-4495, USA

First edition 2011



ISBN: 978-0-12-374794-5
ISSN: 0065-2504

For information on all Academic Press publications
visit our website at elsevierdirect.com

Printed and bound in UK

11 12 13 14 10 9 8 7 6 5 4 3 2 1

# Contents

**Long-Term Dynamics of a Well-Characterised Food Web: Four Decades of Acidification and Recovery in the Broadstone Stream Model System**

KATRIN LAYER, ALAN G. HILDREW, GARETH B. JENKINS, JENS O. RIEDE, STEPHEN J. ROSSITER, COLIN R. TOWNSEND AND GUY WOODWARD

**From Natural to Degraded Rivers and Back Again: A Test of Restoration Ecology Theory and Practice**

CHRISTIAN K. FELD, SEBASTIAN BIRK, DAVID C. BRADLEY, DANIEL HERING, JOCHEM KAIL, ANAHITA MARZIN, ANDREAS MELCHER, DIRK NEMITZ, MORTEN L. PEDERSEN, FLORIAN PLETTERBAUER, DIDIER PONT, PIET F.M. VERDONSCHOT AND NIKOLAI FRIBERG

**Stream Ecosystem Functioning in an Agricultural Landscape: The Importance of Terrestrial–Aquatic Linkages**

SALLY HLADYZ, KAJSA ÅBJÖRNSSON, ERIC CHAUVET, MICHAEL DOBSON, ARTURO ELOSEGI, VERÓNICA FERREIRA, TADEUSZ FLEITUCH, MARK O. GESSNER, PAUL S. GILLER, VLADISLAV GULIS, STEPHEN A. HUTTON, JEAN O. LACOURSIÈRE, SYLVAIN LAMOTHE, ANTOINE LECERF, BJÖRN MALMQVIST, BRENDAN G. MCKIE, MARIUS NISTORESCU, ELENA PREDA, MIIRA P. RIIPINEN, GETA RÎŞNOVEANU, MARKUS SCHINDLER, SCOTT D. TIEGS, LENA B.-M. VOUGHT AND GUY WOODWARD

**A Belowground Perspective on Dutch Agroecosystems: How Soil Organisms Interact to Support Ecosystem Services**

CHRISTIAN MULDER, ALICE BOIT, MICHAEL BONKOWSKI, PETER C. DE RUITER, GIORGIO MANCINELLI, MARCEL G.A. VAN DER HEIJDEN, HARM J. VAN WIJNEN, J. ARIE VONK AND MICHIEL RUTGERS

# Contributors to Volume 44

**KAJSA ÅBJÖRNSSON,** *University College Cork, Department of Zoology, Ecology and Plant Science, Lee Maltings Building, Prospect Row, Cork, Ireland, and Limnology, Department of Ecology, Lund University, Sweden.*

**SEBASTIAN BIRK,** *Faculty of Biology, Department of Applied Zoology/ Hydrobiology, University of Duisburg-Essen, Universitaetsstrasse 2, 45141 Essen, Germany.*

**ALICE BOIT,** *Institute of Biochemistry and Biology, University of Potsdam, Maulbeerallee 2, 14415 Potsdam, Germany.*

**NÚRIA BONADA,** *Freshwater Ecology and Management groups (FEM), Departament d'Ecologia, Universitat de Barcelona, Avda Diagonal 645, 08028 Barcelona, Catalonia/Spain.*

**MICHAEL BONKOWSKI,** *Institute of Zoology, University of Köln, Biozentrum Köln, Zülpicher Straße 47b, 50674 Köln, Germany.*

**DAVID C. BRADLEY,** *APEM Ltd., Riverview, A17 Embankment Business Park, Heaton Mersey, Stockport, SK4 3GN, United Kingdom.*

**ERIC CHAUVET,** *Université de Toulouse; UPS, INP; EcoLab; 118 route de Narbonne, F-31062 Toulouse, France and CNRS; EcoLab; F-31062 Toulouse, France. Present address: Université Paul Sabatier, EcoLab - Laboratoire Ecologie Fonctionnelle et Environnement, Bât. 4R1, 118 route de Narbonne 31 062 Toulouse Cedex 9 France.*

**PETER C. DE RUITER,** *Biometris, Wageningen University, Droevendaalsesteeg 1, 6708PB Wageningen, The Netherlands.*

**MICHAEL DOBSON,** *Manchester Metropolitan University, Environmental and Geographical Sciences, Chester Street, Manchester, M15 6BH, United Kingdom. Present address: Freshwater Biological Association the Ferry Landing, Far Sawrey, Ambleside, Cumbria, LA22 0LP, United Kingdom.*

**MICHAEL J. DUNBAR,** *Centre for Ecology and Hydrology, Maclean Building, Benson Lane, Crowmarsh Gifford, Wallingford, OX10 8BB, United Kingdom.*

**FRANCOIS K. EDWARDS,** *Centre for Ecology and Hydrology, Maclean Building, Benson Lane, Crowmarsh Gifford, Wallingford, OX10 8BB, United Kingdom.*

**ARTURO ELOSEGI,** *Department of Plant Biology and Ecology, Faculty of Science and Technology, University of the Basque Country, Apdo. 644, E-48080 Bilbao, Spain.*

**CHRISTIAN K. FELD,** *Faculty of Biology, Department of Applied Zoology/Hydrobiology, University of Duisburg-Essen, Universitaetsstrasse 2, 45141 Essen, Germany.*

**VERÓNICA FERREIRA,** *Department of Life Sciences, University of Coimbra, PO Box 3046, 3001–401 Coimbra, Portugal.*

**TADEUSZ FLEITUCH,** *Institute of Nature Conservation, Polish Academy of Sciences, Mickiewicza 33 31–120 Kraków, Poland.*

**NIKOLAI FRIBERG,** *Aarhus University, National Environmental Research Institute, Department of Freshwater Ecology, Vejlsøvej 25, DK-8600 Silkeborg, Denmark.*

**MARK O. GESSNER,** *Department of Aquatic Ecology, Eawag: Swiss Federal Institute of Aquatic Science and Technology, 6047 Kastanienbaum, Switzerland, and Institute of Integrative Biology (IBZ), ETH Zurich, Switzerland. Present address: Leibniz Institute of Freshwater Ecology and Inland Fisheries (IGB), Alte Fischerhütte 2, 16775 Stechlin, Germany; Department of Ecology, Berlin Institute of Technology (TU Berlin), Ernst-Reuter-Platz 1, 10587 Berlin, Germany.*

**PAUL S. GILLER,** *University College Cork, Department of Zoology, Ecology and Plant Science, Lee Maltings Building, Prospect Row, Cork, Ireland.*

**JONATHAN GREY,** *School of Biological & Chemical Sciences, Queen Mary University of London, London, E1 4NS, United Kingdom.*

**VLADISLAV GULIS,** *Department of Life Sciences, University of Coimbra, PO Box 3046, 3001–401 Coimbra, Portugal. Present address: Department of Biology, Coastal Carolina University, P.O. Box 261954 Conway, SC 29528–6054, USA.*

**RICHARD B. HAYES,** *School of Biological & Chemical Sciences, Queen Mary University of London, London, E1 4NS, United Kingdom.*

**DANIEL HERING,** *Faculty of Biology, Department of Applied Zoology/Hydrobiology, University of Duisburg-Essen, Universitaetsstrasse 2, 45141 Essen, Germany.*

**ALAN G. HILDREW,** *School of Biological & Chemical Sciences, Queen Mary University of London, London, E1 4NS, United Kingdom.*

**SALLY HLADYZ,** *University College Cork, Department of Zoology, Ecology and Plant Science, Lee Maltings Building, Prospect Row, Cork, Ireland. Present address: Monash University, School of Biological Sciences, Clayton, Melbourne, Victoria, 3800, Australia.*

**STEPHEN A. HUTTON,** *University College Cork, Department of Zoology, Ecology and Plant Science, Lee Maltings Building, Prospect Row, Cork, Ireland.*

**GARETH B. JENKINS,** *School of Biological and Chemical Sciences, Queen Mary University of London, E1 4NS, London, United Kingdom.*

**JOCHEM KAIL,** *Department of Biology and Ecology of Fishes, Leibniz-Institute of Freshwater Ecology and Inland Fisheries, 12587 Berlin, Germany.*

**JEAN O. LACOURSIÈRE,** *Kristianstad University, SE 29188 Kristianstad, Sweden.*

**SYLVAIN LAMOTHE,** *Université de Toulouse; UPS, INP; EcoLab; 118 route de Narbonne, F-31062 Toulouse, France and CNRS; EcoLab; F-31062 Toulouse, France. Present address: Université Paul Sabatier, EcoLab - Laboratoire Ecologie Fonctionnelle et Environnement, Bât. 4R1, 118 route de Narbonne 31 062 Toulouse Cedex 9 France.*

**NICOLAS LAMOUROUX,** *Cemagref, UR BELY, F-69336 Lyon, France.*

**KATRIN LAYER,** *School of Biological and Chemical Sciences, Queen Mary University of London, E1 4NS, London, United Kingdom.*

**ANTOINE LECERF,** *Université de Toulouse; UPS, INP; EcoLab; 118 route de Narbonne, F-31062 Toulouse, France and CNRS; EcoLab; F-31062 Toulouse, France. Present address: Université Paul Sabatier, EcoLab - Laboratoire Ecologie Fonctionnelle et Environnement, Bât. 4R1, 118 route de Narbonne 31 062 Toulouse Cedex 9 France.*

**BJÖRN MALMQVIST,** *Animal Ecology, Department of Ecology and Environmental Science, Umea University, SE 90187 Umeå, Sweden.*

**GIORGIO MANCINELLI,** *University of Salento, ECOTEKNE, Monteroni, 73100 Lecce, Italy.*

**ANAHITA MARZIN,** *Cemagref, Unit HBAN-HEF, Hydrosystems and Bioprocesses Research unit (HBAN), Parc de Tourvoie, 92160 Antony, France.*

**BRENDAN G. McKIE,** *Animal Ecology, Department of Ecology and Environmental Science, Umea University, SE 90187 Umeå, Sweden. Present address: Department of Aquatic Sciences and Assessment, Swedish University of Agricultural Sciences, 75007 Uppsala, Sweden.*

**ANDREAS MELCHER,** *Department of Water, Atmosphere and Environment, University of Natural Resources and Life Sciences (BOKU), A-1180 Vienna, Austria.*

**CHRISTIAN MULDER,** *National Institute for Public Health and Environment RIVM, P.O. Box 1, 3720BA Bilthoven, The Netherlands.*

**DIRK NEMITZ,** *Faculty of Biology, Department of Applied Zoology/Hydrobiology, University of Duisburg-Essen, Universitaetsstrasse 2, 45141 Essen, Germany.*

**MARIUS NISTORESCU,** *Department of Systems Ecology and Sustainability, University of Bucharest, Splaiul Indepedentei 91–95, 050095 Bucharest, Romania.*

**MORTEN L. PEDERSEN,** *Aalborg University, Department of Civil Engineering, Sohngaardsholmsvej 57, DK-9000 Aalborg, Denmark.*

**FLORIAN PLETTERBAUER,** *Department of Water, Atmosphere and Environment, University of Natural Resources and Life Sciences (BOKU), A-1180 Vienna, Austria.*

**DIDIER PONT,** *Cemagref, Unit HBAN-HEF, Hydrosystems and Bioprocesses Research unit (HBAN), Parc de Tourvoie, 92160 Antony, France.*

**ELENA PREDA,** *Department of Systems Ecology and Sustainability, University of Bucharest, Splaiul Indepedentei 91–95, 050095 Bucharest, Romania.*

**JENS O. RIEDE,** *J.F Blumenbach Institute of Zoology and Anthropology, University of Göttingen, Germany.*

**MIIRA P. RIIPINEN,** *Manchester Metropolitan University, Environmental and Geographical Sciences, Chester Street, Manchester, M15 6BH, United Kingdom.*

**GETA RÎŞNOVEANU,** *Department of Systems Ecology and Sustainability, University of Bucharest, Splaiul Indepedentei 91–95, 050095 Bucharest, Romania.*

**STEPHEN J. ROSSITER,** *School of Biological and Chemical Sciences, Queen Mary University of London, E1 4NS, London, United Kingdom.*

**MICHIEL RUTGERS,** *National Institute for Public Health and Environment RIVM, P.O. Box 1, 3720BA Bilthoven, The Netherlands.*

**MARKUS SCHINDLER,** *Department of Aquatic Ecology, Eawag: Swiss Federal Institute of Aquatic Science and Technology, 6047 Kastanienbaum, Switzerland, and Institute of Integrative Biology (IBZ), ETH Zurich, Switzerland.*

**SCOTT D. TIEGS,** *Department of Aquatic Ecology, Eawag: Swiss Federal Institute of Aquatic Science and Technology, 6047 Kastanienbaum, Switzerland and Institute of Integrative Biology (IBZ), ETH Zurich, Switzerland. Present address: Department of Biological Sciences, Oakland University, Rochester Michigan, 48309-4401, USA.*

**COLIN R. TOWNSEND,** *Department of Zoology, University of Otago, Dunedin, New Zealand.*

**MARK TRIMMER,** *School of Biological & Chemical Sciences, Queen Mary University of London, London, E1 4NS, United Kingdom.*

**MARCEL G.A. VAN DER HEIJDEN,** *Agroscope Reckenholz Tänikon, ART Research Station, Reckenholzstraβe 191, 8046 Zürich, Switzerland.*

**HARM J. VAN WIJNEN,** *National Institute for Public Health and Environment RIVM, P.O. Box 1, 3720BA Bilthoven, The Netherlands.*

**PIET F.M. VERDONSCHOT,** *Alterra, Freshwater Ecology, P.O. Box 47, 6700 AA Wageningen, The Netherlands.*

**J. ARIE VONK,** *Institute for Water and Wetland Research, Radboud University, 6525ED Nijmegen, The Netherlands.*

**LENA B.-M. VOUGHT,** *Kristianstad University, SE 29188 Kristianstad, Sweden.*

**GUY WOODWARD,** *University College Cork, Department of Zoology, Ecology and Plant Science, Lee Maltings Building, Prospect Row, Cork, Ireland, and Queen Mary University of London, School of Biological and Chemical Sciences, London, E1 4NS, United Kingdom.*

# Preface

## Linking Ecological Theory to Reality: Measuring and Mitigating Human Impacts in Europe's Heavily Modified Ecosystems

The world's ecosystems have been modified by human activity to such an extent that few can now be considered to be truly "pristine" (Brooks *et al.*, 2006; Vitousek *et al.*, 1997). The European continent, in particular, has experienced profound alterations to its natural systems, and at an increasing rate, for millennia (Baattrup-Pedersen *et al.*, 2008). From the onset of the extensive Neolithic forest clearances, through to the rise of the intensively farmed and urban landscapes that now dominate much of the continent, the human footprint has become ever more prominent. The Agricultural and Industrial Revolutions were born in Europe, and both have had especially far-reaching impacts on the continent's ecosystems and the species they support: the evidence of the changes that these pressures have imposed on natural systems is manifested over many spatial and temporal scales and across multiple levels of biological organisation (Hynes, 1960; Vitousek *et al.*, 1997). Europe therefore represents a microcosm for understanding human impacts on a global scale, as many other parts of the world have also experienced (or are experiencing) marked increases in population growth, agricultural intensification and industrialisation (cf. Bernhardt *et al.*, 2005; Feld *et al.*, 2011): characterising human impacts in Europe thus enables us to glimpse what might happen elsewhere in the near future.

This thematic volume is devoted to the applied ecology of European ecosystems, with a particular emphasis on the higher (i.e. multispecies) levels of organisation. It is focused primarily on freshwater and soil ecosystems, which are both highly susceptible to human perturbations and where biomonitoring approaches and food web, community and ecosystem ecology are well-established traditions (Elton, 1927; Hynes, 1960; Ings *et al.*, 2009). The collection of papers here also explore how new developments in general ecology can be used to form a stronger link between pure and applied branches of the discipline, to the benefit of both. This volume therefore forms a logical extension to the previous two (Volume 42 "Ecological Networks"; Volume 43 "Integrative Ecology: From Molecules to Ecosystems"), in which more fundamental aspects of ecology were addressed. It also forms a bridge to Volume 45 ("The Role of Body Size in Communities and Ecosystems"), which takes this combination of pure and applied ecology further still. Volumes 42–45 essentially represent a quartet of interconnected themes that cut across a range of ecological disciplines, although each thematic volume has its own distinctive flavour. Perhaps inevitably, this volume has an especially broad remit, because much of applied ecology spans the nexus of science, politics and socioeconomics (Moog and Covanec, 2000)—a fact that is highlighted here particularly by Friberg *et al.* (2011), who review the role of biomonitoring in the context of national and international environmental legislation. Nonetheless, we have attempted to retain a strong link to general ecology and recent theoretical advances throughout this volume so that the five papers presented here form part of an interconnected series of threads within a broader theme.

The first of these focuses on the development and application of biomonitoring as a tool for sensing the biota and its responses to environmental pressures, and provides a timely critique of current and emerging techniques in this important field of applied ecology, especially in the light of the imminent implementation of the far-reaching European Union's Water Framework Directive (Friberg *et al.*, 2011). Although biomonitoring has its roots in freshwater ecology (Friberg *et al.*, 2011), the general principles can be applied to all ecosystems, as revealed, for instance, in the final paper in the volume, which focuses on human impacts on soil systems (Mulder *et al.*, 2011).

Layer *et al.* (2011) provide a detailed case study that combines biomonitoring approaches and theoretical ecology, using a unique long-term dataset from a model system, Broadstone Stream in southern England, which complements an earlier paper that explored the impact of pH on stream food webs in a snapshot across multiple systems in a space-for-time comparative approach (Layer *et al.*, 2010). The Layer *et al.* (2011) study documents changes in one of the most intensively studied food webs in the world over four decades of acidification and recovery. This paper considers the consequences of the waves of progressively larger predators that have invaded over successive decades as pH has risen and how the structure and stability of the food web has changed as a result. It also highlights both the scarcity and the value of long-term ecological data, which are key for understanding and ultimately predicting higher-level responses to stress—and, importantly, to the alleviation of that stress. Acidification in European freshwaters was one of the greatest environmental concerns of the day in the 1970s and 1980s—just as climate change now occupies the top of the agenda—and yet we are only now starting to see the fruits of decades of reduced acidifying emissions, as the first signs of biological recovery start to emerge in the wake of longer-term chemical recovery (Hildrew, 2009; Kernan *et al.*, 2010). An intriguing aspect of the Layer *et al.* (2011) study is the suggestion that the biota itself might play an important role in shaping the trajectory of recovery, which does not appear to be simply a straightforward reversal of the changes that occurred during acidification: there is evidence that ecological inertia, modulated via the food web, may slow or even redirect recovery (Ledger and Hildrew, 2005). This has wider implications for biomonitoring and bioassessment schemes because the widespread underlying assumption that the biota can be effectively passively mapped onto the environmental setting or habitat templet is called into question, ideas that are also raised by Friberg *et al.* (2011) in a more general sense. The Layer *et al.* (2011) paper therefore represents a detailed case study, using real data from a model system, of a potentially widespread phenomenon. It also documents a rare environmental success story, as Europe's freshwaters start to show signs of recovery after many years of stringent legislative controls on acidifying emissions, whilst serving as a warning of the growing threats posed by this stressor in other parts of the world, including India and China, where industrialisation is gathering pace and the associated pollution burden is likely to grow substantially in the near future.

The Friberg *et al.* (2011) and Layer *et al.* (2011) papers are complemented by the Feld *et al.* (2011) paper, which focuses on habitat degradation and restoration of the physical (rather than chemical) environment, by assessing what has and has not been learnt over the past two decades of river restoration science. Feld *et al.* (2011) conducted an extensive literature review, in parallel with an investigation of several detailed case studies in Denmark, where river restoration has a particularly long history, to develop and explore a more coherent theoretical framework than currently exists. To date, river restoration in Europe has been carried out on a largely *ad hoc*, piecemeal basis, with little testing or validation of the underlying scientific assumptions as to how the physical environment and the biota interact. The findings of Feld *et al.* (2011) are enlightening, and even startling in some instances: it will probably surprise many readers new to this field that vast financial resources have been devoted to implementing management and restoration practices that are still so poorly understood and often founded on little or no empirical evidence that they actually work. The prevailing view in river restoration ecology has been described as the "Field of Dreams" approach, whereby "if you build it, they will come". This habitats-by-numbers approach to restoring damaged ecosystems is based on the assumption that biological recovery will inevitably follow physical restoration. Feld *et al.* (2011) demonstrate that these ideas, although appealing, are still largely unsubstantiated, at least at the spatiotemporal scales at which most restoration schemes have been conducted, and that we know surprisingly little about the extent to which these communities map onto the physical geomorphological attributes of the habitat (Vaughan *et al.*, 2009).

This mismatch between knowledge, understanding and prediction arises because the scales at which most studies have been conducted are either inappropriate or disconnected in terms of how organisms sense and respond to their physical habitat, and also because traditionally the focus of applied freshwater ecology has been directed almost exclusively towards chemical stressors, such as organic pollution and acidification (Friberg *et al.*, 2011). Few studies have used rigorous experimental designs to make meaningful before-after-control-impact (BACI) comparisons among reference, damaged, and restored sites, and Feld *et al.* (2011) also emphasise the need to consider the impacts of multiple rather than single stressors. In many cases, for restoration schemes to be effective, water quality issues need to be addressed before any potential responses to habitat alteration are likely to be expressed by the biota. It is therefore critical to prioritise conservation and management strategies accordingly if resources are to be used most effectively. This also concurs with Friberg *et al.* (2011), who point out that changing conditions of both reference and impacted conditions need to be incorporated into future assessment and management schemes, because water quality improvements across Europe may be confounded with other long-term changes. This makes it difficult to disentangle potentially confounded cause- and -effect relationships where multiple drivers may be at play (Durance and Ormerod, 2009). Previously secondary environmental constraints, such as habitat quality and the physical environment, are therefore likely to come increasingly to the fore as baseline conditions continue to shift over time, and this needs to be addressed in future biomonitoring research.

Hladyz *et al.* (2011) move beyond the purely aquatic environment and consider the links between the freshwater and terrestrial spheres and how alterations to one can influence the other. Their study is one of the first to assess a range of structural and functional responses of aquatic ecosystems to changes in terrestrial vegetation and includes a continental-scale field bioassay across 100 European streams. This extensive study is followed up by a more intensive set of field and laboratory experiments in Irish pasture streams, which represents one of the dominant land uses across Europe. Hladyz *et al.* (2011) found that human modification of riparian vegetation in general, and clearance for pasture in particular, could have significant impacts on stream ecosystems. For example, although absolute process rates were often not affected, the agents of decomposition changed markedly, with a far greater contribution coming from microbial decomposers, as opposed to invertebrate detritivores, in impacted versus reference sites. In addition, autochthonous feeding pathways within the food web became relatively more important in sites cleared of their native riparian vegetation. This, in turn, has implications for food web dynamics, as the stabilising effects of "slower" detrital pathways may be mitigated by these stronger, "faster" consumer–resource interactions (Rooney *et al.*, 2006). Hladyz *et al.* (2011) make explicit connections between what is happening on the land and in the water and demonstrate that it is not simply alterations to water chemistry (Layer *et al.*, 2011) or in-stream habitats (c.f. Feld *et al.*, 2011) that determine overall ecosystem functioning: it is the links among the component parts of the wider landscape that are also key, but often overlooked, drivers.

Mulder *et al.* (2011) take a further step into the terrestrial biosphere and consider how soil communities, food webs, and ecosystems in the heavily human-modified cultural landscape of the Netherlands are moulded by environmental conditions, as well as by their own internal dynamics. They examine a comprehensive dataset from close to 200 sites that span a range of different land uses and agricultural intensity and explore responses at the higher levels of organisation, including the distribution of functional traits within food webs, to human activity. Soil biodiversity, in both functional and taxonomic terms, is key for provisioning ecosystem goods and services, and Mulder *et al.* (2011) make the first attempt to link these aspects to food web attributes in order to develop a more predictive approach based on first principles derived from ecological theory. This is analogous to the types of *a priori* approaches that are also advocated by Friberg *et al.* (2011) and Feld *et al.* (2011) for freshwaters, but which have yet to be implemented in most national or international biomonitoring and bioassessment schemes.

Clear parallels can therefore be drawn with the Mulder *et al.* (2011) study and the other papers in the volume. In addition exciting new ground opens up from the application of these novel approaches, which offer great promise in other terrestrial and aquatic systems. This potential for cross-fertilisation of ideas across disciplines, as becomes apparent from the recurrent themes emerging from each of the five papers in this volume, is perhaps not so unexpected as it might appear at first sight. After all, it could be argued that many forested stream food webs are essentially very wet soil food webs (or soil webs are "dry" stream food webs, depending upon one's perspective), as both are largely reliant on inputs of terrestrial leaf-litter as a major, often poor-quality, carbon-rich energy source. Consequently, these are systems with similar stoichiometric constraints on consumer–resource interactions at the base of the food web (i.e. fluxes of CNP), where donor-controlled dynamics and their potential stabilising effects are especially prevalent. Both also exhibit strong size structuring, at least in terms of the trophic interactions among their respective consumer assemblages, such that mass–abundance scaling relationships and other size-based allometries therefore offer the potential to provide important insights into both soil (de Ruiter *et al.*, 1995; Mulder *et al.*, 2011) and stream communities (Hladyz *et al.*, 2011; Layer *et al.*, 2011). This echoes some of the findings from marine, freshwater and above-ground systems published recently in sister volumes of *Advances in Ecological Research* (Layer *et al.*, 2010; McLaughlin *et al.*, 2010; O'Gorman and Emmerson, 2010; Reuman *et al.*, 2009; Woodward *et al.*, 2010a) and elsewhere (Cohen *et al.*, 2003; Ings *et al.*, 2009). Further, because size-based approaches are not so strongly linked to taxonomy as is tradition-al freshwater biomonitoring, with its strong focus on indicator species and diversity indices, they could provide a useful means of assessing functional attributes of the community without being subject to the idiosyncratic effects of species identity or ecosystem type *per se* (Petchey and Belgrano, 2010; Yvon-Durocher *et al.*, 2011). This is important because body size is a proxy for many ecological traits, such as trophic status and feeding rates (Petchey and Belgrano, 2010; Reiss *et al.*, 2010; Woodward *et al.*, 2010a), and it should also respond predictably to certain environmental stressors, including environmental warming (Atkinson, 1994; Bonada *et al.*, 2007; Perkins *et al.*, 2010; Woodward *et al.*, 2010b; Yvon-Durocher *et al.*, 2010).

Several major advances in community ecology in the past two decades have been made by the use of size-based approaches to modelling food web structure and dynamics and in assessing biodiversity–ecosystem functioning relationships, with analogous developments occurring in parallel in terrestrial, marine and freshwater ecosystems (Ings *et al.*, 2009). A key challenge now is to integrate these and other recent advances, such as *in situ* metagenomic characterisation of the microbial assemblages that drive the major biogeochemical cycles (He *et al.*, 2010; Purdy *et al.*, 2010), to improve our understanding of the causes and consequences of biodiversity loss (Hector and Bagchi, 2007; Woodward *et al.*, 2010b). If this can be achieved, then the new generation of approaches available to applied ecology will be better able to meet the emerging environmental threats of the twenty-first century (Moss *et al.*, 2009). Hopefully, this thematic volume of *Advances in Ecological Research* represents a small step towards meeting this goal.

Guy Woodward
London

# REFERENCES

Atkinson, D. (1994). Temperature and organism size—A biological law for ectotherms. *Adv. Ecol.Res.* **25**, 1–58.

Baattrup-Pedersen, A., Springe, G., Riis, T., Larsen, S.E., Sand-Jensen, K., and Kjellerup Larsen, L.M. (2008). The search for reference conditions for stream vegetation in northern Europe. *Freshw. Biol.* **53**, 1890–1901.

Bernhardt, E.S., Palmer, M.A., Allan, J.D., Alexander, G., Barnas, K., Brooks, S., Carr, J., Clayton, S., Dahm, C., Follstad-Shah, J., Galat, D., Gloss, S., *et al.* (2005). Synthesizing U.S. River Restoration Efforts. *Science* **308**, 636–637.

Bonada, N., Dolédec, S., and Statzner, B. (2007). Taxonomic and biological trait differences of stream macroinvertebrate communities between Mediterranean and temperate regions: Implications for future climatic scenarios. *Global Change Biol.* **13**, 1658–1671.

Brooks, T.M., Mittermeier, R.A., da Fonseca, G.A.B., Gerlach, J., Hoffmann, M., Lamoreux, J.F., Mittermeier, C.G., Pilgrim, J.D., and Rodrigues, A.S.L. (2006). Global biodiversity conservation priorities. *Science* **313**, 58–61.

Cohen, J.E., Jonsson, T., and Carpenter, S.R. (2003). Ecological community description using the food web, species abundance, and body size. *Proc Natl Acad Sci USA* **100**, 1781–1786.

de Ruiter, P.C., Neutel, A.M., and Moore, J.C. (1995). Energetics, patterns of interaction strengths, and stability in real ecosystems. *Science* **269**, 1257–1260.

Durance, I., and Ormerod, S.J. (2009). Trends in water quality and discharge offset long-term warming effects on river macroinvertebrates. *Freshw. Biol.* **54**, 388–405.

Elton, C.S. (1927). *Animal Ecology*. Sedgewick and Jackson, London.

Feld, C.K., Birk, S., Bradley, D.C., Hering, D., Kail, J., Marzin, A., Melcher, A., Nemitz, D., Pederson, M.L., Pletter Bauer, F., Pont, D., Verdonschot, P.F.M., *et al.* (2011). From natural to degraded rivers and back again: A test of restoration ecology theory and practice. *Adv. Ecol. Res.* **44**, 119–210.

Friberg, N., Bonada, N., Bradley, D.C., Dunbar, M.J., Edwards, F.K., Grey, J., Hayes, R.B., Hildrew, A.G., Lamouroux, N., Trimmer, M., and Woodward, G. (2011). Biomonitoring of human impacts in natural ecosystems: The good, the bad, and the ugly. *Adv. Ecol. Res.* **44**, 1–68.

He, Z., Xu, M., Deng, Y., Kang, S., Kellogg, L., Wu, L., Van Nostrand, J.D., Hobbie, S.E., Reich, P.B., and Zhou, J. (2010). Metagenomic analysis reveals a marked divergence in the structure of belowground microbial communities at elevated CO2. *Ecol. Lett.* **13**, 564–575.

Hector, A., and Bagchi, R. (2007). Biodiversity and ecosystem multifunctionality. *Nature* **448**, 188–190.

Hildrew, A.G. (2009). Sustained research on stream communities: A model system and the comparative approach. *Adv. Ecol. Res.* **41**, 175–312.

Hladyz, S., Åbjörnsson, K., Chauvet, E., Dobson, M., Elosegi, A., Ferreira, V., Fleituch, T., Gessner, M.O., Giller, P.S., Gulis, V., Hutton, S.A., Lacoursière, J.O., *et al.* (2011). Stream ecosystem functioning in an agricultural landscape: The importance of terrestrial-aquatic linkages. *Adv. Ecol. Res.* **44**, 211–276.

Hynes, H.B.N. (1960). *The Biology of Polluted Waters*. Liverpool University Press, Liverpool, England202pp.

Ings, T.C., Montoya, J.M., Bascompte, J., Bluthgen, N., Brown, L., Dormann, C.F., Edwards, F., Figueroa, D., Jacob, U., Jones, J.I., Lauridsen, R.B., Ledger, M.E., *et al.* (2009). Ecological networks—Beyond food webs. *J. Anim. Ecol.* **78**, 253–269.

20 Year Interpretative Report - Recovery of lakes and streams in the UK from acid rain. The United Kingdom Acid Waters Monitoring Network 20 year interpretative report. ECRC Research Report #141. Report to the Department for Environment, Food and Rural Affairs. Kernan, M., Battarbee, R.W., Curtis, C.J., Monteith, D.T. and Shilland, E.M. (Eds.), (2010). http://awmn.defra.gov.uk/resources/interpreports/index.php.

Layer, K., Riede, J.O., Hildrew, A.G., and Woodward, G. (2010). Food web structure and stability in 20 streams across a wide pH gradient. *Adv. Ecol. Res.* **42**, 265–301.

Layer, K., Hildrew, A.G., Jenkins, G.B., Riede, J.O., Rossiter, S.J., Townsend, C.R., and Woodward, G. (2011). Long-term dynamics of a well-characterised food web: Four decades of acidification and recovery in the Broadstone Stream model system. *Adv. Ecol. Res.* **44**, 69–118.

Ledger, M.E., and Hildrew, A.G. (2005). The ecology of acidification and recovery: Changes in herbivore-algal food web linkages across a pH gradient in streams. *Environ. Pollut.* **137**, 103–118.

McLaughlin, O.B., Jonsson, T., and Emmerson, M.C. (2010). Temporal variability in predator-prey relationships of a forest floor food web. *Adv. Ecol. Res.* **42**, 172–264.

Moog, O., and Chovanec, A. (2000). Assessing the ecological integrity of rivers: Walking the line among ecological, political and administrative interests. *Hydrobiologia* **422** (423), 99–109.

Moss, B., Hering, D., Green, A.J., Aidoud, A., Becares, E., Beklioglu, M., Bennion, H., Boix, D., Brucet, S., Carvalho, L., Clement, B., Davidson, T., *et al.* (2009). Climate change and the future of freshwater biodiversity in Europe: A primer for policy-makers. *Freshw. Rev.* **2**, 103–130.

Mulder, C., Boit, A., Bonkowski, M., De Ruiter, P.C., Mancinelli, G., Van der Heijden, M.G.A., van Wijnen, H.J., Vonk, J.A., and Rutgers, M. (2011). A belowground perspective on dutch agroecosystems: How soil organisms interact to support ecosystem services. *Adv. Ecol. Res.* **44**, 277–358.

O'Gorman, E.J., and Emmerson, M.C. (2010). Manipulating interaction strengths and the consequences for trivariate patterns in a marine food web. *Adv. Ecol. Res.* **42**, 301–419.

Perkins, D.M., McKie, B.G., Malmqvist, B., Gilmour, S.G., Reiss, J., and Woodward, G. (2010). Environmental warming and biodiversity-ecosystem functioning in freshwater microcosms: Partitioning the effects of species identity, richness and metabolism. *Adv. Ecol. Res.* **43**, 177–209.

Petchey, O.L., and Belgrano, A. (2010). Body-size distributions and size-spectra: Universal indicators of ecological status? *Biol. Lett.* **6**, 434–437.

Purdy, K.J., Hurd, P.J., Moya-Laraño, J., Trimmer, M., and Woodward, G. (2010). Systems biology for ecology: From molecules to ecosystems. *Adv. Ecol. Res.* **43**, 87–149.

Reiss, J., Bailey, R.A., Cássio, F., Woodward, G., and Pascoal, C. (2010). Assessing the contribution of micro-organisms and macrofauna to biodiversity-ecosystem functioning relationships in freshwater microcosms. *Adv. Ecol. Res.* **43**, 151–176.

Reuman, D.C., Mulder, C., Banasek-Richter, C., Cattin Blandenier, M.F., Breure, A. M., Den Hollander, H., Knetiel, J.M., Raffaelli, D., Woodward, G., and Cohen, J.E. (2009). Allometry of body size and abundance in 166 food webs. *Adv. Ecol. Res.* **41**, 1–46.

Rooney, N., McCann, K., Gellner, G., and Moore, J.C. (2006). Structural asymmetry and the stability of diverse food webs. *Nature* **442**, 265–269.

Vaughan, I.P., Diamond, M., Gurnell, A.M., Hall, K.A., Jenkins, A., Milner, N.J., Naylor, L.A., Sear, D.A., Woodward, G., and Ormerod, S.J. (2009). Integrating ecology with hydromorphology: A priority for river science and management. *Aquatic Conserv. Marine Freshw. Ecosyst.* **19**, 113–125.

Vitousek, P.M., Mooney, H.A., Lubchenco, J., and Melillo, J.M. (1997). Human domination of Earth's ecosystems. *Science* **277**, 494–499.

Woodward, G., Blanchard, J., Lauridsen, R.B., Edwards, F.K., Jones, J.I., Figueroa, D., Warren, P.H., and Petchey, O.L. (2010a). Individual-based food webs: Species identity, body size and sampling effects. *Adv. Ecol. Res.* **43**, 211–266.

Woodward, G., Benstead, J.P., Beveridge, O.S., Blanchard, J., Brey, T., Brown, L., Cross, W.F., Friberg, N., Ings, T.C., Jacob, U., Jennings, S., Ledger, M.E., *et al.* (2010b). Ecological networks in a changing climate. *Adv. Ecol. Res.* **42**, 72–138.

Yvon-Durocher, G., Allen, A.P., Montoya, J.M., Trimmer, M., and Woodward, G. (2010). The temperature dependence of the carbon cycle in aquatic systems. *Adv. Ecol. Res.* **43**, 267–313.

Yvon-Durocher, G., Reiss, J., Blanchard, J., Ebenman, B., Perkins, D.M., Reuman, D.C., Thierry, A., Woodward, G., and Petchey, O.L. (2011). Across ecosystem comparisons of size structure: Methods, approaches, and prospects. *Oikos* (in press).

# Biomonitoring of Human Impacts in Freshwater Ecosystems: The Good, the Bad and the Ugly

NIKOLAI FRIBERG, NÚRIA BONADA, DAVID C. BRADLEY, MICHAEL J. DUNBAR, FRANCOIS K. EDWARDS, JONATHAN GREY, RICHARD B. HAYES, ALAN G. HILDREW, NICOLAS LAMOUROUX, MARK TRIMMER AND GUY WOODWARD



ADVANCES IN ECOLOGICAL RESEARCH VOL. 44

0065-2504/11 $35.00
DOI: 10.1016/B978-0-12-374794-5.00001-8

# SUMMARY

It is critical that the impacts of environmental stressors on natural systems are detected, monitored and assessed accurately in order to legislate effectively and to protect and restore ecosystems. Biomonitoring underpins much of modern resource management, especially in fresh waters, and has received significant sums of money and research effort during its development. Despite this, the incorporation of science has not been effective and the management tools developed are sometimes inappropriate and poorly designed. Much biomonitoring has developed largely in isolation from general ecological theory, despite the fact that many of its fundamental principles ultimately stem from basic concepts, such as niche theory, the habitat template and the *r–K* continuum. Consequently, biomonitoring has not kept pace with scientific advances, which has compromised its ability to deal with emerging environmental stressors such as climate change and habitat degradation. A reconnection with its ecological roots and the incorporation of robust statistical frameworks are key to progress and meeting future challenges.

The vast amount of information already collected represents a potentially valuable, and largely untapped, resource that could be used more effectively in protecting ecosystems and in advancing general ecology. Biomonitoring programmes have often accumulated valuable long-term data series, which could be useful outside the scope of the original aims. However, it is timely to assess critically existing biomonitoring approaches to help ensure future programmes operate within a sound scientific framework and cost-effectively. Investing a small proportion of available budgets to review effectiveness would pay considerable dividends.

Increasing activity has been stimulated by new legislation that carries the threat of penalties for non-compliance with environmental targets, as is proposed, for example, in the EU's Water Framework Directive. If biomonitoring produces poor-quality data and has a weak scientific basis, it may lead either to unjustified burdens placed on the users of water resources, or to undetected environmental damage. We present some examples of good practice and suggest new ways to strengthen the scientific rigour that underpins biomonitoring programmes, as well as highlighting potentially rewarding new approaches and technologies that could complement existing methods.

# I. INTRODUCTION

## A. A Brief Overview of Biomonitoring

Biomonitoring can be broadly defined as the use of the biota to gauge and track changes in the environment (Gerhardt, 2000; Wright *et al.*, 1993, 2000), just as symptoms are used to identify disease in medicine. It has its roots in the microbiology of running waters, following the first bioassays conducted in the polluted rivers of Europe in the late nineteenth century (see Bonada *et al.*, 2006 and references therein), although it subsequently expanded to cover other fresh waters and also marine and terrestrial systems (Ashmore *et al.*, 1978; Hawksworth and Rose, 1970; Mulder *et al.*, 2011; Roberts, 1972; Ruston, 1921).

Biomonitoring now underpins much of the management and conservation of fresh waters, albeit primarily in the developed world, designed to protect the ecosystem goods and services they supply, including the water itself, the production of food, climate regulation and waste processing (Dudgeon *et al.*, 2006; Statzner *et al.*, 1997; Tranvik *et al.*, 2009; Williamson *et al.*, 2009). Over 15 years ago, the former Vice President of the World Bank, Ismail Serageldin, suggested that if the twentieth century's wars were fought over oil, the twenty-first century's wars would be fought over water: despite covering <1% of the Earth's surface, fresh waters are clearly strategically very important (Gleick, 1998; Gleick and Palaniappan, 2010). Within Europe alone almost 80% of the total EU environmental budget involves water-related expenditure, which will rise even further as the EU Water Framework Directive (WFD; Directive 2000/60/EC) comes into full operation (Bonada *et al.*, 2006; European Commission, 1998).

This potent combination of ecological, geopolitical and socioeconomic factors (Cairns, 1988) has resulted in an increase in regional, national and international environmental legislation, including the US Clean Water Act, the Canadian Protection Act, the Urban Wastewater Treatment Directive (91/271/EEC) and the WFD in recent decades (Brooks *et al.*, 2006; Pollard and Huxham, 1998; Rosenberg and Resh, 1993). In order to conserve and manage freshwaters sustainably, however, we first need to be able to measure how they respond to natural and anthropogenic stressors, and biomonitoring is a key tool used for tracking and quantifying impacts (Carter and Resh, 2001; Niemi and McDonald, 2004). Although it is primarily used for supporting legislators and managers and is not strictly a scientific discipline in itself, biomonitoring needs to be supported by strong science to ensure that it is both adaptive and credible, especially since legal challenges to fines imposed for non-compliance with water quality legislation can be extremely costly.

## B. The Emergence of Community and Ecosystem-Level Perspectives

The environmental problems that first emerged on a large scale in the nineteenth century, and which first triggered interest in the degradation of aquatic ecosystems, were manifested primarily through threats to human health: for instance, the great cholera outbreaks of Victorian London in the early nineteenth century were a direct consequence of sewage pollution. Originally, the monitoring of fresh waters was based primarily on chemical measures, but the use of biological parameters was soon to follow (Bonada *et al.*, 2006). The initial focus of biomonitoring was on bacteria, as they were of greatest interest from a health perspective but, by the turn of the twentieth century, the range of taxa had expanded to embrace other microscopic organisms, including algae, fungi and protozoa (Bonada *et al.*, 2006).

It was at this point that biomonitoring first started to enter a new, more holistic, phase, in which 'biocoenoses' (i.e. assemblages of macrophytes, macroinvertebrates and fish associated with particular habitats) were assayed (Hynes, 1960; Kolkwitz and Marsson, 1902, 1909; see review in Bonada *et al.*, 2006 for more detailed coverage). In many respects, this was arguably the single most significant leap forward in biomonitoring, together with the pioneering work of Patrick (1949) on biological measures (using a number taxonomic groups) of stream condition, because it paved the way for the development of biotic indices and advances in statistical analyses based on community and ecosystem ecology. The original approaches have been gradually superseded by newer biological metrics (e.g. Bailey *et al.*, 1998; Chessman and McEvoy, 1998; Kelly, 1998; Kelly *et al.*, 2008; Padisak *et al.*, 2006; Wright *et al.*, 2000), but the same groups of organisms and many notions about biocoenoses still hold sway to this day (e.g. Ashbolt *et al.*, 2001; Attrill and Depledge, 1997). However, the development of this core component of biomonitoring has often lagged some way behind theoretical developments in general ecology.

Many other stressors (beyond organic pollution) have been identified since the first microbial assays, and their potential impacts on ecosystems and people have become the object of burgeoning legislation (Rosenberg and Resh, 1993). This has typically sought to stop the problem at source (e.g. bans on DDT and other organochlorine pesticides) or, more recently, to restore damaged systems (e.g. Feld *et al.*, 2011). The means of doing so has often been via financial sanctions, as in 'the polluter pays' principle or, as anticipated in the EU's WFD, where all surface waters need to achieve 'good ecological status' by 2015 or offending member states can be fined.

Biomonitoring is clearly both necessary and important, and effective schemes based on good ecological understanding can yield invaluable information into how natural systems respond to stressors and provide a basis for effective environmental management. A key advantage it offers is that biological responses to perturbations are assessed directly, rather than relying on inferences from chemical data; after all, it is the biological consequences that are normally of ultimate concern. Further, biological indicators integrate their responses overtime and space, potentially lowering sampling effort and cost compared to the high intensity of sampling often required when relying on chemical variables to detect certain impacts (e.g. the requirement for repeated nutrient concentration measurements to detect organic pollution in running waters; Metcalfe-Smith, 1996). The potential for cross-fertilisation with fundamental ecology is great, yet there remain significant shortcomings in the philosophy and practice of biomonitoring, and the science that underpins it, that impair its development. Here, our aim is to provide an overview of the evolution of biomonitoring, to assess current practices and their respective strengths and weaknesses, to identify new approaches and, finally, to suggest where it might advance in the future and where it seems more likely to stagnate. Above all, it is not our intention to deprecate the invaluable contributions that (good) biomonitoring has already made to environmental management and applied ecology, but rather to identify areas that we feel merit considered (re)evaluation and to propose ways to strengthen its scientific rigour and ability to adapt to future environmental stressors, such as climate change.

## C. Knowledge, Understanding and Prediction

To persist with our medical analogy, a prerequisite for finding a cure to a particular ailment is to be able to assess the condition of the patient, before a treatment can be applied and its efficacy quantified (Bailey *et al.*, 2004). In terms of 'ecosystem health', the stages from diagnosis to cure each raise different sets of challenges and require not only the ability to identify deviations from the desired state (reference condition) but also an understanding of how the system operates and the ability to predict and monitor its responses to future change.

Assessing ecological status objectively is difficult and the challenges faced by modern biomonitoring are many and complex: some are semantic (what *is* 'ecosystem health'?), some are philosophical (do reference conditions even exist in such a human-modified world?), whereas others are more logistical, technological, financial or political (Lancaster, 2000). Many of these difficulties arise because biomonitoring and bioassessment span the interface between natural and social sciences (Feld *et al.*, 2011; Mulder *et al.*, 2011), and its growing prominence arguably reflects the 'greening' of politics and greater public interest in environmental issues, just as much as it does the advances made in the scientific disciplines of applied ecology. This inevitably creates tensions, as the political and legislative decision makers may have little or no scientific training—and, similarly, scientists can be politically naïve or prefer not to engage directly with managers and their difficulties. These challenges are common to much of applied ecology and have been discussed elsewhere (e.g. Downes *et al.*, 2002; Wright *et al.*, 2000) so we will give them only cursory consideration here, whilst focussing more on the scientific issues.

Science progresses from the acquisition of knowledge to developing a mechanistic understanding and, ultimately, to making predictions. Peters (1991) and Rigler (1982), among others, were impatient with this seemingly slow process and advocated that ecologists should make better use of well-known empirical relationships to make specific predictions of practical use, rather than focussing on the reductionist approach of trying to understand the underlying mechanisms. Although their ideas still resonate with much of the current philosophy and practice of biomonitoring, unfortunately serious problems can soon arise in the absence of a firm mechanistic basis. One consequence of this is that biomonitoring is often based on contingent case studies of empirical relationships, such that each monitoring scheme has to be created anew for each set of systems and stressors, and this reflects the wider tension in applied ecology that arises when trying to balance realism with universality (Mulder *et al.*, 2011). Understanding the mechanisms, and their degree of context dependency, allows for greater flexibility across systems (more generality) and extrapolation into less familiar territory (more predictive power). This will become increasingly important in the near future: for instance, as climate change alters current ecological baselines, both reference and impacted conditions that many current biomonitoring schemes are based on could become obsolete (Moss *et al.*, 2009; Woodward, 2009).

## D. What Is a Healthy Ecosystem?

Achieving the seemingly simple prerequisite of identifying "healthy" reference conditions is extremely problematic, especially in the developed world, where pristine ecosystems are virtually non-existent (Feld *et al.*, 2011; Hladyz *et al.*, 2011a,b; Mulder *et al.*, 2011). This is especially true in Europe, where the industrial revolution and urbanisation of the eighteenth and nineteenth centuries have been overlain on millennia of earlier alterations to the rural landscape, including the clearance of much of the native forest and its replacement by agriculture (Hladyz *et al.*, 2011a,b). Unfortunately, biomonitoring itself began only after human impacts were already widespread (indeed, this is the very reason it was initiated), so finding pristine systems against which to gauge the extent of anthropogenic impacts is a considerable challenge (e.g. Bennion *et al.*, 2004). This logistical difficulty also compounds the problem of defining and measuring 'ecosystem health' (Norris and Thomas, 1999).

The need for reference sites is most directly relevant to higher-level (i.e. community or ecosystem) approaches that are carried out *in situ* and it forms the cornerstone of much environmental legislation, including the EU WFD. It is far easier to reconstruct reference communities for standing waters than for most running waters, due to the presence of preserved remains of many components of the biota (e.g. diatom valves, macrophyte pollen, insect chitin, fish scales) in stratified lake sediments, which can also be independently dated (Bennion *et al.*, 2004; Rawcliffe *et al.*, 2010). A common solution for standing waters, therefore, is to use reconstructed community data from circa 1850 to define the reference conditions as a baseline, before large-scale agricultural intensification, population growth, or industrial pollution (e.g. Battarbee *et al.*, 2001; Bennion *et al.*, 2004; Figure 1). Consequently, palaeoecological techniques have long been a key component of limnological biomonitoring, and surface sediments can also be compared with contemporary benthic samples to calibrate past conditions with those in extant communities, providing ways of gauging both temporal and space-for-time gradients in environmental stress (Rawcliffe *et al.*, 2010). The growing challenge to this approach, however, is the problem that reference conditions may themselves be shifting over time, for instance, due to climate change (Bennion *et al.*, 2010; Woodward *et al.*, 2010e).

Palaeoecological techniques are rarely possible in running waters, so space-for-time substitution is often used as an alternative to a truly temporal approach. Unfortunately, contemporary sampling rarely stretches back more than a few years, or decades at most (but see Bradley and Ormerod, 2002; Layer *et al.*, 2010a, 2011; UKAWMN, 2000, 2010). Biomonitoring in streams and rivers thus faces a particular challenge of finding suitable reference conditions (Nijboer *et al.*, 2004; Reynoldson *et al.*, 1997): for instance, it has even been suggested that Latvian, Lithuanian and Polish streams should act as proxy reference conditions for heavily impacted Danish streams (Baattrup-Pedersen *et al.*, 2008). Alternative approaches have therefore emerged that depart from the concept of a fixed, historically based, reference state. For example, the River Management Concept applied in Austria uses a 'Leitbild', or 'target vision' approach, that adjusts the natural potential of a given water course, given existing political and economic constraints (Jungwirth *et al.*, 2005). Though eminently pragmatic, these approaches still require a scientific underpinning to understand and predict the effectiveness of alternative management and conservation scenarios.

Notwithstanding these problems, the use of reference conditions is now widely used in biomonitoring (e.g. Chessman *et al.*, 2008; Wright *et al.*, 2000). In many of these cases, the purportedly impacted site needs to be matched with a suitable paired reference site, or set of sites and, whilst such approaches can be extremely powerful and sensitive, these prerequisites mean that, although the basic principles are global, each scheme has to be revised for each area in which it is applied (Chessman *et al.*, 2008). This

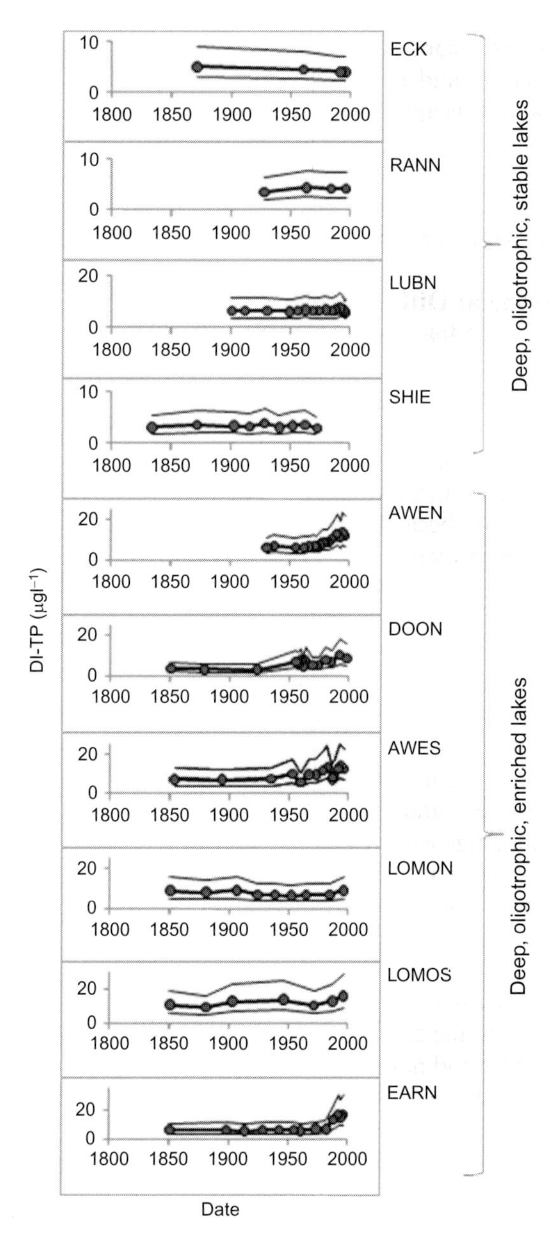


**Figure 1** DI-TP ($\mu$g l-1) reconstructions 1850-present day (solid lines). Values are based on the large lakes model (H. Bennion and N. J. Anderson, unpublished data) and the north-west European model (Bennion *et al.*, 1996). Bounding lines represent the standard error. The lakes are ordered according to increasing current measured TP (redrawn after Bennion *et al.*, 2004). Note: 1850 is used here as the standard reference conditions for the pre-Industrial period, as is commonly the case in palaeolimnological biomonitoring in Europe.

inevitably requires a considerable expenditure of resources that might not appeal to regulators and industry, who are under pressure to economise, although in reality this apparent cost may be relatively small when weighed against the risk of making the wrong decisions based on unreliable data.

## II. CURRENT METHODS—PROS AND CONS

### A. Biomonitoring at Different Levels of Organisation—From Molecules to Ecosystems

A wide range of possible approaches are available in modern biomonitoring, from the use of molecular techniques to the assessment of entire communities and whole-ecosystem metabolism, and each has its own strengths and weaknesses. We will concentrate on those based at the higher (i.e. multispecies) levels of organisation because these are arguably more closely linked to the goods and services provided by ecosystems, particularly at the larger scales where land-use change and catchment management operate, as well as being more realistic approximations as to how natural systems are likely to respond (Kimball and Levin, 1985). Bioassays and classical ecotoxicological approaches (e.g. Maltby *et al.*, 2002), focussed on individuals or small laboratory populations, however, do offer opportunities for replication and control unavailable with other techniques.

Biomonitoring has employed a range of techniques, including the use of molecular biomarkers and bioassays of model organisms (e.g. Heckmann *et al.*, 2005), to investigate biochemical to population level responses to particular stressors (see review by Bonada *et al.*, 2006 and references therein). These responses can be measured in a variety of ways, including perturbations to developmental processes (e.g. as revealed by fluctuating asymmetries; Dobrin and Corkum, 1999; Savage and Hogarth, 1999); behavioural changes; and altered rates of growth, reproduction or mortality. The principal disadvantage of such studies lies in the difficulties of extrapolation and 'scaling up' to multispecies communities and natural ecosystems. In detecting risks to health from novel or poorly understood toxins, however, they may be the method of choice.

### B. Community Structure Metrics: The Dominant Paradigm

The vast majority of biomonitoring is based on measures of community structure rather than ecosystem processes, with the notable exceptions being the long-standing estimates of biological (or biochemical) oxygen demand (BOD), sediment oxygen demand (SOD) and dissolved oxygen measures (DO). This is largely because communities act as remote sensors, integrating the effects of stressors over time and space, thereby enabling impacts to be detected even if the sources are inactive at the time of sampling, and also because they allow impacts to be appraised in an ecological context that cannot be revealed by physicochemical data alone (Bailey *et al.*, 1998; Bradley and Ormerod, 2002; Kowalik *et al.*, 2007).

Modern freshwater biomonitoring is dominated by the use of macroinvertebrates (Carter and Resh, 2001), a situation that has persisted for decades, as a wide range of metrics and indices have emerged, many of which have subsequently been discarded or transformed into new measures (Balloch *et al.*, 1976; Bonada *et al.*, 2006; Cairns and Pratt, 1993; Hellawell, 1986; Rosenberg and Resh, 1993; Woodiwiss, 1964). Macroinvertebrates offer many advantages, including ubiquity, diversity, well-known taxonomy (at least in certain parts of the world), ease of collection, sensitivity to a range of stressors, well-matched lifecycles to the timescales associated with many stressors, importance in key ecosystem processes, and recognised value to freshwater conservation (Chadd and Extence, 2004; Rosenberg and Resh, 1993; Wallace and Webster, 1996).

Other groups commonly employed in biomonitoring include a range of taxa, such as the remaining 'biological quality elements' used to define ecological status under the EU WFD: fish, macrophytes, phytobenthos and phytoplankton. Combinations of simple biotic metrics provide advanced multi-metric indices and these are now commonly used all over the world (e.g. Baptista *et al.*, 2007; Barbour *et al.*, 1999; Hering *et al.*, 2006; Karr and Chu, 1999). The assessment of ecological status also requires that community metrics are related to target values, or linked to models, that detect departures from putative reference conditions. RIVPACS (River InVertebrate Prediction and Classification System; Wright *et al.*, 2000; now superseded by the River Invertebrate Classification Tool—RICT) is such a model, developed on UK river data sets (Figure 2) but with the concept subsequently exported to many other parts of the world (Mason, 2002; Reynoldson *et al.*, 1997; Wright *et al.*, 2000). Unfortunately, the advantages of community-based metrics apply primarily to the developed world, rather than other less well-studied or less wealthy areas (Resh *et al.*, 2004).

The current hegemony of taxonomy-based community metrics will undoubtedly be challenged increasingly, as a range of novel technologies and a renaissance in microbiological approaches appear on the horizon (e.g. Lear *et al.*, 2009). Powerful molecular techniques, based on new sequencing techniques, have started to provide unprecedented new insights into cryptic microscopic biodiversity and, importantly, at costs that have declined exponentially year on year (Figure 3; Hudson, 2008). These new approaches are also appealing from a scientific perspective because they offer, at least theoretically, a means of linking structural measures to key ecosystem processes, such as biogeochemical cycles, and they could provide a powerful microbial complement to current approaches (Purdy *et al.*, 2010).

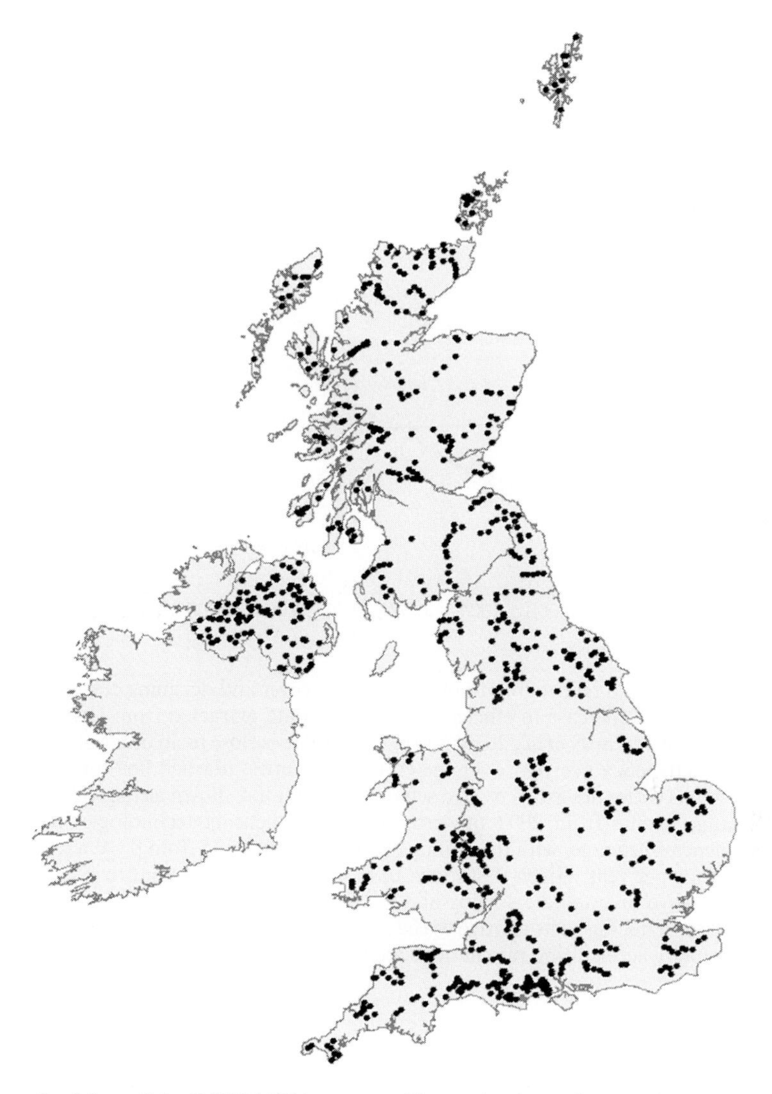

**Figure 2** Map of the RIVPACS long-term biomonitoring reference sites in the UK. These sites were selected primarily to detect and monitor the impacts of organic pollution, via a combination of invertebrate community data and physical habitat attributes.

## C. Putting Nature into Boxes: The Questionable Reliance on 'Typologies'

The deviation of an observed community from what would be predicted at a given site if it were not stressed provides a measure of the response to the stressor, often expressed as a ratio or some form of index. This approach

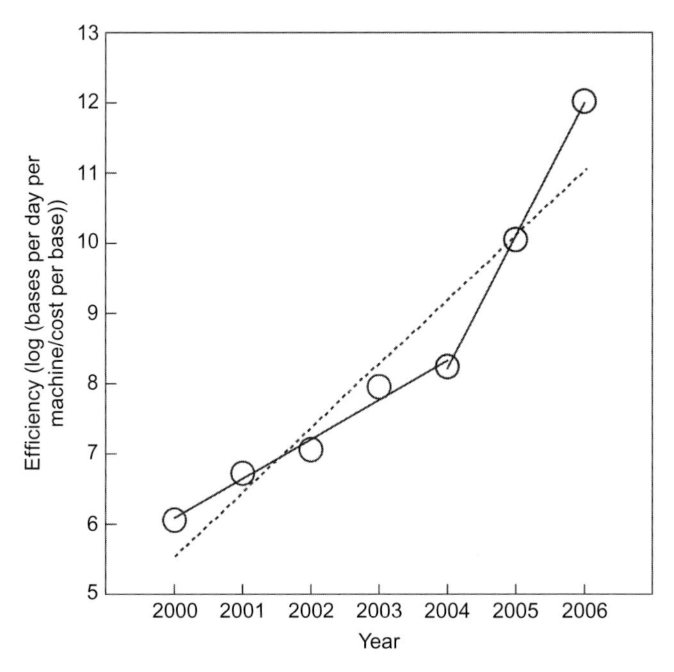


**Figure 3** The rapid recent growth of sequencing power and declining costs, expressed as an exponential increase in efficiency over time (data extracted from Hudson, 2008 and redrawn after Purdy *et al.*, 2010). On average there is close to an order of magnitude increase in efficiency per year over the entire time series: (dashed line): however, the relationship is better described by two separate regressions, shown as the solid lines as a step change occurs from 2004 onwards as new sequencing technologies (e.g. 454-pyrosequencing) emerge, with the scaling coefficient increasing from $\beta = 0.6$ to 2 orders of magnitude per year. Given that much of biomonitoring at present relies heavily on labour-intensive invertebrate sorting and microscope-based identification, there is far less scope for reducing costs relative to the dramatic improvements that are being made at these rapidly accelerating rate in molecular ecology.

generally requires impacted and reference sites to be matched based on their physical characteristics and biogeography, and thus to isolate stressor effects from other natural background differences as far as possible. This has usually been attempted by grouping water bodies or sites into discrete categories, or 'types', so that members of the same type can be compared, one with another.

Forbes and Richardson (1913) proposed a river zonation classification system and Thienemann (1920, 1959) introduced the idea of lake types in the early twentieth century, and typologies have been a persistent concept in freshwater biomonitoring ever since (Bonada *et al.*, 2006). Illies and Botosaneanu (1963) similarly used a complex classification system to divide rivers into distinct zones based on physical characteristics, with each harbouring apparently characteristic and discrete communities. Ecosystem trophic status, community composition or sets of environmental conditions have all been used to define typologies, but all are ultimately based on an *a priori* classification with which the composition of an observed community is then compared. These notions hark back to a long-obsolete concept in plant community ecology (Clements, 1936), in which distinct groups of coevolved species form a 'climax' end-state community or 'super-organism'. Such views are clearly at variance with modern ecology, which is dominated by the concepts of non-equilibrium processes, species turnover, and continua of community structure along environmental gradients, and yet they remain firmly entrenched within many biomonitoring approaches.

Such approaches are not only scientifically questionable, but they are far less effective at predicting reference conditions than more sophisticated multivariate techniques based on continuous gradients that chime more closely with current ecological ideas (e.g. Sandin and Verdonshot, 2006). Community-based approaches based on gradient analysis (e.g. as in RIVPACS; Wright *et al.*, 2000) are more realistic and objectively defined representations of natural systems (Figure 4).

Typologies might serve some useful practical purpose for simplifying our view of complex ecosystems as a basis for defining regional policies, but their value as meaningful and objective means of describing ecosystems is questionable at best. In a comparative study, Snelder *et al.* (2008) found that different typological approaches fitted observed ecological patterns (e.g. large-scale data on fish, invertebrates, water chemistry and hydrology) with similar (low) levels of success, and were little better than a simple neutral model. The persistent use of community 'types' as an integral part of many biomonitoring systems across Europe (e.g. Pottgiesser and Sommerhäuser, 2004), therefore threatens to compromise our understanding and ability to detect and predict impacts, especially as there are better alternatives available that are rooted in modern science rather than ideas that were abandoned by mainstream ecology decades ago.

## D. A More Ecologically Meaningful Alternative to Typologies?

The development of the RIVPACS methodology by the Freshwater Biological Association and then NERC's Centre for Ecology and Hydrology over the past 30 years marked a significant advance in biomonitoring, placing it firmly within the field of general ecology, especially as its statistical approach resonated with niche theory (Clarke *et al.*, 2003; Wright *et al.*, 2000). RIVPACS (and similar approaches based on the same general approach) is a statistically based tool, in which the species composition of

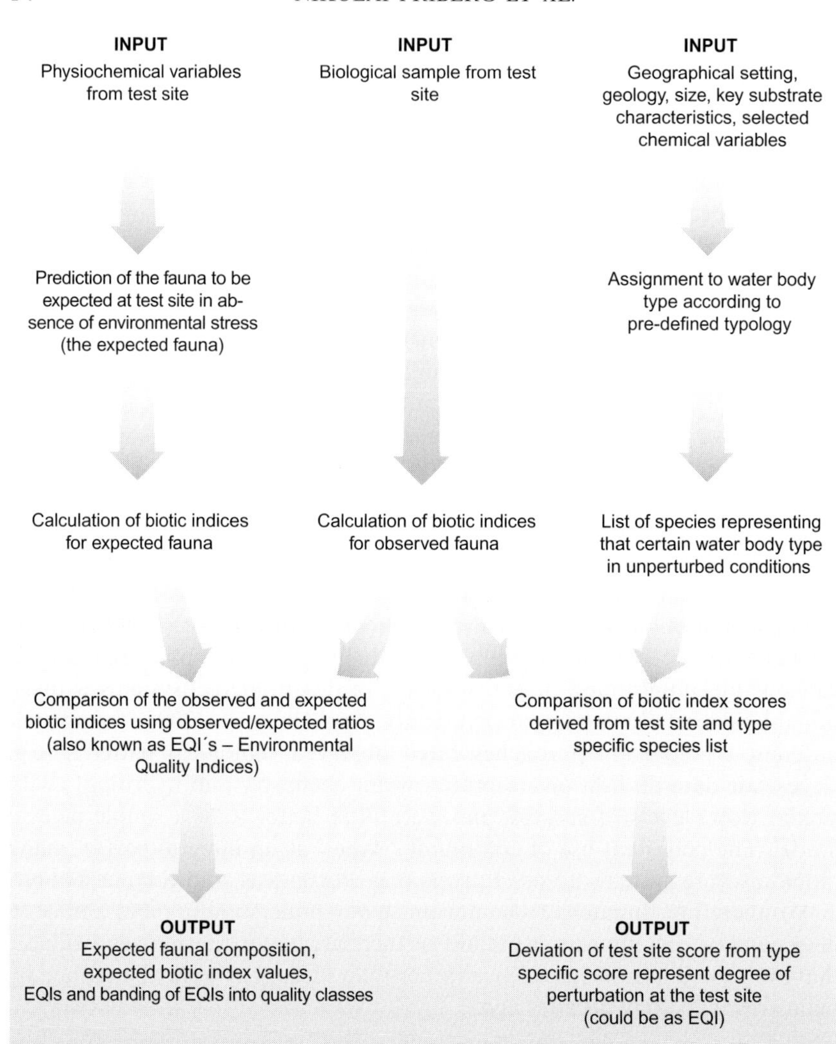


**Figure 4** Schematic highlighting the key differences between Type A and Type B approaches to community-level biomonitoring.

the macroinvertebrate assemblage at a given site can be predicted based on measurements of several key environmental variables (not themselves susceptible to anthropogenic change) and then compared with the assemblage actually observed. A key distinction between these and typology-based approaches is that the former are based on a more realistic gradient-based measure of community change, whereas the latter are based on a reliance on a pre-defined set of discrete types.

Differences between the observed and predicted fauna enable the magnitude of the stressor to be assessed objectively against expected reference conditions along a continuous gradient of community change, a technique that has been adopted in many subsequent schemes around the world, such the Australian AUSRIVAS programme (e.g. Hose *et al.*, 2004; Simpson and Norris, 2000; Sloane and Norris, 2003). Predictions are based on an unconstrained classification (via TWINSPAN) of macroinvertebrate assemblages, followed by predictions derived from multiple discriminant analysis (MDA) derived from a large number of minimally impacted 'reference' stream sites covering all major geographical regions. There are still groupings of sites (types) in RIVPACS, but these are used as a technical step when estimating the probability that observed communities are impacted, rather than assigning them to discrete entities (Figure 4).

The RIVPACS approach is ecologically sound because of two underlying assumptions: (1) it is founded upon analysis of existing communities along environmental gradients, and (2) when sampling a new site, its probability of belonging to each reference group is calculated and these probabilities are combined with the frequency of occurrence of taxa within these groups to derive the list of expected taxa. It is therefore essentially a sophisticated extension of simple discriminant analysis (ter Braak and Prentice, 1988; ter Braak and Šmilauer, 2002; Wright *et al.*, 2000). The extent of the deviation from the expected condition provides the measure of the magnitude of the perturbation.

This system acknowledges that perfect matches are unlikely, even when there is no perturbation, because it is based on probabilities of a site being assigned to a number of different groups (rather than having to conform to a single, fixed typology). RIVPACS offers demonstrably better predictions of expected reference communities than is provided by the typology approach, as revealed by jack-knife tests in which the latter approach provided little improvement on a simple null model based on the average of all reference sites (Figure 5; Davy-Bowker *et al.*, 2006). Other approaches have been based on similar principles as RIVPACS: for example, Oberdorff *et al.* (2001, 2002) and Pont *et al.* (2006) developed bioindication models for fish based on the likelihood of observed communities as estimated from a null 'reference' model of community patterns involving continuous environmental variables, while Ruse (2002) has used multivariate approaches to assess the ecological status of lakes based on chironomid pupal exuviae.

Despite the advantages of RIVPACS-style approaches, it is perhaps surprising that one of the most progressive water policies to date, the WFD, still gives the option of using either a typology (referred to as 'system A') or a predictive modelling approach using site-specific variables (system B) to characterise waterbodies (the latter approach encompasses RIVPACS-style methodologies). This option of being able to use either method has meant that

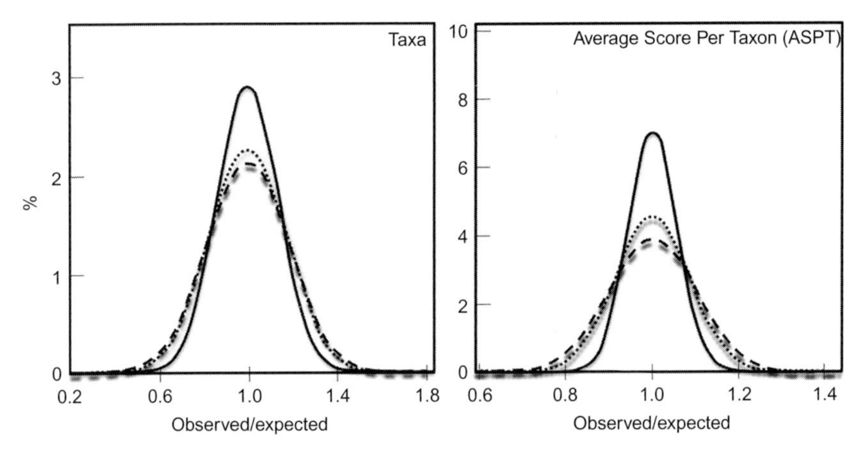


**Figure 5** Relative performance of EU Water Framework Directive *System A* (Fixed Typological Approach, defined by ecoregions and categories of altitude, catchment area and geology: dotted line) and *System B* (Continuous RIVPACS types, defined by a variety of physical and chemical factors: solid line) biomonitoring approaches based on the null model (dashed line). Curves represent observed/expected ratios for the number of taxa (left-hand panel) and ASPT (right-hand panel) for combined spring, summer and autumn data for Great Britain (redrawn after Davy-Bowker *et al.*, 2006). Predictive accuracy increases as the standard deviation of observed/expected ratios declines: note that the typology method represents no marked improvement on the null model. Similar results in the relative performance of *System A* versus *System B*, not shown here, were obtained for Sweden and the Czech Republic (see Davy-Bowker *et al.*, 2006 for details).

fundamentally different statutory biomonitoring approaches exist in parallel across Europe, rather than having a single coherent means of comparing stressor impacts at an international scale, and considerable effort has been devoted to 'intercalibrating' this confusing morass of different methods (Sandin and Hering, 2004). Typology-based approaches currently underpin monitoring programmes in many of the EU Member States in which tools for assessing biological quality elements were not established prior to the implementation of the WFD. The modern revival of typological categorisation might also have arisen partly from misinterpretations of technical steps of some multivariate statistical approaches (e.g. the biological clusters in TWINSPAN), and the desire to compress complex (but realistic) continua into a simpler categories. It should, however, not be confused with the erroneous assumption, based on long-obsolete ecological theory, that real and distinct community 'types' exist as some form of super-organism. There is a real danger that such misunderstandings, if left unchallenged, could become firmly entrenched in the mindset of practitioners.

## E. Functional Approaches

There is a scarcity of clear functional metrics in biomonitoring schemes, especially in running water, as most of the legislation to date has focussed on structural indices. There has been little incentive or impetus to develop new functional metrics, probably reflecting the assumption that both approaches are somehow correlated, and therefore redundant. It has become increasingly apparent, however, that the two approaches often provide complementary information, such that together they are greater than the sum of their parts (Hladyz *et al.*, 2011a,b).

The traditional reliance on community-based perspectives has been questioned recently, as interest in quantifying 'ecosystem function' has grown, especially in light of the debates about biodiversity–ecosystem functioning relationships, conservation and restoration that have gained increasing prominence in recent years (Balvanera *et al.*, 2006; Covich *et al.*, 2004; Hooper *et al.*, 2002, 2005; Loreau *et al.*, 2002; McGrady-Steed *et al.*, 1997; Petchey *et al.*, 2002, 2004; Van der Heijden *et al.*, 1998; Woodward, 2009). The unfortunate term 'function' (which is implicitly teleological) is now firmly embedded in the scientific lexicon and reflects the growing recognition of the need to include dynamical, process-based measures as alternatives to the previous focus on structural metrics (e.g. species richness, Shannon-Wiener indices, indicator taxa, BMWP scores; Metcalfe-Smith, 1996).

In the past decade or so, considerable funds have been allocated to develop functional measures (e.g. European Commission, 1998), but most are still embryonic and have yet to be widely adopted, and of all the functional metrics available to assess or give an integrative measure of the 'health' of running waters, oxygen has undoubtedly been measured the most frequently: its appeal lies in the fact that it is the ultimate integrator of metabolic activity in aquatic ecosystems and is also comparatively easy and cheap to measure. In addition, not only does its consumption provide us with a measure of aerobic respiration coupled to the oxidation of organic carbon compounds but it can also 'integrate' anaerobic respiratory pathways to give an overall measure of ecosystem respiration (ER; Demars *et al.*, 2011; Glud, 2008; Young *et al.*, 2008).

The simple BOD assay gives a measure of the potential of a sample of water to consume oxygen at a standard temperature and forms a staple component of many routine biomonitoring exercises. However, it only gives an indication of organic pollution in the water itself, and provides no information about processes occurring in the system as a whole (Young *et al.*, 2008). Attempts to measure whole system or benthic community respiration as a functional indicator of ecosystem stress have increased in recent years, as the techniques involved have become both cheaper and more refined (Bott *et al.*, 1985; Bunn *et al.*, 1999; Demars *et al.*, 2011; Uzarski *et al.*,

2001; Yvon-Durocher *et al.*, 2010a,b), although it is still necessary to bear in mind the potential confounding effects of advective flow (Cook *et al.*, 2007), light levels and nutrient concentrations (Trimmer *et al.*, 2009).

Young *et al.* (2008) recently carried out an extensive meta-analysis of published gross primary production (GPP) and ER data from both reference ($n = 213$) and impacted streams ($n = 82$) and proposed a robust framework for assessing and distinguishing between streams of 'poor' or 'good' health using such measures of 'ecosystem metabolism'. Recent methodological advances in measuring whole-system metabolism (e.g. Demars *et al.*, 2011; Roberts *et al.*, 2007) have also been complemented by long-term mesocosm scale experiments (Figure 6) and the recent rapid development of theoretical frameworks, which allow responses to stressors to be predicted from first

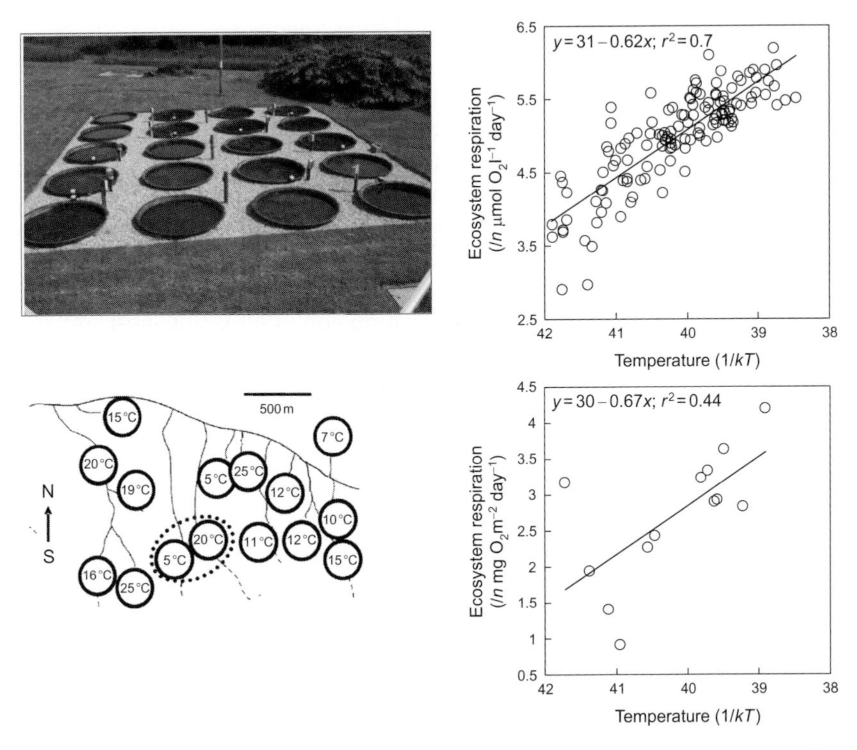


**Figure 6** (A) Experimental ponds used in the long-term environmental warming study at the Freshwater Biological Association River Lab in the UK, and (B) ecosystem respiration as a function of temperature (redrawn after Yvon-Durocher *et al.*, 2010a,b). (C) Icelandic geothermal streams as a model system for measuring higher-level responses to environmental warming in a natural experiment, showing mean summer temperatures, with (D) stream ecosystem respiration as a function of temperature (redrawn after Demars *et al.*, 2011).

principles, and this could open up new possibilities for biomonitoring, especially in response to assessing higher-level responses to climate change (Yvon-Durocher *et al.*, 2010a,b).

To date, most other measures of ecosystem processes of relevance to biomonitoring have focussed mostly on the decomposition of leaf-litter (Figure 7), which is now underpinned by a large and rapidly growing body

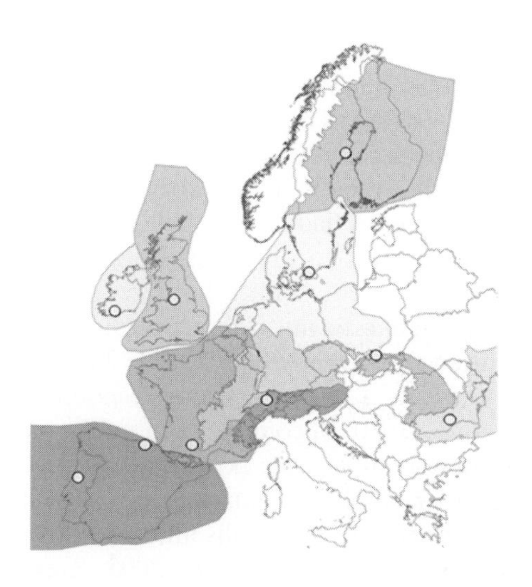

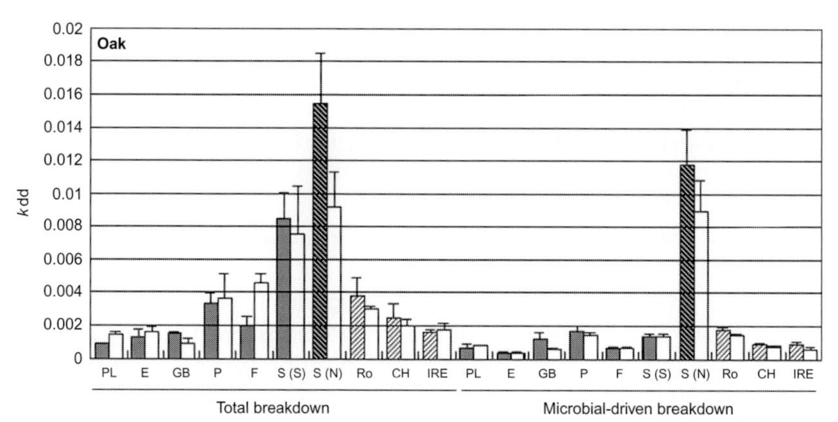


**Figure 7** Functional measures of ecosystem process rates at the continental scale: EU-ecoregions and oak leaf-litter breakdown rates (expressed as $k$-values) in 100 streams in 10 regions across Europe (redrawn after Hladyz *et al.*, 2011b). Fifty streams are reference sites (five per region) bordered by native riparian woodland, whereas the other 50 are impacted sites whose riparian vegetation has been altered by human activity.

of literature (e.g. Dangles *et al.*, 2004; Elosegi *et al.*, 2006; Ferreira and Graça, 2006a,b; Ferreira *et al.*, 2006a,b, 2007; Lecerf and Chauvet, 2008a, b; Lecerf *et al.*, 2006; McKie *et al.*, 2006, 2009; Tiegs *et al.*, 2007, 2008, 2009), much of which derives from the earlier pioneering work of Kaushik and Hynes (1971). This reflects the importance of leaf-litter as a key basal resource in many stream food webs and the fact that its breakdown can be easily measured (e.g. Hladyz *et al.*, 2010, 2011a,b; Riipinen *et al.*, 2010). Energy released from this process fuels the food webs that supply valuable ecosystem goods and services, including the production of fish biomass, and decomposition is also a key step in nutrient cycling.

Despite being advocated as a potentially powerful biomonitoring tool for almost a decade (Gessner and Chauvet, 2002), usable leaf-litter decomposition bioassays are still in their infancy, partly because the process is driven by many different consumers (Hladyz *et al.*, 2011b) and also because the strength and direction of responses to pressures are contingent upon the type of stressor. Nonetheless, recent studies have revealed a general increase in microbial activity relative to invertebrate-mediated decomposition in sites with altered riparian vegetation (Hladyz *et al.*, 2011a,b) and higher temperatures (Boyero *et al.*, 2011), and also strong responses to pH (Dangles *et al.*, 2004; Riipinen *et al.*, 2010). In contrast, and unlike many community structure metrics, responses to organic pollution have been less clear and field bioassays and experimental manipulations of nutrient concentrations have shown conflicting responses (e.g. Gulis and Suberkropp, 2003; Gulis *et al.*, 2006; Krauss *et al.*, 2003; Ramirez *et al.*, 2003; Rosemond *et al.*, 2002). It does, however, offer considerable promise for climate change biomonitoring because altered carbon:nutrient content of litter, which is a strong predictor of breakdown rates (Hladyz *et al.*, 2009), reflects atmospheric $CO_2$ concentrations (Tuchman *et al.*, 2002), and also because decomposition rates are temperature-dependent (Friberg *et al.*, 2009; Perkins *et al.*, 2010a,b).

## F. Species Traits: Linking Structure and Function

Ecosystems processes do not necessarily respond to pressures in the same way, and thus overall 'ecosystem health' cannot be assessed fully by measurements of single rates, such as litter decomposition (Gamfeldt *et al.*, 2008; Reiss *et al.*, 2009, 2010; Ptacnik *et al.*, 2010; Figure 8). In particular, if the species driving these processes are more 'functionally unique' when multiple, rather than single, processes are considered, the consequences of species loss for ecosystem functioning risk being underestimated (Hector and Bagchi, 2007; Ptacnik *et al.*, 2010; Reiss *et al.*, 2009, 2010). Clearly, a better understanding of how functional attributes of organisms map onto ecosystem-level responses is needed (e.g. shredder body mass as a determinant of litter

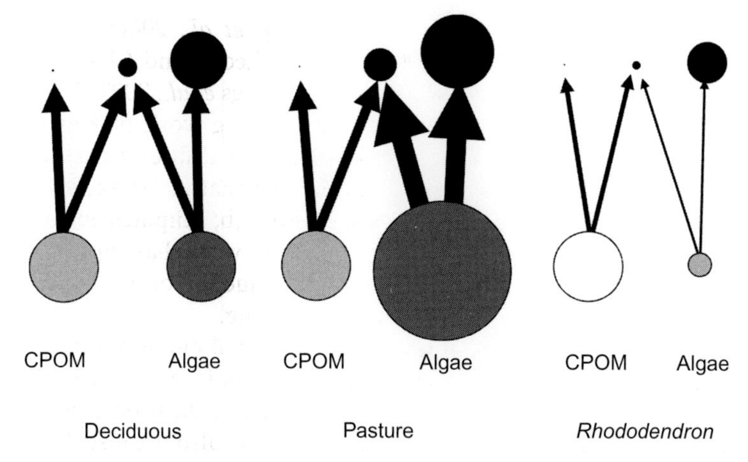


**Figure 8** Functional and structural impacts of an exotic plant (*Rhododendron ponticum*) on stream ecosystems (redrawn after Hladyz *et al.*, 2011a) depicted as simplified food webs in three riparian vegetation types (pasture, native woodland, *Rhododendron* invaded sites). The native woodland sites represent the 'reference' condition, whereas the pasture streams depict a long-term human perturbation (clearance of woodland for pasture) that started in Ireland during the Neolithic, and the *Rhododendron* streams represent a more recent perturbation, resulting from the introduction of this exotic invasive plant in the nineteenth century. Circle diameters are scaled to benthic algal production, the biomass of detrital CPOM per unit area, and primary consumer abundance per unit area, respectively. Specialist detritivore shredders are shown to the left, generalist herbivore–detritivores in the middle, and specialist algal gazers to the right at the top of each web. Circle shading denotes carbon:nitrogen content (black = 0–5; dark grey = 5–10; mid-grey = 10–100; white = 100–1000 [molar ratios]) of consumers and resources. Arrows show whether feeding pathways are degraded in terms of quality (thin arrows), both quality and magnitude (thinnest arrows), or if both are enhanced (thick arrows), relative to the woodland 'reference' streams. Removal of light limitation in pasture streams stimulated algal production, which was consumed by an abundant guild of grazers. *Rhododendron* invasion, however, compromised both algal and detrital pathways in the food web, resulting in marked declines in invertebrate abundance.

breakdown rates) and to understand which species traits are important for particular stressors.

Pioneering work has already been done in assigning species traits to a wide array of freshwater invertebrates, and to measuring the distribution of species traits along environmental gradients using sophisticated multivariate statistics (Dolédec *et al.*, 1999, 2000; Gayraud *et al.*, 2003). Trait-based approaches provide a means of developing and testing *a priori* hypotheses that cannot be addressed using species-based approaches (Statzner *et al.*, 2001a,b; Table 1), and functional trait measures are more stable than structural measures, which can help address some of the problems related to the

**Table 1** Relative performance of invertebrate-based biomonitoring approaches, redrawn after Bonada *et al.* (2006), who selected their 12 grading criteria from Dolédec *et al.* (1999), Lorenz *et al.* (1997), Niemi and McDonald (2004) and Norris and Hawkins (2000)

| | Individuals–populations | | | Community structure | | | Community functional traits | | Ecosystem processes | | |
|---|---|---|---|---|---|---|---|---|---|---|---|
| | Biomarkers | Bioassays | Fluctuating asymmetries | Saprobrian | Multi-metrics | Multivariate approaches | Functional feeding groups | Multitraits approaches | Benthic secondary production | Litter decomposition rates | Σ |
| 1 | ✗ | ✓ | ✓ | ✓ | ± | ✓ | ✓ | ✓ | ± | ✓ | 9 |
| 2 | ✗ | ± | ± | ✗ | ± | ± | ✓ | ✓ | ± | ± | 8 |
| 3 | ✗ | ± | ✗ | ✗ | ± | ? | ✓ | ✓ | ✓ | ✓ | 6 |
| 4 | ± | ? | ✓ | ✗ | ✓ | ✓ | ✗ | ✓ | ✓ | ✓ | 7 |
| 5 | ✓ | ✓ | ± | ✗ | ✓ | ± | ± | ✓ | ± | ± | 9 |
| 6 | ± | ± | ± | ✗ | ± | ✓ | ✓ | ✓ | ✗ | ✓ | 7 |
| 7 | ✓ | ✓ | ✓ | ✗ | ± | ✓ | ? | ✓ | ? | ✗ | 5 |
| 8 | ✓ | ✓ | ✓ | ✗ | ± | ± | ± | ✓ | ± | ✓ | 9 |
| 9 | ? | ± | ? | ✗ | ✗ | ✗ | ± | ✓ | ? | - | 3 |
| 10 | ? | ? | ✗ | ✗ | ± | ± | ✗ | ✓ | ? | ± | 4 |
| 11 | ? | ± | ✗ | ✗ | ✓ | ? | ± | ? | ? | ± | 4 |
| 12 | ± | ± | ? | ✗ | ✗ | ± | ? | ? | ? | ✓ | 4 |
| Σ | 6 | 10 | 7 | 1 | 10 | 9 | 8 | 10 | 6 | 10 | |

Note that the grading criteria are unweighted.

Key to the selection of criteria 1–12.

Rationale (1–5): (1) derived from sound theoretical concepts in ecology; (2) *a priori* predictive; (3) potential to assess ecological processes; (4) potential to discriminate overall human impact (i.e. to identify anthropogenic disturbance); (5) potential to discriminate different types of human impact (i.e. to identify specific types of anthropogenic disturbance).

Implementation (6–8): (6) low costs for sampling and sorting (field approaches) or for standardised experimentation (laboratory approaches); (7) simple sampling protocol; (8) low cost for non-specialist taxonomic identification.

Performance (9–12): (9) large-scale applicability across ecoregions or biogeographic provinces; (10) reliable indication of changes in overall human impact; (11) reliable indication of changes in different types of human impact; (12) human impact indication on linear scale.

Examples of references of each approach highlighted in Table 1:

Biomarkers: Day and Scott (1990), Depledge and Fossi (1994), Karouna-Renier and Zehr (1999), Lagadic *et al.* (2000), Gillis *et al.* (2002), Hyne and Maher (2003).

Bioassays: Crane *et al.* (1995), Maltby *et al.* (2000, 2002), Gerhardt *et al.* (2004).

Fluctuating asymmetries: Groenendijk *et al.* (1998), Dobrin and Corkum, 1999, Drover *et al.* (1999), Hardersen (2000), Savage and Hogarth (1999), Hogg *et al.* (2001), Bonada *et al.* (2005).

Saprobrian: Kolkwitz and Marsson (1902), Rolauffs *et al.* (2004).

Multi-metrics: Thorne and Williams (1997), Mebane (2001), Carlisle and Clements (2005), Vlek *et al.* (2004), Barbour and Yoder (2000), Barbour *et al.* (1995), Pont *et al.* (2006).

Multivariate approaches: Wade *et al.* (1989), Rutt *et al.* (1990), Hämäläinen and Huttunen (1996), Battarbee *et al.* (2001), Boca Raton *et al.* (2000), UKAWMN (2000, 2010), Bradley and Ormerod (2002), Wright *et al.* (2000), Oberdorff *et al.* (2002), Hose *et al.* (2004), Kelly *et al.* (2008), Layer *et al.* (2010a,b, 2011).

Functional feed groups: Palmer *et al.* (1996), Rawer-Jost *et al.* (2000).

Multitraits approaches: Charvet *et al.* (1998, 2000), Bady *et al.* (2005), Bonada *et al.* (2007), Dolédec *et al.* (1996, 1999, 2000), Gayraud *et al.* (2003), Lamouroux *et al.* (2004), Statzner *et al.* (2001a,b, 2004, 2005).

Benthic secondary production: Buffagni and Comin (2000), De Langue *et al.* (2004), Carlisle and Clements (2005), Entrekin *et al.* (2009).

Litter decomposition rates: Carlisle and Clements (2005), Gulis *et al.* (2006), Lecerf *et al.* (2006), Tiegs *et al.* (2009), Hladyz *et al.* (2010, 2011a,b), Riipinen *et al.* (2010).

sensitivity of reference condition to natural variability (Péru and Dolédec, 2010). They could also help overcome one of the major shortcomings of existing community-based approaches, due to their potential lack of biogeographical constraints (Figures 9 and 10), thereby providing a possible means of developing a more standardised, universal approach to biomonitoring. An additional advantage is that calibration can be done *post hoc*, as taxonomic data that have been processed over many decades of intensive biomonitoring could be revisited and traits assigned to invertebrates prior to re-analysis. For instance, the extensive RIVPACS database is now freely available for downloading (http://www.ceh.ac.uk/products/software/RIVPACSDatabase.asp) and could be explored in this way: this would represent an extremely efficient way of investing resources for a potentially dramatic gain in understanding and predictive ability.

Of course, traits need to be defined clearly if they are to be useful, and they are still subject to many of the same constraints as taxonomic-based indices, in that they both needs to be linked clearly to the stressor in question and to provide sufficient accuracy and precision to detect changing conditions. Other non-taxonomic approaches to biomonitoring have been used for a long time in marine systems, and functional approaches are now being embedded in terrestrial biomonitoring schemes (Mulder *et al.*, 2011). In the former, the use of *k*-dominance curves and size-based (as opposed to species-based) analysis have been used to gauge of stressor impacts (Petchey and Belgrano, 2010), and in the latter food web approaches and mathematical modelling of energy fluxes have been used (Mulder *et al.*, 2011). These approaches, like trait-based approaches, also offer the potential advantages of being (largely) freed of taxonomic and biogeographical constraints, and offer clear *a priori* predictive power (Figure 10), although they have yet to be embraced by the freshwater biomonitoring fraternity: as such, these could provide useful grounds for cross-fertilisation across marine, terrestrial and freshwater disciplines, which have largely developed biomonitoring approaches independently and in parallel (Figure 11).

### G. Structural V Functional Approaches: Redundant or Complementary Approaches?

Intriguingly, in the Yvon-Durocher *et al.* (2010a,b, 2011a,b) mesocosm studies of the effects of warming on whole-system metabolism and planktonic assemblages, the community-level structural shifts were far more marked than those of functional measures at the ecosystem level, and there are plenty of other instances of cases in which either structural or functional measures have revealed impacts that have gone undetected by the other (e.g. McKie and Malmqvist, 2009). For example, in the 'clogged' gravels of the River

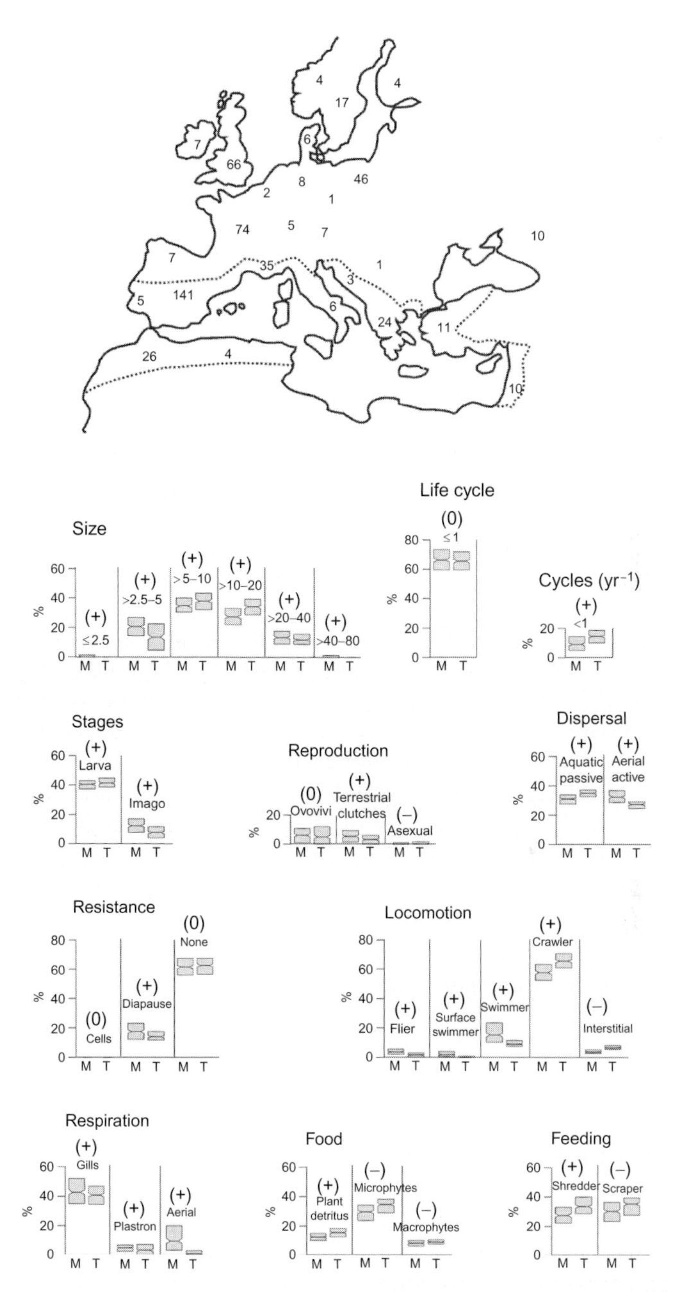


**Figure 9** Number and approximate location of sites and spatial limits of the Mediterranean climate (dotted lines on the map). Box-plots indicating the proportion (in %) of categories per trait for a priori predicted differences in 31 trait categories between Mediterranean (M) and temperate (T) regions. For each trait category, (+) indicates a significant match with the prediction, (–) a significant mismatch, and (0) that no significant differences were found between mediterranean and temperate regions (redrawn after Bonada *et al.*, 2007).

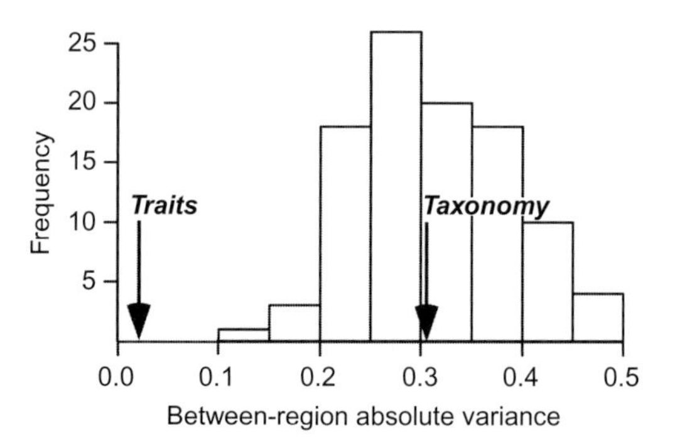


**Figure 10** Comparison of between-region variance (i.e. between Mediterranean and temperate regions) for taxonomic and trait composition. Arrows show the observed values using a dataset of 530 sites. For taxonomic composition, simulations were performed because number of genera did not equal number of trait categories, and frequency of 100 simulated values are presented (redrawn after Bonada *et al.* 2007).

Frome there is a hotspot of anaerobic respiration just 4–6 cm beneath the surface, where nitrate is rapidly reduced and $NH_4^+$, $CO_2$ and $CH_4$ accumulate (Figure 12), and similar observations have been made in otherwise seemingly 'pristine' streams (as suggested by their macroinvertebrate assemblages; Pretty *et al.*, 2006; M. Trimmer, unpublished data). Thus, sampling for surface dwelling communities of invertebrates is unable to detect the clearly abnormal functional state of these rivers, and might also account for the sporadic local extinctions of salmonids, as their spawning grounds are susceptible to anoxia (Malcolm *et al.*, 2004; Walling *et al.*, 2003): our perception of the functional role of the hyporheic zone might need revising, as it may be effectively moribund beneath the sediment surface, even if community structure on the thin skin of the sediment surface appears to be healthy.

## III. EXAMINING THE FOUNDATIONS

### A. The Need for Simplicity and Clarity

Simplicity of approach was a key goal of most biomonitoring and assessment systems developed during the 1970s, whereby sampling and laboratory techniques and taxonomic resolution were kept to the minimum required to address the problem. Cost–benefit analyses and the desire to minimise avoidable errors formed the reasoning behind this parsimonious approach. Since biomonitoring samples are typically collected and processed by a range of individuals with different skills (Rosenberg and Resh, 1993), operator bias can create considerable noise in the data, and the potential for such errors to

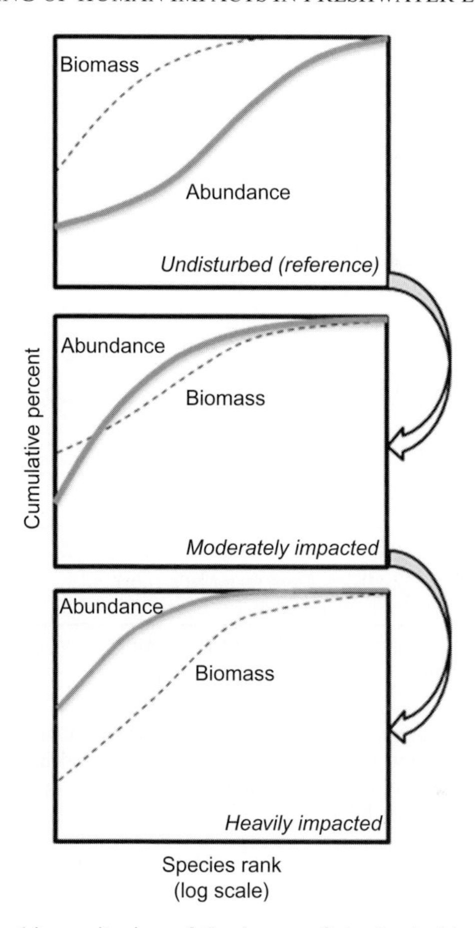


**Figure 11** Freeing biomonitoring of the 'curse of the Latin binomial'? Non-taxonomic metrics of community structure are commonly used in marine biomonitoring: theoretical ABC curves, illustrating patterns of abundance and biomass in response to increasing levels of perturbation (redrawn after Yemane *et al.*'s, 2005 modification of Clarke and Warwick, 1994). Such approaches have yet to be adopted systematically in freshwater biomonitoring, despite their roots in general ecology and their potential for global application due to the absence of contingent biogeographical effects on the taxonomic composition of local communities.

accumulate increases with the complexity of the sampling design: because such errors tend to be confounded with particular sites or dates, they often cannot be corrected without tortuous calibration and fitting of covariate terms that necessitate direct comparisons between operators for a given sample (Carter and Resh, 2001). The UK and many other national environment agencies spend considerable sums auditing the sorting and identification of macroinvertebrates samples in order to quantify operator errors and

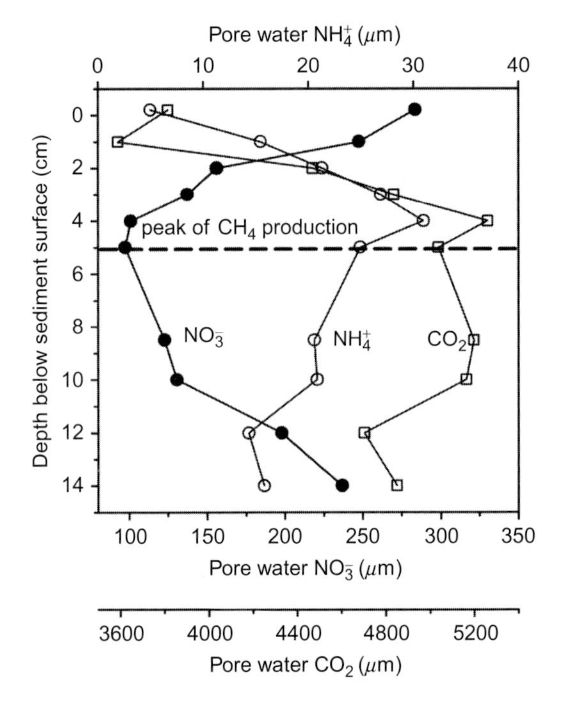


**Figure 12** Sediment pore water profiles for nitrate ($NO_3^-$), ammonium ($NH_4^+$) and dissolved inorganic carbon (as $CO_2$ by GC/FID) measured in the fine sands which accumulate over the gravel beds in the River Frome, Dorset, UK. Data points are means ($n=9$) and error bars have been omitted for clarity but are $<15\%$ of the mean. The sediment depth horizon where the peak in $CH_4$ production was measured is also indicated and coincides with the middle of the $NO_3^-$ reduction zone (reproduced with permission from Trimmer *et al.*, 2009; see Sanders *et al.*, 2007 for production and profile data for $CH_4$).

bias estimates, which are used to quantify uncertainties in assigning sites to quality classes (Dines and Murray-Bligh, 2000) and to ensure that the data are defensible if challenged. Even so, there is often little that can be done to quantify errors associated with the collection of samples from the field (Beisel *et al.*, 1998; Clarke *et al.*, 2002; Vinson and Hawkins, 1996). Consequently, the methods of data collection need to be as simple and as easily replicated as possible to maximise accuracy and precision, and, ideally, at the minimum of cost.

Recently, far more elaborate sampling schemes have started to appear, as attempts are made to develop a standardised pan-European protocol, and these are clearly at variance with the earlier, more pragmatic approaches. The STAR-AQEM method (Hering *et al.*, 2003) is based on Barbour *et al.* (1999) and involves sampling macroinvertebrates from major habitats in relation to their occurrence within a reach (i.e. a logical use of the long-accepted practice of stratified sampling). Twenty sample-units are taken in proportion to the relative area of those habitat 'types' with a 5% minimum coverage, which are then physically pooled and subsamples (aliquots) are subsequently taken, from each of which 700 macroinvertebrates are identified and counted. This process is also very labour intensive and has offers no measurable improvement on the faster and cheaper 1-min kick-sampling method employed previously in a national biomonitoring scheme (Friberg *et al.*, 2006). The STAR-AQEM method is less effective than this earlier and simpler method in several key regards. First, the categorisation of microhabitats (another form of 'typologies') is subjective and assumes that (a) such 'categories' exist and have relevance to natural communities; (b) if they do exist, they can be identified consistently and accurately by different operators, across systems and overtime; and (c) the biota respond differentially to the prescribed microhabitat types. These optimistic assumptions are unlikely to be met in reality, and additional problems further undermine this approach: for instance, the pooling and resampling step destroys potentially useful information related to microhabitat preferences among taxa (Friberg *et al.*, 2006). Indeed, the estimated rates of mis-classifying sites according to the STAR-AQEM method class of ecological status averaged about 40%, raising questions about the ability to detect anything other than gross effects of impacts (Clarke *et al.*, 2006). This also contrasts with the 3-min kick-sampling procedure used for bioassessment in the UK (Dines and Murray-Bligh, 2000) that typically collects over 60% of taxa in the local assemblage in a single sample, which is sufficiently precise to distinguish between impacted and reference sites, as well as detecting long-term change (Bradley and Ormerod, 2002), whilst minimising between-operator variance (Furse *et al.*, 1984; Wright *et al.*, 1994). The rationale for trading greater accuracy, replication and precision, and hence robustness and reliability, when set against an increase in unnecessary detail (and spurious accuracy) is questionable, especially if a key motivation is simply to achieve homogenous approaches across different administrative regions, rather than for practical or theoretically sound scientific reasons.

## B. Intercalibration: Forcing Square Pegs into Round Holes

Failure of EU member states to attain 'good ecological status' for their surface waters by 2015 (WFD, Directive 2000/60/EC) risks incurring severe financial sanctions against member states, unless they are granted exemptions to postpone this for certain water bodies until the end of the next River Basin Management Plans, that is, 2021 or 2027 at the latest. This apparently simple and praiseworthy pursuit of a commonly understood view of 'good ecological status' entails many technical and political challenges: the WFD definition of good ecological status is itself couched in rather nebulous terms (e.g. 'slight...') that are open to interpretation, and which will differ among systems and regions. In addition, the WFD provides ample flexibility for member states to develop their own (different) assessment methods, and this has been embraced enthusiastically. Even in cases where there was no previous national method of assessment, practices in developing a 'WFD compliant' method have often diverged considerably. Thus, the objective that 'good ecological status' should mean the same thing throughout the EU, and therefore that member states should be the same before the law, is problematic to say the least.

Nevertheless, the response to the need to intercalibrate methods and boundaries (particularly the 'good-moderate' boundary) between states has often been both energetic and imaginative (Poikane, 2008; van de Bund, 2008). To standardise measurements of anything, in principle, a common reference point is required (in this case, ideally 'reference conditions' defined generally as 'very low (human) pressure, without the effects of major industrialisation, urbanisation and intensification of agriculture and with only very minor modification of physicochemistry, hydromorphology and biology'; van de Bund, 2008). This serves as is a common means of measuring departure from that point. The difficulty of defining reference conditions (states differ in the standards applied), or simply that there are no reference conditions for some systems (e.g. for very large rivers or estuaries), is perhaps the biggest challenge to intercalibration and to a uniform definition of the legally crucial boundary between good and moderate ecological status.

In practice, groups of member states that shared water categories and water body types formed the Geographic Intercalibration Groups (GIGS) which, in the face of great technical difficulties, have generally done a respectable job of cross-calibrating different approaches. The initial notion was to set up a register of sites that reflect national understanding of good ecological status and on which the performance of national assessment methods could be compared. In the event, such registers were incomplete or of limited use, apparently partly because of inadequate data on pressures (Poikane, 2008; van de Bund, 2008). At this point, three intercalibration options were identified. First, member states in a GIG should use the same, WFD-compliant, method for each 'biological quality element' (phytoplankton, macroinvertebrates, phytobenthos, etc.). This apparently attractive and simple option has not been widely adopted, however. Second, once boundaries between high/good and good/moderate status have been set using national methods, these boundaries are then compared using a common metric. Third, national results are compared using a metric specifically designed for intercalibration (see below). As one can imagine, member states and GIGS have been particularly keen to demonstrate statistically that the various national metrics identify similar thresholds and boundaries.

Given this raft of difficulties and complexities, it is not surprising that the desired objective of reliably defined common standards is still some way off. The correlation between national methods and a common metric is often low, making intercalibration impossible in such circumstances (Poikane, 2008; van de Bund, 2008), and neither have uncertainty or the potentially costly risk of misclassification (if a water body is put into the moderate rather than good class) been well characterised.

The ultimate, Utopian goal is to develop simple pan-European biomonitoring tools (European Commission, 1999), which will likely have to rely on a mixture of four main approaches: (1) some unequivocal measures that can made on all comparable systems (e.g. chlorophyll mass in lake plankton); (2) extensions of RIVPACS-type approaches, avoiding pre-defined typologies but perhaps incorporating expected and observed trait-based community composition offer promise in this regard; (3) measures of ecological status need to combine community structural elements with quantification of ecosystem processes; (4) molecular measures of biodiversity and functional trait composition associated with key processes will need to be developed in the near future.

# IV. MATCHING SYMPTOMS TO STRESSORS

## A. A Case Study of Successful Biomonitoring: Acidification and Recovery in European Freshwaters

Despite the numerous pitfalls described above, there are many examples of successful biomonitoring schemes, and here we highlight the United Kingdom Acid Waters Monitoring Network (UKAWMN; Kernan *et al.*, 2010), and other long-term research into the impacts of acidification on fresh waters (e.g. Hildrew, 2009; Weatherley and Ormerod, 1987), as a case study to highlight potential advantages of the general approach. The study of acidification has been on a far smaller scale than that of many other stressors, and the resources devoted to it are small, but nonetheless invaluable insights into impacts at the higher levels of organisation have been gained, particularly when a range of different approaches has been brought to bear (e.g. experiments, surveys and mathematical models) to provide a more holistic understanding. Viewed as an anthropogenic stressor, acidification has had profound ecological effects of freshwater ecosystems (Hildrew and Ormerod, 1995; Ormerod *et al.*, 1987), leading to dramatic shifts in community structure (Jensen and Snekvik, 1972; Ledger and Hildrew, 1998; Magnuson *et al.*, 1984; Townsend *et al.*, 1983; Rosemond *et al.*, 1992), persistence (Layer *et al.*, 2010a,b; Townsend *et al.*, 1987; Weatherley and

Ormerod, 1990), and ecosystem processes (Hildrew *et al.*, 1984; Winterbourn *et al.*, 1992).

After decades of acidification in many parts of the world, controls on polluting emissions (oxides of sulphur and nitrogen) began in the 1980s through the United Nations Economic Commission for Europe (UNECE) Convention on Long Range Transboundary Air Pollution (LRTAP), aimed at reducing the ecological impact of acid deposition. Over the past 3 decades, there have been substantial reductions in acidifying emissions to the atmosphere in the UK, and across Europe as a whole (Kernan *et al.*, 2010). The UK's AWMN was set up explicitly to assess responses to this reduction, and to compare those responses along a gradient of deposition using acid-sensitive sites in the low-deposition region in the north-west as 'controls'. The 22 AWMN sites (11 lakes and 11 streams) are all located in relatively acid-sensitive regions, in mainly upland areas with catchments underlain by base-poor soils and geology, and largely free form other forms of anthropogenic pollution. The Network was also designed to assess the differences between lakes and streams, especially with respect to their hydrological regimes, and the differences between sites with afforested and moorland catchments.

The Network provides high-quality data on a wide range of chemical determinands (not only those immediately reflecting acidification itself), plus a range of biological assemblages, including epilithic algae, macrophytes, macroinvertebrates and fish. The lake sites all had palaeolimnological records of their acidification status, obtained from coring before the monitoring itself began. It has subsequently been possible to match contemporary ecology with sub-fossil remains in new sediments accumulated over the monitoring period.

A key strength of the Network is founded on its strict quality control of data, consistency of methods and personnel and, above all, that it was based on a set of explicit hypotheses from the outset (broadly, that amelioration of water chemistry would follow reduction in emissions, and that the biota would follow chemical recovery; Figure 13). It has also formed a productive 'outdoor laboratory', greatly increasing its value, as a test-bed for research on environmental change in the uplands. It is thus a case study in which biomonitoring related to policy objectives and long-term data in support of science have gone hand in hand to great advantage. Indeed, as acidification itself as a policy issue has declined in importance in the UK (and consequently the UKAWMN may be threatened by financial cuts in the future), its value in other area of environmental policy, such as climate change, is becoming more evident. In addition, the general approach could be exported to other parts of the world, such as China and India, that are now facing increasing threats of acidification in responses to industrialisation.

Perhaps reassuringly, there is now clear chemical recovery in many of the UKAWMN sites, as in other European systems (Davies *et al.*, 2005; Fowler

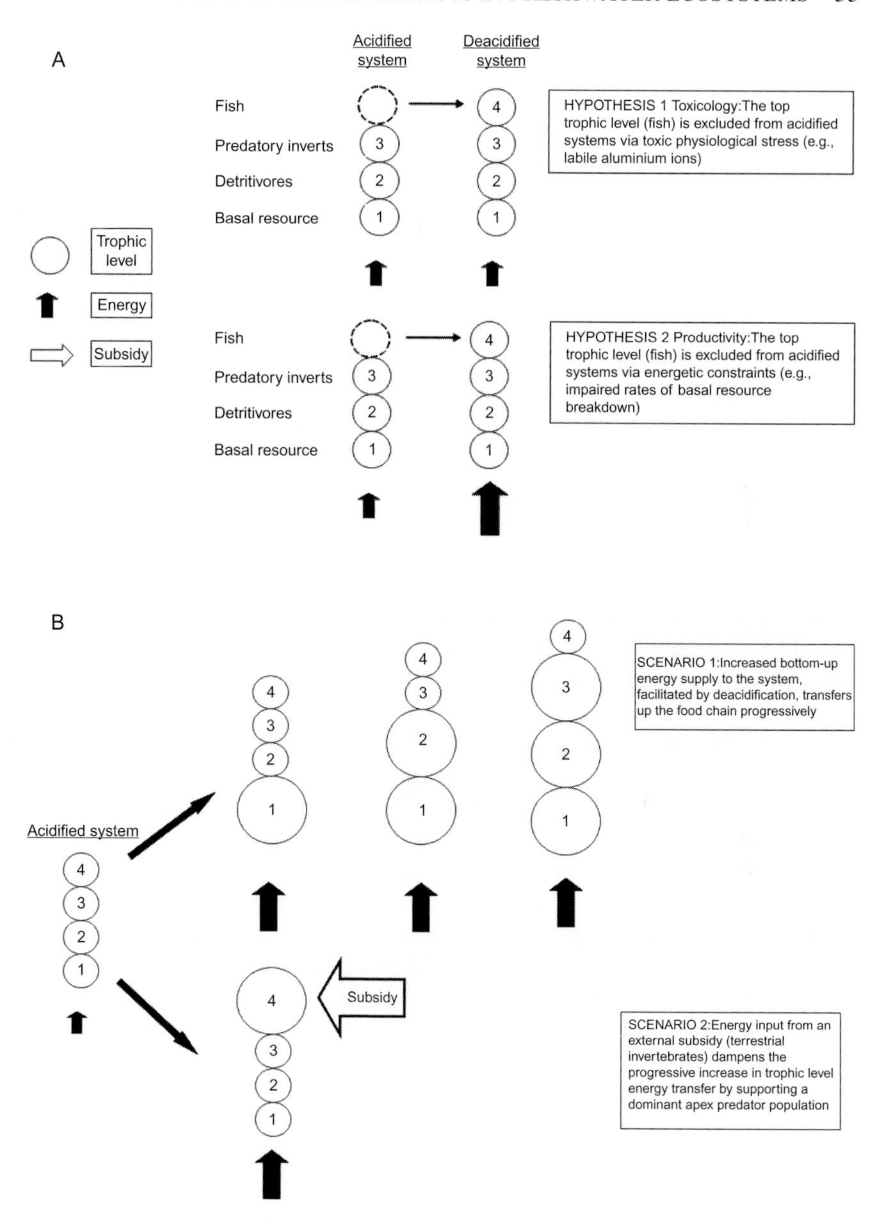


**Figure 13** Schematic outlining the principal general hypotheses underpinning the UKAWMN biomonitoring programme set up to measures responses to and recovery from long acidification.

*et al.*, 2005; Galloway, 1995), and the most recent results from the Network have been summarised in a report to the UK Department of Environment, Food and Rural Affairs (Defra; Kernan *et al.*, 2010). It demonstrates conclusively that acidified rivers and lakes are beginning to recover biologically, although plant and animal communities are far from fully restored (see also Hildrew, 2009; Kowalik *et al.*, 2007; Layer *et al.*, 2010a,b, 2011; Ledger and Hildrew, 2005; Monteith *et al.*, 2005; Rose *et al.*, 2004). This lag between chemical and biological recovery has been ascribed to several possible mechanisms, which are not necessarily mutually exclusive, including repeated acid episodes, dispersal constraints, the intervention of climate change, and 'community closure' due to ecological inertia in the food web (Durance and Ormerod, 2007; Layer *et al.*, 2010a,b, 2011; Lepori *et al.*, 2003; Masters *et al.*, 2007; Yan *et al.*, 2003). Overall, the acidification of fresh waters stands as an excellent example in which science was central to diagnosing an environmental problem, suggesting a solution and (through biomonitoring) assessing the response.

## B. Emerging Stressors and Obsolete Biomonitoring Metrics: The Organic Pollution Time Lag, Habitat Effects and Climate Change

Organic pollution (and increasingly its cousin, eutrophication) has long been the dominant stressor of interest in biomonitoring programmes, with acidification coming second (e.g. Metcalfe-Smith, 1996). Over 50 different macroinvertebrate-based biomonitoring metrics and assessment methods are currently in use (e.g. Friberg *et al.,* 2006; Wright *et al.*, 2000), many of which are based on known or assumed oxygen demands of individual species (Liebmann, 1951; Sladecek, 1973) and are themselves often derived from the earlier saprobic system (Kolkwitz and Marsson, 1902). More recently, there has been an increasing emphasis on multi-metric approaches in many parts of the world and the EU WFD requires that all 'biological quality elements' should be assessed within each type of waterbody, although this does not yet extend to bacteria and other microbial taxa that drive many ecosystem processes. Regardless of which elements of the biota are measured as responses variables, either singly or in combination, organic pollution is still very much the focal predictor variable, as it has been for more than a century.

Despite this pre-eminence in environmental legislation, organic pollution has abated in the past two decades (at least in much of western Europe), as regulation and enforcement, combined with significant investment by the water industry have led to marked improvements in many areas (e.g. Hildrew and Statzner, 2010; Stanners and Bourdeau, 1995). As a result, the effects of previously secondary stressors are now becoming increasingly prominent, raising a new set of challenges and creating the need for new biomonitoring and assessment measures. These apparently newly-emerging stressors are of particular concern to regulatory and monitoring bodies if they prevent the attainment of 'good' or 'high' ecological status in waterbodies that have already undergone significant increases in biological quality. However, many biomonitoring schemes still rely primarily on indices (e.g. ASPT and N-Taxa) designed to measure organic pollution rather than these other, new stressors, and are therefore in danger of becoming increasingly obsolete. This could create mismatches between the drivers and the responses being measured (Woodward, 2009), although there is some evidence that some pressure-specific biological metrics might also be able to detect other previously hidden stressors that they were not originally designed for, including acidification (Acid Waters Indicator Community—AWIC; Davy-Bowker *et al.*, 2005), river flow alterations and habitat modification (Lotic invertebrate Index for Flow Evaluation—LIFE; Extence *et al.*, 1999; Dunbar *et al.*, 2010) and physical habitat degradation (Proportion of Sediment-sensitive Invertebrates, PSI; Extence *et al.*, unpublished data). Further development of such indices is clearly required, particularly if they are to be incorporated into biomonitoring schemes, as a more ecologically meaningful alternative to the current proxy physicochemical measures (e.g. hydrological standards). Such indices will most likely need to be developed in combination and not in isolation, as different pressures are often highly correlated (e.g. Friberg, 2010): for example, LIFE responds not only to low flows but also to some degree to eutrophication, and often these stressors occur together (Dunbar *et al.*, 2010).

Physical habitat degradation has been an important pressure on stream ecosystems for decades and is now a major limiting factor for many rivers achieving their full environmental targets in the light of recent improvements in water quality (Feld *et al.*, 2011). For the first time in any legislation, the EU WFD recognises the importance of river habitats in supporting ecological communities, and introduces the concept of 'hydromorphological quality elements' (a combination of hydrology, geomorphology and connectivity) as a measure of ecosystem status. Despite the development of some standardised protocols (e.g. River Habitat Survey—RHS; Raven *et al.*, 1998), such physical–biological connections remain poorly understood and we are still largely unable to measure habitat characteristics and their degradation (or restoration) effectively and objectively (Feld *et al.*, 2011; Vaughan *et al.*, 2009).

In the UK, river hydromorphology is assessed using the RHS (Raven *et al.*, 1998), which broadly categorises physical features of rivers over 500 m reaches and creates simple indices, such as Habitat Modification Scores and Habitat Quality Assessment scores, to inform river management (Vaughan *et al.*, 2009). It was designed to provide a nationwide inventory of river habitats and the degree of artificial modification. Several studies have related indices derived from the large RHS database to the river biota, using biomonitoring data: one of the most successful of these characterised the slope of the linear response of a macroinvertebrate index (LIFE) to river discharge (Dunbar *et al.*, 2010) Whilst some significant relationships have been reported across the broadest spatial and temporal scales, the strength of these relationships at regional, catchment and reach scales remains uncertain, possibly because RHS surveys provide very detailed information over discrete 500 m reaches, which often do not correspond with macroinvertebrate sampling locations (or scale). Several recent studies and reviews have revealed negligible or weak relationships (Vaughan *et al.*, 2009): Friberg *et al.* (2009), for instance, found a maximum $r^2$ value of just 0.21 between RHS indices and a number of macroinvertebrate indices in streams along a hydromorphological degradation gradient. Further, this study demonstrated that an alternative, more detailed assessment of hydromorphological features suggested that the RHS is not tightly coupled to macroinvertebrate responses at ecologically relevant spatial scales, as more fine-grained measurements (e.g. sediment characteristics at the smaller scales at which macroinvertebrates are sampled) provided a far better fit. The plethora of studies that report sediment conditions as one of the primary drivers of macroinvertebrate community composition lend weight to this contention (e.g. Larsen and Ormerod, 2010; Larsen *et al.*, 2009, 2010; Miyake and Nakano, 2002; Wood and Armitage, 1997).

In the UK, excessive fine sediment in rivers has been reported as a likely reason for failure of meeting WFD standards in ~450 UK waterbodies and is a widespread threat to salmonid spawning grounds. A range of large-scale manipulations to elucidate ecological responses to catchment management initiatives (Defra Demonstration Test Catchments) and recent advances in the monitoring of river habitats, particularly in relation to fish assemblages, have enabled the development of remote sensing methods, based on georeferenced digital aerial imagery combined with, software that can automatically generate river bed grain size maps, water depth estimates and quantified measures of instream mesohabitats of functional importance to the biota (Clough *et al.*, 2010; Figure 14).

In addition to the effects of habitat degradation and flow modification coming increasingly to the fore as the effects of organic pollution subside, climate change is predicted to exert a growing influence in the near future (Perkins *et al.*, 2010a; Walther, 2010), and is likely to interact with these other anthropogenic impacts. In addition to the effects of environmental warming and altered atmospheric composition and carbon cycling, climate change is likely to alter hydrological regimes and the composition of local communities (Perkins *et al.*, 2010a and references therein; also see Bell *et al.*, 2007; Wilby *et al.*, 2008; Johnson *et al.*, 2009). As such, it represents a substantial

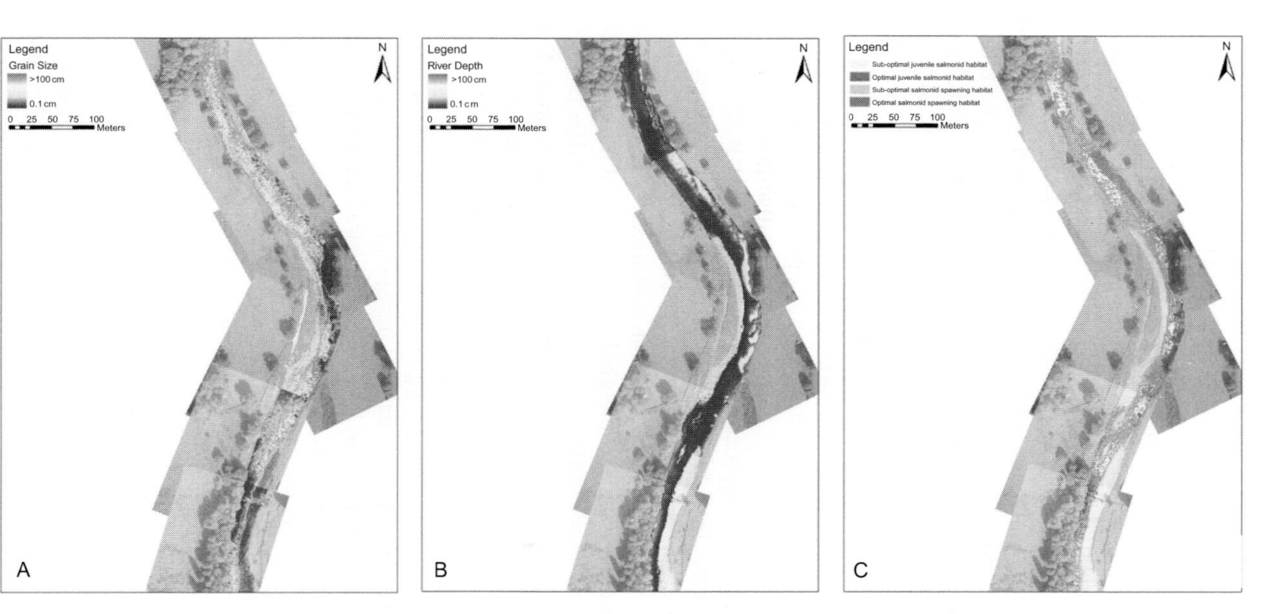


**Figure 14** An example of the outputs the Fluvial Information System (APEM Ltd.) which uses high-resolution digital aerial images to quantify (A) substratum grain size, (B) water depth, and (C) Atlantic salmon functional habitat classification (Clough *et al.*, 2010).

challenge in terms of developing effective biomonitoring strategies to monitor, predict and respond to its effects (Moss *et al.*, 2009), and it is imperative that we start to identify how these components will influence community structure and ecosystem processes, both in isolation and, ultimately, in combination (Perkins *et al.*, 2011).

Most ecological climate change research to date has focussed on the effects of warming and altered atmospheric composition at the lower levels of biological organisation (e.g. individuals, species populations), and biomonitoring has, for instance, already turned its attention to metrics based on temperature-dependent developmental responses (Savage and Hogarth, 1999). Responses to warming at higher levels of organisation are far less well known (Kaushal *et al.*, 2010; Perkins *et al.*, 2010a), although recent surveys, natural experiments and field and laboratory manipulations, in combination with theoretical advances, are starting to provide insights of relevance to biomonitoring (Demars *et al.*, 2011; Friberg *et al.*, 2009; Yvon-Durocher *et al.*, 2010a,b, 2011a,b). *A priori* predictions can be made about the potential impacts of both warming and atmospheric change, based on ecological theory and empirical observations (Perkins *et al.*, 2010a,b, 2011; Sterner and Elser, 2002; Woodward *et al.*, 2010c,e). One recurrent pattern that seems to be emerging from aquatic systems in particular, is that warming favours smaller organisms, both within and across species, as predicted by the temperature size rule and metabolic theory (e.g. Atkinson, 1994; Finkel *et al.*, 2010; Reiss *et al.*, 2010; Yvon-Durocher *et al.*, 2010a,b, 2011a,b,c).

In the absence of more detailed models, coupling of estimates of future high and low flow frequency and magnitude (Bell *et al.*, 2007; Johnson *et al.*, 2009) to existing models of macroinvertebrate community response to flow (Dunbar *et al.*, 2010) may provide a useful line of enquiry. Recent results of long-term mesocosm experiments have also shown that ecosystem processes (secondary production) among certain functional groups of consumers (especially large predators and detritivores) can be severely impaired by intermittent droughts (Ledger *et al.*, 2011). Species invasions or local extinctions due to climate-induced range shifts are likely to be more idiosyncratic and harder to predict, and no strong theoretical *a priori* predictions can be made at present, although this situation is improving (Woodward *et al.*, 2010e).

### C. Biotic Stressors—The Problem of Invasive Species

So far we have focussed on primarily physicochemical stressors, but biotic agents, including invasions of new species associated with climate change, can also exert considerable stress upon natural systems. Invasive species can have particularly strong effects on natural ecosystems (Flecker and Townsend, 1994; Townsend, 2003) and despite their increasing prevalence on a global scale, they have yet to be dealt with in a coherent and consistent way within current biomonitoring schemes. Part of the problem here is that they represent a continuum, such that long-established exotic species may have become such integral components of their recipient ecosystems that their removal would cause a major perturbation, whereas at the other extreme, new arrivals can lead to significant disruption at the invasion front. This is further complicated by the fact that invasions are both natural and anthropogenic-mediated phenomena, with the former generally viewed as being desirable and the latter undesirable. These boundaries are often blurred from a conservation and environmental legislation perspective, as highlighted by the case of *Potamopyrgus jenkinsi*, the widespread and abundant freshwater snail that was long thought to be a native of Europe that was subsequently introduced to New Zealand. More recently it has been identified as an Antipodean species invasive in the UK and in has consequently been renamed as *Potamopyrgus antipodarum* (Ponder, 1988).

Current biomonitoring practice regarding non-native species is highly variable, but in general they are not fully integrated into most assessment schemes (cf. Stoddard *et al.*, 2005, 2006; US EPA, 2006). In parts of the United Kingdom, however, waterbodies with 'a significant presence of invasive non-native species will not meet high ecological status' (Environment Agency (EA), 2009): up to 51% of the 483 water bodies within the Thames catchment could fail to attain 'good' status because of the presence of exotic species (EA, 2009).

The idiosyncratic effects of invasives make them difficult to quantify within traditional biomonitoring programmes. For instance, the zebra mussel (*Dreissena polymorpha*) can suppress phytoplankton biomass by ~85%, increasing water clarity and light availability to the benthos (Caraco *et al.*, 1997) and stimulating primary production as a result (Maguire and Grey, 2006). In contrast, even if the Asian clam (*Corbicula fluminea*) were to replace the native species across the United States, it has been suggested that its impacts on ecosystem processes might be negligible (Vaughn and Hakenkamp, 2001). Similarly, an audit in England conducted in 2005 by English Nature (now Natural England) revealed that, of a total of 2271 non-native species, <1% were identified as having strongly negative environmental impacts (Hill *et al.*, 2005).

Effects are often most severe immediately after establishment, although there are examples of non-native species following a boom-and-bust pattern (Simberloff and Gibbons, 2004; Strayer *et al.*, 2006). For example, the North American water weed (*Elodea canadensis*) initially dominates and excludes most native macrophytes, but within 15–30 years typically diminishes to low levels or disappears; a pattern that has been seen in various countries where it has established (Simberloff and Gibbons, 2004 and references therein). This issue of naturalisation clearly poses problems for biomonitoring schemes: for instance, *P. antipodarum* is no longer considered invasive in the UK, despite its prevalence in many systems (Woodward *et al.*, 2008). Outside its native range it has been recorded at one of the highest levels of production for a stream invertebrate at 194 g $m^{-2}$ $yr^{-1}$, constituting up to 92% of invertebrate production (Hall *et al.*, 2006). It was introduced to the United States just over 20 years ago (Bowler, 1991), where, in contrast to the situation in the U.K., it is still viewed very much as an invasive species that exerts strong impacts on the recipient ecosystems (Hall *et al.*, 2006; Kerans *et al.*, 2005).

There are already too many introduced species for a species-by-species assessment of potential impacts to be feasible in many countries, especially as an insurmountably large number of reference sites would be required. A trait-based functional scheme might therefore be more useful within a biomonitoring context, in which the attributes of both the species and the ecosystem are considered simultaneously to assess overall invasion risk: for instance, what is the likelihood that an invasive species will become a new 'keystone species' (Mills *et al.*, 1993), by exerting a disproportionately strong impact within its recipient ecosystem (Paine, 1995). The warning signs are likely to include whether the organism is an ecosystem engineer (Anderson *et al.*, 2006; Brown and Lawson, 2010), a top predator (potential for trophic cascades; Fritts and Rodda, 1998; Letnic *et al.*, 2009) or a dominant primary producer/consumer (Tuttle *et al.*, 2009).

In terms of identifying traits that make ecosystems vulnerable to invasion, disturbed habitats tend to be especially susceptible, as do those that already have a high number of other invasive species (Lockwood *et al.*, 2007). In particular, urban development, water diversions, aqueducts and agriculture can favour invasions (Hladyz *et al.*, 2011a,b; Marchetti *et al.*, 2004). Such characteristics could be relatively easily incorporated into catchment-level assessments of the risk of the establishment of non-native species, and therefore to guide management.

## D. Multiple Stressors and Their Interactions

Many fresh waters are exposed to multiple stressors (e.g. Ormerod *et al.*, 2010), which can often interact synergistically (Folt *et al.*, 1999; Matthaei *et al.*, 2006; Perkins *et al.*, 2010a; Woodward, 2009): for instance, chemical pollution and low flows frequently combine to exacerbate overall stress levels. Durance and Ormerod (2009) found that improvements in water quality in southern English streams mitigated the effects of warming over a period of 18 years. Since ecological status is an integrated response to multiple pressures, we need to adapt our measurements accordingly. Consequently, an idealised biomonitoring multiple metric should be able to identify not only specific responses to particular stressors but also to gauge general impairment caused by the overall stressor burden (Bonada *et al.*, 2006).

One of the key limitations of existing systems is that they were designed primarily to target a single dominant stressor across a broad gradient (typically from practically pure water to sewage), which inevitably renders them less sensitive to other impacts. In the UK, for example, statutory biomonitoring was originally designed to assess the effects of insanitary water pollution and the current network of sites still reflects this legacy. The multi-metric approaches using macroinvertebrates (e.g. Barbour and Yoder, 2000) and other indicators in running waters (e.g. Fore *et al.*, 1996; Hering *et al.*, 2006; Weigel, 2003) represent important steps towards addressing these problems. Clews and Ormerod (2008) demonstrated that by using simple combinations of standard invertebrate indices, calibrated to detect specific stressors, the diagnostic capabilities of biomonitoring could be improved. Methods sensitive to multiple stressors clearly still need further development, especially in relation to detecting hydromorphological impacts, such as fine sediment deposition (Vaughan *et al.*, 2009).

There are clearly distinct advantages to maintaining consistency within long-term data sets, but these need to be weighed against the appropriateness for answering contemporary and emerging questions in river management. In the UK alone, the quality-assured biomonitoring data set, collected using standardised protocols across circa 20,000 sites over two decades, represents a valuable and still largely untapped resource that could be used more effectively to advance the field and also to feed back into general ecology. The same holds true for Europe as a whole, although data quality and methodologies are often inconsistent both among and within member states, which limits the potential for true continental-scale analyses and reanalyses (Friberg, 2010). Attempts to interrogate existing large-scale biomonitoring data have mostly revealed dominant axes of variation related to organic water pollution, which is not surprising since this is what they were designed to do: however, this can clearly confound the detection of trends due to other stressors, such as climate change or river hydromorphology (Durance and Ormerod, 2009). Whilst acknowledging the potential, as opposed to the realised, value of these vast repositories of data it is timely to review critically the existing body of information, to guide future biomonitoring in tackling contemporary environmental issues in the most scientifically sound and cost-effective manner.

An alternative, but potentially complementary, approach to these hindcasted taxonomic analyses could involve using species traits to assess community-level responses (Statzner *et al.*, 2005; Usseglio-Polatera *et al.*, 2000). Such metrics could, in theory, be applied over areas where there are substantial biogeographical differences, since it is based on functional attributes, to diagnose potential responses to a range of environmental stressors (e.g. Dolédec and Statzner, 2008). At present, these approaches often neglect the natural variation in biological trait distributions among natural reference sites, at least for broad types of streams, so there is clear potential for improvement by developing, as for the RIVPACS approach, environment-dependent reference models, since the traits of invertebrate communities should also vary as a function of multiple environmental variables (e.g. substrate and hydraulics, Lamouroux *et al.*, 2004; Richards *et al.*, 1997).

## V. REASONS FOR STAGNATION: BUREAUCRACY, NEOPHOBIA AND THE INERTIA OF RED TAPE

In addition to the practical and scientific constraints within which biomonitoring schemes operate, legislative and bureaucratic inertia often slow the pace of the development and implementation of new techniques (Moog and Chovanec, 2000), creating significant time lags between current knowledge and operational practices. A lack of desire to introduce new approaches is, of course, understandable in the light of the challenges associated with collecting extensive (new) time series data, staff training and the intercalibration (e.g. Sandin and Hering, 2004) of methods, but as Bonada *et al.* (2006, p. 515) point out, '. . .a long history of use is not a sufficient reason to continue with a biomonitoring tool that is far from ideal'.

For instance, monitoring of restoration schemes often targets physical parameters due to their feasibility and cost-efficiency of measurement (Statzner *et al.*, 1997), whereas current bioindicators are relatively insensitive to hydromorphological changes, so there is a mismatch between what is measured and what needs to be measured (Feld *et al.*, 2011). It must always be kept in mind that notions of 'ecological status' and 'ecological potential' are highly dependent upon somewhat subjective, generally negotiated choices of metrics and thresholds, the definitions of which are constrained by the limits of assessment methods, their interpretability and ability accurately to assign a given system to a particular class. Definitions of ecological status, therefore, need to be viewed more as probability density functions than as discrete contiguous entities (Jones *et al.*, 2010), and this concept needs to be conveyed clearly to policy makers and end-users. The requirement to report not just the membership of a particular sample to a given class, but also a measure of confidence of correct assignment, as can be derived from RIVPACS-style approaches, goes some way towards addressing this issue, and it is critical that the integrity of biomonitoring schemes are not undermined for the sake of political expediency or due to a lack of clarity.

# VI. NEW SOLUTIONS TO OLD PROBLEMS

## A. Reconnecting with Ecological Theory

Although considerable gaps remain, our understanding of the structure and functioning of natural freshwaters and other systems has improved dramatically over the past couple of decades, and this provides greater scope for the development of new approaches to biomonitoring. Unfortunately, many of the key scientific advances made in recent years have outstripped the ability of biomonitoring schemes to keep pace, partly because the latter have become increasingly institutionalised and partly because basic science has been neglected or is still remote from the world of application and implementation. Clearly, a degree of rapprochement is needed, not only to reinvigorate biomonitoring, but also to attract ecologists by offering them access to the vast amounts of empirical data contained in biomonitoring databases.

One key way that these fields can be better linked is for a more *a priori* predictive approach to be embedded in biomonitoring. Pattern-fitting is not mechanistic and even the more sophisticated 'system B' (i.e. RIVPACS-type) approaches are not well suited to *a priori* hypothesis testing. To understand, and hence predict, how ecosystems respond to stressors, ultimately we need to complement biomonitoring schemes with manipulative experiments carried out at appropriate spatiotemporal scales, and also wider use of mathematical models, as is now being done in some of the more recent terrestrial biomonitoring schemes (Mulder *et al.*, 2011). Both approaches are central to ecology but neither has been absorbed fully into the general practice of mainstream biomonitoring, with the exception of a few case studies. One such example comes from the Environment Agency/DEFRA Demonstration Test Catchments project in the UK, in which a large-scale, 4-year manipulation is being undertaken in three large river catchments to test the hypothesis that it is possible, by altering farming practices, to reduce the environmental impacts of diffuse pollution whilst maintaining food security in a cost-effective manner. This is an ambitious, logistically challenging, and long-overdue attempt to link cause and effect at relevant scales, as it involves measuring multiple biological, physical and chemical response variables to track changes in fluxes in response to altered land-use. Whilst this project is essentially research in support of policy, it offers considerable promise for developing future biomonitoring schemes and of reconnecting with basic science.

The community models that underpin biomonitoring tools are generally based on knowledge of species distributions, whose use in the predictive mode requires rigorous testing (Guisan and Thuiller, 2005) due to the complex inter-correlations and spatial autocorrelations of environmental variables (e.g. Dirnböck and Dullinger, 2004). Unfortunately, there is a trend for many models used in large-scale biomonitoring to become more and more diverse, but less and less rigorously tested and there is a clear need to validate models with real data and at appropriate scales, as is now starting to be attempted (e.g. Dunbar *et al.*, 2010).

Identifying the drivers of species presence and abundance is a fundamental ecological goal, yet the focus on pattern-fitting limits the ability of biomonitoring approaches to help develop a mechanistic understanding that would help to inform applied ecology and hence help to improve its ability to adapt to emerging stressors. For instance, it widely assumed that the environment provides the template that in turn determines community composition: if species are 'missing' then the system must be in some way degraded. The problem here is the implicit and axiomatic assumption that communities simply react to the physicochemical environment without any influence of (often indirect) biotic interactions, which are well known as key drivers in general ecology (Woodward, 2009). The shortcomings of this physical-template perspective are revealed by the widespread occurrence of very different suites of species and traits under otherwise identical environmental conditions. A familiar example of this is provided by the alternative equilibria that can exist in mesotrophic shallow lakes: switches between states are driven by biotic rather than abiotic agents (e.g. Scheffer, 1998; Scheffer and Carpenter, 2003; Figure 15).

Identifying cause-and-effect relationships via carefully designed large-scale experiments (Downes, 2010), and the development and testing of *a priori* hypotheses, if integrated carefully into biomonitoring schemes will help in this regard. For instance, the links between biological, geomorphological and hydrological diversity are still poorly understood, largely due to a lack of data collected at relevant spatiotemporal scales (Vaughan *et al.*, 2009) and combining biomonitoring with large-scale experimentation (which is effectively what many restoration schemes are) could provide invaluable new insights of relevance to both pure and applied ecology (Feld *et al.*, 2011). This represents a largely untapped experimental resource on a potentially grand scale: in the United States alone, for instance, only 10% of ~40,000 reported restoration operations have had any form of monitoring (Palmer *et al.*, 2007). If before-and-after data are collected and analysed, ideally in a before-after-control-impact (BACI)-style experimental design, when carrying out future restoration projects, as a form of general best-practice (Souchon *et al.*, 2008), this would equate to a small part of the total budget of such schemes, but the interpretation of the efficacy of restoration schemes would be greatly enhanced. Similar approaches applied to other biomonitoring schemes could also yield significant dividends, and the insights gained could then be used to direct resource in a more targeted and effective manner in future schemes.

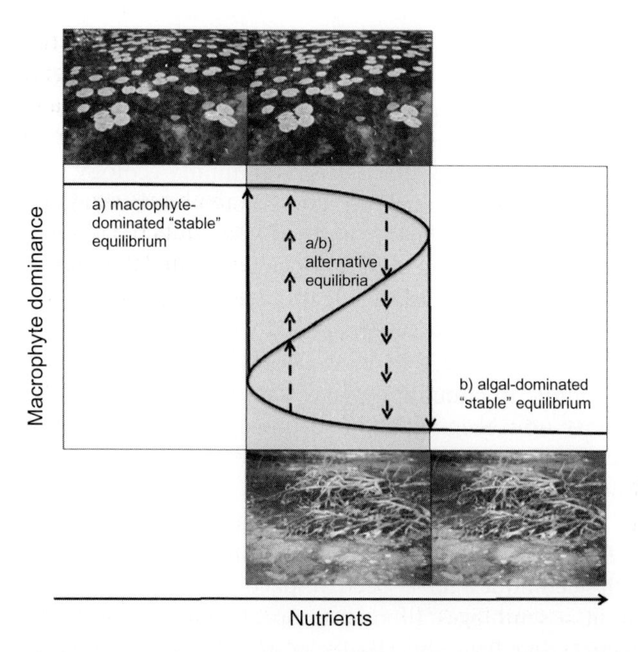


**Figure 15** Biotic and environmental determinants of community structure and ecosystem functioning: alternative equilibria in shallow lakes. At either extreme of the environmental gradient, two different 'stable equilibria' are manifested, with rooted plants (macrophytes) dominating in the clear water state at low nutrient concentrations and phytoplankton dominating under high concentrations, with each state being maintained by a range of positive feedback mechanisms. At intermediate concentrations, denoted by the grey area, alternative states can exist and there is an unstable equilibrium, with the system being able to flip between (A) or (B) if, for instance, it is subjected to a perturbation (dashed arrows), as can occur when fish are added or removed. This demonstrates how biotic drivers can alter ecosystems profoundly, *even* in the absence of any environmental change, via indirect food web effects (redrawn after Jones and Sayer, 2003).

## B. New Technologies: Molecular Microbiology and Functional Genomics

Although biomonitoring has its provenance in microbiology, the field has been dominated by macroinvertebrate indices for much of the previous century, with one of the few notable exceptions being diatom-based indices (e.g. Kelly and Whitton, 1995). This situation could change profoundly in the near future, as new genetic sequencing methods now allow the metagenome of multiple (microbial) species to be characterised *in situ*, rather than having to perform laborious and often unreliable incubations in the laboratory: that is, both taxonomic and functional biodiversity of microbial assemblages can now be measured simultaneously for the first time, without necessarily having to know *which* species are present. These new approaches offer a way to understand how stressors influence the microbial assemblages that drive many ecosystem processes, and it can be used not only to measure microbial biodiversity using traditional community ecology techniques that could previously not be applied (e.g. constructing rank abundance curves of the entire species complement within a given environment), enabling these two fields to become far more closely entwined. In addition, new technologies can also be used simultaneously to identify and quantify key functional genes associated with the ecosystem process of interest (Purdy *et al.*, 2010). A degree of caution still needs to be exercised, however, when extrapolating from metagenomic to community-level data (Kunin *et al.*, 2009), but these techniques have now been applied to a range of microscopic taxa, including viruses, bacteria, fungi, protists and small metazoans.

In addition, variants of these approaches can be used to quantify gene expression, via the even more embryonic field of metatranscriptomics, in which RNA activity is measured *in situ* (Shi *et al.*, 2009; Stewart *et al.*, 2010). Such techniques have been applied recently to characterise gene expression in assemblages of microscopic eukaryotes (including protists and small metazoans) in soil (Bailly *et al.*, 2007) they are still subject to poor coverage and often not reliably reproducible (Gifford *et al.*, 2010; Stewart *et al.*, 2010). If these methodological issues can be resolved, from a biomonitoring perspective, the new generation of techniques that are emerging could potentially provide a means of linking the presence, abundance and expression of genes to whole-ecosystem processes.

The current disconnect between the measures commonly used (i.e. macroinvertebrate community indices and biogeochemical processes) has thus far hindered the development of a useful structural–functional approach. One of the obvious reasons for this mismatch is that macroinvertebrates are not major players in many key ecosystem processes, bar perhaps secondary production and leaf-litter decomposition (Hladyz *et al.*, 2011). Many major biogeochemical cycles are driven primarily by bacteria, and yet we still know almost nothing about their community composition or functional biodiversity in natural systems. This poses a major problem for biomonitoring, as these key groups are often overlooked: for example, although the EU WFD identifies larger organisms (e.g. fishes and invertebrates) as 'biological quality elements', this is not the case for bacteria.

It is conceivable that in the not-too-distant future metagenomic sequencing could be carried out as a fairly routine step in biomonitoring sampling, and potentially for less cost than macroinvertebrate approaches. The newly-emerging approaches, such as 454-pyrosequencing, have not yet been introduced to formal biomonitoring programmes to the best of our knowledge, and so they are still an unknown quantity that requires further investment in research and development. The vast amounts of data that are generated require skilled bioinformaticians to extract the relevant information and this must then be interpreted in an ecological context, but once collected the sequenced data can be revisited repeatedly as bioinformatics databases continue to improve rapidly in size and quality.

If such new techniques are to be introduced in the future it is essential to run them in parallel with traditional metrics, at least initially, in order to be able to assess their relative merits, and to provide a means of potentially calibrating across methods. In addition, as with macroinvertebrate community metrics, the challenge of defining reference conditions remains. Because microbial assemblages have far more rapid turnover rates than macroinvertebrate assemblages, metagenomics could help develop early warning systems in the former, to complement the longer-term integrated signals picked up in the latter. Consequently, we do not recommend simply replacing taxonomic-based with molecular-based approaches, as these could provide even less ecological insight than current schemes unless the functional role of the metagenome (e.g. Lear *et al.*, 2009) can be quantified and linked to ecosystem-level responses to stressors.

## VII. CONCLUSIONS: A LIGHT AT THE END OF THE TUNNEL?

There is an urgent and growing need to anticipate how ecosystems will respond to future environmental change and the importance of new stressors (e.g. climate change and habitat degradation) means that many of the prevailing approaches require a major overhaul, or at least the introduction of additional complementary approaches, to strengthen their ability to provide meaningful and predictive insights. Although biomonitoring is often undertaken primarily for legal compliance, rather than for primarily scientific reasons, with all the political and economic ramifications that this entails (Cairns, 1988), carefully designed schemes can also provide invaluable data to test and inform ecological theories and ultimately feed back into improving management practices.

We are not suggesting that it is necessary to comprehend fully all the component parts of an ecosystem in order to make informed predictions, rather that it is more effective to build on existing knowledge (and data) by applying sound scientific principles to underpin biomonitoring schemes. Resources for basic research have been depleted during the same period that has seen significant increased investment in biomonitoring, even though the main intellectual driver in this field is the aim of furthering understanding and prediction (Battarbee *et al.*, 2005). For instance, major advances have been made in 'biodiversity–ecosystem functioning' research (*sensu* Loreau *et al.*, 2002) and food web ecology (Ings *et al.*, 2009) in the past decade that offer many potentially useful new perspectives, especially as they consider the key role of species interactions that has been largely overlooked in freshwater biomonitoring schemes to date, although they have been incorporated into recent terrestrial schemes (Mulder *et al.*, 2011). In particular, such approaches can be used to make *a priori* predictions about how ecosystem processes or community properties might respond to stressors (Mulder *et al.*, 2011; Perkins *et al.*, 2010b; Petchey and Belgrano, 2010; Reiss *et al.*, 2010). Indeed, many of the biomonitoring programmes employed in marine ecosystems have already adopted such theory-based approaches to predict the likely impacts of perturbations from first principles (Petchey and Belgrano, 2010; Sweeting *et al.*, 2009; Warwick and Clarke, 1991). In freshwaters, for instance, metabolic theory predicts that environmental warming should lead to whole-system respiration rates increasing at a fast rate than primary production, pushing ecosystems towards greater heterotrophy, and favouring smaller bodied organisms, as indeed has been observed in recent long-term mesocosm experiments (Yvon-Durocher *et al.*, 2010a,b, 2011a,b).

Although we have highlighted some of the more major shortcomings, we do not intend in any way to devalue the great importance of well-designed biomonitoring programmes, which provide invaluable information for society as well as a resource for ecologists and others. Recent efforts in biomonitoring activities have certainly contributed to increased methodological exchanges and the development of shared databases at large, and often international, scales. This encouraging new spirit of co-operation and transparency is to be welcomed, especially as these resources can be mined to drive future developments, such as *a posteriori* modelling of functional species traits applied to existing data. However, biomonitoring still needs to become more integrated into adaptive management, based on sound science. To achieve this, more emphasis needs to be placed on understanding mechanisms, particularly via experimentation and modelling, and especially at large scales.

It is also critical to consider carefully both the responses and the drivers that are being measured, and to ensure that they are related to one another in a meaningful way and at appropriate scales. In particular, biomonitoring needs to consider larger scale drivers and responses, and to map these more closely to the relevant scales at which management decisions are made. With new legislation relating to the water environment, such as the WFD, the science is lagging behind the policy objectives, and this increase in scale will help to address this problem. If undertaken using standardised techniques, the use of such large-scale experiments will also improve the prospects for more meaningful meta-analyses and hence interpretation of the effects of management decisions and the efficacy of altered land-use or restoration schemes: the Defra Demonstration Test Catchments provide just one promising example of how this could be done.

Here we argue that it is better to abandon ineffectual approaches that are no longer valid and to replace them with new and more effective methods, but that it is essential that this is done on a rolling schedule, such that new and old methods are run in parallel for sufficient time to make meaningful comparisons of their relative strengths and weaknesses, and also to intercalibrate their results. Given that we are currently standing at a crossroads, it is critical that these issues are given due and careful consideration, as biomonitoring moves into a new and evermore challenging era.

## ACKNOWLEDGEMENTS

The authors would like to thank Iwan Jones and Richard Johnson for providing helpful comments on an earlier version of the manuscript, and the UKAWMN and Helen Bennion for providing data for several of the figures.

## REFERENCES

Anderson, C.B., Griffith, C.R., Rosemond, A.D., Rozzi, R., and Dollenz, O. (2006). The effects of invasive North American beavers on riparian plant communities in Cape Horn, Chile—Do exotic beavers engineer differently in sub-Antarctic ecosystems? *Biol. Conserv.* **128**, 467–474.

Ashbolt, N.J., Grabow, W.O.K., and Snozzi, M. (2001). Indicators of microbial water quality. In: *Water Quality: Guidelines, Standards and Health* (Ed. by L. Fewtrell and J. Bartram), pp. 289–316. IWA Publishing, London.

Ashmore, M.R., Bell, J.N.B., and Reilly, C.L. (1978). A survey of ozone levels in the British Isles using indicator plants. *Nature* **276**, 813–815.

Atkinson, D. (1994). Temperature and organism size—A biological law for ectotherms. *Adv. Ecol. Res.* **25**, 1–58.

Attrill, M.J., and Depledge, M.H. (1997). Community and population indicators of ecosystem health: Targeting links between levels of biological organisation. *Aquat. Toxicol.* **38**, 183–197.

Baattrup-Pedersen, A., Springe, G., Riis, T., Larsen, S.E., Sand-Jensen, K.A., and Kjellerup Larsen, L.M. (2008). The search for reference conditions for stream vegetation in northern Europe. *Freshw. Biol.* **53**, 1890–1901.

Bady, P., Dolédec, S., Fesl, C., Bacchi, M., Gayraud, S., and Schöll, F. (2005). Invertebrate traits for the biomonitoring of large European rivers: Sampling efforts to assess taxa richness or functional diversity. *Freshw. Biol.* **50**(159–73), 9.

Bailey, R.C., Kennedy, M.G., Dervish, M.Z., and Taylor, R.M. (1998). Biological assessment of freshwater ecosystems using a reference condition approach: comparing predicted and actual benthic invertebrate communities in Yukon streams. *Freshw. Biol.* **39**, 765–774.

Bailey, R.C., Norris, R.H., and Reynoldson, T.B. (2004). Bioassessment of Fresh water Ecosystems: Using the Reference Condition Approach. Kluwer, Dordrecht, 184 pp.

Bailly, J., Fraissinet-Tachet, L., Verner, M.C., Debaud, J.C., Lemaire, M., Wesolowski-Louvel, M., and Marmeisse, R. (2007). Soil eukaryotic functional diversity, a metatranscriptomic approach. *ISME J.* **1**, 632–642.

Balloch, D., Davies, C.E., and Jones, F.H. (1976). Biological Assessment of Water Quality in Three British Rivers: The North ESK (Scotland), The Ivel (England) and the TAF (Wales). *Water Pollut. Control* **75**, 92–114.

Balvanera, P., Pfisterer, A.B., Buchmann, N., He, J.S., Nakashizuka, T., Raffaelli, D., and Schmid, B. (2006). Quantifying the evidence for biodiversity effects on ecosystem functioning and services. *Ecol. Lett.* **9**, 1146–1156.

Baptista, D.F., Buss, D.F., Egler, M., Giovanelli, A., Silveira, M.P., and Nessimian, J.L. (2007). A multimetric index based on benthic macroinvertebrates for evaluation of Atlantic streams at Rio de Janeiro State, Brazil. *Hydrobiologia* **575**, 83–94.

Barbour, M.T., and Yoder, C.O. (2000a). The multimetric approach to bioassessment, as used in the United States of America. In: *Assessing the Biological Quality of Freshwaters: RIVPACS and Similar Techniques* (Ed. by J.F. Wright, D.W. Sutcliffe and M.T. Furse), pp. 281–292. Freshwater Biological Association, Ambleside, UK.

Barbour, M.T., and Yoder, C.O. (2000b). The Multimetric Approach to Bioassessment as Used in the United States of America. Lewis 14, Boca Raton, FL. See Ref 149a, pp. 281–292, 15.

Barbour, M.T., Stribling, J.B., and Karr, J.R. (1995). Multimetric approaches for establishing biocriteria and measuring biological condition. In: *Biological Assessment and Criteria: Tools for Water Resource Planning and Decision Making* (Ed. by W.S. Davis and T.P. Simon), pp. 63–77. CRC Press, USA.

Barbour, M.T., Gerritsen, J., Snyder, B.D., and Stribling, J.B. (1999). *Rapid Bioassessment Protocols for Use in Streams and Wadable Rivers: Periphyton, Benthic Macroinvertebrates and Fish.* US EPA, Washington, DC, 202 pp.

Battarbee, R.W., Jones, V.J., Flower, R.J., Cameron, N.G., Bennion, H.B., Carvalho, L., and Juggins, S. (2001). Diatoms. In: *Tracking Environmental Change Using Lake Sediments, Vol. 3: Terrestrial, Algal, and Siliceous Indicators* (Ed. by J.P. Smol and H.J.B. Birks), pp. 155–202. Kluwer Academic Publishers, Dordrecht.

Battarbee, R., Hildrew, A.G., Jenkins, A., Jones, I., Maberly, S., Ormerod, S.J., Raven, P., and Willby, N. (2005). *A Review of Freshwater Ecology in the UK.* Freshwater Biological Association, Ambleside, pp. 1–22.

Beisel, J.N., Usseglio-Polatera, P., Thomas, S., and Moreteau, J.C. (1998). Effects of sampling on benthic macroinvertebrate assemblages in assessment of biological quality of running water. *Ann. De Limnol. Int. J. Limnol.* **34**(4), 445.

Bell, V.A., Kay, A.L., Jones, R.G., and Moore, R.J. (2007). Use of a grid-based hydrological model and regional climate model outputs to assess changing flood risk. *Int. J. Climatol.* **27**, 1657–1671.

Bennion, H., Fluin, J., and Simpson, G.L. (2004). Assessing eutrophication and reference conditions for Scottish freshwater lochs using subfossil diatoms. *J. Appl. Ecol.* **41**, 124–138.

Bennion, H., Juggins, S., and Anderson, N.J. (1996). Predicting epilimnetic phosphorus concentrations using an improved diatom based transfer function, and its application to lake eutrophication management. *Environ. Sci. Technol.* **30**, 2004–2007.

Bennion, H., Battarbee, R.W., Sayer, C.D., Simpson, G.L., and Davidson, T.A. (2010). Defining reference conditions and restoration targets for lake ecosystems using palaeolimnology: A synthesis. *J. Paleolimnol.* **45**, 533–544.

Bonada, N., Vives, S., Rieradevall, M., and Prat, N. (2005). Relationship between pollution and fluctuating asymmetry in the pollution-tolerant caddisfly *Hydropsyche exocellata* (Trichoptera, Insecta). *Arch. Hydrobiol.* **162**, 167–185.

Bonada, N., Prat, N., Resh, V.H., and Statzner, B. (2006). Developments in aquatic insect biomonitoring: A comparative analysis of recent approaches. *Annu. Rev. Entomol.* **51**, 495–523.

Bonada, N., Dolédec, S., and Statzner, B. (2007). Taxonomic and biological trait differences of stream macroinvertebrate communities between mediterranean and temperate regions: Implications for future climatic scenarios. *Glob. Change Biol.* **13**, 1658–1671.

Bott, T.L., Brock, J.T., Dunn, C.S., Naiman, R.J., Ovink, R.W., and Petersen, R.C. (1985). Benthic community metabolism in four temperate stream systems: An interbiome comparison and evaluation of the river continuum concept. *Hydrobiologia* **123**, 3–45.

Boyero, L., Pearson, R.G., Gessner, M.O., Barmuta, L.A., Ferreira, V., Graça, M.A.S., *et al.* (2011). A global experiment suggests climate warming will not accelerate litter decomposition in streams but might reduce carbon sequestration. *Eco. Lett.* **14**, 289–294.

Bowler, P.A. (1991). The rapid spread of the freshwater hydrobiid snail *Potamopyrgus antipodarum* (Gray) in the middle Snake River, Southern Idaho. *Proc. Desert Fish. Counc.* **21**, 173–182.

Bradley, D.C., and Ormerod, S.J. (2002). Long-term effects of catchment liming on invertebrates in upland streams. *Freshw. Biol.* **47**, 161–171.

Brooks, T.M., Mittermeier, R.A., da Fonseca, G.A.B., Gerlach, J., Hoffmann, M., Lamoreux, J.F., Mittermeier, C.G., Pilgrim, J.D., and Rodrigues, A.S.L. (2006). Global biodiversity conservation priorities. *Science* **313**, 58–61.

Brown, B.L., and Lawson, R.L. (2010). Habitat heterogeneity and activity of an omnivorous ecosystem engineer control stream community dynamics. *Ecology* **91**, 1799–1810.

Buffagni, A., and Comin, E. (2000). Secondary production of benthic communities at the habitat scale as a tool to assess ecological integrity in mountain streams. *Hydrobiologia* **422**(423), 183–195.

Bunn, S.E., Davies, P.M., and Mosisch, T.D. (1999). Ecosystem measures of river health and their response to riparian and catchment degradation. *Freshw. Biol.* **41**, 333–345.

Cairns, J., Jr. (1988). Politics, economics, science—Going beyond disciplinary boundaries to protect aquatic ecosystems. In: *Toxic Contaminants and Ecosystem Health: A Great Lakes Focus* (Ed. by M.S. Evans), pp. 1–16. Wiley, New York.

Cairns, J., Jr., and Pratt, J.R. (1993). A History of Biological Monitoring Using Benthic Macroinvertebrates. Chapman and Hall, New York, pp. 10–27.

Caraco, N.F., Cole, J.J., Raymond, P.A., Strayer, D.L., Pace, M.L., Findlay, S.E.G., and Fischer, D.T. (1997). Zebra mussel invasion in a large, turbid river: Phytoplankton response to increased grazing. *Ecology* **78**, 588–602.

Carlisle, D.M., and Clements, W.H. (2005). Leaf litter breakdown, microbial respiration and shredder production in metalpolluted streams. *Freshw. Biol.* **50**, 380–390.

Carter, J.L., and Resh, V.H. (2001). After site selection and before data analysis: Sampling, sorting, and laboratory procedures used in stream benthic macroinvertebrate. *J.N. Am. Benthological Soc.* **20**, 658–682.

Chadd, R., and Extence, C. (2004). The conservation of freshwater macroinvertebrate populations: a community-based classification scheme. *Aquatic Conservation: Marine and Freshwater Ecosystems* **14**, 597–624.

Charvet, S., Kosmala, A., and Statzner, B. (1998). Biomonitoring through biological traits of benthic macroinvertebrates: Perspectives for a general tool in stream management. *Arch. Hydrobiol.* **142**, 415–432.

Charvet, S., Statzner, B., Usseglio-Polatera, P., and Dumont, B. (2000). Traits of benthic macroinvertebrates in semi-natural French streams: An initial application to biomonitoring in Europe. *Freshw. Biol.* **43**, 277–296.

Chessman, B.C., and McEvoy, P.K. (1998). Towards diagnostic biotic indices for river macroinvertebrates. *Hydrobiologia* **364**, 169–182.

Chessman, B.C., Muschal, M., and Royal, M.J. (2008). Comparing apples with apples: use of limiting environmental differences to match reference and stressor exposure sites for bioassessment of streams. *River Res. Appl.* **24**, 103–117.

Clarke, K.R., and Warwick, R.M. (1994). *Changes in Marine Communities: an Approach to Statistical Analysis and Interpretation*. Plymouth Marine Laboratory, Plymouth, England, 144 pp.

Clarke, R.T., Furse, M.T., Gunn, R.J.M., Winder, J.M., and Wright, J.F. (2002). Sampling variation in macroinvertebrate data and implications for river quality indices. *Freshw. Biol.* **47**(9), 1735.

Clarke, R.T., Wright, J.F., and Furse, M.T. (2003). RIVPACS models for predicting the expected macroinvertebrate fauna and assessing the ecological quality of rivers. *Ecol. Modell.* **160**, 219–233.

Clarke, R.T., Lorenz, A., Sandin, L., Schmidt-Kloiber, A., Strackbein, J., Kneebone, N.T., and Haase, P. (2006). Effects of sampling and sub-sampling variation using the STAR-AQEM sampling protocol on the precision of macroinvertebrate metrics. *Hydrobiologia* **566**, 441–459.

Clements, F.E. (1936). Structure and nature of the climax. *J. Ecol.* **24**, 252–284.

Clews, E., and Ormerod, S.J. (2008). Improving bio-diagnostic monitoring using simple combinations of standard biotic indices. *River Res. Appl.* **24**, 1–14.

Clough, S., Campbell, D., Bradley, D., and Hendry, K. (2010). Aerial Photography as a Tool for Salmonid Habitat Assessment, In: *Salmonid Fisheries: Freshwater Habitat Management* (Ed. by P. Kemp), Wiley-Blackwell, Oxford, UK. doi:10.1002/9781444323337.ch13.

Cook, P.L.M., Wenzhofer, F., Glud, R.N., Janssen, F., and Huettel, M. (2007). Benthic solute exchange and carbon mineralization in two shallow subtidal sandy sediments: Effect of advective pore-water exchange. *Limnol. Oceanogr.* **52**, 1943–1963.

Covich, A.P., Austen, M.C., Bärlocher, F., Chauvet, E., Cardinale, B.J., Biles, C.L., Inchausti, P., Dangles, O., Solan, M., Gessner, M.O., Statzner, B., and Moss, B. (2004). The role of biodiversity in the functioning of freshwater and marine benthic ecosystems. *Bioscience* **54**, 767–775.

Crane, M., Delaney, P., Mainstone, C., and Clarke, S. (1995). Measurement by in situ bioassay of water quality in an agricultural catchment. *Water Res.* **29**, 2441–2448.

Dangles, O., Gessner, M.O., Guerold, F., and Chauvet, E. (2004). Impacts of stream acidification on litter breakdown: Implications for assessing ecosystem functioning. *J. Appl. Ecol.* **41**, 365–378.

Davies, J.J.L., Jenkins, A., Monteith, D.T., Evans, C.D., and Cooper, D.M. (2005). Trends in surface water chemistry of acidified UK freshwaters, 1988–2002. *Environ. Pollut.* **137**, 27–39.

Davy-Bowker, J., Murphy, J.F., Rutt, G.R., Steel, J.E.C., and Furse, M.T. (2005). The development and testing of a macroinvertebrate biotic index for detecting the impact of acidity on streams. *Arch. Hydrobiol.* **163**, 383–403.

Davy-Bowker, J., Clarke, R.T., Johnson, R.K., Kokes, J., Murphy, J.F., and Zahradkova, S. (2006). A comparison of the European Water Framework Directive physical typology and RIVPACS-type models as alternative methods of establishing reference conditions for benthic macroinvertebrates. *Hydrobiologia* **566**, 91–105.

Day, K.E., and Scott, I.M. (1990). Use of acetylcholinesterase activity to detect sublethal toxicity in stream invertebrates exposed to low concentrations of organophosphate insecticides. *Aquat. Toxicol.* **18**, 101–114.

De Langue, H.J., De Jonge, J., Den Besten, P.J., Oosterbaan, J., and Peeters, E.T.H.M. (2004). Sediment pollution and predation affect structure and production of benthic macroinvertebrate communities in the Rhine-Meuse delta, The Netherlands. *J. N. Am. Benthol. Soc.* **23**, 557–579.

Demars, B., *et al.* (2011). Temperature and the metabolic balance of streams. *Freshw. Biol.* doi:10.1111/j.1365-2427.2010.02554.x.

Depledge, M.H., and Fossi, M.C. (1994). The role of biomarkers in environmental assessment.2. Invertebrates. *Ecotoxicology* **3**, 161–172.

Dines, R.A., and Murray-Bligh, J.A.D. (2000). Quality assurance and RIVPACS. In: *Assessing the biological quality of fresh waters: RIVPACS and other techniques.* (Eds. by J.F. Wright, D.W. Sutcliffe and M.T. Furse), 373 p. Freshwater Biological Association, Ambleside, UK.

Dirnböck, T., and Dullinger, S. (2004). Habitat distribution models, spatial autocorrelation, functional traits and dispersal capacity of alpine plant species. *J. Veg. Sci.* **15**, 77–84.

Dobrin, M., and Corkum, L.D. (1999). Can fluctuating asymmetry in adult burrowing mayflies (*Hexagenia rigida*, Ephemeroptera) be used as a measure of contaminant stress? *J. Great Lakes Res.* **25**, 339–346.

Dolédec, S., and Statzner, B. (2008). Invertebrate traits for the biomonitoring of large European rivers: an assessment of specific types of human impact. *Freshw. Biol.* **53**, 617–634.

Dolédec, S., Chessel, D., ter Braak, C.J.F., and Champely, S. (1996). Matching species traits to environmental variables: A new three-table ordination method. *Environ. Ecol. Stat.* **3**, 143–166.

Dolédec, S., Statzner, B., and Bournaud, M. (1999). Species traits for future biomonitoring across ecoregions: Patterns along a human-impacted river. *Freshw. Biol.* **42**, 737–758.

Dolédec, S., Olivier, J.M., and Statzner, B. (2000). Accurate description of the abundance of taxa and their biological traits in stream invertebrate communities: Effects of taxonomic and spatial resolution. *Arch. Hydrobiol.* **148**, 25–43.

Downes, B.J. (2010). Back to the future: little-used tools and principles of scientific inference can help disentangle effects of multiple stressors on freshwater ecosystems. *Freshw. Biol.* **55**, 60–79.

Downes, B.J., Barmuta, L.A., Fairweather, P.G., Faith, D.P., Keough, M.J., Lake, P.S., Mapstone, B.D., and Quinn, G.P. (2002). *Monitoring ecological impacts: concepts and practice in flowing water*. Cambridge University Press, New York, NY, USA.

Drover, S., Leung, B., Forbes, M.R., Mallory, M.L., and McNicol, D.K. (1999). Lake pH and aluminum concentration: Consequences for developmental stability of the water strider *Rheumatobates rileyi* (Hemiptera: Gerridae). *Can. J. Zool.* **77**, 157–161.

Dudgeon, D., Arthington, A.H., Gessner, M.O., Kawabata, Z.-I., Knowler, D.J., Lévêque, C., Naiman, R.J., Prieur-Richard, A.-H., Soto, D., Stiassny, M.L.J., and Sullivan, C.A. (2006). Freshwater biodiversity: Importance, threats, status and conservation challenges. *Biol. Rev.* **81**, 163–182.

Dunbar, M.J., Smart, S.M., Murphy, J.F., Clarke, R.T., Baker, R., Edwards, Maskell, L.C., and Scholefield, P. (2010). Chapter 2: Freshwater quality; clean water provision and biodiversity. In: *An integrated assessment of countryside survey data to investigate ecosystem services in Great Britain* (Ed. by S. Smart, M.J. Dunbar, B.A. Emmett, S. Marks, L.C. Maskell, L.R. Norton, P. Rose and I.C. Simpson)., Technical Report No. 10/07 NERC/Centre for Ecology and Hydrology, 230pp. (CEH Project Number: C03259). Centre for Ecology and Hydrology, Wallingford, UK.

Durance, I., and Ormerod, S.J. (2007). Climate change effects on upland stream macroinvertebrates over a 25-year period. *Glob. Change Biol.* **13**, 942–957.

Durance, I., and Ormerod, S.J. (2009). Trends in water quality and discharge offset long-term warming effects on river macroinvertebrates. *Freshw. Biol.* **54**, 388–405.

Elosegi, A., Basaguren, A., and Pozo, J. (2006). A functional approach to the ecology of Atlantic Basque streams. *Limnetica* **25**, 123–134.

Entrekin, S.A., Lamberti, G.A., Tank, J.L., Hoellein, T.J., and Rosi-Marshall, E.L. (2009). Large wood restorations increase macroinvertebrate secondary production in three forested headwater streams. *Freshw. Biol.* **54**, 1741–1748.

Environment Agency (2009). River Basin Management Plan, Thames River Basin District Annex G: Pressures and Risks to waters. Environment Agency, UK.

European Commission (1998). Special report No 3/98 concerning the implementation by the commission of EU policy and action as regards water pollution accompanied by the replies of the commission. *Office J. Eur. Communities* **191**, 2–44.

European Commission (1999). Council decision of 25 January 1999 adopting a specific programme for research, technological development and demonstration on energy, environment and sustainable development (1998–2002). 1999. *Office J. Eur. Communities* **L064**, 58–77.

Extence, C.A., Balbi, D.M., and Chadd, R.P. (1999). River flow indexing using benthic macroinvertebrates: a framework for setting hydrobiological objectives. *Regulated Rivers: Research & Management* **15**, 543–574.

Feld, C.K., Birk, S., Bradley, D.C., Hering, D., Kail, J., Marzin, A., Melcher, A., Nemitz, D., Pederson, M.L., Pletter Bauer, F., Pont, D., Verdonschot, P.F.M., *et al.* (2011). From natural to degraded rivers and back again: a test of restoration ecology theory and practice. *Adv. Ecol. Res.* **44**, 119–210.

Ferreira, V., and Graça, M.A.S. (2006a). Do Invertebrate Activity and Current Velocity Affect Fungal Assemblage Structure in Leaves? *Int. Rev. Hydrobiol.* **91**, 1–14.

Ferreira, V., and Graça, M.A.S. (2006b). Whole-stream nitrate addition affects litter decomposition and associated fungi but not invertebrates. *Oecologia* **149**, 718–729.

Ferreira, V., Graça, M.A.S., de Lima, J.L.M.P., and Gomes, R. (2006a). Role of physical fragmentation and invertebrate activity in the breakdown rate of leaves. *Arch. Hydrobiol.* **164**, 493–513.

Ferreira, V., Elosegi, A., Gulis, V., Pozo, J., and Graça, M.A.S. (2006b). Eucalyptus plantations affect fungal communities associated with leaf-litter decomposition in Iberian streams. *Arch. Hydrobiol.* **166**, 467–490.

Ferreira, V., Gulis, V., and Graça, M.A.S. (2007). Fungal activity associated with decomposing wood is affected by nitrogen concentration in water. *Int. Rev. Hydrobiol.* **92**, 1–8.

Finkel, Z.V., Beardall, J., Flynn, K.J., Quigg, A., Rees, T.A.V., and Raven, J.A. (2010). Phytoplankton in a changing world: Cell size and elemental stoichiometry. *J. Plankton Res.* **32**, 119–137.

Flecker, A.S., and Townsend, C.R. (1994). Community-wide consequences of trout introduction in New Zealand streams. *Ecol. Appl.* **4**, 798–807.

Folt, C.L., Chen, C.Y., Moore, M.V., and Burnaford, J. (1999). Synergism and antagonism among multiple stressors. *Limnol. Oceanogr.* **44**, 864–873.

Forbes, S.A., and Richardson, R.E. (1913). Studies on the biology of the upper Illinois River. *Ill. State Lab. Nat. Hist. Bull.* **9**(10), 481–574.

Fore, L.S., Karr, J.R., and Wisseman, R.W. (1996). Assessing invertebrate responses to human activities: evaluating alternative approaches. *J. N. Am. Benthol. Soc.* **15**, 212–231.

Fowler, D., Smith, R.I., Muller, J.B.A., Hayman, G., and Vincent, K.J. (2005). Changes in the atmospheric deposition of acidifying compounds in the UK between 1986 and 2001. *Environ. Pollut.* **137**(1), 15–25.

Friberg, N. (2010). Pressure-response relationships in stream ecology: introduction and synthesis. *Freshw. Biol.* **55**, 1367–1381.

Friberg, N., Sandin, L., Furse, M.T., Larsen, S.E., Clarke, R.T., and Haase, P. (2006). Comparison of macroinvertebrate sampling methods in Europe. *Hydrobiologia* **566**, 365–378.

Friberg, N., Sandin, L., and Pedersen, M.L. (2009). Assessing the effects of hydromorphological degradation on macroinvertebrate indicators in rivers: examples, constraints and outlook. *Integrated Environmental Assessment and Management (IEAM)* **5**, 86–96.

Friberg, N., Dybkjaer, J.B., Olafsson, J.S., Gislason, G.M., Larsen, S.E., and Lauridsen, T.L. (2009). Relationships between structure and function in streams contrasting in temperature. *Freshw. Biol.* **54**, 2051–2068.

Fritts, T.H., and Rodda, G.H. (1998). The role of introduced species in the degradation of island ecosystems: A case history of Guam. *Annu. Rev. Ecol. Syst.* **29**, 113–140.

Furse, M.T., Moss, D., Wright, J.F., and Armitage, P.D. (1984). The influence of seasonal and taxonomic factors on the ordination and classification of running-water sites in Great Britain and on the prediction of their macro-invertebrate communities. *Freshw. Biol.* **14**, 257–280.

Galloway, J.N. (1995). Acid deposition: Perspectives in time and space. *Water Air Soil Pollut.* **85**, 15–24.

Gamfeldt, L., Hillebrand, H., and Jonsson, P.R. (2008). Multiple functions increase the importance of biodiversity for overall ecosystem functioning. *Ecology* **89**, 1223–1231.

Gayraud, S., Statzner, B., Bady, P., Haybach, A., Schöll, F., *et al.* (2003). Invertebrate traits for the biomonitoring of large European rivers: An initial assessment of alternative metrics. *Freshw. Biol.* **48**, 2045–2064.

Gerhardt, A. (Ed.) (2000). *Biomonitoring of polluted water - reviews on actual topics.* Environ. Science Forum 96; Trans. Tech. Publ., Uetikon-Zürich, Switzerland, 320 pp.

Gerhardt, A., Bisthoven, L., and Soares, A.M.V.M. (2004). Macroinvertebrate response to acid mine drainage: Community metrics and on-line behavioural toxicity bioassay. *Environ. Pollut.* **130**, 263–274.

Gessner, M.O., and Chauvet, E. (2002). A case for using litter breakdown to assess functional stream integrity. *Ecol. Appl.* **12**, 498–510.

Gifford, S., Sharma, S., Rinta-Kanto, J., and Moran, M.A. (2010). Metatranscriptomic analysis of a summer bacterioplankton community in Southeastern U.S. coastal waters. *In* 13th International Symposium on Microbial Ecology. ISME, Seattle, USA.

Gillis, P.L., Diener, L.C., Reynoldson, T.B., and Dixon, D.G. (2002). Cadmium-induced production of a metallothioneinlike protein in *Tubifex tubifex* (Oligochaeta) and *Chironomus riparius* (Diptera): Correlation with reproduction and growth. *Environ. Toxicol. Chem.* **21**, 1836–1844.

Gleick, P.H. (1998). Water in crisis: Paths to sustainable water use. *Ecol. Appl.* **8**, 571–579.

Gleick, P.H., and Palaniappan, M. (2010). Peak water limits to freshwater withdrawal and use. *Proc. Natl. Acad. Sci. USA* **107**, 11155–11162.

Glud, R.N. (2008). Oxygen dynamics of marine sediments. *Mar. Biol. Res.* **4**, 165–179.

Groenendijk, D., Zeinstra, L.W.M., and Postma, J.F. (1998). Fluctuating asymmetry and mentum gaps in populations of the midge Chironomus riparius (Diptera: Chironomidae) from a metal-contaminated river. *Environ. Toxicol. Chem.* **17**, 1999–2005.

Guisan, A., and Thuiller, W. (2005). Predicting species distribution: Offering more than simple habitat models. *Ecol. Lett.* **8**, 993–1009.

Gulis, V., and Suberkropp, K. (2003). Leaf litter decomposition and microbial activity in nutrient-enriched and unaltered reaches of a headwater stream. *Freshw. Biol.* **48**, 123–134.

Gulis, V., Ferreira, V., and Graca, M.A. (2006). Stimulation of leaf litter decomposition and associated fungi and invertebrates by moderate eutrophication: Implications for stream assessment. *Freshw. Biol.* **51**, 1655–1669.

Hall, R.O., Dybdahl, M.F., and Vanderloop, M.C. (2006). Extremely high secondary production of introduced snails in rivers. *Ecol. Appl.* **16**, 1121–1131.

Hämäläinen, H., and Huttunen, P. (1996). Inferring the minimum pH of streams from macroinvertebrates using weighted averaging regression and calibration. *Freshw. Biol.* **36**, 697–709.

Hardersen, S. (2000). The role of behavioural ecology of damselflies in the use of fluctuating asymmetry as a bioindicator of water pollution. *Ecol. Entomol.* **25**, 45–53.

Hawksworth, D.L., and Rose, F. (1970). Qualitative scale for estimating sulphur dioxide pollution in England and Wales using epiphytic lichens. *Nature* **227**, 145–148.

Heckmann, L.-H., Friberg, N., and Ravn, H.W. (2005). Relationship between biochemical biomarkers and pre-copulatory behaviour and mortality in *Gammarus pulex* following pulse-exposure to lambda-cyhalothrin. *Pest Manage. Sci.* **61**, 627–635.

Hector, A., and Bagchi, R. (2007). Biodiversity and ecosystem multifunctionality. *Nature* **448**, 188–190.

Hellawell, J.M. (1986). *Biological Indicators of Freshwater Pollution and Environmental Management*. Elsevier, London, 546 pp.

Hering, D., Buffagni, A., Moog, O., Sandin, L., Sommerhäuser, M., Stubauer, I., Feld, C., Johnson, R.K., Pinto, P., Skoulikidis, N., Verdonschot, P., and Zahrádková, S. (2003). The development of a system to assess the ecological quality of streams based on macroinvertebrates: design of the sampling programme within the AQEM project. *Internat. Rev. Hydrobiol.* **88**, 345–361.

Hering, D., Feld, C.K., Moog, O., and Ofenböck, T. (2006). Cook book for the development of a multimetric index for biological condition of aquatic ecosystems: Experiences from the European AQEM and STAR projects and related initiatives. *Hydrobiologia* **566**, 311–324.

Hildrew, A.G. (2009). Sustained research on stream communities: A model system and the comparative approach. *Adv. Ecol. Res.* **41**, 175–312.

Hildrew, A.G., and Ormerod, S.J. (1995). Acidification: Causes, consequences and solutions. In: *The Ecological Basis for River Management* (Ed. by D.M. Harper and A.J.D. Ferguson), pp. 147–160. John Wiley and Sons Ltd., London.

Hildrew, A.G., and Statzner, B. (2010). European rivers: Perspectives and prospects. In: *European Rivers* (Ed. by K. Tockner and C. Robinson). Elsevier, UK.

Hildrew, A.G., Townsend, C.R., Francis, J.E., and Finch, K. (1984). Cellulolytic decomposition in streams of contrasting pH and its relationship with invertebrate community structure. *Freshw. Biol.* **14**, 323–328.

Hill, M., Baker, R., Broad, G., Chandler, P.J., Copp, G.H., Ellis, J., Jones, D., Hoyland, C., Laing, I., Longshaw, M., Moore, N., Parrott, D., *et al.* (2005). *Audit of non-native species in England.* English Nature Research Reports Number 662. English Nature, UK.

Hladyz, S., Gessner, M.O., Giller, P.S., Pozo, J., and Woodward, G. (2009). Resource quality and stoichiometric constraints in a stream food web. *Freshw. Biol.* **54**, 957–970.

Hladyz, S., Tiegs, S.D., Gessner, M.O., Giller, P.S., Risnoveanu, G., Preda, E., Nistorescu, M., Schindler, M., and Woodward, G. (2010). Leaf-litter breakdown in pasture and deciduous woodland streams: a comparison among three European regions. *Freshw. Biol.* **55**, 1916–1929.

Hladyz, S., Abjornsson, K., Giller, P.S., and Woodward, G. (2011a). Impacts of an aggressive riparian invader on community structure and ecosystem functioning in stream food webs. *J. Appl. Ecol.* **48**, 443–452.

Hladyz, S., Åbjörnsson, K., Chauvet, E., Dobson, M., Elosegi, A., Ferreira, V., Fleituch, T., Gessner, M.O., Giller, P.S., Gulis, V., Hutton, S.A., Lacoursière, J.O., *et al.* (2011b). Stream ecosystem functioning in an agricultural landscape: The importance of terrestrial–aquatic linkages. *Adv. Ecol. Res.* **44**, 211–276.

Hogg, I.D., Eadie, J.M., Williams, D.D., and Turner, D. (2001). Evaluating fluctuating asymmetry in a stream-dwelling insect as an indicator of low-level thermal stress: A large-scale field experiment. *J. Appl. Ecol.* **38**, 1326–1339.

Hooper, D.U., Solan, M., Symstad, A., Diaz, S., Gessner, M.O., Buchman, N., Degrange, V., Grime, P., Hulot, F., Mermillod-Blondin, F., Roy, J., Spehn, E., *et al.* (2002). Species diversity, functional diversity and ecosystem functioning. In: *Biodiversity and Ecosystem Functioning—Synthesis and Perspectives* (Ed. by M. Loreau, S. Naeem and P. Inchausti), pp. 195–208. Oxford University Press, Oxford, UK.

Hooper, D.U., Chapin, F.S., III, Ewel, J.J., Hector, A., Inchausti, P., Lavorel, S., Lawton, J.H., Lodge, D.M., Loreau, M., Naeem, S., Schmid, B., Setälä, H., *et al.* (2005). Effects of biodiversity on ecosystem functioning: A consensus of current knowledge. *Ecol. Monogr.* **75**, 3–35.

Hose, G., Turak, E., and Waddell, N. (2004). Reproducibility of AUSRIVAS rapid bioassessments using macroinvertebrates. *J. N. Am. Benthol. Soc.* **23**, 26–139.

Hudson, M.E. (2008). Sequencing breakthroughs for genomic ecology and evolutionary biology. *Mol. Ecol. Resour.* **8**, 3–17.

Hyne, R.V., and Maher, W.A. (2003). Invertebrate biomarkers: Links to toxicosis that predict population decline. *Ecotox. Environ. Saf.* **54**, 366–374.

Hynes, H.B.N. (1960). *The Biology of Polluted Waters.* Liverpool University Press, Liverpool, England, 202 pp.

Illies, J., and Botosaneanu, L. (1963). Problemes et méthode de la classification et de la zonation ecologique des eaux courantes considerées surtout du pointe de vue faunistique. *Mitteilungen der internationale Vereingigung für theoretische und angewandte Limnologie* **12**, 1–57.

Ings, T.C., Montoya, J.M., Bascompte, J., Bluthgen, N., Brown, L., Dormann, C.F., Edwards, F., Figueroa, D., Jacob, U., Jones, J.I., Lauridsen, R.B., Ledger, M.E., *et al.* (2009). Ecological networks—Beyond food webs. *J. Anim. Ecol.* **78**, 253–269.

Jensen, K.W., and Snekvik, E. (1972). Low pH levels wipe out salmon and trout populations in southernmost Norway. *Ambio* **1**(6), 223–225.

Johnson, A.C., Acreman, M.C., Dunbar, M.J., Feist, S.W., Giacomello, A.M., Gozlan, R.E., Hinsley, S.A., Ibbotson, A.T., Jarvie, H.P., Jones, J.I., Longshaw, M., Maberly, S.C., *et al.* (2009). The British river of the future: How climate change and human activity might affect two contrasting river ecosystems in England. *Sci. Tot. Environ.* **407**, 4787–4798.

Jones, J.I., and Sayer, C.D. (2003). Does the fish-invertebrate-periphyton cascade precipitate plant loss in shallow lakes? *Ecology* **84**, 2155–2167.

Jones, J.I., Davy-Bowker, J., Murphy, J.F., and Pretty, J.L. (2010). Ecological monitoring and assessment of pollution in rivers. In: *Ecology of Industrial Pollution: Remediation, Restoration and Preservation* (Ed. by L. Batty). Cambridge Press, UK.

Jungwirth, M., Haidvogl, G., Hohensinner, S., Muhar, S., Schmutz, S., and Waidbacher, H. (2005). Letbild-specific measures for the rehabilitation of the heavily modified Austria Danube River. *Arch. Hydrobiol.* **15**, 17–36.

Karouna-Renier, N.K., and Zehr, J.P. (1999). Ecological implications of molecular biomarkers: Assaying sub-lethal stress in the midge Chironomus tentans using heat shock protein 70 (HSP-70) expression. *Hydrobiologia* **401**, 255–264.

Karr, J.R., and Chu, E.W. (1999). *Restoring Life in Running Waters: Better Biological Monitoring*. Island, Washington, DC, 200 pp.

Kaushik, N.K., and Hynes, H.B.N. (1971). The fate of the dead leaves that fall into streams. *Arch. Hydrobiol.* **68**, 465–515.

Kaushal, S.S., Likens, G.E., Jaworski, N.A., Pace, M.L., Sides, A.M., Seekell, D., Belt, K.T., Secor, D.H., and Wingate, R.L. (2010). Rising stream and river temperatures in the United States. *Front. Ecol. Environ.* **9**, 461–466.

Kelly, M.G., and Whitton, B.A. (1995). The trophic diatom index: A new new index for monitoring eutrophication. *J. Appl. Phycol.* **7**, 433–444.

Kelly, M.G. (1998). Use of the trophic diatom index to monitor eutrophication in rivers. Water Res., **32**, 236–242.

Kelly, M., Juggins, S., Guthrie, R., Pritchard, S., Jamieson, J., Ripley, B., Hirst, H., and Yallop, M. (2008). Assessment of ecological status in U.K. rivers using diatoms. *Freshw. Biol.* **53**, 403–422.

Kerans, B.L., Dybdahl, M.E., Gangloff, M.M., and Jannot, J.E. (2005). *Potamopyrgus antipodarum*: Distribution, density, and effects on native macroinvertebrate assemblages in the Greater Yellowstone Ecosystem. *J. N. Am. Benthol. Soc.* **24**, 123–138.

Kernan, M., Battarbee, R.W., Curtis, C.J., Monteith, D.T. and Shilland, E.M. (Eds.) (2010). *In* 20 Year Interpretative Report—Recovery of Lakes and Streams in the UK from Acid Rain. http://awmn.defra.gov.uk/resources/interpreports/index.php. The United Kingdom Acid Waters Monitoring Network 20 year interpretative report. ECRC Research Report #141. Report to the Department for Environment, Food and Rural Affairs.

Kimball, K.D., and Levin, S.A. (1985). Limitations of laboratory bioassays: The need for ecosystem-level testing. *Bioscience* **35**, 165–171.

Kolkwitz, R., and Marsson, M. (1902). Grundsatze fur die biologische Beurteilung des Wassers nach seiner Flora und Fauna. *Mitt. Prufungsansalt Wasserversorgung Abwasserreining* **1**, 1–64.

Kolkwitz, R., and Marsson, M. (1909). Ökologie der tierischen Saprobien. *Int. Rev. Gesamten Hydrobiol.* **2**, 126–152.

Kowalik, R.A., Cooper, D.M., Evans, C.M., and Ormerod, S.J. (2007). Acid episodes retard the biological recovery of upland British streams from acidification. *Glob. Change Biol.* **13**, 2439–2452.

Krauss, G., Sridhar, K.R., Jung, K., Wennrich, R., Ehrman, J., and Bärlocher, F. (2003). Aquatic hyphomycetes in polluted groundwater habitats of Central Germany. *Microb. Ecol.* **45**, 329–339.

Kunin, V., Engelbrektson, A., Ochman, H., and Hugenholtz, P. (2009). Wrinkles in the rare biosphere: Pyrosequencing errors can lead to artificial inflation of diversity estimates. *Environ. Microbiol.* **12**, 118–123.

Lagadic, L., Caquet, T., Amiard, J.-C. and Ramade, F. (Eds.) (2000). Use of biomarkers for environmental quality assessment. Enfield/Plymouth: *Science* **84**, pp. 340.

Lamouroux, N., Dolédec, S., and Gayraud, S. (2004). Biological traits of stream macroinvertebrate assemblages: Effects of microhabitat, reach and basin filters. *J. N. Am. Benthol. Soc.* **23**, 449–466.

Lancaster, J. (2000). The ridiculous notion of assessing ecological health and identifying the useful concepts underneath. *Hum. Ecol. Risk Assess.* **6**, 213–222.

Larsen, S., and Ormerod, S.J. (2010). Low-level effects of inert sediment on temperate stream communities. *Freshw. Biol.* **55**, 476–486.

Larsen, S., Vaughan, I.P., and Ormerod, S.J. (2009). Scale-dependent effects of fine sediments on temperate headwater invertebrates. *Freshw. Biol.* **54**, 203–219.

Larsen, S., Pace, G., and Ormerod, S.J. (2010). Experimental effects of sediment deposition on the stucture and function of macroinvertebrate assemblages in temperate streams. *River Res. Appl.* **26**, 1–11.

Layer, K., Hildrew, A.G., Monteith, D., and Woodward, G. (2010a). Long-term variation in the littoral food web of an acidified mountain lake. *Glob. Change Biol.* 10.1111/j.1365-2486.2010.02195.x.

Layer, K., Riede, J.O., Hildrew, A.G., and Woodward, G. (2010b). Food web structure and stability in 20 streams across a wide pH gradient. *Adv. Ecol. Res.* **42**, 265–301.

Layer, K., Hildrew, A.G., Jenkins, G.B., Riede, J.O., Rossiter, S.J., Townsend, C.R., and Woodward, G. (2011). Long-term dynamics of a well-characterised food web: Four decades of acidification and recovery in the broadstone stream model system. *Adv. Ecol. Res.* **44**, 69–118.

Lear, G., Boothroyd, I.K.G., Turner, S.J., Roberst, K., and Lewis, G.D. (2009). A comparison of bacteria and macrobenthic invertebrates as indicators of ecological health in streams. *Freshw. Biol.* **54**, 1532–1543.

Lecerf, A., and Chauvet, E. (2008a). Intraspecific variability in leaf traits strongly affects alder leaf decomposition in a stream. *Basic Appl. Ecol.* **9**, 598–605.

Lecerf, A., and Chauvet, E. (2008b). Diversity and functions of leaf-decaying fungi in human-altered streams. *Freshw. Biol.* **53**, 1658–1672.

Lecerf, A., Usseglio-Polatera, P., Charcosset, J.Y., Lambrigot, D., Bracht, B., and Chauvet, E. (2006). Assessment of functional integrity of eutrophic streams using litter breakdown and benthic macroinvertebrates. *Arch. Hydrobiol.* **165**, 105–126.

Ledger, M.E., and Hildrew, A.G. (1998). Temporal and spatial variation in the epilithic biofilm of an acid stream. *Freshw. Biol.* **40**, 655–670.

Ledger, M.E., and Hildrew, A.G. (2005). The ecology of acidification and recovery: Changes in herbivore-algal food web linkages across a pH gradient in streams. *Environ. Pollut.* **137**, 103–118.

Ledger, M.E., Edwards, F.K., Brown, L.E., Milner, A.M., and Woodward, G. (2011). Impact of simulated drought on ecosystem biomass production: an experimental test in stream mesocosms. *Glob.Change Biol.* **17**, doi: 10.1111/j.1365-2486.2011.02420.x.

Lepori, F., Barbieri, A., and Ormerod, S.J. (2003). Causes of episodic acidification in Alpine streams. *Freshw. Biol.* **48**, 175–189.

Letnic, M., Koch, F., Gordon, C., Crowther, M.S., and Dickman, C.R. (2009). Keystone effects of an alien top-predator stem extinctions of native mammals. *Proc. R. Soc. B Biol. Sci.* **276**, 3249–3256.

Liebmann, H. (1951). Handbuch der Frischwasser und Abwasserbiologie. Verlag v. R. Oldenburg Munchen, Germany, 539 p.

Lockwood, J.L., Hoopes, M.F., and Marchetti, M.P. (2007). Invasion Ecology. Blackwell, Oxford.

Loreau, M., Naeem, S., and Inchausti, P. (2002). Biodiversity and Ecosystem Functioning: Synthesis and Perspectives. Oxford University Press, Oxford, UK.

Lorenz, C.M., Van Dijk, G.M., Van Hattum, A.G.M., and Cofino, W.P. (1997). Concepts in river ecology: Implications for indicator development. *Regul. Rivers* **13**, 501–516.

Magnuson, J.J., Baker, J.P., and Rahel, S.J. (1984). A critical assessment of effects of acidification on fisheries in North America. *Phil. Trans. R. Soc. Lond. B* **305**, 501–516.

Maguire, C.M., and Grey, J. (2006). Determination of zooplankton dietary shift following a zebra mussel invasion, as indicated by stable isotope analysis. *Freshw. Biol.* **51**, 1310–1319.

Malcolm, I.A., Soulsby, C., Youngson, A.F., Hannah, D.M., McLaren, I.S., and Thorne, A. (2004). Hydrological influences on hyporheic water quality: Implications for salmon egg survival. *Hydrol. Process.* **18**, 1543–1560.

Maltby, L., Clayton, S.A., Yu, H., McLoughlin, N., Wood, R.M., and Yin, D. (2000). Using single-species toxicity tests, community-level responses, and toxicity identification evaluations to investigate effluent impacts. *Environ. Toxicol. Chem.* **19**, 151–157.

Maltby, L., Clayton, S.A., Wood, R.M., and McLoughlin, N. (2002). Evaluation of the *Gammarus pulex in situ* feeding assay as a biomonitor of water quality: Robustness, responsiveness, and relevance. *Environ. Toxicol. Chem.* **21**, 361–368.

Marchetti, M.P., Light, T., Moyle, P.B., and Viers, J.H. (2004). Fish invasions in California watersheds: Testing hypotheses using landscape patterns. *Ecol. Appl.* **14**, 1507–1525.

Mason, C.F. (2002). *Biology of Freshwater Pollution*, 4th edn. Prentice-Hall, Pearson Education Limited, Harlow, England, 391 pp.

Masters, Z., Petersen, I., Hildrew, A.G., and Ormerod, S.J. (2007). Insect dispersal does not limit the biological recovery of streams from acidification. *Aquat. Conserv. Mar. Freshw. Ecosyst.* **16**, 1–9.

Matthaei, C.D., Weller, F., Kelly, D.W., and Townsend, C.R. (2006). Impacts of fine sediment addition to tussock, pasture, dairy and deer farming streams in New Zealand. *Freshw. Biol.* **51**, 2154–2172.

McGrady-Steed, J., Harris, P.M., and Morin, P.J. (1997). Biodiversity regulates ecosystem predictability. *Nature* **390**, 162–165.

McKie, B.G., and Malmqvist, B. (2008). Assessing ecosystem functioning in streams affected by forest management: Increased leaf decomposition occurs without changes to the composition of benthic assemblages. *Freshw. Biol.* **54**, 2086–2100.

McKie, B.G., Petrin, Z., and Malmqvist, B. (2006). Mitigation or disturbance? Effects of liming on macroinvertebrate assemblage structure and leaf litter decomposition in the humic streams of northern Sweden. *J. Appl. Ecol.* **43**, 780–791.

McKie, B.G., Schindler, M., Gessner, M.O., and Malmqvist, B. (2009). Placing biodiversity and ecosystem functioning in context: Environmental perturbations and the effects of species richness in a stream field experiment. *Oecologia* **160**, 757–770.

Mebane, C.A. (2001). Testing bioassessments metrics: Macroinvertebrate, sculpin, and salmonid response to stream habitat, sediment, and metals. *Environ. Monit. Assess.* **67**, 293–322.

Metcalfe-Smith, J.L. (1996). Biological water-quality assessment of rivers: Use of macroinvertebrate communities. In: *River Restoration* (Ed. by G. Petts and P. Calow), pp. 17–59. Blackwell Science, Oxford, UK.

Mills, L.S., Soule, M.E., and Doak, D.F. (1993). The keystone-species concept in ecology and conservation. *Bioscience* **43**, 219–224.

Miyake, Y., and Nakano, S. (2002). Effects of substratum stability on diversity of stream invertebrates during baseflow at two spatial scales. *Freshw. Biol.* **47**, 219–230.

Monteith, D.T., Hildrew, A.G., Flower, R.J., Raven, P.J., Beaumont, W.R.B., Collen, P., Kreiser, A.M., Shilland, E.M., and Winterbottom, J.H. (2005). Biological responses to the chemical recovery of acidified fresh waters in the UK. *Environ. Pollut.* **137**, 83–101.

Moog, O., and Chovanec, A. (2000). Assessing the ecological integrity of rivers: Walking the line among ecological, political and administrative interests. *Hydrobiologia* **422**(423), 99–109.

Moss, B., Hering, D., Green, A.J., Aidoud, A., Becares, E., Beklioglu, M., Bennion, H., Boix, D., Brucet, S., Carvalho, L., Clement, B., Davidson, T., *et al.* (2009). Climate change and the future of freshwater biodiversity in Europe: A primer for policy-makers. *Freshw. Rev.* **2**, 103–130.

Mulder, C., Boit, A., Bonkowski, M., De Ruiter, P.C., Mancinelli, G., Van der Heijden, M.G.A., van Wijnen, H.J., Vonk, J.A., and Rutgers, M. (2011). A belowground perspective on dutch agroecosystems: How soil organisms interact to support ecosystem services. *Adv. Ecol. Res.* **44**, 277–358.

Niemi, G.J., and McDonald, M.E. (2004). Application of ecological indicators. *Annu. Rev. Ecol. Evol. Syst.* **35**, 89–111.

Nijboer, R.C., Johnson, R.K., Verdonschot, P.F.M., Sommerhäuser, M., and Buffagni, A. (2004). Establishing reference conditions for European streams. *Hydrobiologia* **516**, 91–105.

Norris, R.H., and Hawkins, C.P. (2000). Monitoring river health. *Hydrobiologia* **435**, 5–17.

Norris, R.H., and Thomas, M.C. (1999). What is river health? *Freshw. Biol.* **41**, 197–209.

Oberdorff, T., Pont, D., Hugueny, B., and Chessel, D. (2001). A probabilistic model characterizing fish assemblages of French rivers: A framework for environmental assessment. *Freshw. Biol.* **46**, 399–415.

Oberdorff, T., Pont, D., Hugueny, B., and Porcher, J.P. (2002). Development and validation of a fish-based index (FBI) for the assessment of 'river health' in France. *Freshw. Biol.* **47**, 1720–1734.

Ormerod, S.J., Boole, P., McCahon, C.P., Weatherley, N.S., Pascoe, D., and Edwards, R.W. (1987). Short-term experimental acidification of a Welsh stream: Comparing the biological effects of hydrogen ions and aluminium. *Freshw. Biol.* **17**, 341–356.

Ormerod, S.J., Dobson, M., Hildrew, A.H., and Townsend, C. (Eds.) (2010). *Multiple Stressors in Freshwater Ecosystems. Freshw. Biol.* **55**, 1–269.

Padisak, J., Brics, G., Grigorszky, I., and Soroczki-Pinter, E. (2006). Use of phytoplankton assemblages for monitoring ecological status of lakes within the Water Framework Directive: the assemblage index. *Hydrobiologia* **553**, 1–14.

Paine, R.T. (1995). A conversation on refining the concept of keystone species. *Conserv. Biol.* **9**, 962–964.

Palmer, C.G., Maart, B., Palmer, A.R., and O'Keeffe, J.H. (1996). An assessment of macroinvertebrate functional feeding groups as water quality indicators in the Buffalo River, eastern Cape Province, South Africa. *Hydrobiologia* **318**, 153–164.

Palmer, M., Allan, J.D., Meyer, J., and Bernhardt, E.S. (2007). River restoration in the twenty-first century: Data and experiential knowledge to inform future efforts. *Restor. Ecol.* **15**, 472–481.

Patrick, R. (1949). A proposed biological measure of stream conditions based on a survey of Conestoga Basin, Lancaster County, Pennsylvania. *Proc. Acad. Natl. Sci. Phila.* **101**, 277–341.

Perkins, D.M., Reiss, J., Yvon-Durocher, G., and Woodward, G. (2010a). Global change and food webs in running waters. *Hydrobiologia* 10.1007/s10750-009-0080-7.

Perkins, D.M., McKie, B.G., Malmqvist, B., Gilmour, S.G., Reiss, J., and Woodward, G. (2010b). Environmental warming and biodiversity–ecosystem functioning in freshwater microcosms: Partitioning the effects of species identity, richness and metabolism. *Adv. Ecol. Res.* **43**, 177–208.

Péru, N., and Dolédec, S. (2010). From compositional to functional biodiversity metrics in bioassessment: A case study using stream macroinvertebrate communities. *Ecol. Indic.* **10**, 1025–1036.

Petchey, O.L., and Belgrano, A. (2010). Body-size distributions and size-spectra: Universal indicators of ecological status? *Biol. Lett.* **6**, 434–437.

Petchey, O.L., Morin, P.J., Hulot, F.D., Loreau, M., McGrady-Steed, J., and Naeem, S. (2002). Contributions of aquatic model systems to our understanding of biodiversity and ecosystem functioning. In: *Biodiversity and Ecosystem Functioning—Synthesis and Perspectives* (Ed. by M. Loreau, S. Naeem and P. Inchausti), pp. 127–138. Oxford University Press, Oxford, UK.

Petchey, O.L., Downing, A.L., Mittelbach, G.G., Persson, L., Steiner, C.F., Warren, P.H., and Woodward, G. (2004). Species loss and the structure and functioning of multitrophic aquatic systems. *Oikos* **104**, 467–478.

Peters, R.H. (1991). *A Critique for Ecology*. Cambridge University Press, Cambridge, UK.

Poikane, S. (2008). Water Framework Directive intercalibration Technical Report: Part 2, Lakes. JRC Scientific and Technical Reports, European Commissison, Institute for Environmental Sustainability, Luxembourg, ISSN: 1018-5593, 183 pp.

Pollard, P., and Huxham, M. (1998). The European Water Framework Directive: A new era in the management of aquatic ecosystem health? *Aquat. Conserv. Mar. Freshwat. Ecosyst.* **8**, 773–792.

Ponder, W.F. (1988). *Potamopyrgus-antipodarum*—A Molluscan Colonizer of Europe and Australia. *J. Molluscan Stud.* **54**, 271–285.

Pont, D., Hugueny, B., Beier, U., Goffaux, D., Melcher, A., Noble, R., Rogers, C., Roset, N., and Schmutz, S. (2006). Assessing river biotic condition at the continental scale: A European approach using functional metrics and fish assemblages. *J. Appl. Ecol.* **43**, 70–80.

Pottgiesser, T., and Sommerhäuser, M. (2004). *VIII-2.1 Fließgewässertypologie Deutschlands: Die Gewässertypen und ihre Steckbriefe als Beitrag zur Umsetzung der EU-Wasserrahmenrichtlinie* (Ed. by C. Steinberg, W. Calmano, H. Klapper and W.-D. Wilken). Handbuch Angewandte Limnologie. 19. Ergänzungslieferung 07/04. ecomed, Landsberg, Germany. 16 pp. + Anhang.

Pretty, J., Hildrew, A.G., and Trimmer, M. (2006). Nitrogen dynamics in relation to surface-subsurface hydrological exchange in a groundwater fed river. *J. Hydrol.* **330**, 84–100.

Ptacnik, R., Moorthi, S.D., and Hillebrand, H. (2010). Hutchinson reversed, or why there need to be so many species. *Adv. Ecol. Res.* **43**, 1–44.

Purdy, K.J., Hurd, P.J., Moya-Laraño, J., Trimmer, M., and Woodward, G. (2010). Systems biology for ecology: From molecules to ecosystems. *Adv. Ecol. Res.* **43**, 87–149.

Ramirez, A., Pringle, C.M., and Molina, L. (2003). Effects of stream phosphorus levels on microbial respiration. *Freshw. Biol.* **48**, 88–97.

Raven, P.J., Holmes, N.T.H., Dawson, F.H., and Everard, M. (1998). Quality assessment using river habitat survey data. *Aquat. Conserv. Mar. Freshw. Ecosyst.* **8**, 477–499.

Rawcliffe, R., Sayer, C.D., Woodward, G., Grey, J., Davidson, T.A., and Jones, J.I. (2010). Back to the future: Using palaeolimnology to infer long-term changes in shallow lake food webs. *Freshw. Biol.* **55**, 600–613.

Rawer-Jost, C., Böhmer, J., Blank, J., and Rahmann, H. (2000). Macroinvertebrate functional feeding group methods in ecological assessment. *Hydrobiologia* **422** (423), 225–232.

Reiss, J., Bridle, J.R., Montoya, J.M., and Woodward, G. (2009). Emerging horizons in biodiversity–ecosystem functioning research. *Trends Ecol. Evol.* **24**, 505–514.

Reiss, J., Bailey, R.A., Cássio, F., Woodward, G., and Pascoal, C. (2010a). Assessing the contribution of micro-organisms and macrofauna to biodiversity–ecosystem functioning relationships in freshwater microcosms. *Adv. Ecol. Res.* **43**, 151–176.

Resh, V.H., Lévêque, C., and Statzner, B. (2004). Long-term, large-scale biomonitoring of the unknown: Assessing the effects of insecticides to control river blindness (onchocerciasis) in West Africa. *Annu. Rev. Entomol.* **49**, 115–139.

Reynoldson, T.B., Norris, R.H., Resh, V.H., Day, K.E., and Rosenberg, D.M. (1997). The reference condition: A comparison of multimetric and multivariate approaches to assess water-quality impairment using benthic macroinvertebrates. *J. N. Am. Benthol. Soc.* **16**, 833–852.

Richards, C., Haro, R.J., Johnson, L.B., and Host, G.E. (1997). Catchment and reach-scale properties as indicators of macroinvertebrate species traits. *Freshw. Biol.* **37**, 219–230.

Rigler, F.H. (1982). Recognition of the possible: An advantage of empiricism in ecology. *Can. J. Fish. Aquat. Sci.* **39**, 1323–1331.

Riipinen, M., Fleituch, T., Hladyz, S., Woodward, G., Giller, P.S., and Dobson, M. (2010). Invertebrate community structure and ecosystem functioning in European conifer plantation streams. *Freshw. Biol.* 10.1111/j.1365-2427.2009.02278.x.

Roberts, T.M. (1972). Plants as monitors of heavy metal pollution. *J. Environ. Planning and Pollution Control* **1**, 43–54.

Roberts, B.J., Mulholland, P.J., and Hill, W.R. (2007). Multiple scales of temporal variability in ecosystem metabolism rates: Results from two years of continuous monitoring in a forested headwater stream. *Ecosystems* **10**, 588–606.

Rolauffs, P., Stubauer, I., Zahrádková, S., Brabec, K., and Moog, O. (2004). Integration of the saprobic system into the European Union Water Framework Directive. *Hydrobiologia* **516**, 285–298.

Rose, N., Monteith, D.T., Kettle, H., Thompson, R., Yang, H., and Muir, D. (2004). A consideration of potential confounding factors limiting chemical and biological recovery at Lochnagar, a remote mountain loch in Scotland. *J. Limnol.* **63**, 63–76.

Rosemond, A.D., Reice, S.R., Elwood, J.W., and Mulholland, P.J. (1992). The effects of stream acidity on benthic invertebrate communities in the south-eastern United States. *Freshw. Biol.* **27**(2), 193–209.

Rosemond, A.D., Pringle, C.M., Ramirez, A., Paul, M.J., and Meyer, J.L. (2002). Landscape variation in phosphorus concentration and effects on detritus-based tropical streams. *Limnol. Oceanogr.* **47**, 278–289.

Rosenberg, D.M. and Resh, V.H. (Eds.) (1993). *In Freshwater Biomonitoring and Benthic Macroinvertebrates*. Chapman and Hall, New York, 488 pp.

Ruse, L. (2002). Chironomid pupal exuviae as indicators of lake status. *Arch. Hydrobiol.* **153**, 367–390.

Ruston, A.G. (1921). The plant as an index of smoke pollution. *Ann. Appl. Biol.* **7**, 390–402.

Rutt, G.P., Weatherley, N.S., and Ormerod, S.J. (1990). Relationships between the physicochemistry and macroinvertebrates of British upland streams: The development of modelling and indicator systems for predicting fauna and detecting acidity. *Freshw. Biol.* **24**, 463–480.

Sanders, I.A., Heppell, C.M., Cotton, J.A., Wharton, G., Hildrew, A.G., Flowers, E.J., and Trimmer, M. (2007). Emission of methane from chalk streams has potential implications for agricultural practices. *Freshw. Biol.* **52**, 1176–1186.

Sandin, L., and Hering, D. (2004). Comparing macroinvertebrate indices to detect organic pollution across Europe: A contribution to the EC Water Framework Directive intercalibration. *Hydrobiologia* **516**, 55–68.

Sandin and Verdonshot (2006). Stream and river typlogies—Major results and conclusions from the STAR project. *Hydrobiologia* **566**, 33–37.

Savage, A., and Hogarth, P.J. 1999. An analysis of temperature-induced fluctuating asymmetry in *Asellus aquaticus* (Linn.). *Hydrobiologia* **411**, 139–143.

Scheffer, M. (1998). *Ecology of Shallow Lakes*. Chapman and Hall, London.

Scheffer, M., and Carpenter, S.R. (2003). Catastrophic regime shifts in ecosystems: Linking theory to observation. *Trends Ecol. Evol.* **18**, 648–656.

Shi, Y.M., Tyson, G.W., and DeLong, E.F. (2009). Metatranscriptomics reveals unique microbial small RNAs in the ocean's water column. *Nature* **459**, 266–269.

Simberloff, D., and Gibbons, L. (2004). Now you see them, now you don't—Population crashes of established introduced species. *Biol. Invasions* **6**, 161–172.

Simpson, J.C., and Norris, R.H. (2000). Biological assessment of river quality: Development of AUSRIVAS models and outputs. In: *Assessing the Biological Quality of Freshwaters: RIVPACS and Similar Techniques* (Ed. by J.F. Wright, D.W. Sutcliffe and M.T. Furse), pp. 125–142. Freshwater Biological Association, Ambleside, UK.

Sladecek, V. (1973). System of water qualification from the biological point of view. *Arch. Hydrobiol. Beih. Ergeb. Limnol.* **7**, 1–218.

Sloane, P.I.W., and Norris, R.H. (2003). Relationship of AUSRIVAS-based macroinvertebrate predictive model outputs to a metal pollution gradient. *J. N. Am. Benthol. Soc.* **22**, 457–471.

Snelder, T.H., Pella, H., Wasson, J.G., and Lamouroux, N. (2008). Definition procedures have little effect on performance of environmental classifications of streams and rivers. *Environ. Manage.* 771–788.

Souchon, Y., Sabaton, C., Deibel, R., Reiser, D., Kershner, J., Gard, M., Katopodis, C., Leonard, P., Poff, N.L., Miller, W.J., and Lamb, B.L. (2008).

Detecting biological responses to flow management: Missed opportunities, future directions. *River Res. Appl.* **24**, 506–518.
Stanners, D., and Bourdeau, P. (1995). *Europe's Environment*. The Dobris Assessment. European Environmental Agency, Copenhagen, Denmark.
Statzner, B., Capra, H., Higler, L.W.G., and Roux, A.L. (1997). Focusing environmental management budgets on non-linear system responses: Potentials for significant improvements to freshwater ecosystems. *Freshw. Biol.* **37**, 463–472.
Statzner, B., Hildrew, A.G., and Resh, V.H. (2001a). Species traits and environmental constraints: Entomological research and the history of ecological theory. *Annu. Rev. Entomol.* **46**, 291–316.
Statzner, B., Bis, B., Dolédec, S., and Usseglio-Polatera, P. (2001b). Perspectives for biomonitoring at large spatial scales: A unified measure for the functional composition of invertebrate communities in European running waters. *Basic Appl. Ecol.* **2**, 73–85.
Statzner, B., Dolédec, S., and Hugueny, B. (2004). Biological trait composition of European stream invertebrate communities: Assessing the effects of various trait filter types. *Ecography* **27**, 470–488.
Statzner, B., Bady, P., Dolédec, S., and Schöll, F. (2005). Invertebrate traits for the biomonitoring of large European rivers: An initial assessment of trait patterns in least impacted river reaches. *Freshw. Biol.* **50**, 2136–2216.
Sterner, R.W., and Elser, J.J. (2002). Ecological Stoichiometry: The Biology of Elements from Molecules to the Biosphere. Princeton University Press, Princeton.
Stewart, F.J., Ottesen, E.A., and DeLong, E.F. (2010). Development and quantitative analyses of a universal rRNA-subtraction protocol for microbial metatranscriptomics. *ISME J.* **4**, 896–907.
Stoddard, J.L., Peck, D.V., Paulsen, S.G., Van Sickle, J., Hawkins, C.P., Herlihy, A.T., Hughes, R.M., Kaufmann, P.R., Larsen, D.P., Lomnicky, G., Olsen, A.R., Peterson, S.A., *et al.* (2005). An Ecological Assessment of Western Streams and Rivers. U.S. Environmental Protection Agency, Washington, DCEPA/620/R-05/005 (66 pp., 4.11MB).
Stoddard, J.L., Peck, D.V., Olsen, A.R., Larsen, D.P., Van Sickle, J., Hawkins, C.P., Hughes, R.M., Whittier, T.R., Lomnicky, G., Herlihy, A.T., Kaufmann, P.R., Peterson, S.A., *et al.* (2006). Environmental Monitoring and Assessment Program (EMAP):Western Streams and Rivers Statistical Summary. U.S. Environmental Protection Agency, Washington, DCEPA/620/R-05/006 (1762 pp., 17.9 MB).
Strayer, D.L., Eviner, V.T., Jeschke, J.M., and Pace, M.L. (2006). Understanding the long-term effects of species invasions. *Trends Ecol. Evol.* **21**, 645–651.
Sweeting, C.J., Badalamenti, F., D'Anna, G., Pipitone, C., and Polunin, N.V.C. (2009). Steeper biomass spectra of demersal fish communities after trawler exclusion in Sicily. *ICES J. Mar. Sci.* **66**, 195–202.
ter Braak, C.J.F., and Prentice, I.C. (1988). A theory of gradient analysis. *Adv. Ecol. Res.* **18**, 271–313.
ter Braak, C.J.F., and Šmilauer, P. (2002). CANOCO Reference Manual and CanoDraw for Windows User's Guide: Software for Canonical Community Ordination (version 4.5). Microcomputer Power, Ithaca, NY, USA.
Thienemann, A. (1920). Die Grundlagen der Biocoenotik und Monards faunistische Prinzipien. *Festschrift Zschokke* **4**, 1–14.
Thienemann, A. (1959). Erinnerungen und Tagebuchblätter eines Biologen. Schweitzerbart, Stuttgart, 499 pp.

Thorne, R.S.J., and Williams, W.P. (1997). The response of benthic macroinvertebrates to pollution in developing countries: A multimetric system of bioassessment. *Freshw. Biol.* **37**, 671–686.
Tiegs, S.D., Langhans, S.D., Tockner, K., and Gessner, M.O. (2007). Cotton strips as a leaf surrogate to measure decomposition in river floodplain habitats. *JNABS* **26**, 70–77.
Tiegs, S.D., Peter, F.D., Robinson, C.T., Uehlinger, U., and Gessner, M.O. (2008). Leaf decomposition and invertebrate colonization responses to manipulated litter quantity in streams. *JNABS* **27**, 321–331.
Tiegs, S.D., Akinwole, P.O., and Gessner, M.O. (2009). Litter decomposition across multiple spatial scales in stream networks. *Oecologia* **161**, 343–351.
Townsend, C.R. (2003). Individual, population, community, and ecosystem consequences of a fish invader in New Zealand streams. *Conserv. Biol.* **17**, 38–47.
Townsend, C.R., Hildrew, A.G., and Francis, J.E. (1983). Community structure in some Southern English streams: The influence of physicochemical factors. *Freshw. Biol.* **13**, 531–544.
Townsend, C.R., Hildrew, A.G., and Schofield, K. (1987). Persistence of stream invertebrate communities in relation to environmental variability. *J. Anim. Ecol.* **56**, 597–613.
Tranvik, L.J., Downing, J.A., Cotner, J.B., Loiselle, S.A., Striegl, R.G., Ballatore, T.J., Dillon, P., Finlay, K., Knoll, L.B., Kortelainen, P.L., Kutser, T., Larsen, S., *et al.* (2009). Lakes and impoundments as regulators of carbon cycling and climate. *Limnol. Oceanogr.* **54**, 2298–2314.
Trimmer, M., Hildrew, A.G., Jackson, M.C., Pretty, J.L., and Grey, J. (2009). Evidence for the role of methane-derived carbon in a free-flowing, lowland river food-web. *Limnol. Oceanogr.* **54**, 1541–1547.
Tuchman, N.C., Wetzel, R.G., Rier, S.T., Wahtera, K.A., and Teeri, J.A. (2002). Elevated atmospheric $CO_2$ lowers leaf litter nutritional quality for stream ecosystem food webs. *Glob. Change Biol.* **8**, 163–170.
Tuttle, N.C., Beard, K.H., and Pitt, W.C. (2009). Invasive litter, not an invasive insectivore, determines invertebrate communities in Hawaiian forests. *Biol. Invasions* **11**, 845–855.
U.S. EPA (Environmental Protection Agency) (2006). Wadable Streams Assessment. EPA 841-B-06-002U.S. Environmental Protection Agency, Office of Research and Development and Office of Water, Washington, DC.
United Kingdom Acid Waters Monitoring Network (2000). 10 year report: Analysis and interpretation of results, April 1988–March 1998. In: *United Kingdom Acid Waters Monitoring Network 10 Year Report.* (Ed. by D.T. Monteith and C.D. Evans), 353 pp. ENSIS Publishing, London.
Usseglio-Polatera, P., Bournaud, M., Richoux, P., and Tachet, H. (2000). Biomonitoring through biological traits of benthic macroinvertebrates: How to use species trait database? *Hydrobiologia* **422/423**, 153–162.
Uzarski, D.G., Burton, T.M., and Stricker, C.A. (2001). A new chamber design for measuring community metabolism in a Michigan stream. *Hydrobiologia* **455**, 137–155.
van de Bund, W. (2008). Water Framework Directive intercalibration Technical Report: Part 1, Rivers. JRC Scientific and Technical Reports, European Commissison, Institute for Environmental Sustainability, Luxembourg, ISSN: 1018-5593, 179 pp.
Van der Heijden, M.G.A., Klironomos, J.N., Ursic, M., Moutoglis, P., Streitwolf-Engel, R., *et al.* (1998). Mycorrhizal fungal diversity determines plant biodiversity, ecosystem variability and productivity. *Nature* **396**, 69–72.

Vaughan, I.P., Diamond, M., Gurnell, A.M., Hall, K.A., Jenkins, A., Milner, N.J., Naylor, L.A., Sear, D.A., Woodward, G., and Ormerod, S.J. (2009). Integrating ecology with hydromorphology: A priority for river science and management. *Aquat. Conserv. Mar. Freshw. Ecosyst.* **19**, 113–125.

Vaughn, C.C., and Hakenkamp, C.C. (2001). The functional role of burrowing bivalves in freshwater ecosystems. *Freshw. Biol.* **46**, 1431–1446.

Vinson, M.R., and Hawkins, C.P. (1996). Effects of sampling area and subsampling procedure on comparisons of taxa richness among streams. *J. N. Am. Benthol. Soc.* **15**(3), 392.

Vlek, H.E., Verdonschot, P.F.M., and Nijboer, R.C. (2004). Towards a multimetric index for the assessment of Dutch streams using benthic macroinvertebrates. *Hydrobiologia* **516**, 173–189.

Wade, K.R., Ormerod, S.J., and Gee, A.S. (1989). Classification and ordination of macroinvertebrate assemblages to predict stream acidity in upland Wales. *Hydrobiologia* **171**, 59–78.

Wallace, J.B. and Webster, J.R. (1996). The role of macroinvertebrates in stream ecosystem function. *Ann. Rev. Entomol.* **41**, 115–139.

Walling, D.E., Collins, A.L., and McMellin, G.K. (2003). A reconnaissance survey of the source of interstitial fine sediment recovered from salmonid spawning gravels in England and Wales. *Hydrobiologia* **497**, 91–108.

Walther, G.-R. (2010). Community and ecosystem responses to recent climate change. *Phil. Trans. R. Soc. B* **365**, 2019–2024.

Warwick, R.M., and Clarke, K.R. (1991). A comparison of some methods for analysing changes in benthic community structure. *J. Mar. Biol. Assoc. UK* **71**, 225–244.

Weatherley, N.S., and Ormerod, S.J. (1987). The impact of acidification on macroinvertebrate assemblages in Welsh streams: Towards an empirical model. *Environ. Pollut.* **46**, 223–240.

Weatherley, N.S., and Ormerod, S.J. (1990). The constancy of invertebrate assemblages in soft water streams: The role of pH and food resources. *J. Appl. Ecol.* **27**, 952–964.

Weigel, B.M. (2003). Development of stream macroinvertebrate models that predict watershed and local stressors in Wisconsin. *J. N. Am. Benthol. Soc.* **22**, 123–142.

Wilby, R.L., Beven, K.J., and Reynard, N.S. (2008). Climate change and fluvial flood risk in the UK: more of the same?. *Hydrol. Process.* **22**, 2511–2523.

Williamson, C.E., Saros, J.E., Vincent, W.F., and Smol, J.P. (2009). Lakes and reservoirs as sentinels, integrators, and regulators of climate change. *Limnol. Oceanogr.* **54**, 2273–2282.

Winterbourn, M.J., Hildrew, A.G., and Orton, S. (1992). Nutrients, algae and grazers in some British streams of contrasting pH. *Freshw. Biol.* **28**, 173–182.

Wood, P.J., and Armitage, P.D. (1997). Biological effects of fine sediment in the lotic environment. *Environ. Manage.* **21**, 203–217.

Woodward, G. (2009). Biodiversity, ecosystem functioning and food webs in freshwaters: Assembling the jigsaw puzzle. *Freshw. Biol.* **54**, 2171–2187.

Woodward, G., Papantoniou, G., Edwards, F.E., and Lauridsen, R. (2008). Trophic trickles and cascades in a complex food web: impacts of a keystone predator on stream community structure and ecosystem processes. *Oikos* **117**, 683–692.

Woodward, G., Perkins, D.M., and Brown, L.E. (2010a). Climate change in freshwater ecosystems: Impacts across multiple levels of organisation. *Philos. Trans. R. Soc. B* **365**, 2093–2106.

Woodward, G., Christensen, J.B., Olafsson, J.S., Gislason, G.M., Hannesdottir, E.R., and Friberg, N. (2010b). Sentinel systems on the razor's edge: Effects of warming on Arctic stream ecosystems. *Glob. Change Biol.* **16**, 1979–1991.

Woodward, G., Benstead, J.P., Beveridge, O.S., Blanchard, J., Brey, T., Brown, L., Cross, W.F., Friberg, N., Ings, T.C., Jacob, U., Jennings, S., Ledger, M.E., *et al.* (2010c). Ecological networks in a changing climate. *Adv. Ecol. Res.* **42**, 72–138.

Woodward, G., Friberg, N., and Hildrew, A.G. (2010d). Science and non-science in the biomonitoring and conservation of fresh waters. In: *Freshwater Ecosystems and Aquaculture Research* (Ed. by F. de Carlo and A. Bassano). 978-1-60741-707-1, Nova Science Publishers, Inc., Hauppauge NY, USA.

Woodward, G., Blanchard, J., Lauridsen, R.B., Edwards, F.K., Jones, J.I., Figueroa, D., Warren, P.H., and Petchey, O.L. (2010e). Individual-based food webs: Species identity, body size and sampling effects. *Adv. Ecol. Res.* **43**, 211–266.

Woodiwiss, F.S. (1964). The biological system of stream classification used by the Trent River Board. *Chem. Ind.* **11**, 443–447.

Wright, J.F., Furse, M.T., and Armitage, P.D. (1993). RIVPACS: A technique for evaluating the biological quality of rivers in the UK. *Eur. Water Pollut. Control* **3**, 15–25.

Wright, J.F., Furse, M.T., and Armitage, P.D. (1994). *Use of macroinvertebrate communities to detect environmental stress in running waters.* Freshwater Biological Association, Ambleside, UK, pp 15–34.

Wright, J.F., Sutcliffe, D.W. and Furse, M.T. (Eds.) (2000). *In* Assessing the Biological Quality of Freshwaters: RIVPACS and Similar Techniques Freshwater Biological Association, Ambleside, UK.

Yan, N.D., Leung, B., Keller, W., Arnott, S.E., Gunn, J.M., and Raddum, G.G. (2003). Developing conceptual frameworks for the recovery of aquatic biota from acidification. *Ambio* **32**, 165–169.

Yemane, D., Field, J.G., and Leslie, R.W. (2005). Exploring the effects of fishing on fish assemblages using Abundance Biomass Comparison (ABC) curves. *ICES J. Mar. Sci.* **62**, 374–379.

Young, R.C., Matthaei, C.D., and Townsend, C.R. (2008). Organic matter breakdown and ecosystem metabolism: Functional indicators for assessing river ecosystem health. *J. N. Am. Benthol. Soc.* **27**, 605–625.

Yvon-Durocher, G., Woodward, G., Jones, J.I., Trimmer, M., and Montoya, J.M. (2010a). Warming alters the metabolic balance of ecosystems. *Philos. Trans. R. Soc. B* **365**, 2117–2126.

Yvon-Durocher, G., Allen, A.P., Montoya, J.M., Trimmer, M., and Woodward, G. (2010b). The temperature dependence of the carbon cycle in aquatic systems. *Adv. Ecol. Res.* **43**, 267–313.

Yvon-Durocher, G., Montoya, J.M., Jones, J.I., Woodward, G., and Trimmer, M. (2011a). Warming alters the fraction of primary production released as methane in freshwater mesocosms. *Glob. Change Biol.* **17**, 1225–1234.

Yvon-Durocher, G., Montoya, J.M., Trimmer, M., and Woodward, G. (2011b). Warming alters the size spectrum and shifts the distribution of biomass in freshwater ecosystems. *Glob. Change Biol.* **17**, 1681–1694.

Yvon-Durocher, G., Reiss, J., Blanchard, J., Ebenman, B., Perkins, D.M., Reuman, D.C., Thierry, A., Woodward, G., and Petchey, O.L. (2011c). Across ecosystem comparisons of size structure: Methods, approaches, and prospects. *Oikos* **120**, 550–563.

# Long-Term Dynamics of a Well-Characterised Food Web: Four Decades of Acidification and Recovery in the Broadstone Stream Model System

KATRIN LAYER, ALAN G. HILDREW, GARETH B. JENKINS, JENS O. RIEDE, STEPHEN J. ROSSITER, COLIN R. TOWNSEND AND GUY WOODWARD



ADVANCES IN ECOLOGICAL RESEARCH VOL. 44 0065-2504/11 $35.00

DOI: 10.1016/B978-0-12-374794-5.00002-X

## SUMMARY

An understanding of the consequences of long-term environmental change for higher levels of biological organisation is essential for both theoretical and applied ecology. Here, we present four decades of data from the well-characterised Broadstone Stream community, detailing biological responses to amelioration of acidification and the recent invasion of a top predator (brown trout, *Salmo trutta* L.) that was previously excluded by low pH. After several decades of reductions in acidifying emissions, species characteristic of less acid conditions have started to invade or recolonise Broadstone and other European freshwaters, but these signs of biological recovery are still patchy and have lagged behind chemical recovery. One possible explanation for slow recovery is ecological inertia arising from the internal dynamics of the food web, a hypothesis we investigate here using a combination of surveys, experiments and mathematical modelling.

The invasion of this hitherto invertebrate-dominated system by a large, generalist vertebrate predator could be expected to alter the structure and stability of the food web. Long-term survey data revealed that the community has experienced waves of invasions or irruptions of progressively larger predators since the 1970s, as pH has risen. Intra-annual fluctuations in prey populations have become increasingly damped and the mean abundance of many species has declined, although none of the previously common taxa have been lost. This suggests that predation, rather than simple chemical tolerance, plays a key role in determining the trajectory of recovery, as the top-down effects of the generalist predators spread diffusely through the reticulate food web. Dynamical simulations indicate that the food web may have become less robust over time as pH has risen and larger predators have become dominant. These results suggest that, though none of the original suite of large invertebrate predators has been driven to local extinction, such an eventual outcome is feasible.

## I. INTRODUCTION

Understanding how complex multispecies systems respond to environmental stress is essential for both theoretical and applied ecology, as we cannot necessarily assume that biological recovery will follow a simple reversal of the trajectory of decline (Feld *et al.*, 2011; Ledger and Hildrew, 2005). Further, responses at the whole community or ecosystem level are unlikely to be predictable by simply scaling up from the component individuals or populations because interactions among species can create emergent effects that may only be understood from, for instance, a food web perspective (Ings *et al.*, 2009; Mulder *et al.*, 2011; Woodward *et al.*, 2010b). Small perturbations often alter relative abundances, but more pronounced disturbances can lead to local extinctions or species invasions and extensive 'rewiring' of the food web, which can ultimately lead to marked changes in community structure and ecosystem processes (Woodward, 2009). These higher-level responses to perturbations can be mediated by food-web properties that influence stability, such as network complexity and the configuration and strength of trophic interactions (Berlow *et al.*, 2009; Emmerson and Yearsley, 2004; May, 1972, 1973; McCann, 2000; McCann *et al.*, 1998; Montoya *et al.*, 2009; Neutel *et al.*, 2002, 2007; Otto *et al.*, 2007). It is critical, therefore, that these features are considered when attempting to assess the community and ecosystem consequences of environmental stress, although most studies remain focussed on the nodes (i.e. species) within food webs, rather than considering the links between them (Hladyz *et al.*, 2011; Mulder *et al.*, 2011; Woodward, 2009).

Acidification provides a particularly useful case study for assessing the higher-level effects of environmental stress because the relevant chemical parameters are relatively straightforward to measure and there are now several high-quality datasets and model systems that can be used to assess biological responses to chemical changes (e.g. Petchey *et al.*, 2004; Townsend *et al.*, 1983). The acidification of freshwaters has affected large areas of Europe and North America for decades (Galloway, 1995; Hildrew and Ormerod, 1995; Mason, 1991) and has had profound ecological impacts, including the loss of many acid-sensitive species, such as salmonids at the top of the food web (Rosemond *et al.*, 1992; United Kingdom Acid Waters Monitoring Network (UKAWMN), 2000). Despite a recent decline in acid deposition in NW Europe, and emerging evidence of chemical recovery in some acidified freshwaters, community responses have so far been modest (Lancaster, 1996; Monteith and Evans, 2005; Monteith *et al.*, 2005; Rundle *et al.*, 1995; UKAWMN, 2010). Several hypotheses have been put forward to explain the more subdued rate of biological recovery (Yan *et al.*, 2003), including dispersal limitations (e.g. Bradley and Ormerod, 2002; Raddum and Fjellheim, 2003; Snucins and Gunn, 2003; but see Masters *et al.*, 2007), incomplete chemical recovery (e.g. Kowalik *et al.*, 2007; Lepori *et al.*, 2003; Rose *et al.*, 2004) and intrinsic characteristics of the community food web that might render it resistant to re-establishment of more acid-sensitive species (e.g. Layer, 2010; Layer *et al.*, 2010a,b; Ledger and Hildrew, 2005; Lunberg *et al.*, 2000; Woodward, 2009; Woodward and Hildrew, 2002a).

The food web provides a conceptual thread that links together individuals, populations, communities and the ecosystem yet, although it is firmly embedded within theoretical ecology, it is only in recent years that it has started to be embraced by applied ecology (Friberg *et al.*, 2011; Hladyz *et al.*, 2011; Ings *et al.*, 2009; Mulder *et al.*, 2011). In terms of understanding the structure and dynamics of food webs, and how they might respond to environmental stress, the key role played by body size has become increasingly recognised, especially in aquatic systems (Cohen *et al.*, 2003; Hildrew *et al.*, 2007; Peters, 1983; Woodward and Warren, 2007). Body size, whether of individuals, species or other biological entities (e.g. guilds), is a major determinant of the occurrence, configuration (Petchey *et al.*, 2008) and strength of feeding interactions (Berlow *et al.*, 2009; Brose *et al.*, 2005; Emmerson and Raffaelli, 2004), as well as being a useful proxy for many other ecological traits, particularly those associated with biological process rates and vulnerability to perturbations (Reiss *et al.*, 2009; Woodward *et al.*, 2005b). It is also easy to measure. Surprisingly, however, it is only in the past few years that it has been recorded routinely in food-web studies. For instance, so-called trivariate food webs that map feeding links onto mass–abundance relationships have revealed remarkably consistent allometric scaling relationships across a wide range of systems (Cohen *et al.*, 2003; Layer *et al.*, 2010a,b; McLaughlin *et al.*, 2010; O'Gorman and Emmerson, 2010; Reuman and Cohen, 2004, 2005; Woodward *et al.*, 2005a,b), and deviations from such recurrent patterns can provide clear evidence that the system has been perturbed (Perkins *et al.*, 2010; Petchey and Belgrano, 2010; Woodward *et al.*, 2010a,d).

The Broadstone Stream food web, like many other aquatic systems, exhibits strong size structuring (e.g. Woodward *et al.*, 2005b, 2010c) and is currently among the best-characterised in the world, having been the subject of numerous studies since the 1970s (e.g. Hildrew, 2009; Hildrew and Townsend, 1976; Olesen *et al.*, 2010; Woodward *et al.*, 2010c). Like other acid freshwaters, it is relatively species-poor and characterised by an abundant guild of macroinvertebrate predators (Hildrew, 2009; Hildrew *et al.*, 2004; Hildrew and Townsend, 1976). The diet of each predator species is effectively a subset of that of the next largest predator, largely irrespective of taxonomic identity (Woodward *et al.*, 2010c). In addition to its influence on topological patterns, body size also determines interaction strengths and this has further implications for biomass flux and dynamic stability of the system (Emmerson *et al.*, 2005; Montoya *et al.*, 2009; Woodward *et al.*, 2005a,b).

From the 1970s to the early 1990s, the Broadstone Stream community was relatively persistent, with no marked species gains or losses, indicating that the food web might be dynamically stable (Speirs *et al.*, 2000; Woodward *et al.*, 2002), a suggested characteristic of acidified freshwaters in general (Layer *et al.*, 2010a,b; Townsend *et al.*, 1987). Despite limited species turnover, however, the relative abundances of species did change, particularly in the mid-1990s when the irruption of the large dragonfly *Cordulegaster boltonii* Donovan restructured the food web, resulting in potentially destabilising increases in food chain length and network complexity (Woodward and Hildrew, 2001). These community responses were attributed to a substantial directed environmental change, reflecting a long-term decrease in acid deposition and a general rise in stream pH (Davies *et al.*, 2005; Fowler *et al.*, 2005).

Since the arrival of *C. boltonii* in the mid-1990s, the stream has been colonised by another new top predator that is also symptomatic of rising pH, the brown trout (*Salmo trutta* L.) (Layer *et al.*, 2010b). These colonists probably arrived from circumneutral reaches further downstream (K. Layer, S.J. Rossiter, C. Mellor and G. Woodward, unpublished data), which have held trout throughout the four decades of research on the acidified upper reaches (Schofield, 1988). This predator is far larger than any other in the food web and is a generalist consumer that is morphologically plastic and ecologically versatile (Baglinière and Maisse, 1999), with a native range that spans the Arctic to North Africa and Iceland to Pakistan (Klemetsen *et al.*, 2003).

Since the brown trout is found not only in the lower reaches of Broadstone but also in many of the less acid streams in the locality, it seems likely that it was a member of the food web of the upper reaches prior to the onset of anthropogenic acidification, although it is not possible to say with certainty in the absence of longer-term historical data. Its (re)colonisation of the headwaters of Broadstone provides further circumstantial evidence that the stream is recovering from acidification, as this species is an important indicator of rising pH (Ledger and Hildrew, 2005; UKAWMN, 2010). Brown trout were not present in the acidified upstream reach (ca. 1 km) of Broadstone when Hildrew and Townsend (1976) first surveyed it in the early 1970s, and it remained fishless at least until 1999, probably due to a combination of low pH and a waterfall created by a log-jam separating it from the more circumneutral reaches (Schofield *et al.*, 1988; Woodward *et al.*, 2005a). Both these physical and chemical barriers to trout colonisation have recently been breached: pH has increased significantly in the upper reaches since the 1970s, with an especially marked rise since the 1990s (Woodward *et al.*, 2002), and the waterfall has partially eroded away in recent years. The waterfall seems insufficiently substantial to have been a complete barrier to trout, but any occasional breach in the past was unlikely to lead to an established population because of the low pH waters above. Note also that the colonisation of Broadstone by trout replicates that in another acid stream nearby (Old Lodge, part of the UKAWMN; Figure 1), the upper reaches of which were fishless until 1990 (Monteith *et al.*, 2005; Rundle and Hildrew, 1992), and where there are no physical barriers but acidity similarly declined after 1990 and fish gradually moved upstream. This, along with the known intolerance of brown trout to low pH, clearly implicates chemical amelioration as the likely prime factor in the appearance of trout (Hildrew, 2009). Trout were first detected in the upper reaches of Broadstone in 2005, and breeding was confirmed for the first time in 2010, by the discovery of fertilised eggs in a spawning redd (G. Woodward, unpublished data).

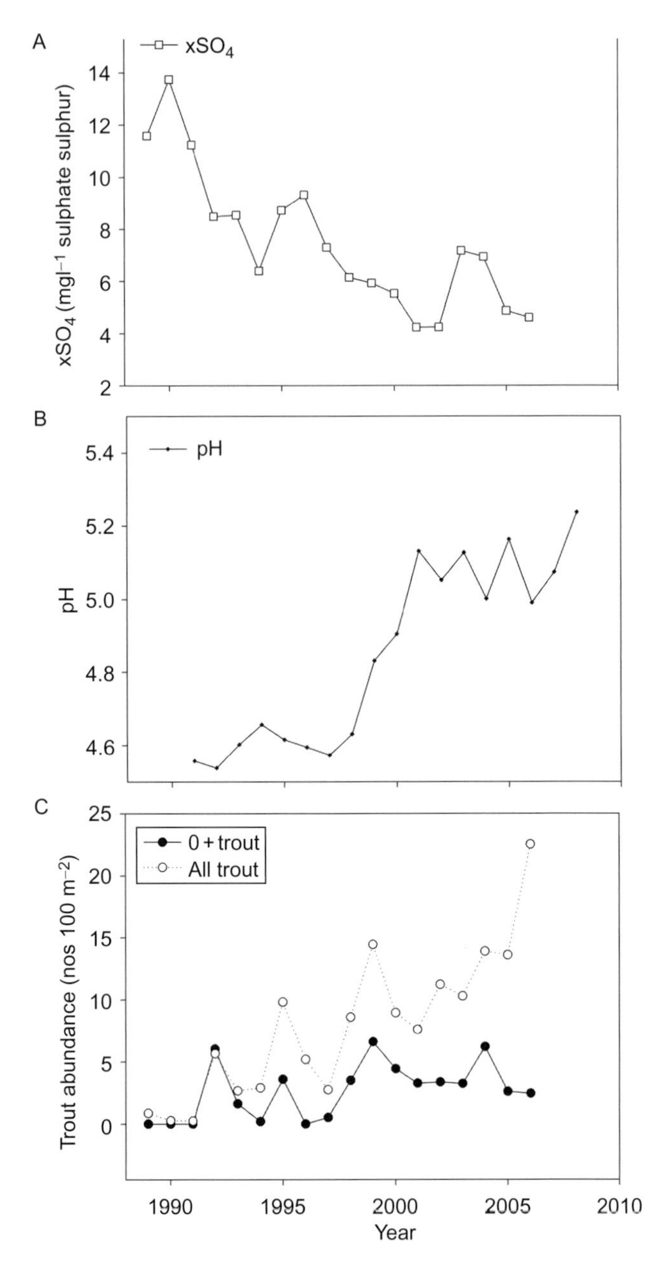


**Figure 1** Long-term chemical and biological recovery from acidification in Old Lodge Stream, adjacent to the Broadstone Stream study site, from 1989 to 2006 (UKAWMN, 2010). Acidifying $xSO_4$ concentrations are shown in (A), mean stream pH in (B) and trout population size in (C), with fry abundance shown as triangles, and total abundance as circles.

Among other effects, the trout invasion has the potential to have made the food web less dynamically stable, by increasing food chain length, consumer–resource body size ratios and, by inference, average interaction strengths (e.g. Emmerson *et al.*, 2005). In addition to size-related increases in interaction strengths, trout can exploit both aquatic and external terrestrial subsidies of invertebrates, potentially allowing it to reach densities beyond those that could be supported by in-stream production alone (e.g. Allen, 1951; Nakano *et al.*, 1999).

In other systems, salmonids have sometimes been reported as having powerful top-down effects, being responsible for local species extinctions (Townsend, 1996; Townsend and Simon, 2006) and trophic cascades (Biggs *et al.*, 2000; Flecker and Townsend, 1994; Nakano *et al.*, 1999; Power, 1990). However, far weaker impacts have been reported from other (mostly experimental) studies (e.g. Allan, 1982; Culp, 1986; Diehl, 1992), and strong impacts may be more likely when the invader is a non-native species (e.g. Flecker and Townsend, 1994). Given this somewhat equivocal evidence, as well as a potential underreporting of weak or negligible effects that could bias our perception of the 'typical' strength of predator impacts in food webs (Woodward, 2009), we sought to assess the impacts of this latest predator invasion in the wider context of long-term change in the Broadstone food web over four decades of research, using several lines of evidence. Strong interactions typically reduce food-web stability; the recent colonisation by a large top predator like trout should reduce system persistence (i.e. the normal number of primary extinctions over time only caused by the basic dynamics of the ecosystem) and, in extreme cases, might lead to widespread species loss and even cascading secondary extinctions. Large-scale space-for-time surveys of other streams in the Ashdown Forest have revealed strong negative relationships between the abundance of one of the dominant large invertebrate predators (the caddis *Plectrocnemia conspersa* Curtis), that characterised Broadstone in previous decades, and brown trout (Figure 2; Hildrew, 2009). In addition, enclosure/exclosure experiments carried out in Broadstone in the 1980s showed that trout can suppress *P. conspersa* abundance. Beyond this single feeding link (Schofield *et al.*, 1988), however, there is little compelling evidence that trout have strong effects in acid streams in general, or Broadstone in particular, suggesting that the fish might not trigger the dramatic rewiring of the food web associated with strong cascading interactions. The invasion of Broadstone Stream thus provides a rare and fortuitous opportunity to explore the possible food-web consequences in a model system and also against a background of ecological responses to environmental stress.

The Broadstone Stream food web possesses many properties that potentially dampen otherwise destabilising influences and strong cascading effects: these include high levels of generalism and omnivory and a prevalence of weak interactions (Woodward *et al.*, 2005a). Such food-web attributes might

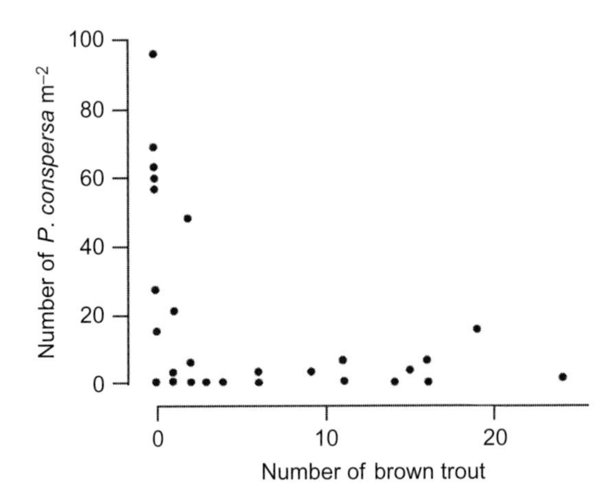


**Figure 2** Negative relationship between the abundance of the large invertebrate predator *Plectrocnemia conspersa* and the abundance of brown trout (total numbers caught in two electrofishing runs through a 50 m reach) across 29 streams in the Ashdown Forest sampled in 1977 (redrawn after Hildrew, 2009).

stabilise the system, thereby slowing the rate of biological recovery or even altering its trajectory, such that it does not simply follow the reverse of the responses to acidification (Woodward, 2009). Indeed, the effects of the earlier *C. boltonii* invasion appeared to diffuse throughout the web, rather than triggering cascades or oscillatory dynamics (Woodward *et al.*, 2005a; Woodward and Hildrew, 2001, 2002b,c,d). There are suggestions that similar ecological inertia, generated by properties of the food web itself, is manifested in other systems (Layer *et al.*, 2010a), although the idea is in need of more rigorous testing with both models and data. Broadstone Stream has an exceptionally well-described food web that has been studied over four decades, making it a uniquely valuable model system for studying long-term responses to environmental change (Hildrew, 2009).

Another potentially important consequence of the recent trout invasion is that, beyond introducing a new species and feeding links, it may also result in behavioural changes in the invertebrate assemblage in response to chemical cues (kairomones) released by the fish (e.g. McIntosh and Peckarsky, 1996). Fish kairomones alter the behaviour of some invertebrates, including *Gammarus pulex* L. (Abjornsson *et al.*, 2001), damselflies (Koperski, 1997), mayflies (McIntosh *et al.*, 1999) and snails (Turner *et al.*, 1999), which respond to predation risk. This could add to potential changes in the food web because the new predator can influence feeding links both directly, via consumption, and indirectly, via behavioural changes that impinge on interactions among the resident species. This has important implications because prey activity is a key determinant of encounter rate, which, in turn, limits predator feeding rates and, hence, interaction strengths in Broadstone Stream (Woodward and Hildrew, 2002c,d).

Because kairomones potentially weaken interactions between the invertebrate predators and their prey, we describe here the results of a new experiment to measure induced behavioural changes in the resident fauna. In laboratory experiments, we gauged the potential for trout to moderate interactions between the three large invertebrate predators, larvae of the dragonfly *C. boltonii*, the alderfly *Sialis fuliginosa* Pictet and the caddisfly *P. conspersa* and two prey species common to them all, the detritivorous stoneflies *Nemurella pictetii* Klapálek and *Leuctra nigra* (Olivier), by quantifying prey survival in the presence/absence of trout kairomones (e.g. after Abjornsson *et al.*, 2001). Each of these three invertebrate predators has occupied the top of the food web at some point over the past four decades (Woodward and Hildrew, 2001), and all three are eaten by trout in other systems (Layer *et al.*, 2010a,b; Woodward *et al.*, 2010d). Of the two prey species we used, *N. pictetii* is highly mobile (Lieske and Zwick, 2007) whereas *L. nigra* is more sedentary (Woodward and Hildrew, 2002b,c) and both are common food items of the Broadstone invertebrate predators (Woodward and Hildrew 2001, 2002b,c) and of trout (Schofield, 1988; Woodward *et al.*, 2010d). Any suppressed activity due to fish kairomones should reduce consumption of the prey species by invertebrate predators indirectly (i.e. 'a behaviourally mediated cascade'), whereas if kairomones do not affect prey survival, then direct consumption by trout is likely to be the primary mode by which the invading fish affects feeding links in the food web.

We have found no evidence that prey avoid the invertebrate predators in Broadstone (Woodward and Hildrew, 2002d), so a similar lack of response to trout would make comparisons over the four decades based on diet data more reliable than if there was an additional behavioural component now operating. Conversely, alterations in prey behaviour in response to chemical cues might suggest that the food web has altered both trophically and behaviourally. In the wider ecological context of biological recovery, it has been suggested that behavioural complexity in freshwater food webs increases as pH rises because it is not only fish that increase in abundance (with a related increase in prevalence of kairomones) but also those species that respond strongly to chemical cues, such as gammarid shrimps, snails and mayflies (Woodward, 2009). In contrast, the invertebrate species that typically dominate in fishless acid waters (e.g. nemourid and leuctrid stoneflies) appear not to have such highly developed chemosensory apparatus for detecting kairomones and they also seem to lack the associated behavioural avoidance responses.

Here, we characterised the feeding links of trout in the invaded Broadstone food web, using a combination of dissected gut contents, stable isotope analyses and supplementary information from literature surveys (after Layer *et al.*, 2010a,b), and we assessed community persistence and long-term shifts in the food web using survey data that span four decades (from 1974 to 2006; Woodward *et al.*, 2002). We then explored the structure and dynamics of the food web in response to a long-term reversal in acidification, including a first assessment of the impacts of the recent trout invasion, via statistical analysis of long-term empirical data, the application of mathematical simulation modelling of food-web persistence and laboratory experiments.

# II. METHODS

## A. Study Site

Broadstone Stream (51°05′N 0°02′E; 120 m above sea level) is a spring-fed headwater of the River Medway in SE England, with a geologically poorly buffered catchment (of Ashdown Sandstone) dominated by heath and woodland vegetation. Stream pH typically ranges seasonally from about 5.2 to 6.5, with the annual peak occurring in summer and the minimum in winter, during acid episodes associated with heavy rain. There has, however, been a general and progressive decline in acidity since the 1970s (also observed in several neighbouring streams; Hildrew, 2009), and particularly so in the summer months when most taxa are in their most acid-sensitive early life-history stages (Woodward *et al.*, 2002). The macroinvertebrate community of Broadstone is species-poor and dominated by acid-tolerant insect larvae, with an abundance of invertebrate predators. The largest of these is the dragonfly *C. boltonii*, which colonised the stream in the 1990s and reached densities of up to 72 nymphs $m^{-2}$ (Woodward *et al.*, 2005a). The two other dominant large macroinvertebrate predators are larvae of the alderfly *S. fuliginosa* and the net-spinning caddis *P. conspersa*, and there are also three common species of much smaller predatory Tanypodinae midge larvae (Hildrew *et al.*, 1985). The macroinvertebrate prey assemblage is composed mainly of detritivorous stoneflies and chironomid midges, reflecting the dominance of basal resources by allochthonous detritus, especially coarse particulate organic matter (CPOM) in the form of terrestrial leaf litter. Algal resources are scarce, due to a combination of acidity and the dense shade cast by the riparian canopy. See Hildrew (2009) for further site details and an in-depth review of the general ecological background of the system.

## B. Community Structure: Characterising the Nodes in the Food Web

To characterise the post-2000 food web, after trout invasion, Broadstone Stream was sampled five times in 2006 (April, June, August, October and December). We compared these results with equivalent datasets obtained from the preceding three decades, as described below. The absolute abundance of macroinvertebrates was quantified by taking 10 Surber samples (sample-unit area 0.0625 $m^2$; mesh aperture 330 $\mu$m) from a 200-m stretch (Woodward *et al.*, 2005a), and samples were preserved in 70% Industrial Methylated Spirit. In the laboratory, macroinvertebrates were sorted from debris, identified to species where possible (except dipteran larvae) using published taxonomic keys (Appendix I) and counted. Head capsule width or body length of each individual was measured with a calibrated ocular micrometre (200–400× magnification) and used to derive individual body mass estimates from published length–mass regression equations (Woodward *et al.*, 2005a). Trout abundance was quantified by depletion electrofishing in a 100-m stretch of the study reach (Junge and Libosvarsky, 1965). Stop nets were installed at each end of the stretch, and three runs were completed with a Smith-Root LR-24 backpack electrofisher, moving upstream and sweeping from one side of the stream to the other. The number of fish caught and the fork-length and body mass of each individual were recorded, prior to release into the stream after the third run.

Data from the 1970s, 1980s and 1990s were collated from A.G. Hildrew and C.R. Townsend (1970s), K. Schofield (1980s), J. Lancaster (1980s and 1990s), J.H. Winterbottom (1990s) and G. Woodward and A.G. Hildrew (1990s): further details of these sampling occasions and methods are supplied in Woodward *et al.* (2002). One of us (A.G.H.) has supervised the collection and identification of macroinvertebrates throughout this four-decade period. To facilitate comparison between sampling decades, taxonomic resolution was standardised to the lowest common denominator among samples after Woodward *et al.* (2002): for instance, all Chironomidae species were grouped into either the predatory ‘Tanypodinae’ or the ‘non-predatory Chironomidae’. This meant that network size, in terms of the absolute number of nodes and links, was smaller than in the most highly resolved versions of the food web (e.g. cf. Schmid-Araya *et al.*, 2002a,b; Woodward *et al.*, 2005a); the standardisation was necessary to facilitate meaningful long-term comparisons. These food webs are species-poor due to the acidity of the stream, and this level of resolution is as good or better than that used in many other studies (e.g. Layer *et al.*, 2010a,b), so the networks were neither artificially nor unusually small. Terrestrial prey were excluded from all the food webs because they included a wide range of often unidentified taxa with unknown body masses.

Data on individual body mass were available for each taxon within the food web from direct measurements made on >40,000 specimens on a subset of sampling occasions in the 1970s, 1990s and 2000s (e.g. see Layer *et al.*, 2010b; Woodward *et al.*, 2005a, 2010c), and these were compiled to calculate species-averaged body masses for each decade. Unfortunately, although abundances were recorded, no direct measurements of body mass for all species were available for the 1980s. Instead, by applying the same mean individual mass derived in the 1990s for a given month to the individuals for each taxon for the corresponding month in the 1980s and then weighting by the observed monthly abundance values to standardise for potential seasonal differences (in the 1980s), hindcasted annual means of body mass were derived for each taxon. This was based on the assumption that although mean abundance (which was measured in each decade) may change markedly, the average body mass of individual species did not change between the 1980s and 1990s, which was supported by the strong correlation in body mass in the 1970s and 1990s among the 11 taxa common to both decades ($r = 0.91$; $P < 0.01$). In addition to species-averaged allometric scaling relationships between mass and abundance and feeding links (e.g. after Cohen *et al.*, 2003), we also constructed individual size distributions (ISDs) by assigning each individual organism, irrespective of species identity, to a series of logarithmic size bins (e.g. after Layer *et al.*, 2010a) to assess potential shifts in size-based allometries over time (e.g. after Brown *et al.*, 2004). These were constructed for 1 month only per decade (August), when the size range and abundance of new recruits are at their peaks for most invertebrate species. Unfortunately, ISDs could not be hindcast for the 1980s because the required individual-level data were not available.

## C. Community Structure: Characterising the Links in the Food Web

Invertebrate diets were characterised via examination of the gut contents under a binocular microscope (gut contents analysis, GCA), following dissection and removal of the foregut (e.g. after Hildrew *et al.*, 1985; Woodward *et al.*, 2005a, 2010c). Feeding links for brown trout were observed directly via gut flushing (e.g. Dineen *et al.*, 2007) of all fish $>$10 cm caught on the five sampling dates in 2006 and by analysis under a binocular microscope at 200$\times$ magnification of the stomach contents of a sample of 10 individuals taken during a fish survey in 2005 (after Woodward *et al.*, 2010c). These direct observations were supplemented by earlier gut contents data collated from a short-term set of field enclosure/exclosure experiments in the 1980s, in which trout were introduced briefly to the acidified portion of Broadstone and subsequently dissected for diet characterisation, and from additional dissections of trout taken from the lower circumneutral reaches (Rundle and Hildrew, 1992; Schofield, 1988; Schofield *et al.*, 1988). In addition, feeding links were also inferred from published data from other studies in which the same species pair of predator and prey were recorded as having a trophic interaction, including the recently described Tadnoll Brook food web, which contains almost the entire invertebrate species pool recorded for Broadstone (data sources: Brose *et al.*, 2005; Schmid-Araya *et al.*, 2002a,b; Warren, 1989; Woodward *et al.*, 2008; Woodward *et al.*, 2010c). This partially inferential approach was necessary because a full characterisation of trout diet in Broadstone would have required the sampling of hundreds of individuals (cf. Ings *et al.*, 2009; Layer *et al.*, 2010b; Woodward *et al.*, 2010c), a number greater than the entire population in the invaded acidified upper reaches.

As with the characterisation of species abundance and body mass data, feeding links were also resolved taxonomically to the lowest common denominator across all sampling occasions, to apply a standardised protocol for comparison among decades. This meant that a necessary compromise was made between replication, sample size and resolution, such that the food webs were comparable with one another, but of a smaller size than the most highly resolved versions constructed from snapshots of particular sampling occasions (cf. Schmid-Araya *et al.*, 2002a,b; Woodward *et al.*, 2005a). Nonetheless, the level of resolution was higher than that of many other studies and was the same as used in two recent large-scale and long-term surveys of the effects of pH on food webs (Layer *et al.*, 2010a,b).

## D. Construction of the Food Webs

A set of four 'animal'-only summary food webs (and also excluding meio- and microfauna) was constructed by pooling data across sampling occasions per decade. To minimise potential temporal biases due to unbalanced numbers of observations among sampling occasions (e.g. Woodward *et al.*, 2002), we restricted the dataset used to construct the webs and to carry out formal statistical analyses by using only data collected for comparable months within a single annual cycle per decade (i.e. to match the recruitment–oviposition cycle as closely as possible). Network structures were compared across the four decades using a suite of food-web metrics, including web size ($S$), number of trophic interactions ($L$), number of interactions including competitive links ($L_c$), linkage density ($L/S$), directed connectance ($L/S^2$) and web complexity ($SC'_{max} = SL_c/S^2 = L_c/S$). Generality ($G$, the number of resources per consumer) and vulnerability ($V$, the number of consumers per resource) as well as standardised generality ($G_k$) and vulnerability ($V_k$), and their standard deviations ($G_{st}$ and $V_{st}$, respectively; see Williams and Martinez, 2000) were also calculated for each decade (after Layer *et al.*, 2010b). The trophic networks were also plotted as trivariate food webs on $\log_{10}$(mean individual body mass)–$\log_{10}$(abundance) scatterplots. The strength of feeding links was inferred by using the species-averaged body mass ratios of predator and prey raised to the 3/4 power as a proxy measure of interaction strength (e.g. after Berlow *et al.*, 2009; Emmerson *et al.*, 2005; Layer *et al.*, 2010b).

## E. Modelling of Food Web Persistence

We explored network persistence by simulating population dynamics within the food web over a period of 1, 5, 10, 20 and 30 years, using a similar approach to that of Layer *et al.* (2010b). Populations of species went extinct if their biomass fell below a critical extinction threshold ($B_i < 10^{-30}$). We defined food-web persistence ($R$) as the fraction of initial species that persisted after species removal: $R = (S_p/S_i)$, where $S_p$ and $S_i$ are the number of persistent and initial species (excluding the species removed at the outset, SR), respectively. Broadstone Stream is dominated by detritivore species, which are not resource limited (Dobson and Hildrew, 1992) and therefore should exhibit a Type I functional response: based on this assumption, we used a logistic growth term for the basal invertebrate prey species in our model.

We used a bioenergetic consumer–resource model (after Yodzis and Innes, 1992) to describe the change of biomass over time ($B'_i$) of the $i$th primary consumer species, while we used food-web structures without algae or any other primary producer species (Eq. (1)) and the $i$th heterotroph consumer species (Eq. (2)) in an $n$-species system:

$$B'_i = r_i(M_i)G_iB_i - \sum_{j=\text{consumers}}^{n} x_j(M_j)y_jB_j, \tag{1}$$

$$B'_i = -x_i(M_i)B_i + \sum_{j=\text{resources}} x_i(M_i)y_iB_i - \sum_{j=\text{consumers}} x_j(M_j)y_jB_j. \tag{2}$$

For each species, $i$, $B_i$ is its biomass, $r_i$ is its mass-specific maximum growth rate, $M_i$ is its average body mass, $G_i$ is its logistic net growth ($G_i = 1 - B_i/K$) with a carrying capacity $K$, $x_i$ is its mass-specific metabolic rate and $y_i$ is its maximum consumption rate relative to its metabolic rate.

The biological rates of production ($W$), metabolism ($X$) and maximum consumption ($Y$) follow negative-quarter power-law relationships with the species' body masses (Enquist *et al.*, 1999; Brown *et al.*, 2004):

$$W_P = a_r M_P^{-0.25}, \tag{3}$$

$$X_C = a_x M_C^{-0.25}, \tag{4}$$

$$Y_C = a_y M_C^{-0.25}, \tag{5}$$

where $a_r$, $a_x$ and $a_y$ are allometric constants, and $C$ and $P$ indicate consumer and producer parameters, respectively (Yodzis and Innes, 1992). The time-scale of the system was defined by normalising the biological rates by the mass-specific growth rate of the smallest producer species $P^*$. Then, the maximum consumption rates, $Y_C$, were normalised by the metabolic rates $X_C$:

$$r_i = \frac{W_P}{W_{P*}} = \left(\frac{M_P}{M_{P*}}\right)^{-0.25}, \tag{6}$$

$$x_i = \frac{X_C}{W_{P*}} = \frac{a_x}{a_r}\left(\frac{M_C}{M_{P*}}\right)^{-0.25}, \tag{7}$$

$$y_i = \frac{Y_C}{X_C} = \frac{a_y}{a_x}. \tag{8}$$

where $W_{P*}$ is the reproduction rate of the smallest primary consumer species. Substituting Eqs. (6)–(8) into Eqs. (1) and (2) yields a population dynamic model with allometrically scaled and normalised parameters. We used constant values for the following model parameters: maximum ingestion rate $y_j = 8$; carrying capacity $K = 1$, half saturation density of the functional response $B_0 = 0.5$ (after Brose *et al.*, 2005), $y_i$ = the maximum ingestion rate relative to its metabolic rate (Brose *et al.*, 2006; Riede *et al.*, 2010). Independent simulations of each food web started with uniformly random initial biomasses ($0.05 < B_i < 1$), and they were run for the equivalent of 1, 5, 10, 20 and 30 years as calculated by inserting $M_{P*}$ in (3) and taking the inverse of $W_{P*}$ (1/years).

## F. Stable Isotope Analysis

GCA provides a detailed snapshot of feeding links on a given occasion, often reflecting the most recent meal of an individual, but does not provide a direct measure of biomass assimilation over a longer timescale (Woodward and Hildrew, 2002a). It is the method of choice when studying taxonomically resolved links (e.g. between individual species), but a large number of samples (often several hundred individuals; e.g. Ings *et al.*, 2009; Woodward and Hildrew, 2001) are needed to describe predator diets fully. Stable carbon and nitrogen isotope analysis (SIA) of consumers and resources provides a more integrated measure of the flux and assimilation of biomass, and this was carried out during the 2006 survey to supplement the GCA in our assessment of trophic interactions in the food web following trout invasion. Despite offering relatively poor taxonomic resolution, SIA provides a more time-integrated and complementary means of tracing broad pathways of biomass flux and of measuring the trophic height of the network as a whole (e.g. Fry, 1988; Grey *et al.*, 2001; Pace *et al.*, 2004; Yoshii, 1999; Yoshioka *et al.*, 1994). We combined both methods to provide the most complete characterization of trout diet that was logistically feasible, while keeping destructive sampling of the small Broadstone population of trout to a minimum.

Macroinvertebrate samples were collected for SIA on each sampling occasion in 2006 using a kick net (mesh size 330 $\mu$m), and the dominant basal resources were also sampled: epilithon (a mixed assemblage of diatoms, fine particulate organic matter [FPOM], fungi and bacteria) was scrubbed off submerged rocks, and CPOM (mainly decomposing plant material of allochthonous origin) was collected by hand from the stream bed. Fish tissue ($<1$ mg per individual) was sampled by fin clipping (McCarthy and Waldron, 2000). All samples were frozen immediately upon return from the field site ($<2$ h) and stored until further processing. After thawing, guts were removed from macroinvertebrates, and all samples dried to constant mass at 60 °C in individual acid-washed glass sample vials. Once dried, samples were ground into a fine powder and approximately 0.6 mg (macroinvertebrates) or 1 mg (basal resources) per replicate was loaded into $5\times7$-mm tin capsules for isotopic analyses. For large macroinvertebrates (e.g. *C. boltonii*, *P. conspersa*), single specimens provided enough material for each replicate. For smaller species, several individuals had to be pooled to obtain sufficient material (e.g. about 12 individuals for non-predatory Chironomidae and for Simuliidae, five to six for leuctrid and nemourid stonefly larvae, and two to three individuals for Tanypodinae). For each sample, three to five replicates were analysed. Stable carbon and nitrogen isotope analyses were performed on the same sample using a ThermoFinnigan Delta$^{\text{Plus}}$ continuous flow isotope ratio mass spectrometer (Thermo Finnigan, Bremen, Germany). The results of estimation of isotopic composition are expressed in standard δ notation:

$$\delta^{\mathrm{I}} = \left[\frac{R_{\text{sample}}}{R_{\text{standard}}} - 1\right]1000,$$

where $I$ is either $^{13}$C or $^{15}$N, and $R$ is the ratio of either one to the respective lighter isotope ($^{12}$C or $^{14}$N). $\delta^{I}$ is expressed as the per mille (‰) deviation of the sample from the recognised isotope standards (Pee Dee Belemnite for $\delta^{13}$C; atmospheric $N_2$ for $\delta^{15}$N).

## G. Laboratory Experiments: Potential Behaviourally Mediated Effects of Trout on Feeding Links

Survival rates of *N. pictetii* and *L. nigra*, two detritivorous stonefly species, were measured in the presence of each of the three dominant large invertebrate predators (*C. boltonii*, *S. fuliginosa* and *P. conspersa*) in the presence/absence of *S. trutta* chemical cues (kairomones). Invertebrate predators and prey were collected from the acidified section of the stream in winter between February and November 2010. Final instars of *S. fuliginosa* and *P. conspersa* and the largest size-class of *C. boltonii* (i.e. instars 12–14) were used, following the protocols of Woodward and Hildrew (2002b,c). Thus, 5 cm depth of washed native gravel substratum was added to plastic aquaria (dimensions 19.5 cm long × 12 cm wide × 13.2 cm deep) (SAVIC, Belgium), with a 60:40 mix of stream water and dechlorinated tap water introduced to a total depth of 10 cm. Stream water was taken from the neighbouring Lone Oak Stream, which has almost identical water chemistry to Broadstone but no fish (e.g. see Layer *et al.*, 2010b), to eliminate the possibility of resident kairomones being present. A single invertebrate predator was placed in each aquarium, with initial prey density being set at 10 *N. pictetii* and *L. nigra*, without replacement of consumed individuals during the 24 h trial. Water temperature was maintained at 10 °C and the light regime was 12 h light/12 h dark. Predators were starved for 5 days prior to introduction to the test aquaria 12 h before the experiment began, to allow individuals to adjust to the environment and especially to enable *P. conspersa* larvae sufficient time to spin nets, as this is their primary means of prey capture (Townsend and Hildrew, 1979). Trials were run for each predator type with/without kairomone addition, with 10 replicates per treatment (i.e. 120 microcosms in total). Sixty millilitres of kairomone-exposed water was added from a containment tank containing one adult *S. trutta*, to achieve a dosage of approximately 3× the concentration that Abjornsson *et al.* (2001) used to demonstrate measurable suppression of activity of *G. pulex*. In other words, to maximise our chances of eliciting any potential behavioural response, we used an amount of kairomone well beyond that likely to be encountered in the stream. Each trial was run for 24 h and the gravel elutriated and sorted to count all surviving prey and establish predation rates, after Woodward and Hildrew (2002c).

## H. Statistical Analysis

Multivariate analyses of invertebrate community structure were performed using PRIMER 6 (v. 6.1.13) and PERMANOVA (v. 1.0.3) (PRIMER-E Ltd., Plymouth, UK) (after Anderson *et al.*, 2008), in which calculations of *P*-values are derived via permutations, thus avoiding the commonly violated assumption of normality in such compositional data in traditional MANOVA. After carrying out these multivariate analyses, linear mixed effects models (LMEM) were used to test for differences in invertebrate abundance for individual (or aggregated) taxa, while accounting for (i) the temporally hierarchical nature of the data structure and (ii) the inclusion of both fixed and random effects in the survey design. Abundances were $\log_{10} x + 1$ transformed to normalise the data and stabilise variances. Sample units (i.e. replicate Surber samples) were nested within individual sampling occasions (i.e. months), months nested within seasons (i.e. summer, May–

October inclusive: winter, November–April inclusive) and seasons and decades as main effects. Season and decade were fitted as fixed effects in the analyses, and nested terms were fitted as random effects. Since our models were unbalanced because of the uneven number of replicates taken in the different decades, even after standardising each decade to include only one annual cycle, we used the restricted maximum likelihood method (REML) to estimate error terms and produce corrected $F$-ratios (Crawley, 2007).

Denominator degrees of freedom (df) were calculated such that the decade main effect was tested against the decade $\times$ month (season) random factor and residuals. The main effect of season was tested against the replicate (month) and month (season) random factors and residuals and the interaction against decade $\times$ month (season) random factor and residuals. Species could differ in their denominator df because some terms among the random factors are redundant and therefore have no variance to partition out of the model: the denominators are calculated without these random terms, and fractions can arise because of the unbalanced design. Several rare taxa (e.g. *Paraleptophlebia submarginata* Stephens, *Halesus radiatus* Curtis) were absent from all sampling occasions in particular decades (i.e. variance was zero) and so could not be analysed formally.

Regression analysis and ANCOVA were performed on mass–abundance scaling relationships within the food web, to compare differences in slopes of the ISDs for August (when the community size spectrum was broadest due to generational overlap) among the three decades for which we had individual-level data. Prey survival rates in the laboratory experiments were arcsine-transformed (after Woodward and Hildrew, 2002c) prior to analysis, using a simple fully factorial ANOVA, with kairomone presence/absence and predator species identity fitted as main effects in a crossed design. Bivariate statistical analyses were performed with PASW statistics v. 18.0 (SPSS Inc., Chicago, IL). The R language for Statistical Computing (R Development Core Team, R Foundation for Statistical Computing, Vienna, Austria (2010); (http://www.R-project.org)) was used for the remaining statistical tests and for graphical visualisation of the food webs.

# III. RESULTS

## A. Community and Food Web Structure

There were significant shifts in the composition of the invertebrate fauna among the four decades (PERMANOVA Pseudo-$F_{3,362}=4.68$, $P=0.0008$) and among seasons (Pseudo-$F_{1,362}=2.94$, $P=0.0248$) but, in contrast to absolute abundances of many of the individual taxa (described below), there was no interaction between the two ($P>0.05$) (Table 1). Within the

**Table 1** Results of PERMANOVA testing for temporal change at different scales in the composition of the Broadstone Stream invertebrate assemblage, from 1974 to 2006 (see Section II for details)

| Source | df | Pseudo-*F* | *P* (perm) |
|---|---|---|---|
| Decade | 3 | 4.68 | 0.0001 |
| Season | 1 | 2.94 | 0.0248 |
| Month (season) | 8 | 4.86 | 0.0001 |
| Decades × season | 3 | 1.06 | 0.4445 |
| Replicate (month (season)) | 203 | 0.911 | 0.8713 |
| Decade × month (season) | 6 | 5.73 | 0.0001 |
| Residual | 138 | | |

large predator guild, there were clear peaks in the absolute abundance of individual predator species, which followed a sequence of increasing body size over the four successive decades (Figure 3): *P. conspersa* was most abundant in the 1970s, *S. fuliginosa* in the 1980s, *C. boltonii* in the 1990s and trout in the 2000s (Figure 4). The annual peak in total macroinvertebrate abundance decreased more than 10-fold between the 1970s and 2000s (Figure 4). Seasonal fluctuations in benthic prey densities were also increasingly damped over this time, with the coefficient of variation of within-year monthly averages declining markedly in the latter two decades relative to the two earlier decades (1970s = 89%; 1980s = 83%; 1990s = 49%; 2000s = 53%). Two of the less acid-tolerant species, the detritivorous stonefly *Leuctra hippopus* Kempny and *C. boltonii*, however, increased significantly in mean annual abundance in the 2000s, after the trout invasion, whereas most taxa and the invertebrate assemblage as a whole declined in abundance over the same period, despite the increase in pH (Table 2).

Brown trout abundance peaked in April in 2006, reaching a density of 0.13 individuals $m^{-2}$, and this new top predator was several orders of magnitude bigger than the largest individual of any other species in the system (Figure 3). The trout population was dominated by 1+ fish, and GCA in 2006 revealed that terrestrial invertebrates, the adult stages of aquatic insects, and the resident large invertebrate predators were prominent in the diet (Figure 5), in addition to a wide range of other prey detected in the diet during the field experiments in the 1980s (Appendix III). These data and additional dietary information obtained from the literature showed that trout were predators of 81% of the invertebrate taxa in the food web. Stable isotope analysis of consumers and resources in 2006 further confirmed that brown trout, with the highest $\delta^{15}N$ values (5.17‰ ± 0.04 SD), occupied the top of the food web and that all members of the macroinvertebrate assemblage analysed were feasible prey items (i.e. none had a higher $\delta^{15}N$; Figure 6). The trout invasion represented the addition of a full trophic level

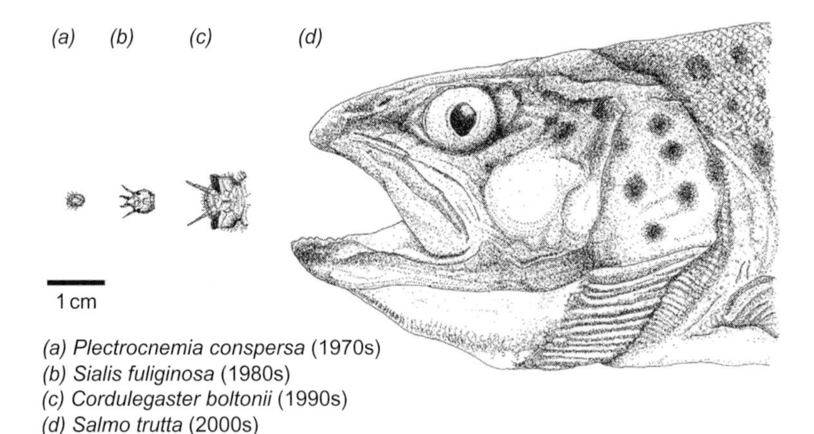


**Figure 3** Scale drawings of heads of the largest individuals of the dominant macroinvertebrate predator species (dorsal elevation) in Broadstone Stream (*Cordulegaster boltonii*, *Sialis fuliginosa* and *Plectrocnemia conspersa*), and of the invading top predator, *Salmo trutta* (side elevation). The respective peak abundances of each species (see Figure 4) are shown in parentheses. For example, during the April 2006 survey, the average brown trout dry body mass (12.8 g) was 388 times greater than that of the next largest predator, *C. boltonii* (33 mg). Macroinvertebrate head-capsules are redrawn after Woodward and Hildrew (2001).

(assuming 3.4‰ fractionation, after Minagawa and Wada, 1984), as revealed by its $\delta^{15}N$ isotopic signature being 3.8‰ higher than that of the predatory macroinvertebrates.

The 2006 Broadstone Stream summary food web contained 17 trophic elements (Table 3), with brown trout at the top. The primary consumer assemblage through all four decades was dominated by detritivorous nemourid and leuctrid stoneflies and chironomid midge larvae, although relative abundances changed considerably both among and within decades. Summary network statistics of the predatory food webs, constructed for all four decades of sampling (Table 4), revealed that linkage density, connectivity and complexity all increased over time, and especially so after the trout invasion. For both pre- and post-trout invasion trivariate food webs, the general direction of energy flux was from smaller and more abundant to a smaller number of larger and rarer taxa, but the body mass–abundance scaling relationships of the network changed markedly over time, as abundances declined and body mass increased (Figure 7). This trend was even stronger after the trout invasion, which reshaped the food web to such an extent that it occupied a very different constraint space in 2006 relative to the 1970s.

The ISDs for July–August (i.e. during the peak of invertebrate recruitment) also revealed marked shifts in the size structure of the food web, with the log $M$ – log $N$ regression slope being significantly steeper in the 2000s than

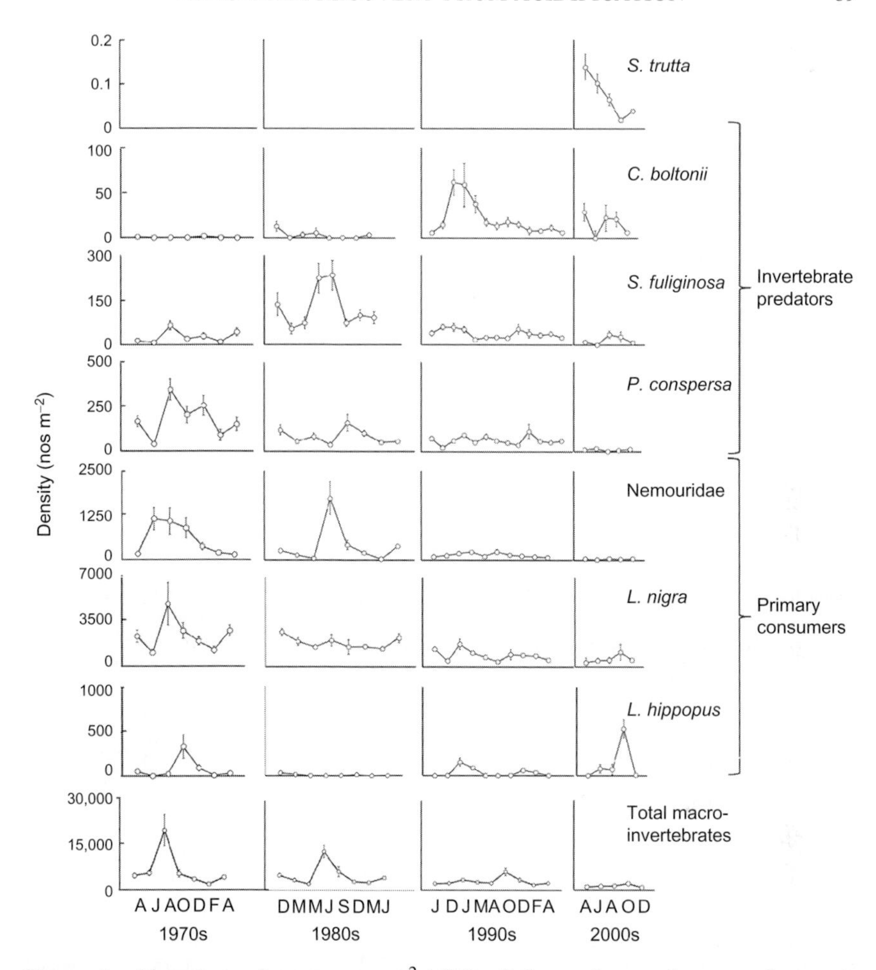


**Figure 4** Abundance (mean nos. $m^{-2} \pm SE$) of the major predators and prey in Broadstone Stream over four decades between 1974 and 2006. *Note*: for the formal statistical LMEM analyses, as described in Section II, data were standardised by restricting the sampling period for each decade to a single annual cycle of recruitment-emergence for the univoltine invertebrates (e.g. in the 1990s, only the months from June 1996 to April 1997 were included).

in the 1970s and 1990s, which did not differ significantly from one another (ANCOVA: $F_{1,25} = 7.80$; $P = 0.01$). Essentially, mean abundance of the smaller members of the food web declined, while mean individual body size increased, principally because the new dominant predators occupied existing (or created new) size bins at the upper end of the size spectrum (Figure 8).

The frequency distributions of inferred interaction strengths among consumer–resource pairs (Figure 9) shifted during the four decades

**Table 2** LMEM GLM Type III nested mixed model output of temporal change in abundance of the dominant predator and prey taxa within the Broadstone Stream food web over 4 decades (analysis was restricted to one annual cycle per decade in the model, to minimise sampling bias; see Figure 4 for full time series data)

| | Numerator df | Denominator df | *F* | *P* |
|---|---|---|---|---|
| *Cordulegaster boltonii* | | | | |
| Decade | 3 | 358 | 34.064 | <0.001 |
| Season | 1 | 358 | 0.358 | 0.550 |
| Decade × season | 3 | 358 | 1.122 | 0.340 |
| *Sialis fuliginosa* | | | | |
| Decade | 3 | 155.7 | 18.252 | <0.001 |
| Season | 1 | 5.5 | 4.587 | 0.080 |
| Decade × season | 3 | 155.7 | 3.212 | 0.025 |
| *Plectrocnemia conspersa* | | | | |
| Decade | 3 | 142.2 | 36.678 | <0.001 |
| Season | 1 | 4.6 | 0.077 | 0.794 |
| Decade × season | 3 | 142.2 | 0.853 | 0.467 |
| Nemouridae | | | | |
| Decade | 3 | 157.3 | 52.432 | <0.001 |
| Season | 1 | 2.7 | 2.478 | 0.224 |
| Decade × season | 3 | 157.3 | 7.676 | <0.001 |
| *Leuctra nigra* | | | | |
| Decade | 3 | 145.8 | 35.474 | <0.001 |
| Season | 1 | 5.7 | 0.048 | 0.834 |
| Decade × season | 3 | 145.8 | 5.322 | 0.002 |
| *Leuctra hippopus* | | | | |
| Decade | 3 | 229.5 | 25.935 | <0.001 |
| Season | 1 | 8.1 | 0.052 | 0.825 |
| Decade × season | 3 | 229.5 | 7.828 | <0.001 |
| *Non-predatory Chironomidae* | | | | |
| Decade | 3 | 327.0 | 47.139 | <0.001 |
| Season | 1 | 7.2 | 1.983 | 0.201 |
| Decade × season | 3 | 327.0 | 21.971 | 0.000 |
| *Tanypodinae* | | | | |
| Decade | 3 | 183.7 | 17.533 | <0.001 |
| Season | 1 | 7.2 | 11.966 | 0.010 |
| Decade × season | 3 | 183.7 | 1.556 | 0.202 |
| *All invertebrates* | | | | |
| Decade | 3 | 169.8 | 27.698 | <0.001 |
| Season | 1 | 6.7 | 10.797 | 0.014 |
| Decade × season | 3 | 169.8 | 0.626 | 0.599 |

Values shown are for the fixed effects only in the model; variances due to random effects (sample-units, month of sampling) have been partitioned in the model and are therefore not shown (see Section II for a more detailed explanation of the derivation of the terms in the table.

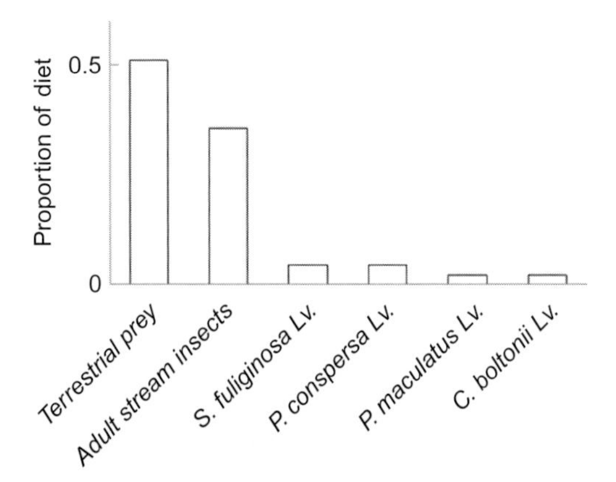


**Figure 5** Gut contents analysis of trout ($n=10$) collected from Broadstone Stream in 2006, which consisted of 'Terrestrial prey' (mainly woodlice and litter mites) and adult aquatic insects (mostly nemourid stoneflies, *P. conspersa* and *C. boltonii*) as well as larval (Lv.) *S. fuliginosa*, *P. conspersa*, *P. maculatus* and *C. boltonii*. Additional dietary data for constructing the food webs were taken from earlier experimental studies in the 1980s in Broadstone (Schofield, 1988) and from the wider literature (see Section II).

($\chi^2=29.764$ df$=6$, $P<0.001$) with a significant pairwise difference between the 1970s and 1980s (Kruskal–Wallis, $H_3=11.36$, $P=0.01$; 1970s<2000s Mann–Whitney pairwise comparison, $P=0.02$) and with an especially pronounced rise after the trout invasion. Here, a bimodal distribution appeared, with the additional 'peak' of especially strong interactions representing those arising from the large body mass ratios of *S. trutta* to its macroinvertebrate prey.

## B. Food Web Dynamics: Persistence

In the simulations, the Broadstone webs in the 1970s and 1980s were less prone to species loss than was the case for the 1990s and 2000s webs (Figures 10 and 11). The establishment of the two largest carnivores, the dragonfly *C. boltonii* and trout, as the new top predators in the 1990s and 2000s, respectively, was associated with a marked decline in persistence in the simulations. Species loss was typically fastest within the first 5 years of the simulation, and followed a nonlinear function, such that the rate slowed subsequently and in many cases the size of the network eventually appeared to stabilise (Figure 10). Intriguingly, the same numbers of species (8) were lost in the 30-year simulation of the trout-invaded food web of the 2000s as for the 1990s food web, although most extinctions occurred much faster in the 2000s food web.

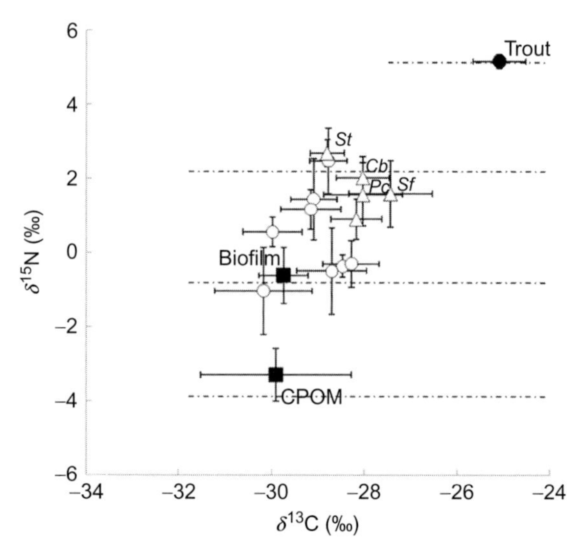


**Figure 6** Stable Isotope Analysis biplot showing mean (±SD) of $\delta^{13}$C and $\delta^{15}$N (‰) in the Broadstone Stream food web after trout invasion in 2006. The two main basal resources (detrital CPOM and algal biofilm) are denoted as solid squares, invertebrate primary consumers as white circles, invertebrate predators (in order of decreasing body size: Cb = *Cordulegaster boltonii*; Sf = *Sialis fuliginosa*; Pc = *Plectrocnemia conspersa*; St = *Siphonoperla torrentium*) as triangles and the apex predator (trout) as a black circle. The horizontal dashed lines are equivalent to an entire 'trophic level', assuming fractionation of $\delta^{15}$N at 3.4‰: that is, there are four full levels in the food web, with intermediate species representing omnivores. The web is based primarily on detrital inputs of terrestrial leaf litter, rather than epilithic biofilm.

**Table 3** The Broadstone Stream food-web taxa (1970–2000s), as defined in the current study

| Trophic position | Taxon |
|---|---|
| Top predator | *Salmo trutta* L. |
| Predators | *Cordulegaster boltonii* Donovan, *Sialis fuliginosa* Pictet, *Siphonoperla torrentium* Pictet, *Plectrocnemia conspersa* Curtis, Tanypodinae |
| Primary consumers | Nemouridae, *Leuctra hippopus* Kempny, *Leuctra nigra* Olivier, *Halesus radiatus* Curtis, *Potamophylax cingulatus* Stephens, *Asellus meridianus* L., *Niphargus aquilex* Schiödte, *Paraleptophlebia submarginata* Stephens, *Pisidium* sp., Tipulidae, Chironomidae (non-predatory), Coleoptera, Simuliidae |

Other more intensive, but less extensive studies have used more highly resolved data (and hence a greater number of nodes within the food web), but here taxa were aggregated to a common denominator across all sampling occasions to standardise the resolution of the food webs from each decade, as described in Section II.

**Table 4** Summary statistics for the Broadstone Stream predatory food web are shown for the decades before (1970–1990s) and after (2000s) brown trout invasion

| Decade | pH | $S$ | $L$ | $L_c$ | $L/S$ | $L/S^2$ | $SC'_{max}$ | $G$ | $V$ | $G_k$ | $V_k$ | $G_{st}$ | $V_{st}$ |
|---|---|---|---|---|---|---|---|---|---|---|---|---|---|
| 1970s | 5 | 13 | 29 | 29 | 2.23 | 0.17 | 2.23 | 9.67 | 2.42 | 3.89 | 0.94 | 1.73 | 0.48 |
| 1980s | 5.2 | 13 | 31 | 31 | 2.38 | 0.18 | 2.38 | 9.67 | 2.42 | 3.91 | 0.92 | 1.72 | 0.34 |
| 1990s | 5.4 | 17 | 46 | 47 | 2.71 | 0.16 | 2.76 | 9.2 | 2.88 | 3.1 | 1.02 | 1.58 | 0.66 |
| 2000s | 5.6 | 17 | 60 | 60 | 3.5 | 0.2 | 3.5 | 10 | 3.75 | 2.57 | 1.01 | 1.42 | 0.53 |

Shown are mean pH (pH), number of taxa ($S$), number of links ($L$); number of links including competitive links (shared resources, $L_c$); linkage density ($L/S$); directed connectance ($L/S^2$); web complexity ($SC'_{max}$); generality ($G$; number of resources per consumer); vulnerability ($V$; number of consumers per resource); standardised generality ($G_k$), vulnerability ($V_k$) and standard deviation of $G_k$ and $V_k$ ($G_{st}$ and $V_{st}$, respectively; see Williams and Martinez, 2000). Taxonomic resolution was standardised to the lowest common denominator across samples before web construction and analysis. Additional pH data for Broadstone prior to the trout invasion were taken from Hildrew (2009) and Woodward *et al.* (2002).

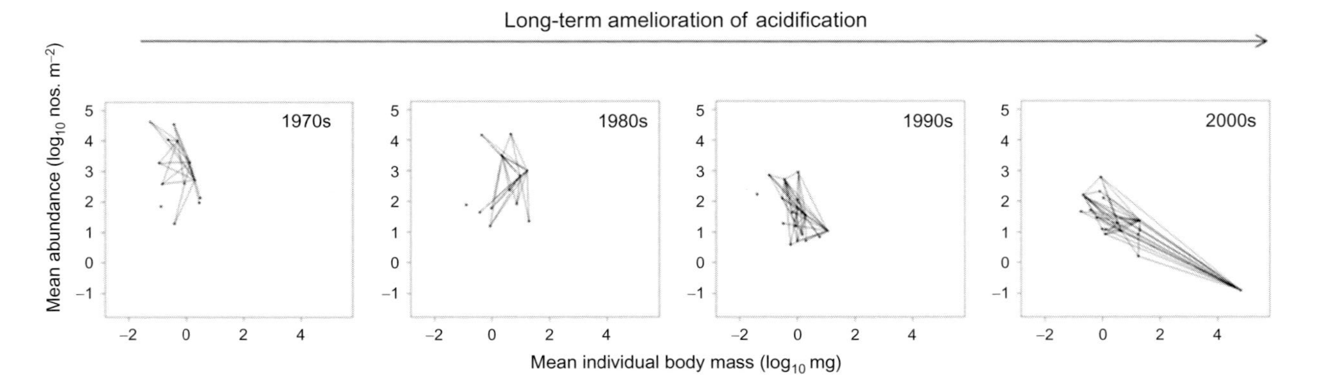


**Figure 7** Broadstone Stream trivariate food webs pre- and post-trout invasion depicted using species-averaged data of $log_{10}$-transformed abundance–body mass relationships and food-web structure. The R language for Statistical Computing (R Development Core Team, R Foundation for Statistical Computing, Vienna, Austria (2010); (http://www.R-project.org)) was used for graphical visualisation of the food webs.

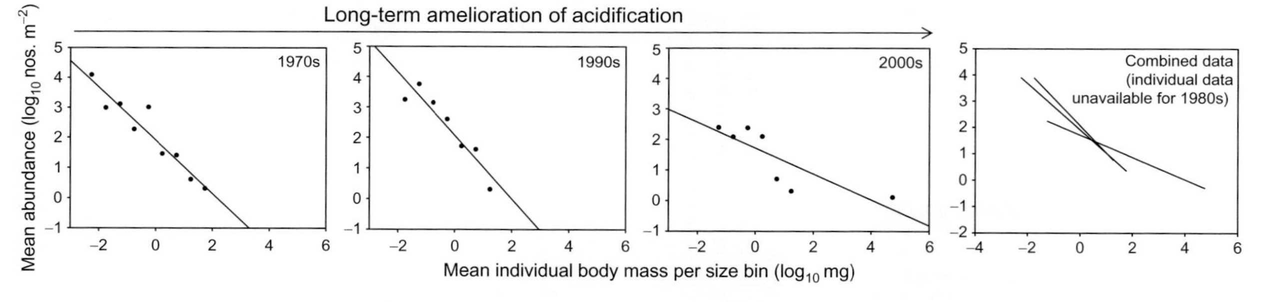


**Figure 8** Individual size distribution (ISD) mass–abundance plots for the Broadstone Stream assemblage (excluding meiofauna) for August 1974, August 1997, August 2006 and all three decades combined. Regression equations as follows: (a) $y = -0.8x + 2.01$, $F = 19.7$, $P < 0.01$ (b) $y = -0.7x + 2.01$, $F = 16.3$, $P < 0.01$ and (c) $y = -0.4x + 1.7$, $F = 10.7$, $P < 0.05$. Individual body mass data were unavailable for the 1980s dataset and therefore no ISD plot could be produced (see Section II).

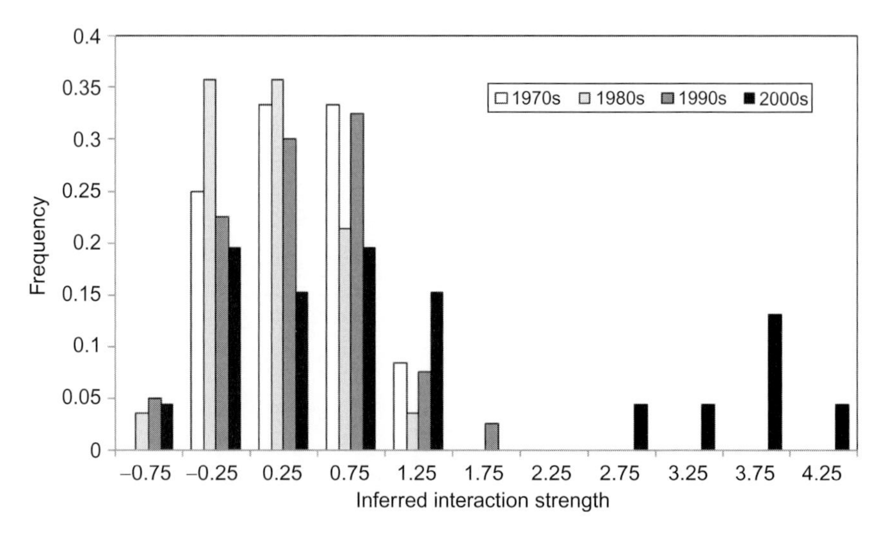


**Figure 9** The frequency distribution of inferred interaction strengths derived from $\log_{10}$-transformed predator–prey body mass ratios ($M_{\mathrm{pred}}/M_{\mathrm{prey}}$) before *C. boltonii* invasion (1970s, white bars; 1980s, light grey bars), and before (1990s, dark grey bars) and after (2000s, black bars) brown trout invasion, calculated for all predatory interactions in Broadstone Stream.

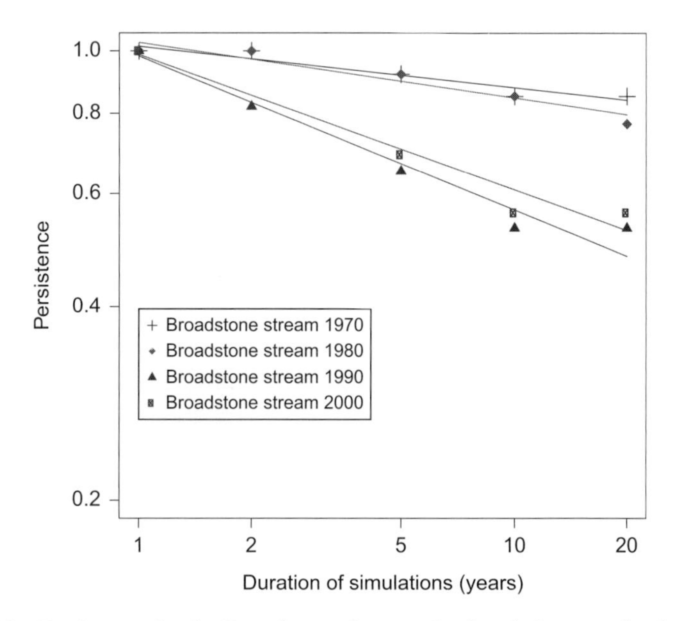


**Figure 10** Persistence in the Broadstone Stream food web from each of the four decades after different time periods of simulated dynamical modelling.

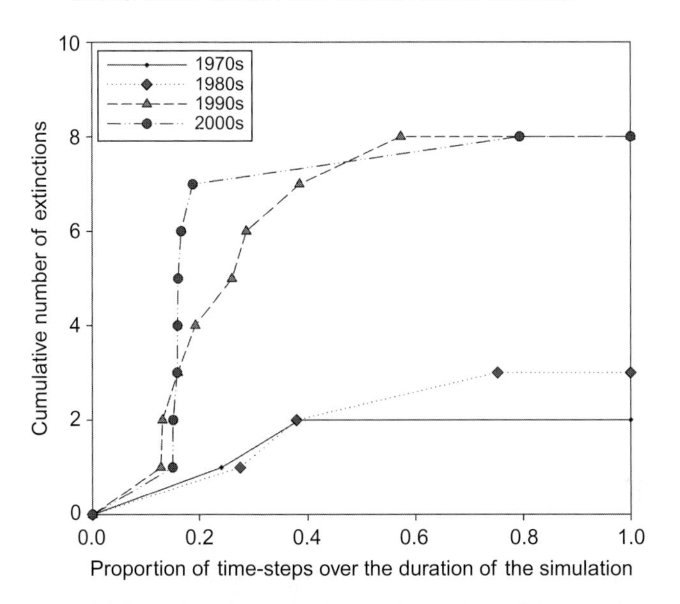


**Figure 11** Trajectories of species extinctions in Broadstone Stream over 30 years of simulated population dynamics for food webs from each of the four decades. Predatory taxa are denoted by solid circles, non-predatory taxa by open circles. Note the *x*-axis has been normalised by the maximum number of time steps in each simulation, to facilitate comparisons. The identity and timing of each species extinction are given in Appendix IV.

Among the species most vulnerable to extinction in the simulations, *P. conspersa* was lost after 30 years in each of the four decades, despite always being present in the empirical data over an equivalent time span. Overall, predators were far more likely to be lost than prey species: after 30 years of simulated dynamics, there were 11 cases of predator extinctions, compared to 10 among the prey assemblage, despite the fact that the former accounted for only about one-quarter to one-third of total species (Figure 11). In the 2000s, all four of the largest predators, including trout, were lost within the first 5 years of the 30-year simulation, whereas in the other three decades, *P. conspersa* was the only one of them to go extinct (Appendix IV).

## C. Behaviourally Mediated Indirect Food Web Effects

In the laboratory trials, trout kairomones did not affect predation by the three large resident invertebrate predators on their stonefly prey. There were, however, significant differences in predation rate among the predators (*C. boltonii* > *P. conspersa* > *S. fuliginosa*), with the largest, *C. boltonii*, exerting far stronger predation pressure than the two smaller predators (Table 5; Figure 12). There was a significant two-way interaction between predator species and prey species, with *N. pictetii* being especially vulnerable to the larger sit-and-wait predators (i.e. *C. boltonii*), whereas the mortality of the less mobile *L. nigra* was lower and at a comparable level for all three predators. The lack of a kairomone main effect and its non-significance in two-way or three-way interaction terms revealed that chemical cues did not alter prey (or predator) behaviour sufficiently to affect direct predatory feeding links (Table 5; Figure 12).

# IV. DISCUSSION

## A. Changes in Community and Food-Web Structure

Complexity can stabilise food webs if most trophic interactions are weak (e.g. Borrvall *et al.*, 2000; McCann *et al.*, 1998; Rooney *et al.*, 2006). In other topological-based modelling studies, persistence (described as robustness) has been observed to decline as connectance increases (Dunne *et al.*, 2002), whereas in our dynamical modelling, we found the opposite to be the case. The key differences between the current study and these earlier modelling approaches are that we included body masses and functional responses into our model, and that we considered both bottom-up effects (which were also considered in the topological models of Dunne *et al.*, 2002) and top-down effects (which were not considered in the Dunne *et al.*, 2002 study). The presence or absence of the brown trout in particular had strong effects on both types of effect, and this accounted for the seemingly contradictory results when compared with previous studies that used different modelling approaches but the same response variable. The Broadstone Stream

**Table 5** Analysis of variance testing for effects of predator species identity and presence/absence of fish kairomone on prey survival rates in laboratory experiments ($n=10$ replicates per treatment: that is, 120 aquaria were used in total)

| | df | $F$ | $P$ |
|---|---|---|---|
| Predator | 2 | 13.06 | <0.001 |
| Prey | 1 | 1.71 | 0.193 |
| Kairomone | 1 | 0.27 | 0.606 |
| Predator × prey | 2 | 6.76 | 0.002 |
| Prey × kairomone | 1 | 0.03 | 0.857 |
| Predator × kairomone | 2 | 1.22 | 0.298 |
| Predator × kairomone × prey | 2 | 2.32 | 0.103 |

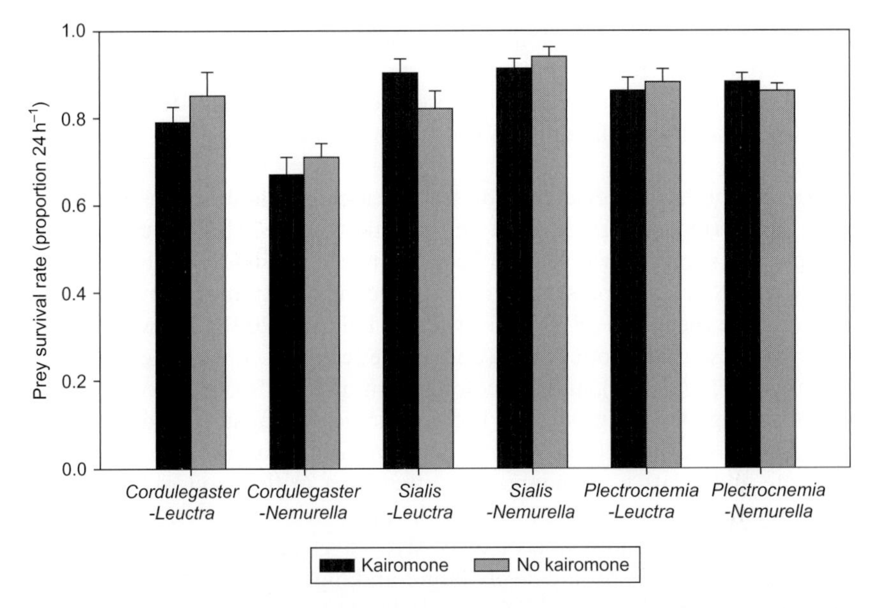


**Figure 12** Mean survival rates (bars = 1 SE) of two common prey species, the detritivorous stoneflies *N. pictetii* and *L. nigra*, after 24-h exposure to the three previously dominant large invertebrate predators (*Cordulegaster boltonii*, *S. fuliginosa* and *P. conspersa*) prior to the trout invasion, in the presence/absence of trout kairomones under controlled laboratory conditions. Statistical analysis was performed on arcsin-transformed survival data: although predation rates were higher for *C. boltonii* than for the two smaller species, the presence of trout kairomones had no significant effect on the predator–prey feeding interaction for any of these six common feeding links within the food web.

community has, at least until now, often been perceived as being dynamically stable. It was, for instance, resilient to the invasion of the dragonfly *C. boltonii* (Woodward and Hildrew, 2001) and in response to the large-scale and long-term manipulation of another previous top predator, the alderfly *S. fuliginosa* (Hildrew *et al.*, 2004). Dynamical simulations have also revealed its food web to be a relatively robust, at least when compared with those at higher pH (Layer *et al.*, 2010b), echoing the results of earlier population and network modelling of interactions within the invertebrate assemblage (Emmerson *et al.*, 2005; Speirs *et al.*, 2000). Several factors, potentially operating in concert, could have contributed to this apparent stability, including the dominance of detritus as the basal resource (Dobson and Hildrew, 1992), the high degree of generalism, the large number of weak interactions in the web (Woodward *et al.*, 2005a), density-dependent population regulation (Hildrew *et al.*, 2004) and the particular configuration of links within the feeding matrix (Emmerson *et al.*, 2005; Layer *et al.*, 2010a; Woodward *et al.*, 2005a). As the system has become less acid and the biota has responded by the (re)colonisation of acid-sensitive species, however, the food web appears to have become less stable. This is potentially due to the recent arrival of the first vertebrate predator that may have further eroded food-web stability, although not yet to the extent of triggering widespread species loss.

Despite the lack of evidence for strong cascading interactions, we found clear long-term trends in the food web over the four decades of study. During this period, species turnover has been relatively low (cf. Woodward *et al.*, 2002), with *S. trutta* representing only the second addition of a new dominant species since the 1970s (cf. Woodward and Hildrew, 2001). Brown trout are typically present at about pH 5.4 and above (e.g. Jensen and Snekvik, 1972; Magnuson *et al.*, 1984), and the colonisation of Broadstone, also seen in the neighbouring Old Lodge stream, is clearly indicative of sustained reductions in acidity (Hildrew, 2009; Monteith *et al.*, 2005), as are the changes in absolute and relative abundances of other taxa in the food web (Hildrew, 2009; Woodward *et al.*, 2002). One of the most striking patterns in the long-term data is the phenomenon of successive waves of invasions or irruptions of predators that are progressively larger and higher in the food web, as pH has risen, and this has coincided with declines in system persistence.

Elsewhere, fish invasions have sometimes triggered strong cascading effects (Baxter *et al.*, 2004; Biggs *et al.*, 2000; Flecker and Townsend, 1994) or local species extinctions (Townsend, 1996; Townsend and Simon, 2006) lower in the food web, so it is intriguing that this has not been observed, at least not yet, in Broadstone. Most such studies have focussed on predator–herbivore–algal food chains in systems with a greater reliance on autochthonous production than on allochthonous detritus (cf. Dahl, 1998; Power, 1992a,b; Townsend, 2003): interaction strengths at the base of detrital food webs, such as Broadstone, are effectively zero, as consumers do not control the supply rate of resources, whereas algal–herbivore links can be very strong in systems that are more reliant on in-stream primary production. Trophic cascades and top-down effects induced by fish predators in detritus-based systems are less well known (but see Greig and McIntosh, 2006; Konishi *et al.*, 2001; Woodward *et al.*, 2008), but they can occur and strong effects on prey populations have been reported, although these have not necessarily cascaded through the web to affect other species indirectly. Indeed, Schofield *et al.* (1988) found that trout predation reduced *P. conspersa* abundance in field experiments conducted in Broadstone, but with negligible effects on other species. There is also a strong negative correlation between the abundances of trout and *P. conspersa* across a wide pH gradient among 29 other streams within the Ashdown Forest, but this was not the case for other invertebrate species, again suggesting that the trout-*P. conspersa* link might be an exceptionally strong interaction embedded within a network of generally much weaker effects exerted by trout. Our long-term data also revealed that *P. conspersa* has decreased significantly following the arrival of trout, as have some of the other invertebrate predators, in line with general expectations based on our modelling results and observations from other systems. However, this again did not translate into what might be considered a cascade. Although we found no compelling evidence of the release of the prey assemblage as a whole from invertebrate predation, following trout invasion, there was an overall suppression of both means and annual variability of species populations across the food web, which suggests strong but diffuse density-dependent control of the prey assemblage by the predator guild (cf. Hildrew *et al.*, 2004). The effects of the predators in Broadstone thus appeared to be manifested relatively directly, rather than by triggering oscillatory dynamics that are 'felt at a distance', possibly because there are so few links separating any pair of species in the small world of the food web (Montoya *et al.*, 2009).

Given the generalist nature of all the Broadstone predators (within the size range of prey they can handle), there is considerable scope for apparent competition among prey for enemy-free space (Woodward and Hildrew, 2001). Rare prey species are therefore unlikely to avoid predation because the predators are supported by other more abundant species. This might account for the loss of some of the rarest taxa (e.g. Coleoptera, *Asellus meridianus* Racovitza, *Siphonoperla torrentium* Pictet) during dynamical simulations, and the fact that some of these taxa have only appeared sporadically in the stream over the four decades of study. In addition, brown trout is by far the largest species in the system, and its energy gain and simulated abundance will be influenced by its body mass in our bioenergetic model. Such a large species could cause substantial redistributions of the energy flow through the food web, and since the brown trout is extremely generalist, it can have dual negative interactions with other predators in the food web: firstly, via direct consumption and secondly by competing for the same prey species.

The only prey species to have become significantly more abundant (on average) since the trout invasion is *L. hippopus*, and this may be linked more directly to a reduction in acid stress (Hildrew, 2009; Larrañaga *et al.*, 2010). The generalist nature of all the predators, and the increased overall predation pressure exerted over time as this guild has expanded, might therefore make it extremely challenging for new prey species to become established as pH rises: new colonists are likely to be rare, but as this is unlikely to confer any density-dependent advantage upon them if apparent competition is strong, it will be difficult to establish a foothold for building a larger, and more viable, population.

Profoundly acid streams do not usually support fish, due to the combination of toxic effects of pH and associated high concentrations of labile aluminium (e.g. Magnuson *et al.*, 1984), in addition to low prey production (Hildrew, 2009). Among the trout guts sampled in Broadstone in 2006, 30%

contained only terrestrial prey, emphasising the importance of this external subsidy in such an unproductive system and the potential for competition for this resource, which is also shared with the large invertebrate predators (Townsend and Hildrew, 1979; Woodward *et al.*, 2005a). It is conceivable that external subsidies to the predator guild have become relatively more important in the food web overall after the trout invasion, and the decreased slope of the body mass–abundance scaling relationships in the 2000s lends support to this suggestion (Figure 9). The terrestrial invertebrate subsidy must at least partially explain how a chronically unproductive acid system like Broadstone Stream can support fish, and this explanation, often referred to as the 'Allen paradox', has often been invoked elsewhere (Allen, 1951; Huryn, 1996; Nakano *et al.*, 1999). Note that the exploitation of terrestrial items by trout might exacerbate top-down effects on its aquatic prey (cf. Nakano *et al.*, 1999) by partially decoupling the invader's reliance on in-stream invertebrate production. Even if trout do eventually exclude the large predators via the dual effects of predation and resource competition, however, the generalist diet of trout suggests that it might simply eventually replace them without any indirect effects cascading down the food web.

Various lines of evidence, when taken together, suggest that the recent brown trout invasion has reduced the dynamic stability of the food web by increasing mean interaction strengths: indeed, increases in top predator size in the 1990s and 2000s were associated with reduced persistence relative to the webs from the two earlier decades. Perhaps paradoxically, however, the increased damping of fluctuations in prey populations throughout the food web might hint at a general stabilising top-down effect (cf. Hildrew *et al.*, 2004). In addition, the long-term lengthening of food chains suggests an overall bottom-up effect of increasing total secondary production in Broadstone, possibly mediated via reduced acidity and faster rates of decomposition of the detrital resource base (see Chamier, 1987; Dangles *et al.*, 2004; Hildrew *et al.*, 1984). This, in turn, might lead to diffuse suppression of prey standing stocks by the array of generalist predators, with both top-down and bottom-up effects operating simultaneously (Hildrew, 2009). Further long-term monitoring and experimental manipulations are clearly needed to track the consequences of the ongoing invasion as they ramify through the community, and the Broadstone model system provides a rare opportunity to observe this phenomenon in a well-characterised food web.

## B. Caveats, Limitations and Future Directions

Because we were investigating a natural system in a shifting environment, there are of course several possible explanations for the change in the food web over time, and these are not necessarily mutually exclusive.

In particular, it is unlikely that the changes we observed were solely due to top-down or bottom-up effects, since both probably operate simultaneously. This could arise via, for instance, the establishment of populations of progressively larger predators at the top of the web and the stimulation of secondary production due to reduced physiological stress and improved basal resource quality (e.g. Groom and Hildrew, 1989; Winterbourn *et al.*, 1992). Since all of these are related directly or indirectly to pH, it is impossible to completely disentangle their effects. However, the fact that prey abundance has declined, rather than increased, as pH has risen over time suggests that overall top-down effects have taken precedence, despite the likely concurrent enrichment of resources at the base of the food web (e.g. Groom and Hildrew, 1989).

Taxonomic resolution can influence many network properties, including connectance and other measures of complexity (Martinez, 1991, 1992, 1993), so we should interpret the long-term structural changes in the food web with some caution (e.g. Dunne, 2006; Hall and Raffaelli, 1993; Schoener, 1989; Solow, 1996) because, as in almost all other food-web studies, the meiofauna were not included, due to logistic challenges (e.g. many taxa can only be identified alive). An earlier study resolved the Broadstone Stream meiofaunal sub-web (Schmid-Araya *et al.*, 2002b) in the 1990s, but comparable data were not available from the other decades, nor were they within the scope of the current study. Such microscopic animals are numerous and diverse in stream food webs, and they can influence estimates of connectance and mass–abundance scaling relationships (Schmid-Araya *et al.*, 2002a,b; Stead *et al.*, 2005; Woodward *et al.*, 2005a), but our analysis was designed to compare taxonomically standardised food webs rather than attempting to include all the species in the system. The chironomid taxa were also lumped into two taxa in our study due to inconsistent levels of resolution of these other small taxa over the four decades of sampling, but this is common to most freshwater food-web studies which, unlike our study, tend to combine the predatory tanypods with the other non-predatory taxa (see examples of webs provided in Ings *et al.*, 2009 and references therein). Further, within each of the two groups, there are only a small number of species, and most are trophically very similar or identical in Broadstone (cf. Woodward and Hildrew, 2001). As such the webs we constructed here, although not as resolved as other versions of the Broadstone web, were neither atypical of acid systems in general nor unusually small in comparison with other studies (e.g. Layer *et al.*, 2010a,b; Woodward *et al.*, 2008).

Interaction strengths were inferred from allometric scaling relationships applied to empirically measured consumer–resource body mass ratios (after Emmerson *et al.*, 2005), due to logistic constraints and the considerable difficulties of characterising them empirically. In a detailed study in which links were quantified for the 1990s food web, there was clear allometric scaling between the strength of links and predator and prey size (Woodward *et al.*, 2005b), so we have assumed similar patterns applied in the other three decades. Only a relatively small number of fish were available for the direct characterisation of trout diet, and these were all relatively large fish, due to the size limits imposed by the non-destructive method of gut flushing. Large trout are frequently reported as interacting strongly with large macroinvertebrates (e.g. Woodward *et al.*, 2010d), but the smaller individuals whose diet could not be characterised directly were identified from the literature sources, and from dissection of individuals from the lower reaches of Broadstone in the 1980s (Rundle and Hildrew, 1992), as potential predators of 81% of the benthic prey assemblage, and even this is likely to be a conservative estimate. This partial reliance on inferred feeding links might lead to overestimates of network complexity, but a diametrically opposing argument could equally be levelled at studies based solely on direct observation of gut contents because these are prone to undersampling, even when sample sizes run into the hundreds for single consumer species, as is the case for trout in the Tadnoll Brook food web, which contains most of the Broadstone taxa (cf. Ings *et al.*, 2009; Woodward *et al.*, 2010c). A realistic compromise must be made, as we have attempted here, for instance, by using data extracted from earlier surveys and field experiments conducted in Broadstone (e.g. Schofield *et al.*, 1988) and from other food webs that contain many of the same species (e.g. Woodward *et al.*, 2010c).

Part of the current study involved the exploration of possible trait-mediated behavioural effects exerted by the invader on existing feeding links between pairs of dominant resident species of predators and prey. We found no evidence of any effect of chemical cues on these simple food-web modules, suggesting that trout had primarily direct, consumptive effects in the food web. It is possible that, while foraging in the benthos, the physical presence of trout (rather than simply chemical cues) could elicit avoidance responses among the prey. At present, we do not have any data to test this, but previous studies with invertebrate predators in Broadstone have not detected any such responses among the prey assemblage (Woodward and Hildrew, 2002c,d). It is an intriguing possibility that these more sophisticated types of behaviour, which have been reported primarily from systems at high pH and for species with highly developed chemosensory apparatus and/or a propensity to enter the drift, such as mayflies and freshwater shrimps, might not be evident in acid streams. Further research is needed to address this hypothesis, but if true it implies that food webs at higher pH are more complex not only in terms of the number of species and feeding links they contain (Layer *et al.*, 2010b; Woodward, 2009) but also in terms of the potential scope for subtle indirect behavioural effects to modulate the strength of those links.

# V. CONCLUSIONS

Many European and North American freshwaters have been exposed to decades of acidification, which has led to widespread species loss and restructuring of food webs. It is only in more recent years, after several decades of reduced acidifying emissions, that chemical recovery has started to be translated into biological recovery. We have presented the first empirical and simulated data on one of the best-described food webs in the world, which has shown several clear signs of recovery over four decades, including (re) colonisation by less acid-tolerant species and a progressive lengthening of food chains, as both chemical (pH) and physical (the waterfall below the upper reach) barriers to establishing a persistent population have been breached. The recent colonisation of this hitherto invertebrate-dominated system by a large, generalist, vertebrate predator had the potential to alter the food web dramatically, and yet, although declines in some species have been observed and local extinctions simulated, most species that were recorded in the first surveys are still present in the stream four decades later.

These data demonstrate that biological recovery does not necessarily simply follow the reverse trajectory of response to acidification. Many other potential colonists from the regional species pool that are associated with less acid conditions, particularly those at the lower trophic levels, did not become (re)established over the study period, and this phenomenon has been reported in other long-term community studies (UKAWMN, 2010). The internal dynamics of the food web could therefore play an important role in delaying recovery or imposing a hysteresis, as has been suggested as a possible explanation for the apparent time lag seen between chemical and biological recovery (e.g. Hildrew, 2009; Layer *et al.*, 2010a,b). This mismatch between observation and expectation undermines to some extent the widespread reliance on simple space-for-time substitutions in many biomonitoring schemes to predict true temporal change in response to environmental stress and emphasises the value of combining long-term data series with experimental and modelling approaches (Friberg *et al.*, 2011).

Empirical before-and-after datasets that track species invasions in real time, such as those presented here, are extremely rare, and it will remain an important task to monitor future changes in Broadstone. This is especially pertinent in the light of the long-term amelioration of acidification seen elsewhere in Europe, as chemical recovery has moved on apace while biological recovery has lagged behind after several decades of reduced acidifying deposition. It might not be possible to simply turn back the clock after all (Friberg *et al.*, 2011). Understanding how and why these systems respond to pH change will also prove invaluable when extrapolating to other regions of the world, such as parts of India and China where the threat of acidification is currently growing, rather than receding (Layer *et al.*, 2010a,b). Acidification will undoubtedly re-emerge as a major global environmental stressor in the very near future.

## ACKNOWLEDGEMENTS

We thank Paul Fletcher and Dan Harris for help with fieldwork, and the landowners and the Board of Conservators of the Ashdown Forest for giving us access to the field site over many years. K. L. was supported by a Queen Mary University of London postgraduate studentship with additional support from the University of London Central Research Fund. G. B. J. was supported by a Natural Environment Research Council doctoral training grant (Ref: NE/I528069/1). We also thank Bo Ebenman, Chris Mellor, Eoin O'Gorman, Mark Emmerson and an anonymous referee for providing helpful comments on earlier versions of the chapter, and Sally Hladyz for assistance with the statistical analyses. This study also relied partly on the United Kingdom Acid Waters Monitoring Network (UKAWMN), established and supported by the Department for Environment, Food and Rural Affairs (DEFRA), and now co-funded by the Scottish Government, the Welsh Assembly Government and the Natural Environment Research Council (NERC). Additional water chemistry data for the UKAWMN were provided by the Freshwater Laboratory, Fisheries Research Services, Pitlochry and the NERC Centre for Ecology and Hydrology, with special thanks to Don Monteith and Lynne Irvine.

## APPENDIX I

Identification keys used for construction of the Broadstone food web:
Brooks (2004)
Edington and Hildrew (1995)
Friday (1988)
Hynes (1993)
Wallace *et al.* (2003)

## APPENDIX II

Additional LMEM GLM Type III nested mixed model output of temporal change in abundance of three of the less-abundant taxa found in each decade not shown in Table 1 (where only the dominant taxa are presented) within the Broadstone Stream food web over four decades. Values shown are for the fixed effects only in the model; variances due to random effects (sample-units, month of sampling) have been partitioned in the model and are therefore not shown. See Section II for a more detailed explanation of the derivation of the terms in the table.

| | | | | |
|---|---|---|---|---|
| *Asellus meridianus* | | | | |
| Decade | 3 | 182.7 | 0.915 | 0.435 |
| Season | 1 | 9.5 | 0.079 | 0.784 |
| Decade × season | 3 | 182.7 | 4.517 | 0.004 |
| *Simulium* spp. | | | | |
| Decade | 3 | 176.6 | 1.951 | 0.123 |
| Season | 1 | 8.3 | 2.544 | 0.148 |
| Decade × season | 3 | 176.6 | 0.004 | 1.000 |
| *Pisidium* spp. | | | | |
| Decade | 3 | 191.2 | 24.878 | 0.000 |
| Season | 1 | 11.0 | 0.626 | 0.445 |
| Decade × season | 3 | 191.2 | 3.742 | 0.012 |

# APPENDIX III

Diet of *Salmo trutta* in Broadstone Stream from two earlier studies. Data are presented in absolute numbers and as proportion of diet, as described from gut contents analysis performed by Schofield (1988) following (a) surveys conducted in the lower (circumneutral) reaches of Broadstone Stream in April and August 1986 and (b) short-term field enclosure/exclosure experiments conducted in Broadstone Stream in August 1985.
Schofield data:

| Location of sampling reach | Date of study | Taxon | 0+ Trout | | 1+ Trout | |
|---|---|---|---|---|---|---|
| | | | Number | Proportion | Number | Proportion |
| Lower reaches (circumneutral) | April 1986 | *Plectrocnemia* | 8 | 0.012 | 28 | 0.025 |
| | | Trichoptera | 61 | 0.090 | 45 | 0.041 |
| | | Coleoptera | 1 | 0.001 | 31 | 0.028 |
| | | Chironomidae | 64 | 0.095 | 183 | 0.167 |
| | | Other Diptera | 8 | 0.012 | 37 | 0.033 |
| | | Plecoptera | 378 | 0.559 | 582 | 0.532 |
| | | Other Aquatic | 0 | 0.000 | 19 | 0.017 |
| | | Terrestrial | 156 | 0.231 | 169 | 0.154 |
| | | Total | 520 | | 1094 | |
| | August 1986 | *Plectrocnemia* | 6 | 0.004 | 17 | 0.009 |
| | | Trichoptera | 9 | 0.006 | 64 | 0.033 |
| | | Coleoptera | 12 | 0.008 | 85 | 0.044 |
| | | Chironomidae | 871 | 0.575 | 1530 | 0.800 |
| | | Other Diptera | 192 | 0.127 | 39 | 0.020 |
| | | Plecoptera | 417 | 0.275 | 166 | 0.087 |
| | | Other Aquatic | 7 | 0.005 | 6 | 0.003 |
| | | Terrestrial | 57 | 0.038 | 6 | 0.003 |
| | | Total | 1514 | | 1913 | |

(*continued*)

| Location of sampling reach | Date of study | Taxon | 0+ Trout | | 1+ Trout | |
|---|---|---|---|---|---|---|
| | | | Number | Proportion | Number | Proportion |
| Upper reaches (acidified section) | August 1985 | *Plectrocnemia* | 6 | 0.0065862 | 1 | 0.047619 |
| | | Trichoptera | 9 | 0.0098793 | 1 | 0.047619 |
| | | Coleoptera | 20 | 0.0219539 | 1 | 0.047619 |
| | | Chironomidae | 794 | 0.8715697 | 14 | 0.6666667 |
| | | Other Diptera | 7 | 0.0076839 | 0 | 0 |
| | | Plecoptera | 46 | 0.050494 | 3 | 0.1428571 |
| | | Other aquatic | 20 | 0.0219539 | 1 | 0.047619 |
| | | Terrestrial | 9 | 0.0098793 | 0 | 0 |
| | | Total | 911 | | 21 | |

Rundle (1988) data: Numerical percentages of prey items in the guts of small (0+) trout caught from the lower circumneutral reaches of Broadstone Stream in the 1980s.

| Size range (mm) | Trout (*n*) | Percentage of items in gut | | |
|---|---|---|---|---|
| | | Cyclopoids (%) | Chironomids (%) | Other (%) |
| 20–25 | 12 | 25 (±6) | 54.6 (±7.5) | 20.4 (±8.3) |
| 26–30 | 18 | 12.2 (±4.7) | 72.3 (±6.8) | 15.5 (±5.7) |
| 31–35 | 20 | 0.3 (±0.2) | 82.4 (±3.4) | 17.3 (±3.5) |
| 36–40 | 13 | 0.7 (±0.5) | 69.8 (±6.2) | 29.5 (±6.2) |
| 41–50 | 11 | 0.6 (±0.5) | 71.3 (±6.7) | 28.1 (±6.4) |
| 51–60 | 12 | – | 70.3 (±8.8) | 29.7 (±8.8) |

# APPENDIX IV

Extinction sequences from modelling simulations of 30 years of population dynamics within the Broadstone Stream food web for each of 4 decades from the 1970s to the 2000s (see Section II)

| Decade | Extinction sequence number | Proportion of total time steps within the 30-year simulation | Species lost |
|---|---|---|---|
| 1970s | 1 | 0.24 | *Asellus meridianus* |
| | 2 | 0.38 | *Plectrocnemia conspersa* |
| 1980s | 1 | 0.28 | Simuliidae |
| | 2 | 0.38 | Chironomids |
| | 3 | 0.75 | *P. conspersa* |

| Decade | Extinction sequence number | Proportion of total time steps within the 30-year simulation | Species lost |
|---|---|---|---|
| 1990s | 1 | 0.13 | *Potamophylax cingulatus* |
| | 2 | 0.13 | Tanypods |
| | 3 | 0.16 | *P. conspersa* |
| | 4 | 0.19 | Coleoptera |
| | 5 | 0.26 | *Siphonoperla torrentium* |
| | 6 | 0.29 | Chironomids |
| | 7 | 0.38 | *A. meridianus* |
| | 8 | 0.57 | Simuliidae |
| 2000s | 1 | 0.15 | *Cordulegaster boltonii* |
| | 2 | 0.15 | *Halesus radiatus* |
| | 3 | 0.16 | *P. conspersa* |
| | 4 | 0.16 | *Salmo trutta* |
| | 5 | 0.16 | *Sialis fuliginosa* |
| | 6 | 0.17 | *Siphonoperla torrentium* |
| | 7 | 0.19 | Tanypods |
| | 8 | 0.79 | Tipulidae |

# REFERENCES

Abjornsson, K., Dahl, J., Nystrom, P., and Bronmark, C. (2001). Influence of predator and dietary chemical cues on the behaviour and shredding efficiency of Gammarus pulex. *Aquat. Ecol.* **34**, 379–387.

Allan, J.D. (1982). The effects of reduction in trout density on the invertebrate community in a mountain stream. *Ecology* **63**, 1444–1455.

Allen, K.R. (1951). The Horokiwi Stream: A study of a trout population. *N. Z. Mar. Depart. Fish. Bull.* **10–10a**, 1–231.

Anderson, M.J., Gorley, R.N., and Clarke, K.R. (2008). *PERMANOVA+for PRIMER: Guide to Software and Statistical Methods*. PRIMER-E, Plymouth, UK.

Bagliniére, J.L., and Maisse, G. (1999). *Biology and Ecology of the Brown and Sea Trout*. Springer-Verlag Ltd., London.

Baxter, C.V., Fausch, K.D., Murakami, M., and Chapman, P.L. (2004). Fish invasion restructures stream and forest food webs by interrupting reciprocal prey subsidies. *Ecology* **85**(10), 2656–2663.

Berlow, E.L., Dunne, J.A., Martinez, N.D., Stark, P.B., Williams, R.J., and Brose, U. (2009). Simple prediction of interaction strengths in complex food webs. *Proc. Natl. Acad. Sci. USA* **106**, 187–191.

Biggs, B.J.F., Francoeur, S.N., Huryn, A.D., Young, R., Arbuckle, C.J., and Townsend, C.R. (2000). Trophic cascades in streams: Effects of nutrient enrichment on autotrophic and consumer benthic communities under two different fish predation regimes. *Can. J. Fish. Aquat. Sci.* **57**, 1380–1394.

Borrvall, C., Ebenman, B., and Jonsson, T. (2000). Biodiversity lessens the risk of cascading extinction in model food webs. *Ecol. Lett.* **3**, 131–136.

Bradley, D.C., and Ormerod, S.J. (2002). Long-term effects of catchment liming on invertebrates in upland streams. *Freshw. Biol.* **47**, 161–171.

Brooks, S. (2004). *Field Guide to the Dragonflies and Damselflies of Great Britain and Ireland.* British Wildlife Publishing, Gillingham.

Brose, U., Cushing, L., Berlow, E.L., Jonsson, T., Banasek-Richter, C., Bersier, L.F., Blanchard, J.L., Brey, T., Carpenter, S.R., Cattin Blandenier, M.F., Cohen, J.E., Dawah, H.A., *et al.* (2005). Body sizes of consumers and their resources. *Ecology* **86**, 2545–2546.

Brose, U., Williams, R.J., and Martinez, N.D. (2006). Allometric scaling enhances stability in complex webs. *Ecol. Lett.* **9**, 1228–1236.

Brown, J.H., Gillooly, J.F., Allen, A.P., Savage, V.M., and West, G.B. (2004). Toward a metabolic theory of ecology. *Ecology* **85**, 1771–1789.

Chamier, A.C. (1987). Effect of pH on microbial degradation of leaf litter in seven streams of the English Lake District. *Oecologia* **71**, 491–500.

Cohen, J.L., Jonsson, T., and Carpenter, S.R. (2003). Ecological community description using the food web, species abundance, and body size. *Proc. Natl. Acad. Sci. USA* **100**, 1781–1786.

Crawley, M.J. (2007). *The R Book*. John Wiley and Sons Ltd., Chichester, UK.

Culp, J.M. (1986). Experimental evidence that stream macroinvertebrate community is unaffected by different densities of coho salmon fry. *J. N. Am. Benthol. Soc.* **5**, 140–149.

Dahl, J. (1998). Effects of a benthivorous and a drift-feeding fish on a benthic stream assemblage. *Oecologia* **116**, 426–432.

Dangles, O., Gessner, M.O., Guérold, F., and Chauvet, E. (2004). Impacts of stream acidification on litter breakdown: Implications for assessing ecosystem functioning. *J. Appl. Ecol.* **41**, 365–378.

Davies, J.J.L., Jenkins, A., Monteith, D.T., Evans, C.D., and Cooper, D.M. (2005). Trends in surface water chemistry of acidified UK freshwaters, 1988-2002. *Environ. Pollut.* **137**, 27–39.

Diehl, S. (1992). Fish predation and benthic community structure: The role of omnivory and habitat complexity. *Ecology* **73**, 1646–1661.

Dineen, G., Harrison, S.S.C., and Giller, P.S. (2007). Diet partitioning in sympatric Atlantic salmon and brown trout in streams with contrasting riparian vegetation. *J. Fish Biol.* **71**, 17–38.

Dobson, M., and Hildrew, A.G. (1992). A test of resource limitation among shredding detritivores in low order streams in southern England. *J. Anim. Ecol.* **61**, 69–77.

Dunne, J.A., Williams, R.J., and Martinez, N.D. (2002). Network structure and biodiversity loss in food webs: robustness increases with connectance. *Ecol. Lett.* **5**, 558–567.

Dunne, J.A. (2006). The network structure of food webs. In: *Ecological Networks: Linking Structure to Dynamics in Food Webs* (Ed. by M. Pascual and J.A. Dunne), pp. 27–86. Oxford University Press, Oxford.

Edington, J.M., and Hildrew, A.G. (1995). *Caseless caddis larvae of the British Isles.* Freshwater Biological Association Scientific Publication No. 53. Titus Wilson and Son Ltd, Kendal.

Emmerson, M.C., and Raffaelli, D. (2004). Predator–prey body size, interaction strength and the stability of a real food web. *J. Anim. Ecol.* **73**, 399–409.

Emmerson, M.C., and Yearsley, J.M. (2004). Weak interactions, omnivory and emergent food-web properties. *Proc. R. Soc. Lond. B* **271**, 397–405.

Emmerson, M.C., Montoya, J.M., and Woodward, G. (2005). Body size, interaction strength and food web dynamics. In: *Dynamic Food Webs: Multispecies Assemblages, Ecosystem Development and Environmental Change* (Ed. by P. de Ruiter, V. Wolters and J.C. Moore), pp. 167–178. Academic Press: Theoretical Ecology Series, San Diego, USA.

Enquist, B.J., West, G.B., Charnov, E.L., and Brown, J.H. (1999). Allometric scaling of production and life-history variation in vascular plants. *Nature* **401**, 907–911.

Feld, C.K., Birk, S., Bradley, D.C., Hering, D., Kail, J., Marzin, A., Melcher, A., Nemitz, D., Pedersen, M.L., Pletterbauer, F., Pont, D., Verdonschot, P.F.M., *et al.* (2011). From natural to degraded rivers and back again: a test of restoration ecology theory and practice. *Adv. Ecol. Res.* **44**, 119–210.

Flecker, A.S., and Townsend, C.R. (1994). Community-wide consequences of trout introduction in New Zealand streams. *Ecol. Appl.* **4**, 798–807.

Fowler, D., Smith, R.I., Muller, J.B.A., Hayman, G., and Vincent, K.J. (2005). Changes in the atmospheric deposition of acidifying compounds in the UK between 1986 and 2001. *Environ. Pollut.* **137**(1), 15–25.

Friberg, N., Bonada, N., Bradley, D.C., Dunbar, M.J., Edwards, F.K., Grey, J., Hayes, R.B., Hildrew, A.G., Lamouroux, N., Trimmer, M., and Woodward, G. (2011). Biomonitoring of human impacts in natural ecosystems: the good, the bad and the ugly. *Adv. Ecol. Res.* **44**, 1–68.

Friday, L.E. (1988). *A key to the adults of British water beetles*. Fields Studies, Vol. 7. Dorset Press, Dorset No. 1.

Fry, B. (1988). Food web structure on Georges Bank from stable C, N, and S isotopic compositions. *Limnol. Oceanogr.* **33**, 1182–1190.

Galloway, J.N. (1995). Acid deposition: Perspectives in time and space. *Water Air Soil Pollut.* **85**, 15–24.

Greig, H.S., and McIntosh, A.R. (2006). Indirect effects of predatory trout on organic matter processing in detritus-based stream food webs. *Oikos* **112**, 31–40.

Grey, J., Jones, R.I., and Sleep, D. (2001). Seasonal changes in the importance of the source of organic matter to the diet of zooplankton in Loch Ness, as indicated by stable isotope analysis. *Limnol. Oceanogr.* **46**, 505–513.

Groom, A.P., and Hildrew, A.G. (1989). Food quality for detritivores in streams of contrasting pH. *J. Anim. Ecol.* **58**, 863–881.

Hall, S.J., and Raffaelli, D. (1993). Food webs: Theory and reality. *Adv. Ecol. Res.* **24**, 187–239.

Hildrew, A.G. (2009). Sustained Research on Stream Communities: A model system and the comparative approach. *Adv. Ecol. Res.* **41**, 175–312.

Hildrew, A.G., and Ormerod, S.J. (1995). Acidification: Causes consequences and solutions. In: *The Ecological Basis for River Management* (Ed. by D.M. Harper and A.J.D. Ferguson), pp. 147–160. John Wiley and Sons Ltd., London.

Hildrew, A.G., and Townsend, C.R. (1976). The distribution of two predators and their prey in an iron-rich stream. *J. Anim. Ecol.* **45**, 41–57.

Hildrew, A.G., Townsend, C.R., Francis, J.E., and Finch, K. (1984). Cellulolytic decomposition in streams of contrasting pH and its relationship with invertebrate community structure. *Freshw. Biol.* **14**, 323–328.

Hildrew, A.G., Townsend, C.R., and Hasham, A. (1985). The predatory Chironomidae of an iron-rich stream: Feeding ecology and food web structure. *Ecol. Entomol.* **10**, 403–413.

Hildrew, A.G., Woodward, G., Winterbottom, J.H., and Orton, S. (2004). Strong density-dependence in a predatory insect: Larger scale experiments in a stream. *J. Anim. Ecol.* **73**, 448–458.

Hildrew, A.G., Raffaelli, D. and Edmonds-Brown, R. (Eds.) (2007). Body Size *The Structure and Function of Aquatic Ecosystems*. Ecological Reviews. Cambridge University Press, United Kingdom.

Hladyz, S., Åbjörnsson, K., Chauvet, E., Dobson, M., Elosegi, A., Ferreira, V., Fleituch, T., Gessner, M.O., Giller, P.S., Gulis, V., Hutton, S.A., Lacoursière, J.O., *et al.* (2011). Stream ecosystem functioning in an agricultural landscape: the importance of terrestrial-aquatic linkages. *Adv. Ecol. Res.* **44**, 211–276.

Huryn, A.D. (1996). An appraisal of the Allen paradox in a New Zealand trout stream. *Limnol. Oceanogr.* **41**, 243–252.

Hynes, H.B.N. (1993). *Adults and nymphs of the British Stoneflies (Plecoptera)*. Freshwater Biological Association Scientific Publication No. 17. Titus Wilson and Son Ltd., Kendal.

Ings, T.C., Montoya, J.M., Bascompte, J., Bluthgen, N., Brown, L., Dormann, C.F., Edwards, F., Figueroa, D., Jacob, U., Jones, J.I., Lauridsen, R.B., Ledger, M.E., *et al.* (2009). Ecological networks—Beyond food webs. *J. Anim. Ecol.* **78**, 253–269.

Jensen, K.W., and Snekvik, E. (1972). Low pH levels wipe out salmon and trout populations in southernmost Norway. *Ambio* **1**, 223–225.

Junge, C.O., and Libosvarsky, J. (1965). Effects of size selectivity on population estimates on successive removals with electrical fishing gear. *Zool. Listy* **14**, 171–178.

Klemetsen, A., Amundsen, P.A., Dempson, J.B., Jonsson, B., Jonsson, N., O'Connell, M.F., and Mortensen, E. (2003). Atlantic salmon *Salmo salar* L., brown trout *Salmo trutta* L. and Arctic charr *Salvelinus alpinus* (L.): A review of aspects of their life histories. *Ecol. Freshw. Fish* **12**, 1–59.

Konishi, M., Nakano, S., and Iwata, T. (2001). Trophic cascading effects of predatory fish on leaf litter processing in a Japanese stream. *Ecol. Res.* **16**, 415–422.

Koperski, P. (1997). Changes in feeding behaviour of the larvae of the damselfly Enallagma cyathigerum in response to stimuli from predators. *Ecol. Entomol.* **22**(2), 167–175.

Kowalik, R.A., Cooper, D.M., Evans, C.M., and Ormerod, S.J. (2007). Acid episodes retard the biological recovery of upland British streams from acidification. *Global Change Biol.* **13**, 2439–2452.

Lancaster, J. (1996). Scaling the effects of predation and disturbance in a patchy environment. *Oecologia* **107**, 321–331.

Larrañaga, A., Layer, K., Basaguren, A., Pozo, J., and Woodward, G. (2010). pH alters consumer biochemistry and community structure in a stream food webs. *Freshw. Biol.* **55**, 670–680.

Layer, K. (2010). *Responses of freshwater food webs to spatial and temporal pH gradients*. Ph.D. Thesis, Queen Mary & Westfield College, University of London, 236pp.

Layer, K., Hildrew, A.G., Monteith, D., and Woodward, G. (2010a). Long-term variation in the littoral food web of an acidified mountain lake. *Global Change Biol.* **11**, 3133–3143.

Layer, K., Riede, J.O., Hildrew, A.G., and Woodward, G. (2010b). Food web structure and stability in 20 streams across a wide pH gradient. *Adv. Ecol. Res.* **42**, 265–301.

Ledger, M.E., and Hildrew, A.G. (2005). The ecology of acidification and recovery: Changes in herbivore-algal food web linkages across a pH gradient in streams. *Environ. Pollut.* **137**, 103–118.

Lepori, F., Barbieri, A., and Ormerod, S.J. (2003). Causes of episodic acidification in Alpine streams. *Freshw. Biol.* **48**, 175–189.

Lieske, R., and Zwick, P. (2007). Food preference, growth and maturation of *Nemurella pictetii* (Plectoptera: Neumoridae). *Freshw. Biol.* **52**, 1187–1197.
Lunberg, P., Ranta, E., and Kaitala, V. (2000). Species loss leads to community closure. *Ecol. Lett.* **3**, 465–468.
Magnuson, J.J., Baker, J.P., and Rahel, S.J. (1984). A critical assessment of effects of acidification on fisheries in North America. *Philos. Trans. R. Soc. B* **305**, 501–516.
Martinez, N.D. (1991). Artifacts or attributes? Effects of resolution on the Little Rock Lake food web. *Ecol. Monogr.* **61**, 367–392.
Martinez, N.D. (1992). Constant connectance in community food webs. *Am. Nat.* **139**, 1208–1218.
Martinez, N.D. (1993). Effects of resolution on food web structure. *Oikos* **66**, 403–412.
Mason, C.F. (1991). *Biology of freshwater pollution*, 2nd edn. Longman Scientific & Technical, New York, USA.
Masters, Z., Petersen, I., Hildrew, A.G., and Ormerod, S.J. (2007). Insect dispersal does not limit the biological recovery of streams from acidification. *Aquat. Conserv. Mar Freshw. Ecosyst.* **16**, 1–9.
May, R.M. (1972). Will a large complex system be stable? *Nature* **238**, 413–414.
May, R.M. (1973). *Stability and Complexity in Model Ecosystems*. Princeton University Press, Princeton.
McCann, K.S. (2000). The diversity-stability debate. *Nature* **405**, 228–233.
McCann, K.S., Hastings, A., and Huxel, G.R. (1998). Weak trophic interactions and the balance of nature. *Nature* **395**, 794–798.
McCarthy, I.D., and Waldron, S. (2000). Identifying migratory Salmo trutta using carbon and nitrogen stable isotope ratios. *Rapid Commun. Mass Spectrom.* **14**, 1325–1331.
McIntosh, A.R., and Peckarsky, B.L. (1996). Differential behavioural responses of mayflies from streams with and without fish to trout odour. *Freshw. Biol.* **35**, 141–148.
McIntosh, A.R., Peckarsky, B.L., and Taylor, B.W. (1999). Rapid size-specific changes in the drift of *Baetis bicaudatus* (Ephemeroptera) caused by alterations in fish odour concentration. *Oecologia* **118**, 256–264.
McLaughlin, O.B., Jonsson, T., and Emmerson, M.C. (2010). Temporal variability in predator-prey relationships of a forest floor food web. *Advances in Ecological Research* **42**, 172–264.
Minagawa, M., and Wada, E. (1984). Stepwise enrichment of $^{15}N$ along food chains: Further evidence and the relation between $d^{15}N$ and animal age. *Geochim. Cosmochim. Acta* **48**, 1135–1140.
Monteith, D.T., and Evans, C.D. (2005). The United Kingdom Acid Waters Monitoring Network: A review of the first 15 years and introduction to the special issue. *Environ. Pollut.* **137**, 3–13.
Monteith, D.T., Hildrew, A.G., Flower, R.J., Raven, P.J., Beaumont, W.R.B., Collen, P., Kreiser, A.M., Shilland, E.M., and Winterbottom, J.H. (2005). Biological responses to the chemical recovery of acidified fresh waters in the UK. *Environ. Pollut.* **137**, 83–101.
Montoya, J.M., Woodward, G., Emmerson, M.C., and Solé, R.C. (2009). Indirect effects propagate disturbances in real food webs. *Ecology* **90**, 2426–2433.
Mulder, C., Boit, A., Bonkowski, M., De Ruiter, P.C., Mancinelli, G., Van der Heijden, M.G.A., van Wijnen, H.J., Vonk, J.A., and Rutgers, M. (2011). A below-ground perspective on Dutch Agroecosystems: how soil organisms interact to support ecosystem services. *Adv. Ecol. Res.* **44**, 277–358.

Nakano, S., Miyasaka, H., and Kuhara, N. (1999). Terrestrial-aquatic linkages: Riparian arthropod inputs alter trophic cascades in a stream food web. *Ecology* **80**, 2435–2441.

Neutel, A.-M., Heesterbeek, J.A.P., and de Ruiter, P.C. (2002). Stability in real food webs: Weak links in long loops. *Science* **296**, 1120–1223.

Neutel, A.-M., Heesterbeek, J.A.P., de Koppel, J.V., Hoenderboom, G., Vos, A., Kaldeway, C., Berendse, F., and de Ruiter, P.C. (2007). Reconciling complexity with stability in naturally assembling food webs. *Nature* **449**, 559–603.

O'Gorman, E.J., and Emmerson, M.C. (2010). Manipulating interaction strengths and the consequences for trivariate patterns in a marine food web. *Adv. Ecol. Res.* **42**, 301–419.

Olesen, J.M., Dupont, Y.L., O'Gorman, E., Ings, T.C., Layer, K., Melián, C.J., Troejelsgaard, K., Pichler, D.E., Rasmussen, C., and Woodward, G. (2010). From Broadstone to Zackenberg: space, time and hierarchies in ecological networks. *Adv. Ecol. Res.* **42**, 1–71.

Otto, S.B., Rall, B.C., and Brose, U. (2007). Allometric degree distributions facilitate food-web stability. *Nature* **450**, 1226–1230.

Pace, M.L., Cole, J.J., Carpenter, S.R., Kitchell, J.F., Hodgson, J.R., Van De Bogert, M., Bade, D.L., Kritzbewrg, E.S., and Bastviken, D. (2004). Whole-lake carbon-13 additions reveal terrestrial support of aquatic food webs. *Nature* **427**, 240–243.

Perkins, D.M., Reiss, J., Yvon-Durocher, G., and Woodward, G. (2010). Global change and food webs in running waters. *Hydrobiologia* **657**, 181–198.

Petchey, O.L., and Belgrano, A. (2010). Body-size distributions and size-spectra: Universal indicators of ecological status? *Biol. Lett.* **6**, 434–437.

Petchey, O.L., Downing, A.L., Mittelbach, G.G., Persson, L., Steiner, C.F., Warren, P.H., and Woodward, G. (2004). Species loss and the structure and functioning of multitrophic aquatic systems. *Oikos* **104**, 467–478.

Petchey, O.L., Beckerman, A.P., Riede, J.O., and Warren, P.H. (2008). Size, foraging, and food web structure. *Proc. Natl. Acad. Sci. USA* **105**, 4191–4196.

Peters, R.H. (1983). *The Ecological Implications of Body Size*. Cambridge University Press, Cambridge, UK.

Power, M.E. (1990). Effects of fish in river food webs. *Science* **250**, 811–814.

Power, M.E. (1992a). Top down and bottom up forces in food webs: Do plants have primacy? *Ecology* **73**, 733–746.

Power, M.E. (1992b). Habitat heterogeneity and the functional significance of fish in river food webs. *Ecology* **73**, 1675–1688.

Raddum, G.G., and Fjellheim, A. (2003). Liming of River Audna, southern Norway: A large-scale experiment of benthic invertebrate recovery. *Ambio* **32**, 230–234.

Reiss, J., Bridle, J., Montoya, J.M., and Woodward, G. (2009). Emerging horizons in biodiversity and ecosystem functioning research. *Trends Ecol. Evol.* **24**, 505–514.

Reuman, D.C., and Cohen, J.E. (2004). Trophic links' length and slope in the Tuesday Lake food web with species' body mass and numerical abundance. *J. Anim. Ecol.* **73**, 852–866.

Reuman, D.C., and Cohen, J.E. (2005). Estimating relative energy fluxes using the food web, species abundance, and body size. *Adv. Ecol. Res.* **36**, 137–182.

Riede, J.O., Rall, B.C., Banasek-Richter, C., Navarrete, S.A., Wieters, E.A. and Brose, U. (2010). Scaling of food web properties with diversity and complexity across ecosystems. *Adv. Ecol. Res.* **42**, 139–170.

Rooney, N., McCann, K., Gellner, G., and Moore, J.C. (2006). Structural asymmetry and the stability of diverse food webs. *Nature* **442**, 265–269.

Rose, N., Monteith, D.T., Kettle, H., Thompson, R., Yang, H., and Muir, D. (2004). A consideration of potential confounding factors limiting chemical and biological recovery at Lochnagar, a remote mountain loch in Scotland. *J. Limnol.* **63**, 63–76.

Rosemond, A.D., Reice, S.R., Elwood, J.W., and Mulholland, P.J. (1992). The effects of stream acidity on benthic invertebrate communities in the south-eastern United States. *Freshw. Biol.* **27**(2), 193–209.

Rundle, S.D. (1988). *The micro-arthropods of some southern English streams*. Ph.D. Thesis. Queen Mary and Westfield College (University of London).

Rundle, S.D., and Hildrew, A.G. (1992). Small fish and small prey in the food webs of some southern English streams. *Arch. Hydrobiol.* **125**, 25–35.

Rundle, S.D., Weatherley, N.S., and Ormerod, S.J. (1995). The effects of catchment liming on the chemistry and biology of upland Welsh streams: Testing model predictions. *Freshw. Biol.* **34**, 165–175.

Schmid-Araya, J.M., Hildrew, A.G., Robertson, A., Schmid, P.E., and Winterbottom, J.H. (2002a). The importance of meiofauna in food webs: Evidence from an acid stream. *Ecology* **83**, 1271–1285.

Schmid-Araya, J.M., Schmid, P.E., Robertson, A., Winterbottom, J.H., Gjerløv, C., and Hildrew, A.G. (2002b). Connectance in stream food webs. *J. Anim. Ecol.* **71**, 1056–1062.

Schoener, T.W. (1989). Food webs from the small to the large. *Ecology* **70**, 1559–1589.

Schofield, K. (1988). *Predation and the prey community of a headwater stream*. Ph.D. Thesis, Queen Mary and Westfield College, University of London.

Schofield, K., Townsend, C.R., and Hildrew, A.G. (1988). Predation and the prey community of a headwater stream. *Freshw. Biol.* **20**, 85–96.

Snucins, E., and Gunn, H.M. (2003). Use of rehabilitation experiments to understand the recovery dynamics of acid-stressed fish populations. *Ambio* **32**, 240–243.

Solow, A.R. (1996). On the goodness of fit of the cascade model. *Ecology* **77**, 1294–1297.

Speirs, D.C., Gurney, W.S.C., Hildrew, A.G., and Winterbottom, J.H. (2000). Long-term demographic balance in the Broadstone Stream insect community. *J. Anim. Ecol.* **69**, 45–58.

Stead, T.K., Schmid-Araya, J.M., and Hildrew, A.G. (2005). Distribution of body size in a stream community: One system, many patterns. *J. Anim. Ecol.* **74**, 475–487.

Townsend, C.R. (1996). Invasion biology and ecological impacts of brown trout *Salmo trutta* in New Zealand. *Biol. Conserv.* **78**(1–2), 13–22.

Townsend, C.R. (2003). Individual, population, community and ecosystem consequences of a fish invader in New Zealand streams. *Conserv. Biol.* **17**, 38–47.

Townsend, C.R., and Hildrew, A.G. (1979). Form and function in the prey-catching net of *Plectrocnemia conspersa* (Trichoptera: Polycentropodidae). *Oikos* **33**, 412–418.

Townsend, C.R., and Simon, K.S. (2006). Consequences of brown trout invasion for stream ecosystems. In: *Biological Invasions in New Zealand* (Ed. by R.B. Allen and W.G. Lee), pp. 213–225. Springer Verlag, Berlin.

Townsend, C.R., Hildrew, A.G., and Francis, J.E. (1983). Community structure in some Southern English streams: The influence of physicochemical factors. *Freshw. Biol.* **13**, 531–544.

Townsend, C.R., Hildrew, A.G., and Schofield, K. (1987). Persistence of stream invertebrate communities in relation to environmental variability. *J. Anim. Ecol.* **56**, 597–613.

Turner, A.M., Fetterolf, S.A., and Bernot, R.J. (1999). Predator identity and consumer behavior: Differential effects of fish and crayfish on the habitat use of a freshwater snail. *Oecologia* **118**, 242–247.

UK Acid Waters Monitoring Network: 20 year Interpretative Report. 2010. (Ed. by M. Kernan). ENSIS, London (in press).

United Kingdom Acid Waters Monitoring Network (2000). In: *10 Year Report: Analysis and Interpretation of Results* Monteith, D.T. and Evans, C.D. (Eds.) (2000) ENSIS, London (April 1988–March 1998).

Wallace, I.D., Wallace, B., and Philipson, G.N. (2003). *Keys to the case-bearing caddis larvae of Britain and Ireland.* Freshwater Biological Association Scientific Publication No. 61. Titus Wilson and Son Ltd., Kendal.

Warren, P.H. (1989). Spatial and temporal variation in the structure of a freshwater food web. *Oikos* **55**, 299–311.

Williams, R.J., and Martinez, N.D. (2000). Simple rules yield complex food webs. *Nature* **409**, 180–183.

Winterbourn, M.J., Hildrew, A.G., and Orton, S. (1992). Nutrients, algae and grazers in some British streams of contrasting pH. *Freshw. Biol.* **28**, 173–182.

Woodward, G. (2009). Biodiversity, ecosystem functioning and food webs in freshwaters: assembling the jigsaw puzzle. *Freshw. Biol.* **54**, 2171–2187.

Woodward, G., and Hildrew, A.G. (2001). Invasion of a stream food web by a new top predator. *J. Anim. Ecol.* **70**, 273–288.

Woodward, G., and Hildrew, A.G. (2002a). Food web structure in riverine landscapes. *Freshw. Biol.* **47**, 777–798.

Woodward, G., and Hildrew, A.G. (2002b). Body-size determinants of niche overlap and intraguild predation within a complex food web. *J. Anim. Ecol.* **71**, 1063–1074.

Woodward, G., and Hildrew, A.G. (2002c). Differential vulnerability of prey to an invading top predator: Integrating field surveys and laboratory experiments. *Ecol. Entomol.* **27**, 732–744.

Woodward, G., and Hildrew, A.G. (2002d). The impact of a sit and wait predator: Separating consumption and prey emigration. *Oikos* **99**, 409–418.

Woodward, G., and Warren, P.H. (2007). Body size and predatory interactions in freshwaters: Scaling from individuals to communities. In: *Body size: The Structure and Function of Aquatic Ecosystems* (Ed. by A.G. Hildrew, D. Raffaelli and R. Edmonds-Brown), pp. 98–117. Cambridge University Press, Cambridge.

Woodward, G., Jones, J.I., and Hildrew, A.G. (2002). Community persistence in Broadstone Stream (UK) over three decades. *Freshw. Biol.* **47**, 1419–1435.

Woodward, G., Speirs, D.C., and Hildrew, A.G. (2005a). Quantification and temporal resolution of a complex size-structured food web. *Adv. Ecol. Res.* **36**, 85–135.

Woodward, G., Ebenman, B., Emmerson, M., Montoya, J.M., Olesen, J.M., Valido, A., and Warren, P.H. (2005b). Body size in ecological networks. *Trends Ecol. Evol.* **20**, 402–409.

Woodward, G., Papantoniou, G., Edwards, F.E., and Lauridsen, R. (2008). Trophic trickles and cascades in a complex food web: Impacts of a keystone predator on stream community structure and ecosystem processes. *Oikos* **117**, 683–692.

Woodward, G., Perkins, D.M., and Brown, L.E. (2010a). Climate change in freshwater ecosystems: Impacts across multiple levels of organisation. *Philos. Trans. R. Soc. B* **365**, 2093–2106.

Woodward, G., Christensen, J.B., Olafsson, J.S., Gislason, G.M., Hannesdottir, E. R., and Friberg, N. (2010b). Sentinel systems on the razor's edge: Effects of warming on Arctic stream ecosystems. *Global Change Biol.* **16**, 1979–1991.

Woodward, G., Benstead, J.P., Beveridge, O.S., Blanchard, J., Brey, T., Brown, L., Cross, W.F., Friberg, N., Ings, T.C., Jacob, U., Jennings, S., Ledger, M.E., *et al.* (2010c). Ecological networks in a changing climate. *Adv. Ecol. Res.* **42**, 72–138.

Woodward, G., Blanchard, J., Lauridsen, R.B., Edwards, F.K., Jones, J.I., Figueroa, D., Warren, P.H., and Petchey, O.L. (2010d). Individual-based food webs: Species identity, body size and sampling effects. *Adv. Ecol. Res.* **43**, 211–266.

Yan, N.D., Leung, B., Keller, W., Arnott, S.E., Gunn, J.M., and Raddum, G.G. (2003). Developing conceptual frameworks for the recovery of aquatic biota from acidification. *Ambio* **32**, 165–169.

Yodzis, P., and Innes, S. (1992). Body size and consumer-resource dynamics. *Am. Nat.* **139**, 1151–1175.

Yoshii, K. (1999). Stable isotope analysis of benthic organisms in Lake Baikal. *Hydrobiologia* **411**, 145–159.

Yoshioka, T., Wada, E., and Hayashi, H. (1994). A stable isotope study on seasonal food web dynamics in a eutrophic lake. *Ecology* **75**, 835–846.

# From Natural to Degraded Rivers and Back Again: A Test of Restoration Ecology Theory and Practice

CHRISTIAN K. FELD, SEBASTIAN BIRK, DAVID C. BRADLEY, DANIEL HERING, JOCHEM KAIL, ANAHITA MARZIN, ANDREAS MELCHER, DIRK NEMITZ, MORTEN L. PEDERSEN, FLORIAN PLETTERBAUER, DIDIER PONT, PIET F.M. VERDONSCHOT AND NIKOLAI FRIBERG



ADVANCES IN ECOLOGICAL RESEARCH VOL. 44

0065-2504/11 $35.00
DOI: 10.1016/B978-0-12-374794-5.00003-1

## SUMMARY

Extensive degradation of ecosystems, combined with the increasing demands placed on the goods and services they provide, is a major driver of biodiversity loss on a global scale. In particular, the severe degradation of large rivers, their catchments, floodplains and lower estuarine reaches has been ongoing for many centuries, and the consequences are evident across Europe. River restoration is a relatively recent tool that has been brought to bear in attempts to reverse the effects of habitat simplification and ecosystem degradation, with a surge of projects undertaken in the 1990s in Europe and elsewhere, mainly North America. Here, we focus on restoration of the physical properties (e.g. substrate composition, bank and bed structure) of river ecosystems to ascertain what has, and what has not, been learned over the last 20 years.

First, we focus on three common types of restoration measures—riparian buffer management, instream mesohabitat enhancement and the removal of weirs and small dams—to provide a structured overview of the literature. We distinguish between abiotic effects of restoration (e.g. increasing habitat diversity) and biological recovery (e.g. responses of algae, macrophytes, macroinvertebrates and fishes).

We then addressed four major questions: (i) Which organisms show clear recovery after restoration? (ii) Is there evidence for qualitative linkages between restoration and recovery? (iii) What is the timescale of recovery? and (iv) What are the reasons, if restoration fails?

Overall, riparian buffer zones reduced fine sediment entry, and nutrient and pesticide inflows, and positive effects on stream organisms were evident. Buffer width and length were key: 5–30 m width and >1 km length were most effective. The introduction of large woody debris, boulders and gravel were the most commonly used restoration measures, but the potential positive effects of such local habitat enhancement schemes were often likely to be swamped by larger-scale geomorphological and physico-chemical effects. Studies demonstrating long-term biological recovery due to habitat enhancement were notable by their absence. In contrast, weir removal can have clear beneficial effects, although biological recovery might lag behind for several years, as huge amounts of fine sediment may have accumulated upstream of the former barrier.

Three Danish restoration schemes are provided as focal case studies to supplement the literature review and largely supported our findings. While the large-scale re-meandering and re-establishment of water levels at River Skjern resulted in significant recovery of riverine biota, habitat enhancement schemes at smaller-scales in other rivers were largely ineffective and failed to show long-term recovery.

The general lack of knowledge derived from integrated, well-designed and long-term restoration schemes is striking, and we present a conceptual framework to help address this problem. The framework was applied to the three restoration types included in our study and highlights recurrent cause–effect chains, that is, commonly observed relationships of restoration measures (cause) and their effects on abiotic and biotic conditions (effect). Such conceptual models can provide useful new tools for devising more effective river restoration, and for identifying avenues for future research in restoration ecology in general.

# I. INTRODUCTION

## A. Why Is River Restoration Necessary?

For thousands of years humans have changed and degraded landscapes significantly and this has had a profound effect on all natural ecosystems (Vitousek *et al.*, 1997). At the global scale, extensive ecosystem degradation, combined with the increasing exploitation of and reliance upon ecosystem services is a major threat to biodiversity (MA, 2005). The loss of species and pristine ecosystems continue at an alarming rate in Europe and globally. By the conversion of entire forest ecosystems into arable land, habitat complexity and functioning have been fundamentally changed in the past: indeed, in Europe's agricultural landscape many anthropogenic habitats are now so familiar to us they are now widely perceived as being 'natural' (Hladyz *et al.*, 2011b; Mulder *et al.*, 2011). The agricultural use of large river floodplains, for instance of River Rhine, together with a severe regulation of main river courses, decreased the ecosystem's water retention capacity over centuries (Dister *et al.*, 1990). International conventions, such as the 2010 biodiversity target set by a pan-European initiative to 'halt the loss of biodiversity by 2010' (EEA, 2007), have so far not had the desired effect in reversing this trend, which poses a serious future threat to human society if essential goods and ecosystem services are irreversibly lost.

It is evident now in all parts of Europe that freshwater and coastal ecosystems have suffered from the severe degradation affecting entire rivers, their catchments, floodplains and estuaries over centuries (Tockner *et al.*, 2009). In particular, the increasing human demands on water resources in heavily urbanised and agricultural landscapes are among the major drivers of ecosystem degradation (Millennium Ecosystem Assessment [MA], 2005), and are associated with a multitude of pressures, including pollution, the modification of instream and riparian habitat and the regulation of flow. Within the river catchment, lake ecosystems are mainly affected by eutrophication (agricultural land use) and physical habitat modification of their shoreline. Estuaries and wetlands are mostly impaired because they act as sinks for nutrients and contaminants originating from river basins (Cloern, 2001; Diaz and Rosenberg, 2008), and many transitional and coastal waters are heavily physically modified for flood protection purposes (e.g. Pollard and Hannan, 1994) or navigation (e.g. van der Wal *et al.*, 2002). These and other pressures (*sensu* EEA, 2007) might occur individually, but more often they act in combination and pose a serious threat to the ecological status of aquatic ecosystems. They ultimately impose environmental stress on communities and can cause a severe deterioration of ecosystem ecology. Even the large River Rhine, for instance, is impacted by a concert of human-induced stress (e.g. navigation, regulation, contamination, alien species invasions), which have altered its food web (van Riel *et al.*, 2006).

The majority of European river basins and their sub-catchments suffer from such combined impacts on both water quality (organic pollution, eutrophication, toxic compounds) and physical habitat degradation, including simplified habitat structure, barriers to dispersal and biologically unsuitable flow regimes (Friberg, 2010a). These pressures can have severe impacts on aquatic communities of plants and animals (e.g. Johnson and Hering, 2009). Freshwaters are essential for providing a wide array of ecosystem services (MA, 2005), in particular, provisional services (e.g. drinking water, food) and regulating services (e.g. nutrient spiralling, self-purification, water regulation). If these services are to be maintained, the level of degradation that we experience today and the current and predicted rates of habitat loss from climate change (Moss *et al.*, 2009) constitute serious future challenges for human societies (Vörösmarty *et al.*, 2010), in addition to the threats they pose to natural systems in their own right.

The evaluation of freshwater pollution and degradation, and the resulting impacts on ecosystem properties is by no means a novel discipline (e.g. Hynes, 1960). Since the 1970s, various policies and specific rehabilitation schemes have been introduced to improve the status of the environment and, hence, to counteract the decline in environmental quality. As early as 1972, the United States of America launched the Clean Water Act in order to protect and improve the nation's freshwater resources (U.S. Senate, 1972). Almost 30 years later, the European Parliament passed the Water Framework Directive in order to achieve and maintain a 'good ecological quality' for surface waters in Europe—rivers, lakes, estuaries and coasts (2000/60/EC). The concept of river restoration is embedded in these efforts, which in its strict sense aims at restoring degraded river sections to their pre-impacted (natural) state, although in reality this optimistic Utopian vision is often compromised (Palmer *et al.*, 2010). Whilst there are some notable examples of the restoration of water chemistry in river catchments (e.g. Bradley and Ormerod, 2002; Layer *et al.*, 2010, 2011), the dominant paradigm in river restoration has been the rehabilitation of the system, that is, the manipulation of habitat structure and water flow to mitigate adverse environmental and human impacts and ultimately to enhance habitat heterogeneity and biodiversity. These works range from minor reach-scale rehabilitation measures carried out over a few metres, such as the introduction of gravel bars and large woody debris (LWD), to larger-scale projects aimed at attaining near natural conditions in entire (sub-) catchments (e.g. Hansen and Baattrup-Pedersen, 2006; Palmer *et al.*, 2007, 2010).

There has been a surge of river restoration projects undertaken since the 1990s in Europe and elsewhere, mainly North America (Palmer *et al.*, 2007). In this chapter, we focus on restoration of the physical properties of river ecosystems (e.g. bed substrate enhancement, improvement of bank structures, riparian revegetation) to ascertain what has and has not been learnt over 20 years of practical river restoration, and to develop a more coherent theoretical and practical framework for the future.

## B. Rivers Under Siege: Years of Physical Abuse

The physical structure of rivers and the diversity of biological communities are closely linked with past and present human activities in river basins, which influence riverine ecosystems on multiple scales, ranging from changes of the instream environment at the reach scale (channelisation, removal of LWD, etc.), to altering the landscape and land use at the catchment scale, influencing hydrological pathways and geomorphological structure (Allan, 2004; Fitzpatrick *et al.*, 2001; Vaughan *et al.*, 2009). Past and present disturbances impact stream ecosystems in various ways, often making it difficult to distinguish the exact disturbance from individual pressures on the biotic communities (Allan, 2004; Harding *et al.*, 1998; Ormerod *et al.*, 2010).

Natural river morphology depends on catchment-scale structural controls, differences in channel patterns at the reach-scale and micro-scale variations in the form and composition of the river, all of which vary over different time scales (Frissell *et al.*, 1986). The natural physical attributes of rivers are therefore often classified hierarchically based on descriptions of morphological features such as dominant bed type, entrenchment ratio, sinuosity, width:depth ratio and water surface slope (Rosgen, 1994), which can respond differently to degradation (Downs, 1995). In a purely ecological context, however, the important issue relates to how spatial and temporal heterogeneity creates a range of habitats and niches for the biota. Understanding how the templet provided by the physical environment affects the biota offers a means of gaining better insight into why certain species are lost and what features are a prerequisite for recovery, although such links are still surprisingly poorly understood (Vaughan *et al.*, 2009).

Because rivers are so closely interlinked with the terrestrial ecosystems through which they flow, channel degradation cannot be separated entirely from the effects on the riparian area and the floodplain. The former constitutes the transitional zone between the riverbanks and the floodplain that connects the river ecosystem with the wider terrestrial ecotone via fluxes of energy, nutrients and organisms. Both riparian areas and floodplains are highly dynamic, characterised by many habitats, and often support very high species diversity (Hladyz *et al.*, 2011a,b; Naiman and Decamps, 1997). Riparian zones are also donors of organic matter, the main source of energy to most stream ecosystems, at least in their upper reaches (Allan, 1995). River engineering has had a substantial impact not only on-channel plan form, but also on the riparian zones and the entire floodplain, as exemplified by the Skjern River, Denmark (Pusch *et al.*, 2009). Six years after the onset of flow regulation of the lower 20 km of the river in the 1960s, the main river channel was straightened and moved its course on the floodplain. River length and channel complexity were reduced by the removal of most meanders and a number of lakes. Flooding was prevented by large dykes and the groundwater table was lowered by establishing pumping stations and drainage ditches, and 40 km$^2$ of arable farm land could be claimed from what were previously meadows and wetlands. This had a significant impact on the extent of wetlands on the floodplain (Figure 1), which covered an area of only approx. 4.3 km$^2$ after regulation (Svendsen and Hansen, 1997).

## C. The Confounding Influence of Multiple Pressures

Rivers are not alone in terms of their vulnerability to human activity: by changing natural land *cover* in pristine ecosystems to a number of land *uses*, considerable anthropogenic stress is imposed on all kinds of aquatic and

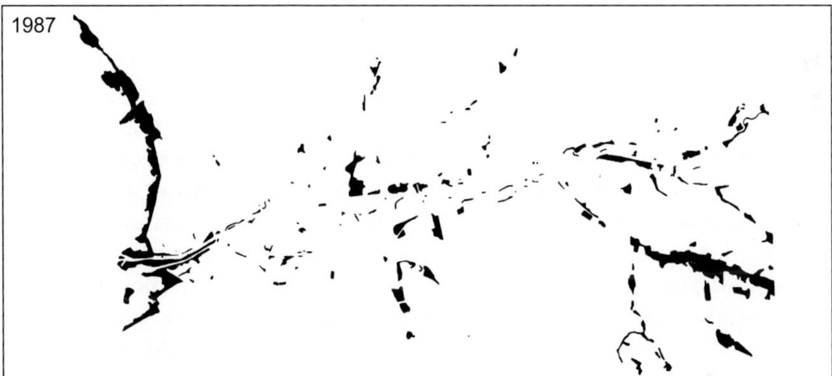


**Figure 1** Floodplain area of River Skjern in Denmark, covered by wetlands in 1871 and 1987.

terrestrial ecosystems. Rivers, are however, especially susceptible to perturbations because they integrate the adverse effects of various types of land use in the surrounding catchment, including agriculture, deforestation, urbanisation, stormwater treatment, flow regulation and water abstraction (Palmer *et al.*, 2010). Overall, human water and land use have constituted the major cause of degradation of aquatic ecosystems on a global stage (MA, 2005), but the scale and diversity and the myriad resultant physico-chemical effects mean that separating individual pressures and their specific impact on biota is extremely challenging (Ormerod *et al.*, 2010). This poses a major problem for river managers responsible for making decisions on the most environmentally sensitive and cost-effective management schemes and restoration/rehabilitation measures: the challenge is to identify and prioritise the main drivers and responses at appropriate scales for implementing effective management of our natural resources.

A general review of the principal mechanisms by which land use may influence stream ecosystems was compiled by Allan (2004), who identified six groups of pressures: 'sedimentation', 'nutrient enrichment', 'contaminant pollution', 'hydrologic alteration', 'riparian degradation' and 'loss of LWD'. Runoff of insoluble fine particles from agricultural land increases turbidity in the water column (Henley *et al.*, 2000) and sedimentation of the riverbed, impacting benthic algae, interstitial macroinvertebrates and gravel-spawning fishes, such as many salmonids (Wood and Armintage, 1997). Excessive fine sediment entry has been recognized a major and growing limitation to achieving ecological improvements (e.g. Larson *et al.*, 2001), and has been considered an important reason for failure of restoration at over 450 river sections in England alone (Bradley *et al.*, 2009). Nutrient enrichment typically stimulates autotrophic production and biomass, at least until the associated toxic effects of organic pollution are manifested, resulting in shifts of algal or macrophyte community composition (e.g. Bakker *et al.*, 2010; Tilman *et al.*, 1982). Decomposition processes, if enhanced by an increasing production of instream plant biomass, can lead to a severe decline of dissolved oxygen concentration or water quality and sensitive taxa (e.g. oxygen-demanding fishes and insects) may then be replaced by tolerant taxa (e.g. Furse *et al.*, 2006; Hering *et al.*, 2004). In particular, macroinvertebrates and fishes are affected by contaminant pollution associated with low flow and reduced oxygen availability (Schulz, 2004; Woodward and Foster, 1997). Individual growth rates may be suppressed, reproduction may fail and the endocrine systems may be disrupted, leading to reduced overall fitness (e.g. Feist *et al.*, 2005), and this could in turn have repercussions at the population level and beyond.

Hydrologic alteration affects surface waters worldwide and causes impacts on aquatic organisms primarily by reduced water quantity from abstractions and unsuitable flow regimes (Bunn and Arthington, 2002). Besides the removal of riparian shade and the increase in water temperature, the loss of riparian woody vegetation can also increase bank erosion while sediment and nutrient trapping from surface runoff will decrease (Allan, 2004). Further, the instream capacity to cycle nutrients and toxic compounds can decrease when woody riparian vegetation is lost (Sweeney *et al.*, 2004), and decomposition of leaf litter becomes increasingly driven by microbial, rather than macroinvertebrate, activity (Hladyz *et al.*, 2010, 2011b). Finally, the loss of LWD causes a loss of habitat and organic matter storage, all of which can have adverse effects on the taxonomic and functional diversity of fishes and benthic macroinvertebrate communities (Gurnell *et al.*, 1995, 2002; Stauffer *et al.*, 2000).

The impacts of urbanisation on rivers are especially related to increased impervious surface cover (Allan, 2004; Paul and Meyer, 2001), with severe implications for both hydrology and morphology (Allan, 2004; Arnold and Gibbons, 1996; Booth and Jackson, 1997). Increased surface runoff and peak discharge (Arnold and Gibbons, 1996; Booth and Jackson, 1997) enlarge the channel, leading to increased water temperature due to decreased shading (Galli, 1991), which could impose metabolic stress on the biota (Perkins *et al.*, 2010; Woodward *et al.*, 2010a,b,c). This hydromorphological (physical) degradation reduces habitat suitability and impacts the diversity and integrity of riverine fish communities (Wang *et al.*, 1997; Yoder *et al.*, 1999) and of benthic macroinvertebrates (Horner *et al.*, 1997; Yoder *et al.*, 1999).

These potential pressures and impacts on river ecosystems is by no means exhaustive, but they are all relevant in the context of river restoration as each individually poses a threat on the integrity of the instream community. In addition, they are likely to occur in combination in densely populated river basins, leading to potential synergies (Friberg *et al.*, 2011). Such synergies are still little studied (Friberg, 2010a,b; Ormerod *et al.*, 2010), but among the studies conducted there is strong evidence that pressures often act in this manner (Folt *et al.,* 1999; Matthaei *et al.*, 2006; Ormerod *et al.,* 2010).

In the context of river restoration, as an example, rainfall-driven runoff from agricultural land can wash fine sediments and chemical contaminants into river channels, and the fate of these components and their biological consequences are subsequently determined by an interaction with the quantity and dynamics of water flow (e.g. Merz *et al.*, 2004). The first challenge in undertaking effective river restoration in this case is to diagnose the primary constraint to ecological recovery, and then to identify a likely hierarchy of stressors in order to target management activities most effectively: for instance, is the principal problem due to too much sediment and too many contaminants entering the river channel or an unsuitable flow regime due to abstraction or impoundment? The next task is to prioritize available resources, repairing first those problems most likely to be limiting ecological recovery (i.e. tackle the land-users or the water company?).

The typical ecological consequence of multiple pressures is that streams and rivers lose biodiversity and normal ecosystem processes are, in turn, impaired. Many sensitive species quickly disappear even under moderate impact levels, while basic ecosystem processes (such as self-purification of polluted water, biomass production and decomposition) may change significantly under severe degradation. Studies on the implications of biodiversity loss on ecosystem service provision in river ecosystems are still scarce, but the degradation or even loss of rivers, riparian areas and floodplains pose a clear and present threat to both, as suggested by Tockner *et al.* (2009), who reported that nearly 90% of the former floodplains in Europe are degraded and therefore already functionally compromised.

Biodiversity, ecosystem functioning, and community characteristics often track gradients of environmental stress or degradation closely and are, therefore, frequently used as bioindicators within assessment and monitoring schemes (Borja *et al.*, 2009a,b; Feld and Hering, 2007; Feld *et al.*, 2009; Hering *et al.*, 2004; Huryn *et al.*, 2002). Their relation to the components of ecosystem degradation can be based on ecological theory (e.g. Friberg *et al.*, 2011; Lake *et al.*, 2007; Woodward *et al.*, 2010a) and has often been tested and discussed in degradation ecology: for example, Clews and Ormerod (2008) demonstrated how using simple combinations of standard biotic (macroinvertebrate) indices, calibrated to detect specific pressures (organic pollution, eutrophication, acidification and low flows), successfully diagnosed reasons for ecological impairment in different parts of the River Wye catchment in Wales. This approach shows promise as a practical tool in guiding specific management needs in different locations, which are subject to multiple pressures, but further development is needed to include more pressure-specific indices for hydromorphological stress (e.g. Feld, 2004) and indices from more than one organism group (Vaughan *et al.*, 2009).

## D. Restoration as an Active Cure or Just a Placebo?

River restoration can be very expensive, and yet its scientific foundations are often weak or of questionable validity (Friberg *et al.*, 2011; Woodward *et al.*, 2010b). In order to judge the effectiveness of restoration measures, knowledge is required about the pre- and post-measurement conditions (e.g. hydromorphological variables, biological community samples). In addition, untreated (i.e. unrestored) 'control' sites and natural (i.e. untreated and unimpacted) 'reference' sites should be included to account for natural changes, provided that these prerequisites are in place. This experimental design represents the classic before–after-control-impact (BACI) approach that is widely used in general ecology, and constitutes a scientifically sound sampling design to detect changes due to restoration and to separate them from the noise implied by natural variability (e.g. seasonal and annual differences) (Friberg *et al.*, 2011). Unfortunately, despite the statistical power that such designs provide, the huge body of restoration literature comprises only a few BACI studies that can be used to measure the success of restoration schemes objectively (Downes *et al.*, 2002; but see Bradley and Ormerod, 2002). Consequently, there is surprisingly little compelling evidence for the effectiveness of many of the common and widespread restoration techniques, which have mostly been applied at the (small) reach scale where such designs could be relatively easily implemented (e.g. Harrison *et al.*, 2004; Palmer *et al.*, 2007; Pretty *et al.*, 2003). For instance, the enhancement of 'flow diversity' is a common goal of restoration projects, but the small artificial structures, such as the 'vanes', 'deflectors', and artificial riffles that are typically used to this end have no measurable impact on either macroinvertebrate or fish communities (Harrison *et al.*, 2004; Pretty *et al.*, 2003). It is therefore critical in future studies to use monitoring and restoration operations as opportunities for learning about the mechanisms driving community responses and guiding future investment, which has rarely been the case to date. As an example, in the U.S. only 10% of ~40,000 reported restoration operations included any form of monitoring (Palmer *et al.*, 2007). Therefore, it is largely unknown if river restoration is a true cure for degraded physical habitat conditions; current practice might be compared with treating a broken leg with a band-aid or, more worryingly, with a placebo.

Restoration ecology is largely based on the theory that habitat heterogeneity promotes biodiversity (King and Hobbs, 2006; McCoy and Bell, 1991; Tews *et al.*, 2004). The concept of influencing habitat structure is understandably more tractable for river managers and stakeholders compared with dealing with other, less tangible or easily measured factors that might be operating to impact biodiversity, such as productivity and disturbance regimes (Palmer *et al.*, 2010). It is not surprising, therefore, that an almost universal article of faith within the field of river restoration is that this 'habitats-by-numbers' approach to introducing physical heterogeneity will lead to an increase in biodiversity. This belief, although intuitively appealing, is supported by few published studies and evidence suggests that habitat heterogeneity should not be the single driving force behind the choice of river restoration options if ecological recovery is the goal (Palmer *et al.*, 2010). Rather, the processes that should be subjected to restoration are those that are closely coupled to meaningful habitat scales, such as erosion and deposition controlled by a natural and dynamic flow regime that generates the classic pool-riffle sequence over the river's length. The links between biological, geomorphological and hydrological diversity are still poorly understood, largely due to lack of data collected at relevant spatio-temporal scales and a paucity of experimental manipulations that can detect causal, rather than correlational, relationships (Vaughan *et al.*, 2009). With regard to streams and rivers, monitoring the effects of restoration measures in the few cases where this is done has revealed that communities do not necessarily always match the anticipated outcomes (e.g. Jähnig *et al.*, 2009a,b; Palmer *et al.*, 1997). Similarly ambiguous results have been reported from lakes (e.g. Jeppesen *et al.*, 2005) and transitional/coastal waters (Duarte *et al.*, 2009), so these phenomena are by no means unique to river ecosystems. The relationships of restoration and its ecological impacts seem to (at least partly) differ from those identified for degradation: restoration is unlikely to be 'simply' the opposite of degradation (Moerke *et al.*, 2004). Palmer *et al.* (1997) described this as the 'field of dreams' approach, such that 'if you build it, they will come', that is, if you restore the habitat, it is axiomatic that the organisms will automatically follow. This hypothesis all too often is not met as rehabilitation and restoration schemes often fall short of the biological expectations (compare Sections III–V), and much of this mismatch is almost certainly due to our currently poor understanding of the real drivers and responses (Palmer *et al.*, 2010; Roni *et al.*, 2008).

For example, in the United Kingdom, attention has been focused recently on the role of habitat heterogeneity in buffering the impacts of water abstraction and flow variation on macroinvertebrate communities in rivers (Bradley *et al.*, 2009; Dunbar *et al.*, 2010a,b). Examination of broad-scale and long-term biomonitoring data in the United Kingdom has indicated that macroinvertebrate communities in rivers are more sensitive to the effects of discharge variation in modified (i.e. habitat-poor) reaches compared to more natural (i.e. diverse, habitat-rich reaches) (Dunbar *et al.*, 2010a,b). Whilst reasonably significant linear relationships were reported across wide spatial and temporal scales, they were weak and inconsistent at the individual waterbody scale. This has nevertheless resulted in proposals to undertake habitat manipulations in degraded river channels to offset the ecological impacts of water abstraction to meet legislative targets in a simple and cost-effective manner. Many restoration projects are based on the underlying, but largely untested, assumption that geomorphological habitat heterogeneity promotes biodiversity. The widespread preference for (local and instream) habitat enhancement may also be driven by practical constraints: specific limitations due to water and land use practices (fishing, agriculture) often inhibit the implementation of more effective, larger-scale, ecological restoration. Recently, Bradley *et al.* (2009) undertook an appraisal of the ecological relevance of geomorphologically driven approaches to guiding river restoration options to mitigate the impacts of water abstraction in a southern English chalk river. They suggested that indices of habitat diversity and patchiness derived from geomorphological/fluvial audits correlated reasonably well with an ecological habitat index, but were not adequate on their own for guiding habitat manipulation works to achieve the desired ecological outcomes. This was mainly due to the fact that geomorphological indices did not consider the appropriate location of ecologically functional mesohabitats (e.g. salmonid spawning habitat located upstream of nursery and adult habitats), some habitats scored low in the geomorphological scheme, but were functionally important to priority taxa (e.g. fine sediment is critical habitat for juvenile lampreys) and the scheme was designed only to guide the location of individual reaches in need of specific habitat manipulations. This chapter emphasises that river restoration in degraded river channels to mitigate the impacts of water abstraction must be guided by ecological criteria and encompass the water body as whole, rather than just smaller-scale geomorphological criteria.

Our knowledge about the temporal extent of river ecosystem recovery from degradation, although improving, is still limited (e.g. Moerke *et al.*, 2004; Nilsson *et al.*, 2005) and one important gap relates to the link between 'endpoint of restoration' (i.e. the ultimate goal of restoration/rehabilitation measures) and its deviance from the ecological status prior to degradation. Another refers to the time scale needed for an ecosystem to recover from degradation; recent research in acidified fresh waters has revealed there is a time lag between chemical and biological recovery, possibly due to ecological inertia within the food web itself (e.g. Layer *et al.*, 2010, 2011), and there is no reason to assume that similar lags or hystereses are not operating in restoration schemes.

The ultimate aim of restoration ecology still is to identify and test the relationships between degradation and ecology and to transfer the findings to practical restoration (King and Hobbs, 2006). Given the numerous studies on biologically ineffective restoration measures (see Palmer *et al.*, 2010 for a review), this statement could be extended to also embrace the relationships between restoration and ecology. A more direct focus on restoration studies could also help better identify the effects of restoration, or lack thereof (Palmer *et al.*, 2010).

We attempt to do this by reviewing and synthesizing existing knowledge regarding three frequent types of restoration measures (re-establishment of riparian buffer strips, instream habitat improvements and weir removal). Additionally, we demonstrate successes and failures by revisiting a number of case studies on a fourth type of the larger-scale restoration measures that are now increasingly favoured over the earlier, smaller-scale approaches, that is, re-meandering lowland rivers. We then apply our findings to propose a novel conceptual framework to elucidate cause–effect chains that links restoration efforts with structural and functional effects on the river ecosystem. Finally, we highlight shortcomings in the current implementation of many river restoration schemes and suggest options for moving the field forward, within the general restoration paradigm to restore ecological characteristics and enhance ecological processes and functioning of river ecosystems. Inevitably, we were not able to address practical constraints that may constrain the implementation of ecologically effective restoration measures (e.g. conflicting stakeholder interests and water/land uses), because such information is largely lacking in the reviewed body of literature. Therefore, ecologically effective restoration was considered the common 'benchmark' for the evaluation of restoration success and failure in this study.

## II. REVIEW AND SYNTHESIS OF THE RESTORATION LITERATURE

*Water quality improvement by riparian buffers* primarily aims at mitigating the adverse impacts of intensive agricultural land use adjacent to streams and rivers. A sufficiently wide and ideally mixed riparian vegetation strip at both sides of a stream is considered to retain plant nutrients (e.g. nitrogen and phosphorous components), fine sediments and toxic substances (e.g. pesticides) that enter streams via surface runoff from adjacent agricultural areas (Barton *et al.*, 1985; Castelle *et al.*, 1994). Riparian trees provide shade and organic material (leaf litter, wood) and thus supply both food and habitats for instream biota (Davies-Colley and Quinn, 1998; Davies-Colley *et al.*, 2009; Parkyn *et al.*, 2005). *The enhancement of instream mesohabitat structures* aims at increasing structural physical diversity and is often considered, by extension if not direct observation, to promote biological diversity (Palmer *et al.*, 2010). In particular, the introduction (or omission of the removal) of LWD provides a key habitat for fishes and benthic macroinvertebrates (Kail *et al.*, 2007; Roni and Quinn, 2001a,b) and also stimulates habitat diversity (e.g. creation of pools and flow refugia) by diversifying hydraulic conditions (Baillie *et al.*, 2008). Besides LWD, we evaluated the mitigation effects of the introduction of boulders, deflectors, fish spawning substrates and the removal of bank enforcement (e.g. 'sheet piling' or 'rip-rap'). *iii. The removal of weirs and dams primarily aims at r*estoring the longitudinal connectivity of streams and rivers. Weir removal is considered to facilitate the migration of fishes and benthic macroinvertebrates, which can be critical for certain species, especially those that are catadromous or anadromous (Doyle *et al.*, 2005; Gregory *et al.*, 2002), and there are positive effects on flow conditions and sediment particle size upstream and water temperature up- and downstream (Bednarek, 2001; Hart *et al.*, 2002).

The literature survey was conducted in October 2010 using the ISI Web of Knowledge and SCOPUS, focussing on publications in peer-reviewed journals (and references therein), which was then extended by selected peer-reviewed reports, grey literature and other publications using Google Scholar and further web search engines. Strict focus was on references from the restoration literature, that is, either publications that specifically address active restoration, or reviews thereof. The following search terms were used: (restoration OR rehabilitation) AND (riparian vegetation OR riparian buffer OR large wood OR LWD OR habitat structure OR bed structure OR channel structure OR weir removal OR dam removal) AND (fish OR invertebrates OR macrophytes OR phytobenthos OR algae) AND (river OR stream). The search terms were used in different combinations to make sure that all relevant restoration/rehabilitation literature was found. Publications on general ecological relationships, such as studies comparing natural (control) streams possessing an intact riparian vegetation with degraded (impact) streams lacking riparian buffers, were only considered if they provided strong and generally applicable evidence based on comprehensive data and representing a broad geographical extent (e.g. entire catchments). All studies had to provide information on further criteria, including full reference details, study type (restoration, general ecological study), river name and size, altitude, region and continent, type of restoration measure, abiotic and biological effects studied, qualitative and quantitative effects detected (i.e. real recovery), period between restoration and monitoring, length of study reache(s), brief summary of core findings, estimated strength of the study (subjective judgement based on number of replicate samples, geographical extent of samples and statistical analysis applied), and variables that potentially may have limited restoration success. The distinction between qualitative and quantitative effects was used to help identify potential mechanistic or otherwise quantifiable relationships and separate them from rather descriptive (qualitative) trends. The criteria include those recently reported by Miller *et al.* (2010) in a recent meta-analysis.

Of the roughly 1000 hits reported back by the Web literature databases, altogether 370 publications (including nine 'grey' reports) were selected for a detailed review, 168 of which fulfilled our criteria (Appendix B). The final selection of references was analysed according to the attributes listed above and converted to a database for further descriptive analysis.

Studies of the effects of restoration involving instream habitat improvements, buffer strips and weir removals originated primarily from North America (USA) followed by Europe, Australia and New Zealand (Figure 2A), whereas little was reported from South America and Asia. The majority of studies (> 70%) were published within the last decade (Figure 2B).

## III. WHAT HAS BEEN ACHIEVED BY RESTORING BUFFER STRIPS?

Out of a total of 57 papers reviewed on this topic, 70% were published after 2000. The restoration of riparian vegetation either refers to active measures, that is, the establishment of riparian buffers (Northington and Hershey, 2006; Schultz *et al.*, 1995; Sutton *et al.*, 2009) or to passive restoration by allowing riparian buffer strips to (re)establish either with fencing to exclude large herbivores (Opperman and Merenlender, 2004) or without fencing (McBride *et al.*, 2008; Pedraza *et al.*, 2008). Mixed riparian buffers consisting of trees, shrubs and grass strips are generally considered to be most effective in the retention of fine sediments and nutrients from both surface runoff and the upper groundwater layer (Correll, 2005). Results and guidelines on the minimum width and length of effective riparian buffer strips vary markedly (e.g. Castelle *et al.* (1994) reported a width range of 3–200 m) although a minimum width of 15 m on either side of a stream appeared sufficient to protect streams under most conditions, while a minimum buffer width of at least 30 m on either side has been found to provide also shading comparable to old-growth riparian forest. Buffers of 30 m width were successful in maintaining macroinvertebrate background levels in Californian streams adjacent to logging activities (Castelle *et al.*, 1994). A similar minimum width was suggested by Wenger (1999), who in addition developed a function to calculate the minimum buffer width based on the riparian slope (compare Table 1 for a selection of example studies).

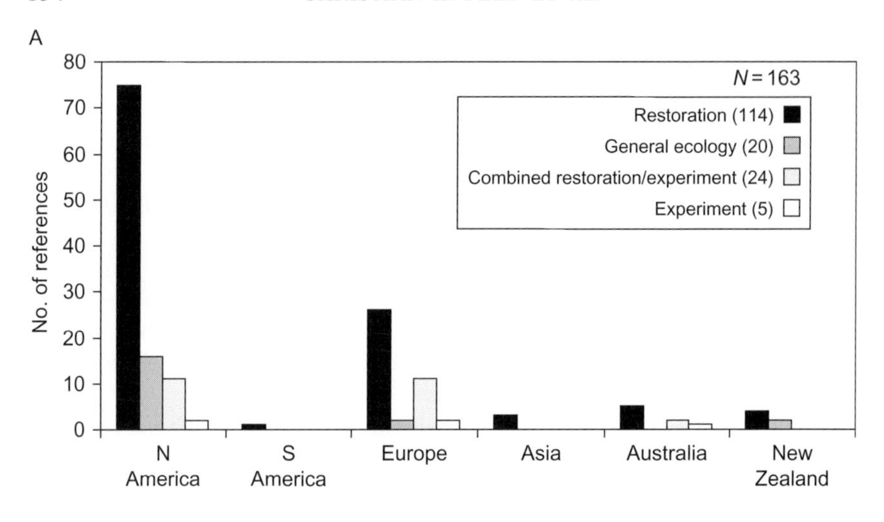


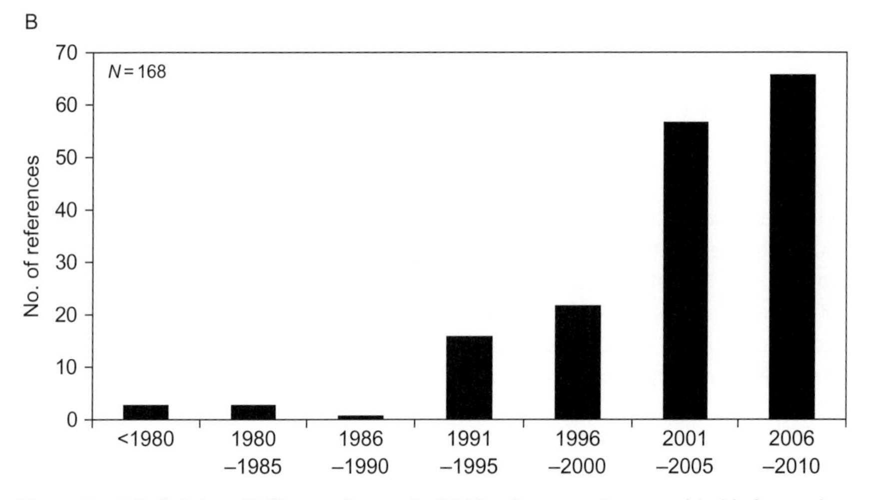


**Figure 2** (A) Origin of 163 out of a total of 168 references that provided information on the origin, grouped according to the character of the studies included in our detailed literature review. (B) Time of publication of 168 references analysed.

Results on the minimum length of a buffer strip were less frequent. As far as reductions in water temperature are targeted by riparian restoration, Parkyn *et al.* (2003) concluded from modelling studies that at least 1–5 km was required for first-order streams versus 10–20 km for fifth-order streams to reduce water temperature to more typical reference conditions. Based on a total of 20 studies that also provided information on the length of the studied sites or reaches (Figure 3), this was less than 1000 m for 70% of the studies; only two

**Table 1** Examples of qualitative and quantitative evidence for the effectiveness of riparian buffer management and related habitat improvement

| Reference | Type | Qualitative | Quantitative |
|---|---|---|---|
| Dosskey (2001) | Review | Buffers can retain pollutants from surface runoff, filter surface and groundwater runoff, stabilise eroding banks, contribute to processes that remove pollutants from the stream water flow. | |
| Broadmeadow and Nisbet (2004) | Review | Removal of riparian woodland can lead to a temperature increase of up to 4 °C, sufficient LWD and CPOM inputs into the river require buffers of 25–100 m width. | |
| Lester and Boulton (2008) | Review (meta-analysis) | Addition of LWD leads to increases in fish and macroinvertebrate richness and abundance, macroinvertebrate diversity, bank stability, sediment and organic matter storage, habitat diversity (greater diversity of depths, velocities and habitat elements). | |
| Miller *et al.* (2010) | Review (meta-analysis) | Addition of LWD and boulders increased instream habitat structure and lead to increased macroinvertebrate richness but not density. | |
| Opperman and Merenlender (2004) | Passive restoration | Restored reaches had higher frequency of LWD, lower temperature and improved habitat characteristics. | |
| Moustgaard-Pedersen *et al.* (2006) | Active restoration | Macrophyte species richness was significantly higher in restored reaches, but plant coverage was not. | |
| Castelle *et al.* (1994) | Review | Buffer width of 25–60 m was found to retain 75–95% of fine sediments, buffer widths of 4.5–10 m can retain up to 95% of plant nutrients, buffers of at least 30 m widths have been found to provide shading comparable to old-growth riparian forest and proved successful in maintaining macroinvertebrate background levels. | |

(*continued*)

**Table 1** (*continued*)

| Reference | Type | Qualitative | Quantitative |
|---|---|---|---|
| Osborne and Kovacic (1993) | Review and active restoration | 10–30 m forested riparian buffers maintain ambient stream temperatures, 9–45 m vegetated buffers retain a substantial portion of sediment in overland, 5–50 m forested riparian buffer retain 60–100% of N and P, riparian forests are more effective in removing nitrate-N from shallow ground waters than are grass strips. | |
| Wenger (1999) | Review | 30 m buffer width sufficient to trap sediments under most circumstances, absolute minimum width is 9 m, 30 m buffers should provide good control of N, 10–30 m native forested riparian buffers should be preserved or restored along all streams to maintain the aquatic habitat/ maintain aquatic habitats, protecting diverse terrestrial riparian wildlife communities requires a number of buffers up to 100 m width. | $W = k\ (s\^0.5)$<br>$W$ = buffer width<br>$k$ = constant (50 ft)<br>$s$ = slope |
| Warren *et al.* (2009) | Experiment (no restoration) | Volume ($V$) and frequency ($F$) of LWD and wood accumulations (wood jams) in streams were most closely associated with the age of the dominant canopy trees in the riparian forest. | $\log_{10} V$ ($m^3$/100 m) = (0.0036 × stand age) − 0.2281; $p < 0.001$, $r^2 = 0.80$; $F$ (No./100 m) = (0.1326 × stand age) + 7.3952; $p < 0.001$, $r^2 = 0.63$ |

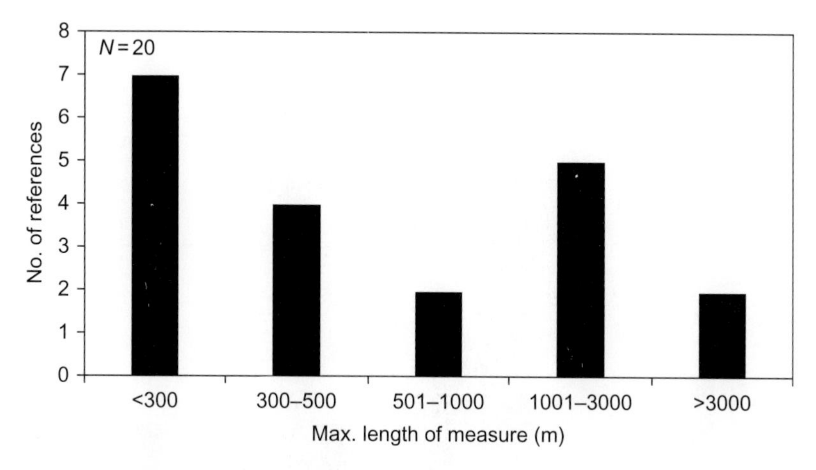


**Figure 3** Maximum lengths of study reaches reported in the 20 restoration studies on riparian buffer restoration that provided this information.

references had study reaches > 3 km length. This pinpointed an important up-scaling issue when extrapolating scientific studies to management tools: the reviewed studies did not address the question of minimum buffer lengths required to restore shading or fine sediment impacts at a river section.

There is considerable evidence that riparian buffer establishment increases instream water quality and habitat complexity (Broadmeadow and Nisbet, 2004; Dosskey, 2001; Mankin *et al.*, 2007), decreases fine sediments and water temperature (Broadmeadow and Nisbet, 2004; Opperman and Merenlender, 2004) and provides LWD (Opperman and Merenlender, 2004). The latter was frequently identified as a key structure or habitat that not only provided direct habitat to benthic macroinvertebrates and shelter to young fish (Brooks *et al.*, 2004), but also played a major role in structuring the stream bottom by enhancing the depth and frequency of pools and flow refugia (Brooks *et al.*, 2004; Larson *et al.*, 2001).

## A. Which Organism Groups and Group Attributes Have Shown Evidence of Recovery After Restoration?

The majority of studies addressed effects on benthic macroinvertebrates (70%) and fishes (16%), and composition and richness measures (as opposed to absolute abundance) dominated the indicators used to monitor these effects (Figure 4). A few studies reported an increase of benthic macroinvertebrate richness after riparian restoration (Becker and Robson, 2009; Broadmeadow and Nisbet, 2004; Castelle *et al.*, 1994; Jowett *et al.*, 2009; Pedraza *et al.*, 2008; Quinn *et al.*, 2009), or the increase of fish richness and biomass after passive

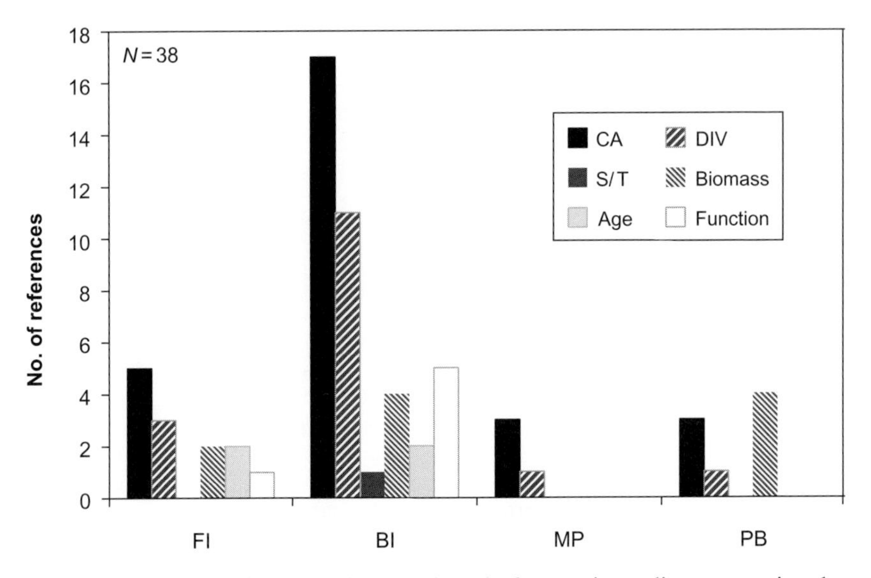


**Figure 4** Riparian buffer restoration: number of references in our literature review that addressed community attributes composition/abundance (C/A), sensitivity/tolerance (S/T), age structure (Age), diversity (Div), biomass and function of fish (FI), benthic macroinvertebrates (BI), macrophytes (MP) and phytobenthos (PB). As a study may refer to more than one community attribute, the overall number of references exceeds the number of the 38 restoration references that provided this information.

buffer restoration Penczak (1995), although the specific mechanisms of recovery were not identified. Five studies addressed functional aspects of the community, while six reported effects on benthic macroinvertebrate biomass and age structure (larval development; e.g. Entrekin *et al.*, 2009; Lester and Boulton, 2008). The most important variables responsible for changes in macroinvertebrate communities were fine sediment, water temperature, food/energy supply (particulate organic matter) and LWD (Figure 5). Changes of the fish community were most often related to water temperature and the amount of LWD (Chen *et al.*, 2008; Hilderbrand *et al.*, 1997), which was reported a key habitat to young fish (Cederholm *et al.*, 1997).

## B. Was There Evidence for Strong Qualitative or Quantitative Linkages?

There was clear and, in many cases, strong evidence for the role of riparian buffers in controlling nutrient and sediment retention, water temperature and instream habitat structure (Table 1). Nevertheless, there was a clear lack of evidence for strong relationships with the aquatic biota. Only two studies

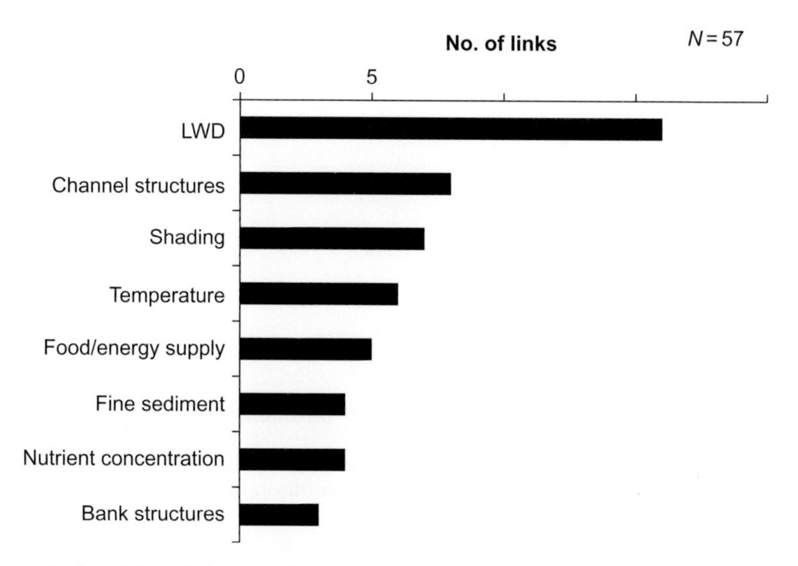


**Figure 5** Ranking of the most important environmental state variables based on the number of linkages (arrow thickness in Figure A1), derived from 57 references on riparian buffer restoration and experimental studies.

reported effects on richness of benthic macroinvertebrate (Miller *et al.*, 2010) and aquatic macrophytes (Moustgaard-Pedersen *et al.*, 2006), while other organism groups and community attributes remained unaddressed. Only three studies (Entrekin *et al.*, 2009; Warren *et al.*, 2009; Wenger, 1999) included quantitative results and provided formulas on the relationships of restoration causes and impacts on the biota (Table 1). Hence, the restoration literature rarely reported information on the mechanisms of restoration that could be used to model and predict the effects of management.

In summary, there is sufficient evidence for the ecological effectiveness of mixed riparian buffers in the retention of fine sediments, plant nutrients and toxic contaminants from agricultural areas adjacent to rivers. Although rarely based on quantifiable, mechanistic relationships, this evidence from the restoration literature is sufficient to develop best-practice guidance for riparian buffer restoration and related instream habitat improvements.

## C. What is the Timescale of Recovery?

A satisfactory answer to this question was not possible based on the existing literature, which is striking given the relative large number of studies and the common use of re-establishment of riparian buffers as a management tool. Two thirds of the studies referred to monitoring activities of $<10$ years

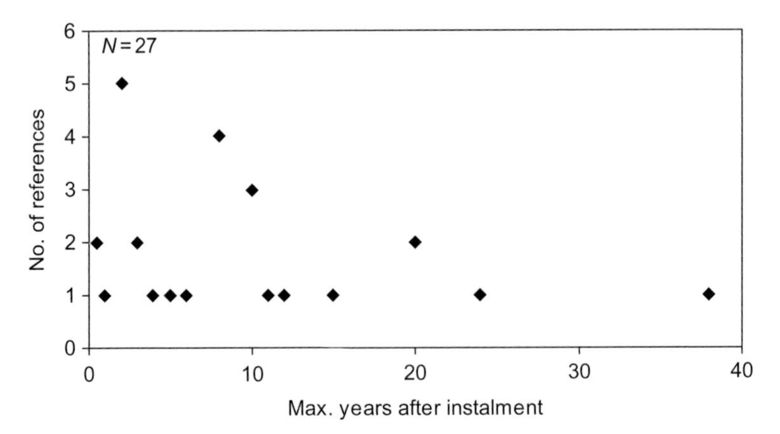


**Figure 6** Timing of investigations of 27 reviewed restoration studies (that provided this information), relative to the time of implementation of measures.

(Figure 6), a timescale that may be sufficient to detect the direct effects of instream habitat improvements, but not to detect major indirect effects on overall ecosystem functioning. Important processes such as wood recruitment and the supply of energy to the instream food web would require larger and older stands of riparian trees (Warren *et al.*, 2009). The timescale required for riparian buffers to achieve maturity and provide all relevant ecological functions may be considerable: riparian trees, such as black alder and willow, can require 30–40 years to mature and reach their final height and maximum canopy cover (Jowett *et al.*, 2009) needed to provide nutrient and sediment retention, and temperature control. Longer time spans are required to provide natural amounts of LWD and influence the instream hydromorphology (e.g. by stimulating bed form processes and habitat improvement), so full recovery may take centuries (Davies-Colley *et al.*, 2009).

## D. Reasons for Failure and Limiting Factors When Restoring Buffer Strips

Only three studies revealed (negligible) effects with regard to the anticipated restoration effects hypothesised *a priori*. Larson *et al.* (2001) expected positive changes after the addition of LWD but found no detectable effects upon the benthic macroinvertebrate communities up to 10 years later, although pool frequency consistently increased in all streams. The authors attributed their negative findings to watershed-scale disturbances, in particular to increased loads of fine sediments that continued to affect the restored section after restoration.

Sutton *et al.* (2009) investigated the effects of active riparian buffer restoration on nutrient retention up to 8 years after planting streamside vegetation (trees and managed grassland). The authors observed only insignificant changes in nutrient concentrations, although the mean forested buffer density in the 15 stream reaches increased from 33% to 44%. Sutton *et al.* (2009) attributed their finding to insufficient buffer age, size, and to gaps within the buffer strips that largely reduced their effectiveness.

Becker and Robson (2009) investigated the effects of riparian buffer revegetation on instream benthic macroinvertebrate communities in Southern Victoria, Australia. Non-native willows had been removed and replaced by native tree species, but up to 8 years after revegetation, there was no measurable effect on the benthic macroinvertebrate community, which was ascribed to a lack of sufficient time for recovery.

Warren *et al.* (2009) quantified LWD loading in 28 streams in the northeastern United States and documented the current volume and frequency of occurrence of wood in streams with riparian forests varying in their stage of stand development as well as stream size and gradient. The authors developed linear regression models to predict the amount ($r^2 = 0.80$; $p < 0.001$) and volume ($r^2 = 0.63$; $p < 0.001$) of wood using riparian forest age as a single descriptor, but these could not be transferred to other systems because of regionally different forest characteristics and the legacy of land use.

## IV. ENHANCEMENT OF INSTREAM HABITAT STRUCTURES

The introduction of substrates (LWD, boulders and gravel) is often aimed at providing key habitats to macroinvertebrates or fishes, while artificial structures (deflectors) and the removal of artificial bank enforcements (e.g. rip-rap) aimed at the re-establishment of dynamic processes, such as flow diversity and erosion/deposition of finer material, at small scales of a few metres. The effects of such structures and their removal, respectively, were comprehensively reviewed in the recent studies of Roni *et al.* (2008), Miller *et al.* (2010) and Palmer *et al.* (2010). Roni *et al.* (2008) primarily addressed fish and biotic production, while the latter two studies investigated the effects of various forms of local and reach-scale habitat improvement (including measures of channel realignment/re-meandering) on benthic macroinvertebrate richness and abundance. Overall, mesohabitat enhancement was found to have only limited positive effects on the instream biota. Miller *et al.* (2010) found significant and positive effects of LWD addition on macroinvertebrate richness but not abundance, in the 24 studies reviewed for their meta-analysis. Palmer *et al.* (2010) did not support this finding and found no evidence that enhanced habitat heterogeneity (including the addition of LWD) was the primary factor controlling macroinvertebrate diversity. Roni *et al.* (2008) found that firm conclusions were impossible because of the limited information provided by the reviewed literature.

From the 75 studies reviewed here, the majority have been conducted in North America (60%), while roughly a third originated from Europe (Figure 7). Many dealt with the introduction of LWD (60% of all studies) and boulders (32%); the introduction of spawning gravel and deflector structures were less often applied (12% and 9%, respectively). Most studies also revealed that restoration and rehabilitation studies on local habitat enhancement are frequently designed prior to the implementation, either following a comparison of monitoring results Before and After the measure (BA), a comparison of untreated Control sites with Impact (treated) sites (CI), or combinations of both (full BACI design). The full BACI design was most common (44%), followed by a CI design (32%). This reflected the more classically experimental character of these studies, which were mostly conducted at relatively small spatial scales (Figure 8A) and short time spans (< 5 years between introduction and monitoring, Figure 8B). As such, they rarely provided much insight into the large-scale and long-term effects of restoration.

Introducing substrates and deflectors primarily influenced instream hydromorphology by increasing habitat (substrate) diversity and pool frequency/area, and by providing habitat cover for fish (Figure 9). Both pool frequency

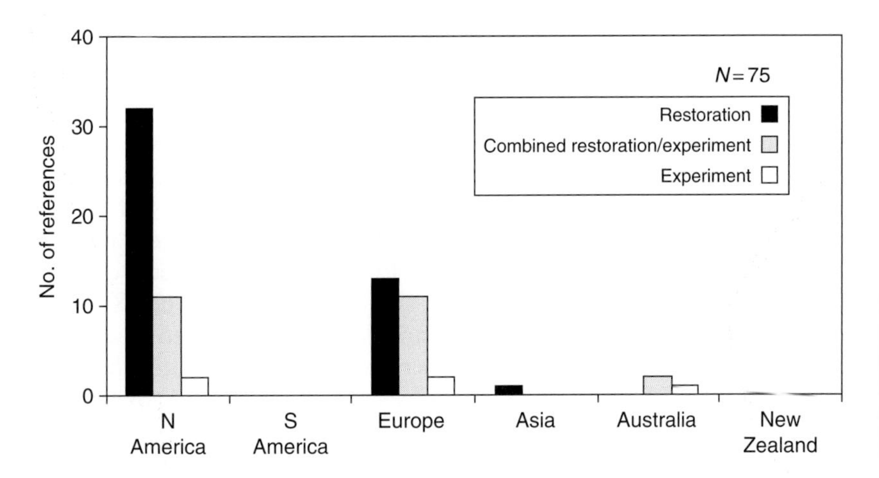


**Figure 7** Origin of 75 mesohabitat enhancement studies (including removal of bank enforcement structures), ordered by study type.

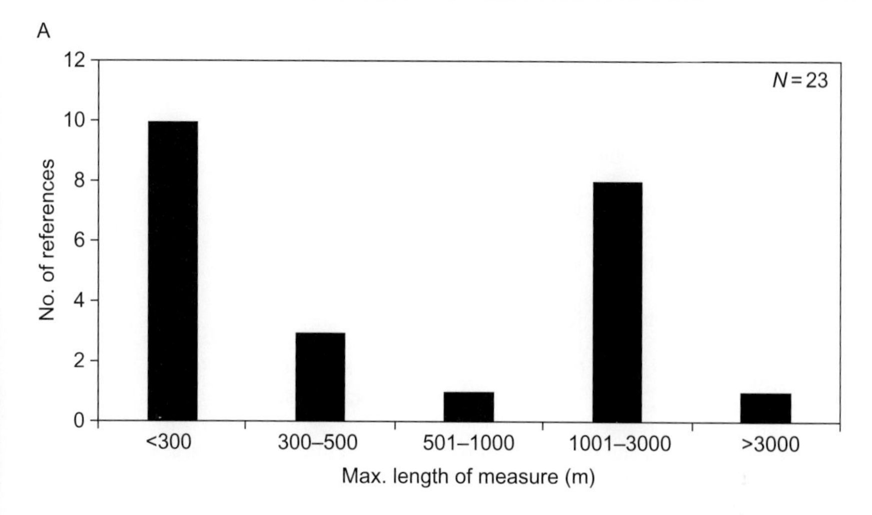


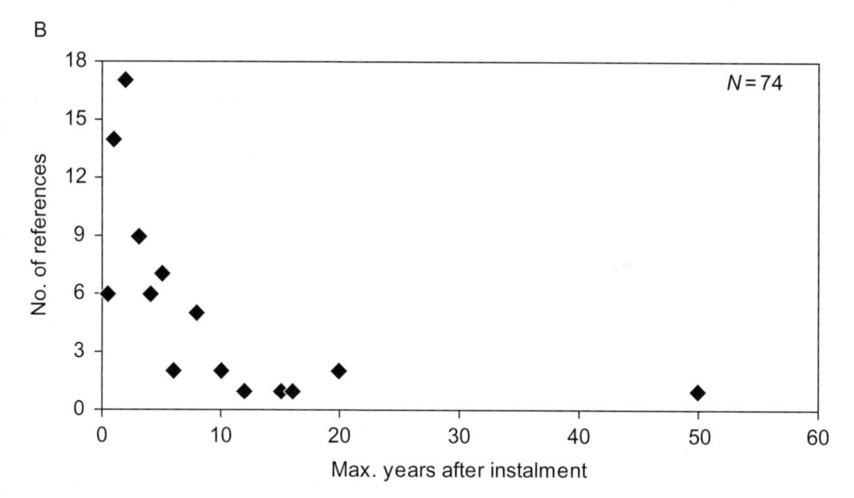


**Figure 8** (A) Maximum lengths of study reaches reported by 23 mesohabitat enhancement studies (that provided this information) and (B) timing of 74 mesohabitat enhancement studies relative to the time of implementation of measures.

and area were frequently used indicators of habitat enhancement in restoration studies that aimed at recovery of sensitive and economically important fishes, particularly salmonids (e.g. *Salmo trutta* Linnaeus, 1758, *Salvelinus fontinalis* [Mitchill, 1814], Avery, 2004; Baldigo *et al.*, 2008; *Oncorhynchus* spp., Cederholm *et al.*, 1997; Johnson *et al.*, 2005). Overall substrate diversity favoured recovery of both fish and benthic macroinvertebrate taxa and communities (Baldigo *et al.*, 2008; Brooks *et al.*, 2004; Jungwirth *et al.*, 1995). Substrate diversity was directly enhanced by the addition of LWD

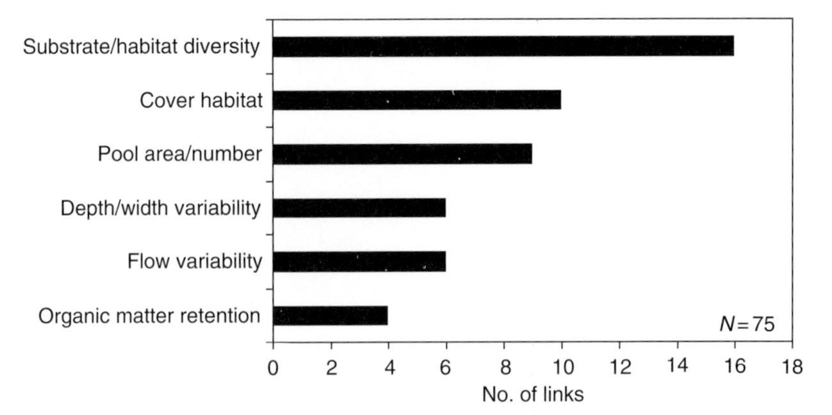


**Figure 9** Ranking of the most important environmental state variables based on the linkages (arrow thickness in Figure A2), derived from 75 references on mesohabitat enhancement.

(e.g. Brooks *et al.*, 2006; Harrison *et al.*, 2004) and the placement of boulders (Gerhard and Reich, 2000; Jungwirth *et al.*, 1995), both of which increased flow heterogeneity (e.g. Pretty *et al.*, 2003), on-channel sediment retention (e.g. Quinn and Kwak, 2000) and bank stability (e.g. Levell and Chang, 2008).

## A. Which Organism Groups and Group Attributes Have Shown Evidence of Recovery After Restoration?

Most ecological effects were reported for fish species or communities (21 references) followed by benthic macroinvertebrates (7), macrophytes and phytobenthos (2 each). Community attributes were referred to in several studies on both fishes and benthic macroinvertebrates (Figure 10). In addition to well-referenced measures of richness and abundance, some studies included effects on sensitive taxa (e.g. greyling *Thymallus thymallus* [Linnaeus 1758]: Muhar *et al.*, 2008), age structure (e.g. Jungwirth *et al.*, 1995; Roni *et al.*, 2006), biomass (e.g. Coe *et al.*, 2009; Edwards *et al.*, 1984) and functional guilds (e.g. Larson *et al.*, 2001; Tullos *et al.*, 2006).

Roni and Quinn (2001a,b) found that a 10-fold increase of the density of LWD resulted in a sixfold rise in of Coho salmon (*Oncorhynchus kisutch* Walbaum, 1792) abundance during summer, but a decline in rainbow trout (*Oncorhynchus mykiss* [Walbaum, 1792]) along the gradient of increasing pool area was also observed. This highlights potentially conflicting effects of specific instream enhancements on different species, even within the same genus (although other mechanisms, such as resource competition, may also explain the opposed response of *O. kisutch* and *O. mykiss* in the study of

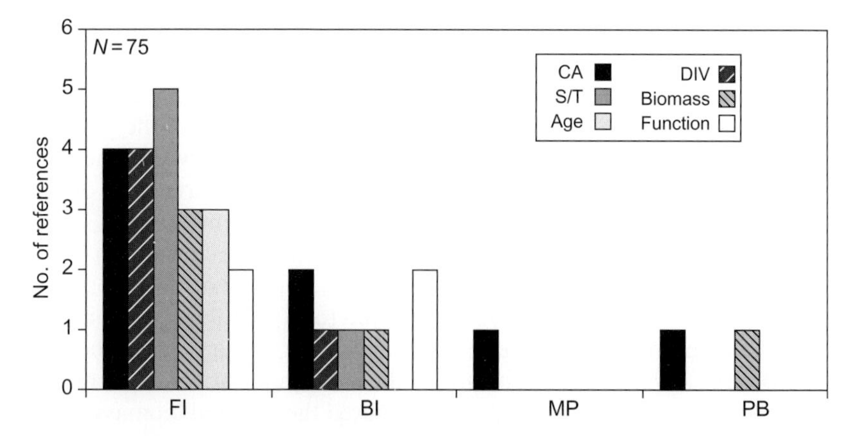


**Figure 10** Mesohabitat enhancement: number of references addressing the community attributes composition/abundance (C/A), sensitivity/tolerance (S/T), age structure (Age), diversity (Div), biomass and function of fish (FI), benthic macroinvertebrates (BI), macrophytes (MP) and phytobenthos (PB). As a study may refer to more than one community attribute, the overall number of references exceeds the number of 75 restoration references reviewed.

Roni and Quinn, 2001a,b). Our review did not reveal stream size- or stream type-dependent natural thresholds for pool frequency or pool area beyond which the lack of other habitats (e.g. riffles, runs) might have adverse effects on these and other species.

A notable example of restoration success was described by Jungwirth *et al.* (1995) in a large 5th order river in Austria, where multiple small scale measures employed along a stretch 1.5kn stretch yielded clear and measurable improvements of the fish fauna. Both species richness and diversity were highly correlated with the variance of maximum depth ($r > 0.86$). A 30-fold increase of this variance resulted in twice as many fish species and enhanced the overall community's diversity. According to Jungwirth *et al.* (1995) these strong relationships '... can also be used to forecast the effects of river restoration plans' (p. 205). Muhar *et al.* (2008) reported similar success from another large Austrian river, the Drau, in which treatments also included a range of instream measures at the local and reach-scale over several kilometres, resulting in a marked improvement in ecological status (according to the national fish assessment system). The level of enhancement reflected the large spatial dimension of the measure: the most extensive measure (total length: 2 km) resulted in a threefold enlargement of the active channel dimension and improved the habitat availability for a key fish species, grayling (Muhar *et al.*, 2008).

The references on macroinvertebrate responses to restoration revealed significant effects on the community abundances, despite the generally high spatio-temporal variability of this parameter. After the addition of LWD, the density of Ephemeroptera preferring pool habitats increased, while abundances of Coleoptera, Trichoptera, Plecoptera and Oligochaeta decreased due to the decreasing proportion of faster-flowing riffle habitats (Hilderbrand *et al.*, 1997). Gerhard and Reich (2000) observed highest species richness and abundance on LWD, twigs and coarse particulate organic matter—microhabitats that only occurred after log placement.

In general, phytobenthos and macrophyte communities have been largely neglected in past restoration studies (Figure 10). Only two studies provided evidence for a significant increase of periphyton (phytobenthos) biomass, expressed as chlorophyll *a* concentration and/or ash free dry mass (Coe *et al.*, 2009; Moerke *et al.*, 2004). According to the latter publication the increase of habitat surface area after log placements caused the elevated biomass values.

## B. Was There Evidence for Strong Qualitative or Quantitative Linkages?

The linkages between instream habitat restoration and biological recovery have been described in primarily qualitative terms, and very few studies provided even quasi-quantitative relationships (Table 2).A positive relationship between maximum depth variance and fish species richness and diversity was derived in the Jungwirth *et al.* (1995) study, but real quantitative (i.e. quasi-mechanistic) relationships between habitat enhancement and biological recovery were otherwise missing from this body of literature. Whether such relationships can be detected without sampling a huge population of restored streams, given the high degree of variability of treated stream reaches, remains to be tested rigorously. Several examples of non-effect studies concluded that the absence of biological recovery was owed to continuing pressures at larger scales that were not mitigated by restoration, such as water quality problems (e.g. Pretty *et al.*, 2003) or fine sediment entries due to intensive land use upstream (e.g. Larson *et al.*, 2001; Levell and Chang, 2008) that 'spoiled' and limited restoration effects, rather than a true absence of an effect.

## C. What Is the Timescale of Recovery?

The time span between restoration and monitoring of effects ranged from one to 50 years with an average (median) value of 2.5 years (Figure 8B). On average, monitoring was performed only twice after restoration and was then compared against before and/or control values (e.g. Gerhard and Reich, 2000; Muhar *et al.*, 2008; Roni *et al.*, 2006). Other studies sampled in

**Table 2** Examples of qualitative and quantitative evidence for the effectiveness of instream mesohabitat enhancement measures

| Publication | Qualitative | Quantitative |
|---|---|---|
| Baldigo *et al.* (2008) | Unspecified measures (acc. to "Natural Channel Design" approach) led to shift in dominant species and increase of intolerant species richness of the macroinvertebrate community. | |
| Binns (2004) | Trout response to habitat manipulation varied among projects, but acceptable responses occurred across all sizes of streams. Mean increases of wild trout abundance and biomass among different stream orders ranged from 30% to 250%. | |
| Cederholm *et al.* (1997) | Salmon abundance increased in winter season after treatment by LWD. | |
| Gerhard and Reich (2000) | Increase of macroinvertebrate richness and diversity in sections treated with LWD. | |
| Herbst and Kane (2009) | An increase of EPT taxa richness by 7 taxa followed after the complete relocation/recreation of 150 m channel. | |
| Hilderbrand *et al.* (1997) | Systematic placement of 50 logs at 225 m channel length increased the pool area by 150% and decreased the riffle area by 40%, but no significant changes in the macroinvertebrate community were observed. | |
| Jungwirth *et al.* (1995) | Measures led to increased heterogeneity of water depth and current velocity, and added sandy and muddy in-channel microhabitats. One year after restructuring, the number of fish species increased from 10 to 19. Fish density and biomass tripled during the period of investigation. The abundance of individual species changed considerably (decrease of *Leuciscus cephalus*, *Gobio gobio*), resulting in a more balanced fish community structure. | $NFS = 0.00927 \times VMD + 6.12$, $r = 0.86$; $n = 15$; NFS: number of fish species; VMD, variance of maximum depths $FSD = 0.0007014 \times VMD + 1.28$; $r = 0.897$, $n = 15$; FSD: fish species diversity |
| Moerke *et al.* (2004) | Increase in abundance of macroinvertebrates and periphyton in sections treated with LWD. | |

(*continued*)

**Table 2** (*continued*)

| Publication | Qualitative | Quantitative |
|---|---|---|
| Muhar *et al.* (2008) | Riverbed widening and re-construction of former side channel at 1900 m river length yield improvement of habitat and fish assessment scores by one quality class. Other restored reaches/sites showed minor improvements. | % aquatic habitat area and fish ecological status highly correlated ($R^2 = 0.81$; $n = 6$) |
| Riley and Fausch (1995) | Abundance and biomass of adult trout (age-2 and older), and often juveniles (age 1) as well, increased significantly in the treatment sections of each of the six streams after log drop structures were established/ introduced. Patterns of change in trout biomass were similar to abundance changes in all streams. | |
| Roni and Quinn (2001a,b) | Juvenile Coho salmon densities were 1.8 and 3.2 times higher in treated reaches compared with reference reaches during summer and winter, respectively. The response of Coho density to LWD placement was correlated with the number of pieces of LWD forming pools during summer and total pool area during winter. | Summer: CDR = 0.59 × LWD − 0.01; $R^2 = 0.25$, $n = 27$, CDR: Coho salmon density response<br>SDR = −0.83 × PAR + 0.15; $R^2 = 0.45$, $n = 20$; SDR: age 1+ steelhead trout density response<br>Winter: JDR = 0.25 × PAR + 0.04; $R^2 = 0.27$, $n = 24$; JDR: juvenile Coho salmon density response; TFR = −0.42 × PAR + 0.21; $R^2 = 0.20$, $n = 20$; TFR: trout fry density response |
| Roni *et al.* (2006) | Both Coho salmon and trout response to boulder weir placement were positively correlated with difference in pool area ($p < 0.10$), while the response of dace and young-of-year trout response to boulder weir placement was negatively correlated with difference in LWD ($p < 0.05$). | Pearson's $r$ significant at $p < 0.1$ (*) and $p < 0.05$ (**)<br>% pool area/Coho abundance: 0.51*; % pool area/trout abundance (>100 mm length): 0.54*; LWD/dace: −0.77**; LWD/trout abundance (<100 mm length): −0.70** |

subsequent years to record the biological succession or return to assumed equilibrial conditions in the treatment reaches (Herbst and Kane, 2009; Riley and Fausch, 1995). Three references reported sampling during different seasons to gain information about within-year (seasonal) variability of the fish communities (Cederholm *et al.*, 1997; Jungwirth *et al.*, 1995; Roni and Quinn, 2001a,b).

Moerke *et al.* (2004) observed initial improvements in habitat quality, which then declined over the following 5 years due to insufficient sediment trapping upstream: that is, the restoration itself was transient. Roni *et al.* (2008) stated that the potential benefits of most instream structures will be short-lived (<10 years) unless coupled with riparian planting or other process-based restoration activities supporting long-term recovery of key ecological and physical processes. Merz *et al.* (2008) found aquatic macrophytes to immediately (i.e. within several months) cover up to 70% of spawning gravel that was introduced to support Chinook salmon (*Oncorhynchus tshawytscha* [Walbaum, 1792]); spawning use of the section by salmon was significantly reduced. This highlights that different elements of the biota will respond with very different rates, and is suggestive that bottom-up recovery via basal resources occurs more rapidly than recovery at the higher trophic levels.

In summary, the available literature shows two main trends. First, the vast majority of restoration studies on instream mesohabitat improvements are short-term and, thus, cannot provide insight into long-term, intergenerational effects. Second, the few references including long-term monitoring results largely imply that habitat enhancement and related biological effects of the most frequent restoration measures (introduction of LWD and boulders) are vulnerable to environmental impacts beyond the scale of restoration. Such impacts include hydrological and geomorphological processes and are largely controlled by catchment land use and land cover upstream (Larson *et al.*, 2001; Shields *et al.*, 2008) and, thus, these often overlooked factors require further consideration in future long-term reach-scale restoration schemes.

## D. Reasons for Failure and Limiting Factors When Restoring Instream Habitat Structures

As perceived 'failures' (i.e. non-significant or opposite-to-expected statistical results) rarely reach the status of publication, existing studies are inevitably biased towards 'positive' findings. It is therefore striking that even the published peer-reviewed literature contained so few clear examples of ecological benefits from undertaking instream mesohabitat restoration: roughly half of the reviewed references imply failures of measurable habitat improvement, biological recovery or both, when it comes to long-term enhancement and recovery. This does not necessarily mean that these studies clearly reported failures, but we simply cannot judge on the level of success of many studies because of their limited monitoring efforts.

The list of potential limiting factors of mesohabitat improvement and biological recovery is comprehensive. Six main aspects were frequently proposed: (i) inappropriate scaling of restoration, (ii) inappropriate timing of monitoring, (iii) inappropriate implementation of restoration (which may also be driven by practical constraints, for example, conflicting stakeholder interests), (iv) inappropriate indicators and indicator groups, (v) confounding effects of natural variability and (vi) presence of multiple pressures not addressed by restoration.

Palmer *et al.* (2010) and Entrekin *et al.* (2008) pointed at the overall level of watershed deterioration, which was considered particularly relevant to determining the success of mesohabitat enhancement in studies on the addition of LWD. Surrounding land use, substrate composition, temperature and the method of LWD placement are variables that interact and influence the recovery of stream biota: thus, only the watershed perspectives can provide insight into whether a project will succeed (Entrekin *et al.*, 2008). In addition there are also biotic aspects that require consideration. In densely populated, intensively used areas the establishment and recovery of populations at restored sites is likely to be governed by complex interactions between tolerant species already present at a site, newly arriving sensitive (targeted) species and possibly also invasive alien species not targeted by restoration. These biotic interactions are poorly understood and currently render prediction of community-level restoration effects challenging.

According to Miller *et al.* (2010), who published a comprehensive meta-analysis of the effects of mesohabitat enhancement on benthic macroinvertebrates, there is a 'myriad of weakly replicated, inconclusive, and even conflicting published studies' (p. 16). The authors pointed at some general flaws in restoration science (e.g. lack of sound study design including inappropriate replication, or publication bias) and questioned the methods to evaluate treatment effects (see also Shields, 2003). Study designs lacking pre-restoration data render impacts on macroinvertebrates questionable as these communities vary naturally at small (and larger) spatial scales. Further, conclusions about restoration significance remain inconclusive if only unrestored (still impacted), but not undisturbed (pristine) controls are being used to detect the effects, at least where control sites would be available. Such undisturbed control sites might be located in the same river upstream of a restored site or in a different river of the same stream type. Only the inclusion of undisturbed control sites in the sampling design (i.e. BACI design, see above) will allow for a more logically defensible partitioning of restoration effects from those arising from natural or other sources of variation (e.g. seasonal and inter-annual variability).

# V. RESTORATION BY REMOVAL OF WEIRS AND DAMS (<5 M HEIGHT)

Among 31 restoration studies (and five additional general ecological papers), the majority examined the effects of weir removal at North American streams, and only one restoration study (Tszydel *et al.*, 2009) and two reviews originated from Europe (de Leaniz, 2008; Schmitt, 2005). Most of the studies sampled several stretches per measure, each of which was several hundred metres long and together spanned stream sections of one to a few kilometres in length. The comparison of conditions before and after restoration was common to all restoration studies, while roughly half of the studies applied a full BACI design. The ecological effects of weir removal were comprehensively reviewed by Bednarek (2001), who also presented a series of case studies to underpin their review with empirical data.

Bunn and Arthington (2002) stressed the role of flow as a major determinant of physical habitat in streams which, in turn, was a major determinant of biotic composition. Acreman and Dunbar (2004) referred to the flow regime required in a river to achieve desired ecological objectives, that is, the 'environmental flow', and multiple elements of which are important (Poff, 1997). Low flows provide minimum habitat for resident species, medium flows sort river sediments and stimulate fish migration and spawning, and floods maintain channel structure and allow movement between floodplain habitats (Acreman and Dunbar, 2004). Occasional floods reconnect the aquatic and riparian habitat (Shuman, 1995), and backwaters are refilled. Fine materials (e.g. sand, silt, mud) erode and uncover coarser substrata (e.g. gravel, pebble and cobbles), which enhances the overall habitat diversity (Born *et al.*, 1996; Kanehl *et al.*, 1997) and dissolved oxygen and water quality improve (Hill *et al.*, 1993). Bednarek (2001), however, also referred to some negative effects, such as the downstream transport of contaminated sediments or the overall abrasive effect of fine sediment movement, although these adverse effects are typically short-term, whereas overall improvement is more likely to occur in the longer term.

The changing abiotic conditions may improve biodiversity and reproduction of fish: the spawning grounds for salmonid species increase (Iversen *et al.*, 1993), while fish passage is more likely because of the restored longitudinal connectivity. Migratory fishes depend on the re-establishment of hydrological connectivity, which was a common key argument for restoration (Iversen *et al.*, 1993; Poff, 1997; but see also de Leaniz, 2008 for a more recent summary of findings during the past decade). The maintenance not only of the longitudinal but also of the lateral connectivity with the floodplain is essential to the viability of populations of many riverine species (Bunn and Arthington, 2002; Bushaw-Newton *et al.*, 2002; Maloney *et al.*, 2008).

Stanley and Doyle (2003) suggested weir removals may be best considered as ecological disturbances, as removal of small dams generally results in the transformation from lentic to lotic river systems upstream, leading to the reservoir sediment release and a pulse downstream, which could cause short-term reductions in productivity and possibly diversity (Bednarek, 2001). In addition, effects of restoration could be very variable depending on the hydrologic nature of the river (Chaplin, 2003). As a result, the effectiveness of a dam removal, that is, the recovery of a river from the induced disturbance, is likely to vary widely among systems and to be contingent upon both temporal and spatial scales of the restoration scheme.

The literature provided little information on the effectiveness of such restorations: it was rarely measured and the judging criteria were usually vague. In most cases, negative impacts of weir removal were rather short-term (e.g. increase in suspended sediments) while the assumed beneficial changes were likely to act in the longer term (e.g. increase in flow diversity, connectivity); the natural free-flowing state of the river was always regained whereas recovery of the biota following this habitat shift was more uncertain. Five consistent effects of weir removal were identified relating to: morphology (width and depth); substrate particle size (and gravel bars); flow diversity (and turbidity); temperature and connectivity (Figure 11). Several studies

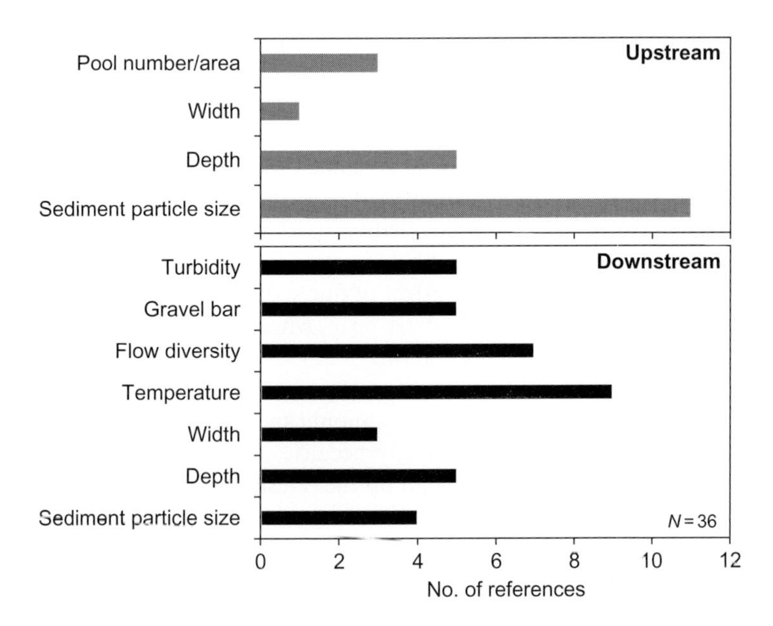


**Figure 11** Ranking of the most important environmental state variables based on the linkages (arrow thickness in Figure A3), derived from 36 references on weir removal.

(e.g. Chaplin, 2003; Cheng and Granata, 2007) found an increase in sediment particle size upstream and a decrease downstream of the former weir and other studies commonly reported an increase in flow diversity upstream (e.g. Hill *et al.*, 1993), decreased water temperature upstream (e.g. Hill *et al.*, 1993; Kanehl *et al.*, 1997) and restoration of the hydro-ecological connectivity (e.g. Gregory *et al.*, 2002; Poff, 1997).

In summary, the removal of weirs and small dams re-establishes natural physical river characteristics with some evidence of positive effects for the instream fauna, in particular for migrating fishes for whom it is arguably a general prerequisite for restoring their populations. The full beneficial effects of weir removal (e.g. Gregory *et al.*, 2002), however, may take decades to be manifested, while some adverse effects are likely to dominate in the short-term due to the mobilisation of fine sediments (e.g. Orr *et al.*, 2006; Pollard and Reed, 2004; Thomson *et al.*, 2005) accumulated in the former stagnant section (i.e. upstream of the weir). The deposition of this material further downstream on gravel areas, included artificial riffles introduced in other restoration schemes, may limit the availability of spawning habitat for fish in the short-term (i.e. up to several years), so spatial and temporal aspects of this form of restoration are again important (e.g. weirs should not be removed close to the spawning season).

## A. Which Organism Groups and Group Attributes Have Shown Evidence of Recovery After Restoration?

The biological impact of weir removal was studied most often for benthic macroinvertebrates (83% of all references), and, to a lesser extent, aquatic macrophytes and fishes (58% and 50%, respectively); phytobenthos was rarely considered (Figure 12). Most commonly, community composition and abundance measures were used to indicate changes due to restoration, irrespective of the organism group. Some papers also considered effects on community functional metrics such as benthic macroinvertebrate feeding habits (e.g. Maloney *et al.*, 2008) or fish growth (e.g. Harvey and Stewart, 1991; Schlosser, 1982). Twelve papers studied the effects of weir removal on sensitive and tolerant benthic macroinvertebrates (mainly EPT taxa: Ephemeroptera–Plecoptera–Trichoptera) and the effects of water quality improvement, such as the abatement of turbidity and oxygen enrichment (Bushaw-Newton *et al.*, 2002; Orr *et al.*, 2006). Changes in the macrophyte community were most often associated with changes in channel morphometry (depth, width) and connectivity (e.g. Shafroth *et al.*, 2002), while the composition and abundance of benthic algae responded to changes in sediment size and turbidity (e.g. Baattrup-Pedersen and Rijs, 1999; Orr *et al.*, 2006).

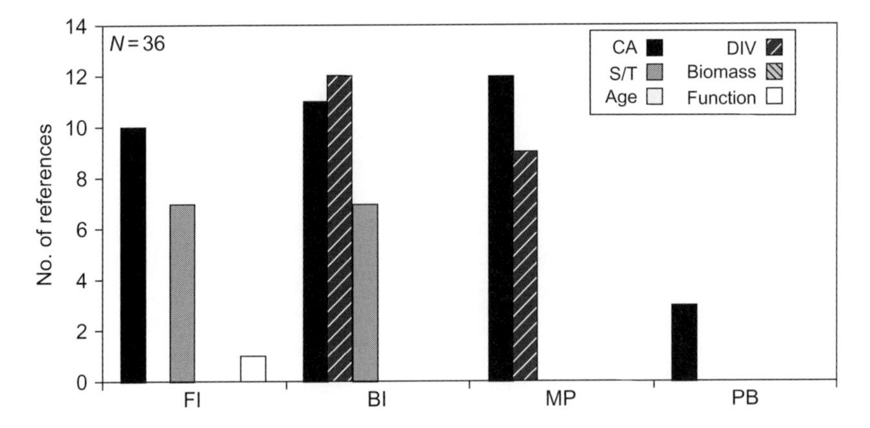


**Figure 12** Weir removal: number of references addressing the community attributes composition/abundance (C/A), sensitivity/tolerance (S/T), age structure (Age), diversity (Div), biomass and function of fish (FI), benthic macroinvertebrates (BI), macrophytes (MP) and phytobenthos (PB). As a study may refer to more than one community attribute, the overall number of references exceeds the number of 36 restoration references reviewed.

## B. Was There Evidence for Strong Qualitative or Quantitative Linkages?

All restoration studies provided qualitative analyses (Table 3). Nonetheless, sound statistical approaches (ANOVA, ordination) were frequently used to detect and identify patterns of biological impact (e.g. Bushaw-Newton *et al.*, 2002; Pollard and Reed, 2004; Thomson *et al.*, 2005). Cheng and Granata (2007) showed that following removal of a dam, bed deposition and scouring caused a 30% decrease in bed slope and a 40% decrease in bed material size downstream, compared with pre-removal conditions. These impacts reflect the similarly 'strong' linkages reported by Hill *et al.* (1993), Bushaw-Newton *et al.* (2002) and Stanley *et al.* (2002), who revealed a consistent decrease in water temperature upstream, leading to an increase of dissolved oxygen conditions that are likely to favour certain benthic macroinvertebrate and fish species.

In summary, considerable effort has been devoted to investigating the effects of weir removal on riverine systems during the past decade, although there remains a general lack of quantitative results that might help elucidate the mechanistic relationships and provide the means for predictions of the impact of weir removal not only on community recovery but also on community status and recovery further downstream.

**Table 3** Examples of qualitative evidence for the effectiveness of weir removal and related instream modifications (no quantitative evidence found in the reviewed literature)

| Reference | Type | Qualitative | Quantitative |
|---|---|---|---|
| Kanehl *et al.* (1997) | Active restoration | After dam removal depth varied considerably following flow variations, rocky bottom increased upstream, bank stability increased upstream and decreased downstream, habitat quality index scores increased dramatically. Short-term effects on fish biomass: increase upstream/long-term effects: general increase. | |
| Bushaw-Newton *et al.* (2002) | Active restoration | Increased sediment transport has led to major changes in channel form in the former impoundment and downstream reaches, leading benthic macroinvertebrate and fish assemblages to shift dramatically from lentic to lotic taxa. No significant upstream–downstream differences in dissolved oxygen, temperature, or most forms of nitrogen (N) and P, were observed either before or after dam removal. | |
| Hart *et al.* (2002) | Review | The overall objectives of this article are to assess the current understanding of ecological responses to dam removal and to develop a new approach for predicting dam removal outcomes based on stressor–response relationships. | |
| Pizzuto (2002) | Review | If the impoundment contains relatively little sediment and is significantly wider than equilibrium channels upstream and downstream of the dam, then the primary processes above the dam are likely to be deposition and floodplain construction rather than erosion and incision. Increased sediment supply at the reach scale could destroy alternate bars, pools and riffles, and armoured beds. | |
| Shafroth *et al.* (2002) | Review | Following dam removal, large areas of former reservoir bottom are exposed upstream and may be colonised by riparian plants. Transport of upstream sediment may lead to a pulse of sediment deposition downstream, which combined with increased flooding, may both stress existing vegetation and create sites for colonisation and establishment of new vegetation. | |
| Pollard and Reed (2004) | Active restoration | Cobble habitat without silt generally supports higher taxonomic diversity than do silted areas. | |

(*continued*)

**Table 3** (*continued*)

| Reference | Type | Qualitative | Quantitative |
|---|---|---|---|
| Doyle *et al.* (2005) | Review | Changes in channel form affect riparian vegetation, fish, macroinvertebrates, mussels, and nutrient dynamics. | |
| Thomson *et al.* (2005) | Active restoration | Downstream sedimentation following dam removal can reduce densities of macroinvertebrates and benthic algae and may reduce benthic diversity, but for small dams such impacts may be relatively minor and will usually be temporary; benthic macroinvertebrate density was significantly lower at downstream sites after complete removal than during pre-removal or partial removal stages, but remained relatively constant at upstream sites (ANOVA); benthic macroinvertebrate assemblages were studied using the NMDS ordination method. | |
| Cheng and Granata (2007) | Active restoration | After weir removal, net sediment deposition occurred downstream of the dam, and net erosion occurred in the reservoir resulting in bed deposition, and scouring in the reservoir accounted for a decrease in the bed slope of 30% and bed material sizes downstream at least 40% finer than pre-removal conditions; bed deposition and scouring in the reservoir accounted for a decrease in the bed slope of 30% and bed material sizes downstream were at least 40% finer than under pre-removal conditions. | |
| Maloney *et al.* (2008) | Active restoration | Following the breach, relative abundance of Ephemeroptera, Plecoptera and Trichoptera (largely due to hydropsychid caddisflies) increased upstream, probably because the increased flow and particle size in former impoundments favour filter feeding taxa that cling to substrate (e.g. hydropsychid caddisflies). | |
| Burroughs *et al.* (2009) | Active restoration | Sediment fill incision resulted in a narrower and deeper channel upstream, with higher mean water velocity and somewhat coarser substrates. Downstream deposition resulted in a wider and shallower channel, with little change in substrate size composition. Water velocity also increased downstream because of the increased slope that developed. | |
| Tszydel *et al.* (2009) | Active restoration | Riparian and land plants developed intensively at the bottom of the Drzewieckie Reservoir immediately after it was emptied. Short-term flow fluctuations usually diminish the quality and quantity of benthos. | |

### C. What Is the Timescale of Recovery?

Recovery of the longitudinal connectivity after removal of a weir or dam is immediate, as is the effect on water temperature: it will immediately start changing back to natural (free flowing) conditions. In contrast, biological recovery in general requires several years or even decades after removal and is expected to occur once the fine sediments have been transported farther downstream (e.g. Thomson *et al.*, 2005). This effect depends largely on the quantity of sediments accumulated above the barrier, water velocity, the gradient of the riverbed, and also on the specific technique of weir (dam) removal (Bednarek, 2001). According to Bednarek (2001), full recovery may take up to 80 years, but the literature rarely includes *post hoc* monitoring for longer than 5 years. The timescale and trajectory of recovery after weir removal thus remains speculative, in the absence of long-term monitoring data.

### D. Examples of Failure and Limiting Factors When Removing Weirs

Many organisms are limited in their recovery by restricted habitat availability and potential barriers within the river channel. A re-establishment of habitat variability requires geomorphological processes similar to pre-damming conditions (Doyle *et al.*, 2005), which may be key for facilitating fish reproduction, which is often limited by a shortage of suitable habitats to complete their life cycle (i.e. habitat for spawning, nursery, foraging). If geomorphological degradation, however, is irreversible, ecological recovery will hardly be possible without the management of natural geomorphological and hydrological processes (e.g. sediment and flow dynamics). The size of the barrier is also critical: Orr *et al.* (2006) concluded that the initially negative effects of the removal of small dams were trivial and short-lived relative to the natural variability of the entire system (see also Thomson *et al.*, 2005).

## VI. CONCEPTUALISING RESTORATION EFFORTS

### A. The General Conceptual Framework

Throughout our review, we frequently encountered common limitations in the restoration literature that often prevented clear conclusions from being drawn, in order to develop and design more ecologically effective restoration schemes. This was because most studies either focused on abiotic effects (e.g.

habitat enhancement) or on biological/ecological effects (mainly species richness and abundance), but rarely integrated both. Studies that reported entire cause–effect chains, for instance, from a specific restoration measure, to altered habitat variables, and ultimately up to the effects on community characteristics of the instream fauna and flora, were scarce (e.g. Jungwirth *et al.*, 1993, 1995). Such comprehensive cause–effect chains would clearly provide far better mechanistic understanding of the links between the biota and the physical environment and, more practically, would help water managers gain deeper insight into the processes and dynamics needed to achieve measurable and meaningful ecological improvements at a given site. Moreover, if key habitat variables and processes could be identified *a priori*, practitioners could then focus on restoration measures specifically tailored to their particular management objectives. At present, there is no coherent framework to help guide restoration towards replicable, ecologically effective cause–effect solutions.

We addressed this current shortfall by attempting to conceptualise and structure the results of our review to develop a theoretical framework that could also be applied usefully in the real world. The task involved linking management practices with hydrology, morphology and eventually with ecology, as far as was possible within the constraints of our current knowledge. In this section, we aim to develop such a general framework and then apply it to the reviews of the three different restoration measures presented in Sections III–V.

In order to keep the framework as parsimonious and straightforward as possible, we broadly followed the general *Driver-Pressure-State-Impact-Response* scheme (DPSIR; EEA, 2007) that is commonly used to link causes and effects in socio-economic and ecological contexts (e.g. Elliott, 2002; Karageorgis *et al.*, 2005). The main advantage of this approach is its simplicity, which facilitates communication with non-scientists and the wider end-user and stakeholder communities (Stanners *et al.*, 2007). The following two examples illustrate the general way in which the scheme is applied in our study:

1. Society's food demand, for instance, is a *Driver* of agricultural land use. In order to increase food production, fertilisers and pesticides are applied to the crops, which, through stormwater and groundwater runoff, are partly flushed into adjacent rivers and lakes, where they cause pollution and eutrophication (*Pressure*), leading to water quality deterioration (*State*). Eutrophication has a stimulating, direct effect on the growth of instream flora, but can also negatively affect the aquatic fauna (fishes, benthic macroinvertebrates) when decomposers start depleting oxygen (*State*). In parallel with eutrophication and contamination, rivers in agricultural landscapes are morphologically modified and hydrologically

regulated (*Pressure*). As a result, microhabitats and flow regimes may change negatively (*State*) and affect the flora and fauna (*Impact*).
2. Following high population density and its demand for food (*Driver*) weirs and dams (*Pressure*) are built to control the ground water levels (*State*), but disrupt the longitudinal connectivity of the system (*State*). The flow conditions are altered and sections of the river may become stagnant (*State*). Land use is often extended to the river banks and inhibits the development of a natural (vegetated) riparian buffer. Solar radiation is increased and stagnant sections may heat up exceptionally during summer. As a consequence of changing *States*, the riverine fauna and flora are being disrupted, sensitive taxa disappear (*Impact*) and a few tolerant taxa become dominant in the system (*Impact*).

Restoration and mitigation measures are required to reverse degradation and to improve ecological status. In example 1, for instance, best-practice agriculture (*Response*) might reduce the amount of fertilisers applied per area to the amount that is equivalent to the plant biomass produced per area. In example 2, hydromorphological conditions might be actively restored (*Response*) to a more diverse habitat and flow regime and land use in the riparian zone might be abandoned (*Response*) to promote the natural development of a diverse riparian corridor, that is, a mixed buffer strip with grasses, shrubs and trees.

The *Response* component is represented by the three restoration measures reviewed in Sections III–V and thus constitutes the starting point of these sections, from which we attempted to construct the cause–effect links, that is, *Response-State-Impact* chains (Figure 13). A specific restoration measure or any other kind of ecosystem management was considered to have a positive effect on environmental conditions (*State*), which in turn should have had a positive *Impact* on the biota, that is, *Recovery*, which, in its strict sense refers to the full recovery of community structure and ecosystem functioning to the condition prior to degradation (Henry and Amoros, 1995). Hence, this extension results in the DPSIRR scheme, that is, the *Driver-Pressure-State-Impact-Response-Recovery* chain, which underpins our new conceptual framework. Our focus here is on the part of the scheme illustrated in Figure 13, that is, the initial degradation part of the chain (*Driver-Pressure-Impact*) is not considered further, as these are far more widely known than the latter part (Friberg *et al.*, 2011; Layer *et al.*, 2011).

In particular, we were interested in the effects of river restoration and management measures on physico-chemical, hydrological, and morphological conditions (*State*) and eventually on the *Recovery* of the instream flora and fauna (*Impact*). This information was derived from our literature database and summarised in three conceptual models (see Appendix A), with one model derived per restoration measure (as defined in Sections III–V).

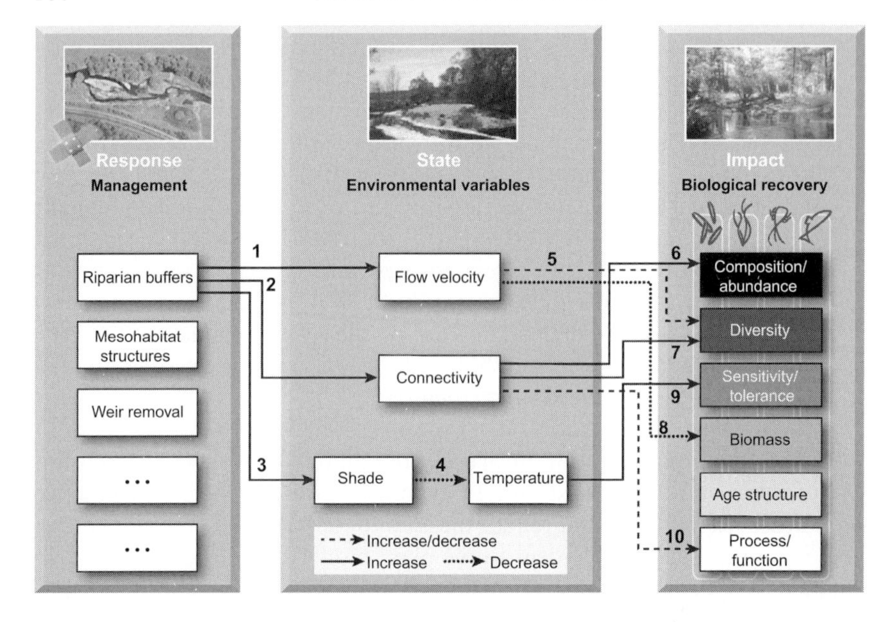


**Figure 13** General conceptual framework of *Response–State–Impact* chains, designed to help structure the cause–effect relationships between management and biological recovery. These links are likely to be of indirect nature (i.e. they connect via environmental *State* variables), but direct linkages may be possible, for instance, with biomanipulation and fish stock management. Each link refers to a relationship referenced in the reviewed restoration literature (see Appendixes A and B for conceptual models and respective literature references). Relationships may be either positive (solid lines), negative (dashed lines) or equivocal/ambiguous (dotted lines). The *Recovery* characteristics 'composition/abundance', 'diversity', 'sensitivity/tolerance', 'biomass', 'age structure' refer to those defined by the WFD, amended by the group of biological measures of 'functions' (and processes). See the text for further explanations.

## B. Response–State–Recovery Variables

The *Response* variables (restoration/management measures) considered here are riparian buffer zones, instream habitat enhancement and weir removal.

The *State* variables are those abiotic (physico-chemical, hydrological, morphological) variables that change due to the effects of a management measure. The most common *States* referred to in the reviewed literature are presented in Figures 5, 9 and 11 for the three types of restoration measures we investigated, respectively. Management measures and recovery might be linked via a single *State* variable (e.g. chain 1–5 in Figure 13) or along several interrelated *State* variables (e.g. chain 3–4–9 in Figure 13).

Six variable types were applied to the four organism groups (fishes, benthic macroinvertebrates, macrophytes and phytobenthos) to describe biological *Recovery*:

1. composition and abundance (e.g. composition of specific taxa, total community abundance),
2. sensitive and tolerant taxa (e.g. number of salmonid fish species, number of Ephemeroptera–Plecoptera–Trichoptera [EPT] taxa among benthic macroinvertebrates, abundance of red-bodied chironomid taxa),
3. diversity (e.g. diversity indices, taxon richness),
4. age structure (e.g. relative abundance of young-of-the-year, larval development in benthic macroinvertebrates),
5. biomass (e.g. fish catch biomass, phytoplankton biomass/biovolume expressed as chlorophyll *a*)
6. processes and functions (e.g. species traits such as feeding types or body size, or measures of primary production and decomposition).

## C. Linking Components of the Conceptual Model

### *1. Linkage Type*

Each cause–effect linkage, that is, between two objects in the conceptual model, is illustrated by an arrow (Figure 13), which relates to a literature reference (coded as arrow number). Solid arrows indicate a positive relationship, that is, the effect variable at the tip of the arrow increases if the cause variable increases. Dashed arrows indicate a negative relationship, while dotted arrows reflect ambiguous or unknown relationships.

The linkages were further distinguished according to their quantitative or qualitative nature.

- A quantitative linkage refers to relationships with information on the degree of change (e.g. an increase of $x$ by 10% causes an increase of $y$ by 20%). Linkages that are based on either empirical statistical or mechanistic relationships (e.g. correlation coefficient and significance level provided; $y$ is $a$ times $x$) were also considered quantitative and rated stronger than qualitative relationships.
- A qualitative linkage was given if only the direction of a trend was reported (e.g. an increase of $x$ causes $y$ to decrease). Though unlikely to be useful for quantitative/mechanistic modelling, such information is considered important for the development of hypotheses. Qualitative relationships were also rated as being strong if supported by $>3$ references based on a sound sampling design that included replicate sampling.

### *2. Linkage Strength*

The strength of each linkage was estimated based on the number of supporting references that included this specific relationship, as well as on a 'rating' of the references (see Section VI.C.1), which was based on the number of site or sample replicates, the study design (before–after, control/impact or BACI), the statistical analysis applied, the significance of results, and the geographical coverage and representativeness of the study. If strong negative and positive implications were reported for a specific linkage, the relationship was considered to be equivocal and coded with a grey/dotted line, unless there was a clear dominance of supportive studies. We are aware of the subjectivity behind this approach, but given the low number of equivocal linkages (7), referred to by a comparatively low number of inconclusive studies, this is not considered important for the conclusions drawn from our models. Rating was done by three researchers only (Anahita Marzin, Christian K. Feld, Jochem Kail) based on cross-checking in order to keep the researcher-dependent variability low.

The conceptual models (Appendix A) show all the possible relationships of management (*Response*) measures within our scheme, their effects on different environmental *State* variables and eventually the *Recovery* of instream organisms as derived from the reviewed literature. It should be stressed, however, that no reference in the literature provided statistically significant evidence of an entire cause–effect chain from the *Response* measure via one or several environmental *States* to the biological *Impact*, and this highlights an obvious gap in our currently piecemeal, rather than holistic, understanding that clearly needs to be addressed in future studies. The majority of studies were limited to measuring environmental effects of a *Response* measure, while comparatively few studies measured the biological *Recovery*. Moreover, studies on biological effects often did not attribute their findings to changing environmental *States*, which renders the construction of cause–effect relationships from these studies difficult. Consequently, the conceptual models reflect an overall constructed summary of the state-of-the-art of the effects of river restoration derived from the literature (for details compare Section II). Indeed, a key output of the development of the models is to highlight precisely those research areas that need to be prioritised to redress these important gaps in our current knowledge and understanding.

## D. Application of the Conceptual Framework

Each of the three conceptual models consists of two main components: a graphical illustration of cause–effect chains based on the general conceptual framework shown in Figure 13, and a corresponding tabular compilation of the characteristics of the reviewed body of literature with references allocated to link numbers (arrow numbers) in the conceptual models (compare Appendixes A and B). Consequently, each number represents a link (arrow) that can be referenced by one or more restoration studies. Both the conceptual models and the corresponding literature are provided in Appendixes A and B, respectively. The arrow thickness in the complete models is equivalent to the number of references that address the individual linkage (arrow). This illustrative approach is designed to facilitate an intuitive and immediate identification of well-referenced cause–effect chains, as reported in the literature.

## E. Are Cause–Effect Chains Detectable from the Conceptual Model?

The conceptual model derived for buffer strip effects reveals some complex relationships between the restoration of riparian vegetation, its environmental effects and eventually its impact on instream plant and animal communities. Three major chains are evident, via (i) enhanced nutrient/sediment retention, (ii) increased shading effects and subsequent temperature decrease and (iii) increased amounts of LWD on the stream bed.

The conceptual model of mesohabitat enhancement, that is, the introduction of substrates (LWD, boulders, spawning gravel) and deflectors, highlights the importance of three key hydromorphological *State* variables: habitat (substrate) diversity, the provision of cover habitat (for fishes) and pool availability (frequency and area). Flow heterogeneity, sediment retention and bank stability were also frequently enhanced. The studies, however, did not provide further information to link these *States* to biology and, thus, constituted only partial chains that ultimately resulted in dead ends in our conceptual model. This revealed not only important knowledge and research gaps but also raises the possibility that such assumed relationships might not exist, at least not at the spatio-temporal scales investigated. For example, the mesohabitat enhancement case exhibited many clear links between the measure 'placement of LWD' and the *State* 'sediment storage', but there was no link to the biological effects. Although there was no compelling evidence in the restoration literature that sediment storage does not enhance biological *Recovery*, it is not inconceivable that it is not, in fact, among the biologically effective (key) variables influenced by the introduction of LWD. Hence, these findings do not necessarily only represent knowledge gaps, and it may be that additional research continues to be unable to identify strong relationships here.

According to the conceptual model on weir removal this restoration type is different from the two other cases, since it is necessary to distinguish between upstream and downstream effects to understand the different processes involved. Indeed, effects of weir removal are often diametrically opposed upstream and downstream of the barrier, such as changes in substrate particle size. Altogether, four major cause–effect relationships were frequently found: (i) erosion of fine sediment at the former impounded section and deposition further downstream; (ii) increase of flow diversity upstream; (iii) decrease of water temperature upstream; (iv) restoration of hydro-ecological connectivity.

Overall, the application of the conceptual framework presented in Figure 13 provided a simple, but powerful tool for reviewing and assessing the existing literature in a more structured way than has been attempted previously. The criteria defined for the review, together with the datasheets compiled to illustrate the conceptual models, provide a sound and transparent basis for subsequent, more quantitative reviews, including formal meta-analyses (cf. Miller *et al.*, 2010). Further, the adoption and extension of the DPSIR scheme (in its more recent version published by EEA, 2007), with a strong focus on the known linkages between *Response*, *State* and *Impact*, facilitates the communication of results to practitioners, which has often been problematic in the past. Even moderately complex models can be used to illustrate well-known cause–effect chains as well as identifying key knowledge and research gaps. Recommendations can be made based on the descriptive analysis of objects (boxes) and relationships (arrows) in the conceptual models. For example, the conceptual model on riparian buffers (Figure A1) reveals strong evidence for riparian wooded vegetation to retain nutrients and fine sediment, to provide shade and decrease the instream water temperature, and to structure the instream habitat via the provision of LWD.

These effects on the *State* variables often have positive effects on the richness, diversity and abundance of instream fishes and benthic macroinvertebrates (*Impact*). The effects on aquatic macrophytes and benthic algae (*Impact*) were, in contrast, less well studied. Nevertheless, a clear focus on measures of richness and abundance was evident for all organism groups, which suggested a general suitability of taxa counts and densities as useful measures to indicate the effects of riparian buffer restoration. But presumably, this finding is also owed to the lack of suited indicators in restoration ecology that account for other aspects of the species and communities. Very often, indicators represent disturbance-sensitive species or taxonomic groups (e.g. salmonid fishes or stoneflies) as opposed to measures of structural and functional integrity (e.g. relationship of sensitive to tolerant species compared to natural conditions or the proportion of macroinvertebrate feeding types). That is, restoration monitoring focuses on the presence or absence of certain taxa rather than on the structural and functional characteristics of the environment and the communities within the system.

Restoration monitoring must not solely focus on the structural attributes (biodiversity, richness, abundance) of aquatic communities, and a more general goal must be to improve and maintain natural ecosystem processes (Palmer *et al.*, 2005). In addition, riverine habitat structures are largely controlled by processes such as erosion and deposition of sediments, and riparian organic matter (leaves, terrestrial insects, LWD) supplies allochthonous energy to the system's nutrient flux and contributes to the riverine food web (e.g. by supporting shredders and detrital feeders among the benthic macroinvertebrates; Hladyz *et al.*, 2011a,b). These and other functional aspects, however, remain largely ignored in ecological restoration, presumably because the complex array of processes makes them difficult and costly to implement as indicators of ecological status (Palmer, 2009; Palmer *et al.*, 2005), although their value is being recognised increasingly in biomonitoring schemes (Friberg *et al.*, 2011). Future research therefore needs to include both structural and functional measures that are capable of indicating whether natural processes and their dynamics do change towards the targeted restoration endpoints, or not, as the case may be.

The discussion of the advantages of conceptual modelling, as presented in this chapter, should not conceal its limitations. For instance, numerous cause–effect chains can be identified in the models that direct from the *Response* measure (e.g. buffer restoration) to biological *Impacts* (e.g. macroinvertebrate composition and abundance), over one or several *State* variables. Such 'complete' chains, however, are rarely supported by single restoration references, rather they are assembled piecemeal via a weight of evidence approach across studies, instead of via unequivocal experimental manipulations. Most studies focus only on a short part (e.g. one or two linkages) of such chains and, hence, the overall conceptual model can be considered a puzzle in which each study refers only to one, or a few pieces at most. Nonetheless, we believe that the overall models provide sufficient evidence for the effectiveness of such constructed cause—effect chains to help guide future restoration efforts, and this in itself represents a marked improvement on previous, more anecdotal approaches. Moreover, the models do not present a final state; future monitoring of, and scientific studies on restoration will help test, adapt and improve the models to refine the framework even further.

# VII. RE-MEANDERING LOWLAND STREAMS IN DENMARK: LARGE SCALE CASE STUDIES

## A. River Restoration: Trial and Error?

In Denmark restorative mitigation measures were employed as early as the late 1980s to fulfil what were, at that point in time, very progressive legislative demands on river habitat quality (Iversen *et al.*, 1993). Since then, at least 200 million EUR have been spent on restoring spawning grounds for salmonids, removing obstacles for migration and re-meandering entire reaches, simultaneously with a >90% reduction of sewage inflow across the country. One of the world's largest restoration projects involving re-meandering of the River

Skjern was initiated in 2000 and completed in 2002 (Pedersen *et al.*, 2007; Pusch *et al.*, 2009). The comprehensive efforts were a strong signal to society of the importance placed on environmental issues by the Danish Parliament, and many of these are now enshrined in even more far-reaching international legislation, such as the EU WFD.

Despite being one of the smallest countries in Europe (total area: 43,000 km$^2$), Denmark is very densely populated and the landscape has, as a result, been heavily modified for generations: it is therefore a useful microcosm for studying human impacts on a larger European and even global scale. The extent and long history of river restoration projects (covering almost 30 years) in Denmark mean that the experiences gained regarding societal decision-making and the outcome of river restoration provide a template that can be applied in many other countries, as has already happened in the United States (Palmer *et al.*, 2007). Re-meandering projects (i.e. the re-introduction of natural riffle/pool sequences an re-construction of the original river course) at the scale of hundreds of metres to tens of kilometres are also well known from other regions (e.g. in the United States: Palmer *et al.*, 2005; Walter and Merritts, 2008) and are now the main restoration target in Denmark: this is the only natural channel plan form type occurring in a geology solely consisting of sandy and loamy moraine soils and yet for decades rivers had been straightened, deepened and channelised to remove these features, often as part of a system of engineering responses to the need for land drainage and flood protection (Sand-Jensen *et al.*, 2006). Here we discuss several comparable case studies (the 'good', the 'bad' and the 'ugly') in order to gain insight into the restoration-monitoring-appraisal process.

## B. The Good: River Skjern

River Skjern is the largest river in Denmark in terms of both discharge and drainage area (2490 km$^2$), and the total length of all river channels including tributaries is 1526 km. In 1987, the Danish Parliament decided to restore the lower reaches of the river and its valley, which included re-meandering of the river and re-establishment of the natural water levels and water level fluctuations. The new meandering river course increased the channel length by roughly a third to 26 km and involved removal of existing dykes and filling in of the old channelised reaches. Whenever possible, the restored river channel was positioned at the original river course as defined by old maps and aerial photographs.

Before first regulation works in 1901/1902, the River Skjern meandered dynamically across its floodplain. The channel width ranged between 65 and 100 m with a mean slope of 20 cm/km in the lower reaches (Rambusch, 1900). After the large-scale regulation in 1968, the width of the river channel was fixed at 45 m at the downstream reach below the confluence with the River Omme, and 30 m at the reach below Borris (Figure 14), and bank height was raised to 3.5–3.8 m along the regulated reach.

According to surveys of the macroinvertebrate communities in the decade prior to implementation of the restoration project in 1999, the river contained one of the most taxon-rich macroinvertebrate communities in Denmark, despite a pronounced alteration of the species composition during the twentieth century. Mayflies (Ephemeroptera) and stoneflies (Plecoptera), and to a lesser extent caddisflies (Trichoptera) exhibited a high number of species that were rare or even completely absent from other parts of the country. Several species, such as the mayfly *Brachycercus harrisella* Curtis, 1834 and the stonefly *Isoptena serricornis* (Pictet, 1841) had abundant populations in the River Skjern, but were threatened by extinction at a national level (Jensen, 1995).

After restoration in 2003, 46 new meanders were added to the river. The average cross-sectional area was reduced by 20–30%, the dominant depth interval decreased from 100–160 to 40–140 cm (Figure 15), and current velocities increased by up to 100% to around 30–60 cm $s^{-1}$ in the mid-channel. Initial effects on river habitats, macrophytes and macroinvertebrates were examined at survey reaches along the restored section using unaltered upstream reaches as controls. Just 2 years after restoration, macroinvertebrate diversity and abundance had reached pre-restoration levels (Figure 16). Overall taxon richness increased from 63 (in 2000) to 76 after restoration in 2003. Before restoration, black flies (Simuliidae indet.) were dominant and about twice as abundant as the non-biting midges (Chironomidae), whereas by 2003 three taxa dominated: Orthocladiinae indet., the caddisfly *Brachycentrus maculatus* (Fourcroy, 1785), and the mayfly *Heptagenia sulphurea* (Müller, 1776) (Table 4), and other insect species become more abundant after restoration (Pedersen *et al.*, 2007).

Restoration of the Skjern was a large-scale project involving both a substantial stretch of the river itself and 19 km of the lower valley. This allowed natural processes to recover and develop and facilitated the reversion of the river towards its pre-regulation state, with significant improvements occurring within just a few years, highlighting the potential resilience and responsiveness of the biota to effective restoration schemes. Prior to restoration, water quality was high and the potential existed for recolonisation by macroinvertebrates from remnant (source) populations further upstream of the restored section. Therefore, the ecological improvement was not significantly threatened by other pressures acting in parallel on the system, or by barriers limiting dispersal and recolonisation of benthic macroinvertebrates: that is,, habitat degradation was the rate-limiting step, whereas in other (e.g. highly polluted) rivers this might not be the case. Since restoration monitoring at River Skjern followed a BACI study design and included a true undisturbed

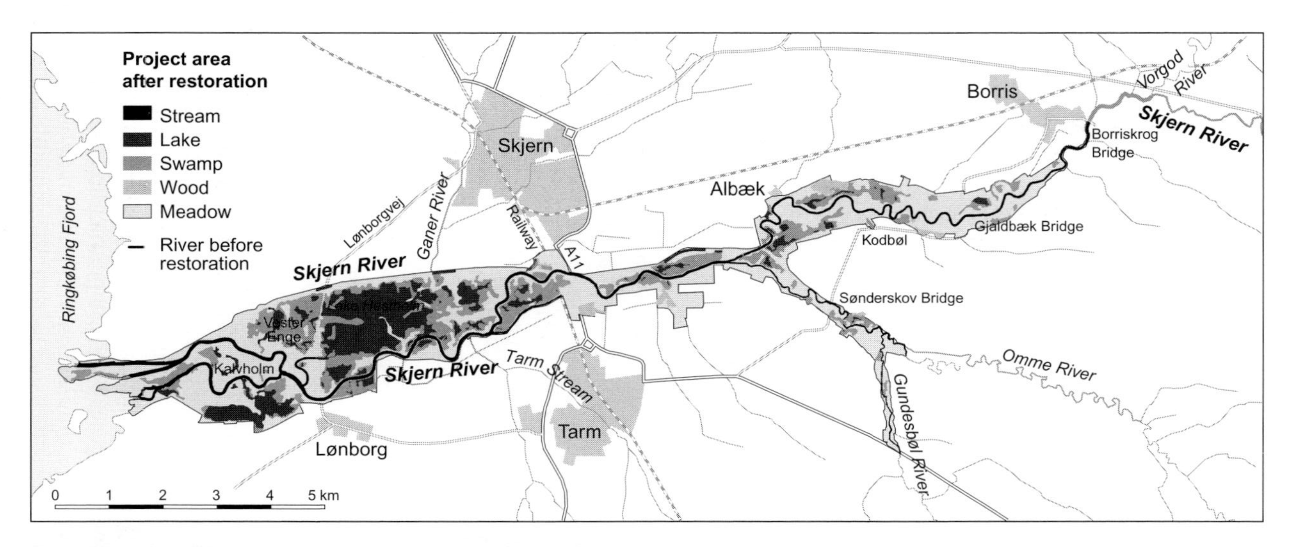


**Figure 14** The River Skjern restoration area, including main landscape/vegetation types, before restoration and after the restoration was finalised in 2002.

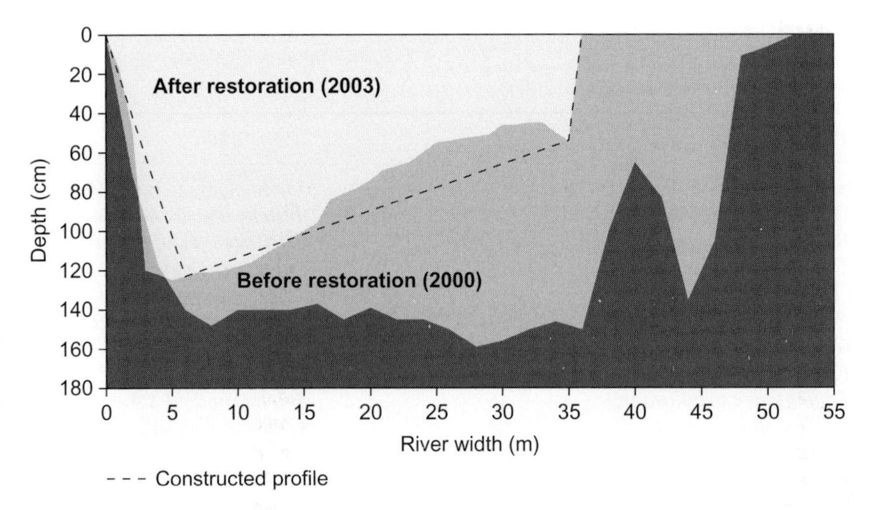


**Figure 15** Example of the changes to the cross-sectional profiles in River Skjern. The cross-sectional area has generally decreased by approximately 30%. The morphology of the profiles has changed from the constructed rectangular shape to a more natural physical appearance (from Pedersen *et al.*, 2007).

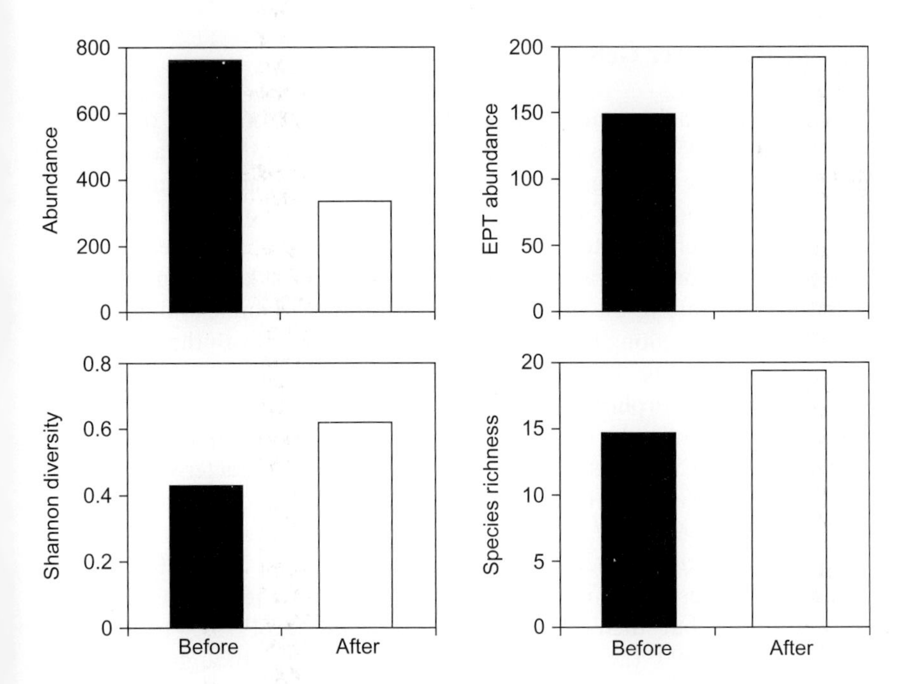


**Figure 16** Macroinvertebrate community variables before and after restoration of the River Skjern.

**Table 4** Frequency (percentage) of occurrence of the 10 most abundant macroinvertebrate taxa in River Skjern before the restoration in 2000 and after the restoration in 2003 (from Pedersen *et al.*, 2007)

| Taxa 2000 | Taxa 2003 |
|---|---|
| **Simuliidae indet. 45.2** | **Orthocladinae indet 16.7** |
| **Orthocladinae indet. 21.1** | ***Brachycentrus maculatus* 13.6** |
| ***Brachycentrus maculatus* 14.3** | *Heptagenia sulphurea* 11.7 |
| Tanytarsini indet. 5.5 | ***Gammarus pulex* 6.0** |
| **Oligochaeta indet. 1.5** | ***Elmis aenea* 5.7** |
| *Pisidium* sp. 1.5 | **Oligochaeta indet. 4.9** |
| ***Gammarus pulex* 1.3** | **Simuliidae indet. 3.5** |
| ***Elmis aenea* 1.0** | Corixinae indet. 1.8 |
| *Taniopteryx nebulosa* 0.9 | *Baetis rhodani* 1.4 |
| *Asellus aquaticus* 0.6 | Chironomini indet. 1.4 |

Bold taxa occurred in both samples.

control reach upstream, the changes and effects implied by restoration could be clearly assigned to the measures implemented.

## C. The Bad: River Gelså

River Gelså is a lowland river located in the southern part of Jutland, Denmark. In 1952, parts of the river were straightened and channelised in order to increase drainage efficiency and discharge capacity, thereby reducing the risk of flooding of agricultural land. In 1989, a restoration project was carried out to rehabilitate a 1.3 km straightened and channelised course of the river to a more natural meandering course over 1.8 km (Figure 17).

Before restoration, the instream substratum was unstable and dominated by sand transport along the riverbed, due to erosion of both the bed and the banks (Friberg *et al.*, 1994). The creation of 16 new meanders changed the stream channel morphology markedly (Figure 18), decreasing the channel width by 3–4 m and reducing discharge by almost 50% (from 6.6 to 3.5 $m^3 s^{-1}$). A similar reach 0.5 km upstream of the restored section characterised by similar physical, chemical and biological characteristics served as control reach (Friberg *et al.*, 1998).

After restoration, macroinvertebrate abundance in the restored reach immediately increased and peaked at >11,000 individuals $m^{-2}$ in 1991 (Figure 19A), but then subsequently decreased until 1995 to the level measured in 1990 (i.e. shortly after restoration) although abundance remained constant in the upstream control reach during the same period (Friberg *et al.*, 1994, 1998). In general, the taxonomic richness increased

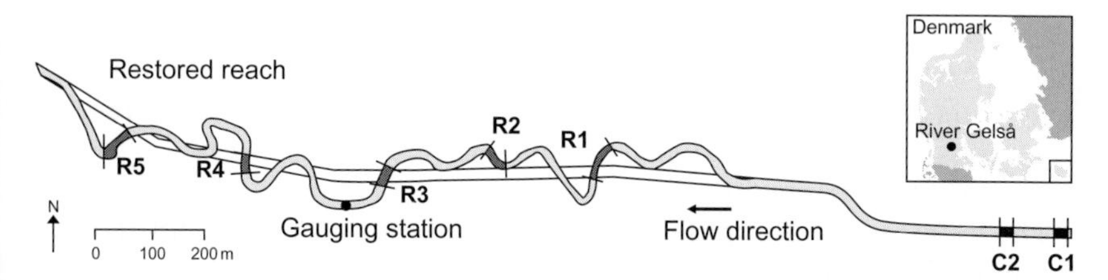


**Figure 17** Restoration of River Gelså at Bevtoft. Restored reaches included in the sampling are labelled R1 to R5. Upstream control reaches are labelled C1 and C2.

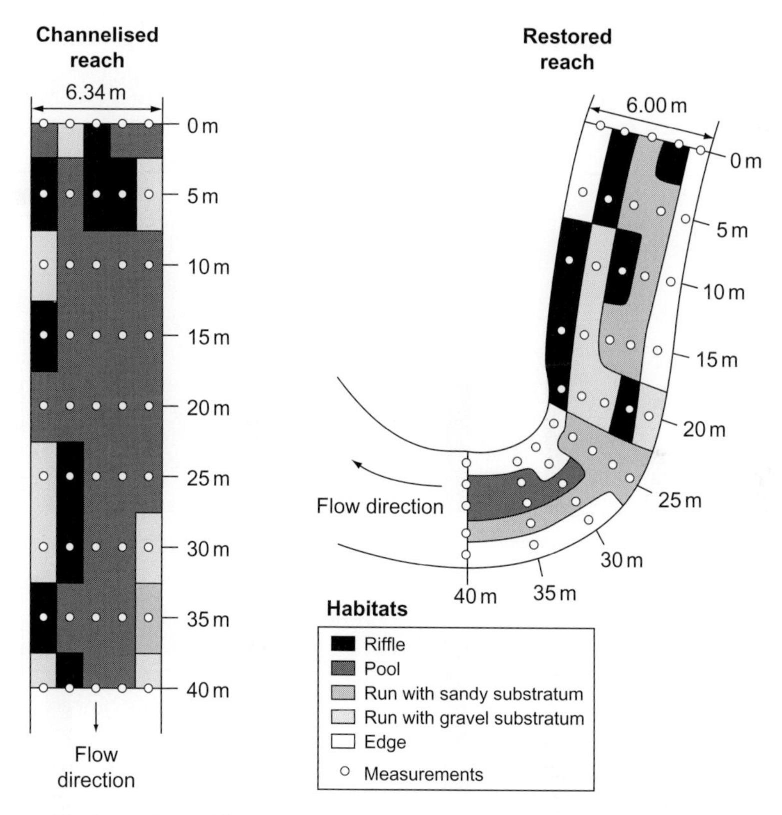


**Figure 18** Location of instream habitats in a channelised and restored section of River Gelså.

markedly in both restored and control reaches during the study period (Figure 19B), hence indicating only minor restoration effects on community abundance. Some species associated with coarse substrates and stones, and not recorded prior to the restoration, were only found in the restored reach, for example, the mayfly *Heptagenia sulphurea*. An initial colonisation lag phase of several years was observed for species with poor dispersal abilities, but 6 years after restoration (1995) the species composition of the two reaches was similar, indicating that the regional species pool was fully dispersed across the local communities within both reaches (Friberg *et al.*, 1998).

The results of the River Gelså restoration indicate that the restored reach had stabilised physically by 1993 and that the macroinvertebrate community seems to have largely recovered by 1995, as no major changes in community composition, density and diversity were recorded subsequently. At this point the community resembled those found in other parts of the river, as indicated in a comparison with the upstream reach. Since many invertebrates were univoltine this implies that it recovery was achieved within six generations of most taxa.

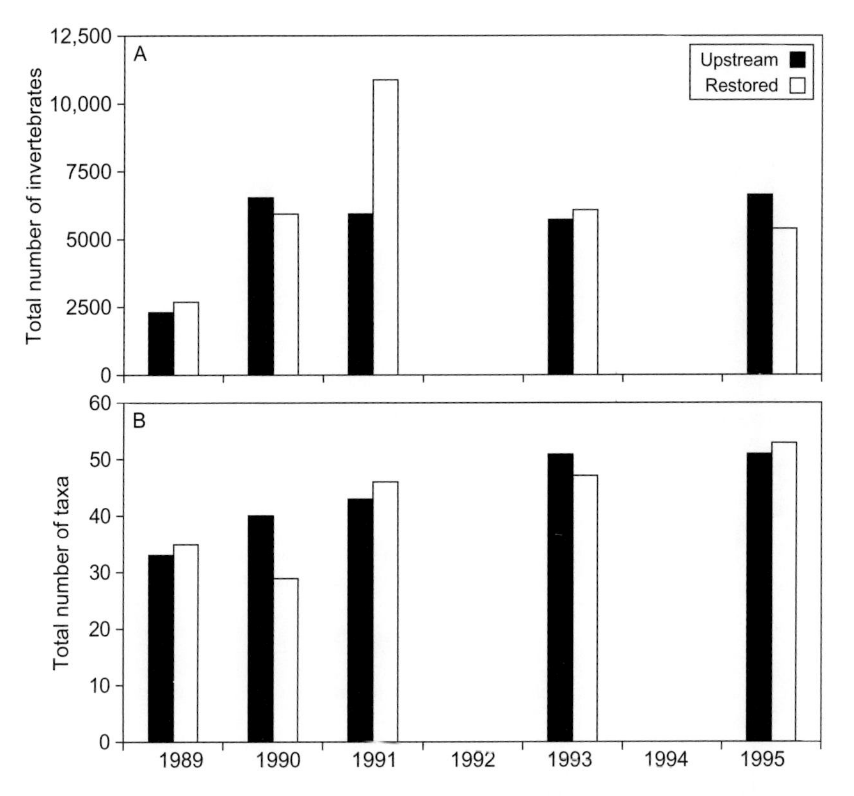


**Figure 19** Total number of individuals (A) and taxa (B) of macroinvertebrates found in kick-samples from the upstream and restored reaches of the River Gelså before (1989) and after restoration (1990–1995). No sampling was undertaken in 1992 or 1994 (from Friberg *et al.*, 1998).

Remarkably, the upstream ('control') reach also displayed signs of recovery during the study period too, because the reach was not confined and natural morphological processes were allowed to dynamically shape the physical conditions. Moreover, the local river authority stopped cutting of submerged plants (weeds) in both the restored and control reach in 1989. The aquatic macrophyte community significantly affected and improved instream habitat conditions (e.g. a more natural riffle/pool sequence) and served as 'ecological engineers' (Sand-Jensen *et al.*, 1996), improving the physical heterogeneity of the instream environment by increasing the variation in flow conditions. Thus, the abandonment of weed cutting at River Gelså, with respect to biological recovery, turned out to be as effective as restoration by re-meandering the river course - and for a fraction of the cost (Figure 20). Softer engineering measures (i.e. less and more scattered engineering works instead of constructing an entire new section) may help avoided biological disruption caused by the mechanical disturbance and excess sediment

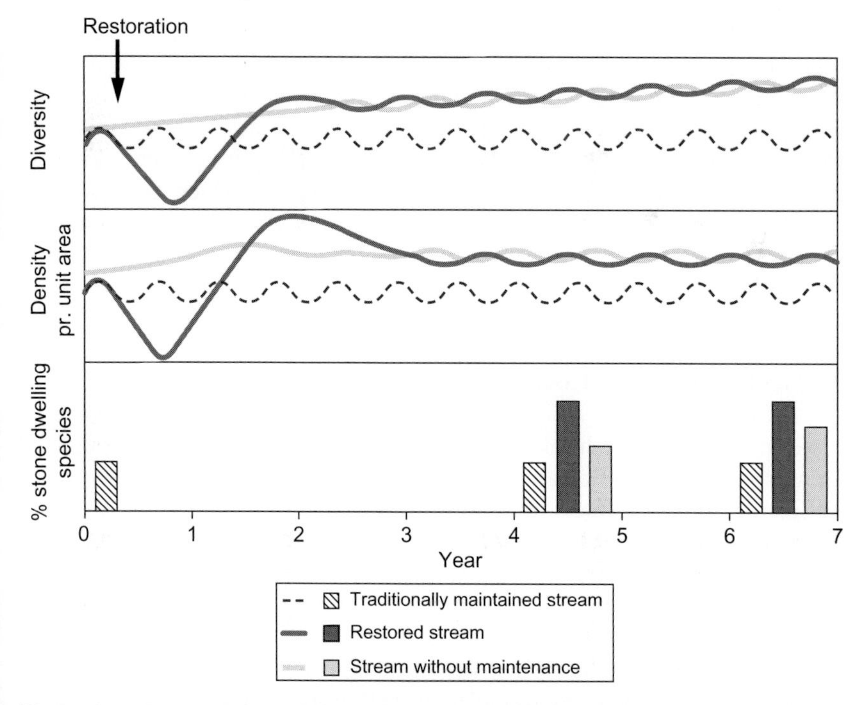


**Figure 20** Conceptual model of development of macroinvertebrate communities within a decade in an actively restored (re-meandered) stream, in a stream in which channel management (e.g. weed cutting) has ceased, and in a stream reach in which management is continued.

transport associated with construction works (Doeg and Koehn, 1994; Kronvang *et al.*, 1998). Further, despite the generally high resilience of many stream macroinvertebrates (e.g. Niemi *et al.*, 1990; Reice, 1985), removal of large areas of the riverbed could impair recovery.

The River Gelså restoration was one of the first re-meandering projects of its kind not only in Denmark but also internationally. As the effects were thoroughly monitored, important lessons were learned from this project, and many outcomes arose that were not anticipated nor necessarily desired. The new channel was extremely confined by boulders in the meanders and hence did not allow bed and bank erosion in the confined areas. The restored section hence rapidly turned into a static 'canal' without further morphological change. Eight years after restoration, biological recovery was insignificant, despite some earlier encouraging initial signs of biological improvement (Carl, 2000; Friberg *et al.*, 2000), suggesting that the entire project might be considered primarily 'cosmetic landscape gardening'.

In general, despite a vast financial outlay, like so many other restoration schemes, the conclusions from this study were compromised by an imperfect experimental design, with only one sampling occasion before restoration and the lack of a proper control site: the lesson that became apparent was that it is essential in future studies to also include either an unperturbed site, if possible, or at least a channelised reach that was maintained throughout the study period in a similar manner to the pre-restoration situation. Despite this early warning as to what should constitute best-practice, these points are still largely ignored in most current restoration schemes, which compromises their scientific value and undermines arguments about their (presumed) efficacy.

## D. The Ugly: Adding Coarse Substrates to Lowland Streams

Restoration studies on the effect of physical habitat enhancement constituted a major portion (45%) of all restoration studies reviewed and presented in Sections III–V. The effects of habitat enhancement were also studied using comprehensive data from six natural (reference), six channelised (control) and six re-meandered (restored) stream sections in Jutland, Denmark. The sections were located in different stream of comparable size (stream width: 2–5 m, depth: 0.5–0.7 m). Each section was 100 m long and covered 10–20 riffle/pool sequences, depending on the channel width. At re-meandered sections, coarse substrates (cobble, gravel) were added, the banks were re-profiled and, in some cases, the bed level was raised. All restoration measures aimed at creating a more 'natural' stream reach with riffle pool sequences, in line with historical maps; the measures were completed >3 years prior to the study. Point source pollution (e.g. fish farms, waste water treatment plants) was absent from all surveyed and upstream stream sections.

The restored stream sections had a higher proportion of gravel and stones relative to the natural stream sections, while sand cover was lower at restored sites (Figure 21). The abundance of Ephemeroptera, Plecoptera and Trichoptera (EPT taxa) was significantly lowest in channelised streams, intermediate in natural streams and highest in re-meandered streams (Figure 22). Many other taxa did not show a significant or systematic difference among natural, restored and channelised streams and overall macroinvertebrate abundance and diversity varied little among the stream types. However, several taxa associated with stone and gravel substrates were clearly favoured in re-meandered streams (Figure 23), including the river limpet *Ancylus fluviatilis* O.F. Müller, 1774 the mayfly genus *Baetis* and the caddis genus *Hydropsyche*.

Our results suggest that re-meandering schemes involving the excessive introduction of gravel and stone substrates created an artificially high percentage of such habitats, such that although to the naked eye they might appear natural, that is, meandering across the stream valley, a closer inspection revealed substrate conditions more typical of upland streams at far higher latitudes in Northern Scandinavia. Typically, coarse substrates are introduced to increase the availability of spawning grounds, particularly for commercially valuable salmonid fishes. Unfortunately, this blanket approach all too often ignores the hydraulic and morphological conditions that govern the natural occurrence and dynamics of coarse mineral

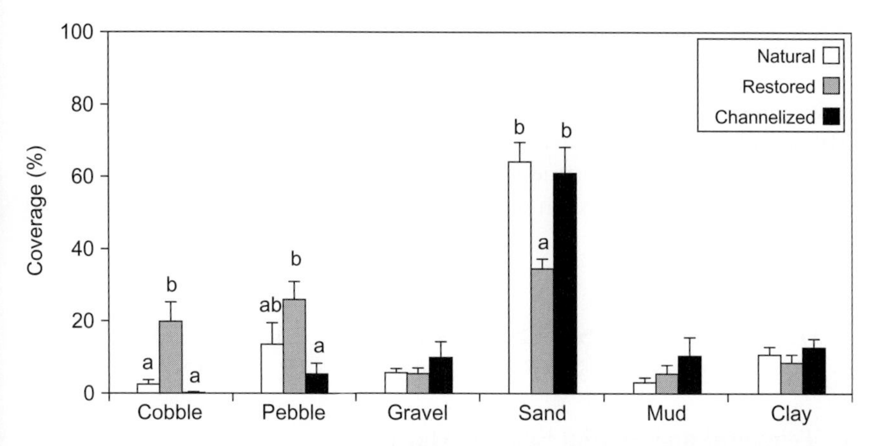


**Figure 21** Substrate composition in natural (i.e. unrestored and unimpacted), channelised (control) and restored streams. Differences in coverage of individual substrates were tested using a one-way ANOVA and pair wise differences were tested using Bonferroni-corrected *post hoc* tests. Lower case letter indicate significant differences.

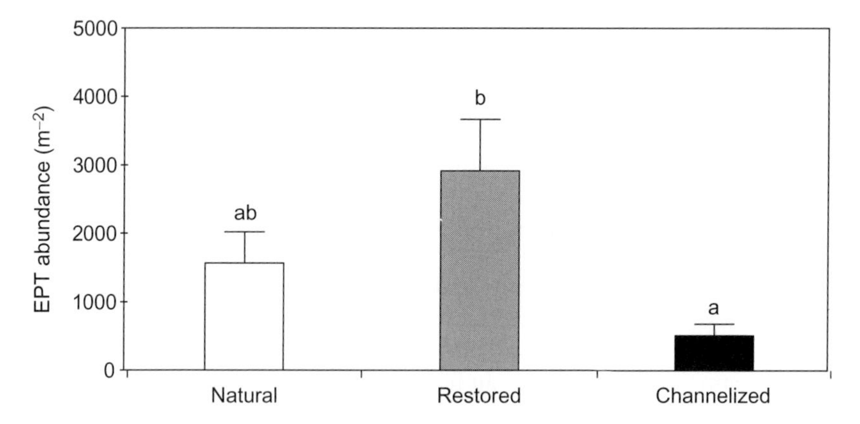


**Figure 22** Abundance of Ephemeroptera–Plecoptera–Trichoptera (EPT) taxa in natural, restored and channelised streams. Differences in EPT abundance were tested using one-way ANOVA and pair wise differences were tested using Bonferroni-corrected *post hoc* tests. Lower case letter indicate significant differences.

substrates, with the consequence that aberrant responses can arise among the biota (compare Section IV). For instance, benthic macroinvertebrates associated with gravel and stones are favoured, creating a skewed community structure dominated by EPT and other oxygen-demanding taxa. Our results clearly reflect the need for more focused studies designed to understand the structure of natural lowland streams and how the biota maps onto the physical environment (Vaughan *et al.*, 2009). This knowledge is required to inform river restoration projects *before* committing vast resources to what could turn out to be a very expensive lottery. Our results also demonstrate that water managers often still ignore the morphological processes and underlying dynamics that control these processes. Rather, they tend to create artificial physical conditions, which may be erroneously perceived as being 'natural', beneficial and attractive and, thus, skewing biological communities towards strong structural and functional bias favoured by such artificial conditions.

# VIII. WHAT LESSONS HAVE BEEN LEARNED AFTER 20 YEARS OF RIVER RESTORATION?

## A. Temporal and Spatial Scaling Matter

The huge body of literature in general supports the view that restoration can alter both the environmental variables that make up aquatic habitats and the biological communities that recover and occupy these habitats, although the

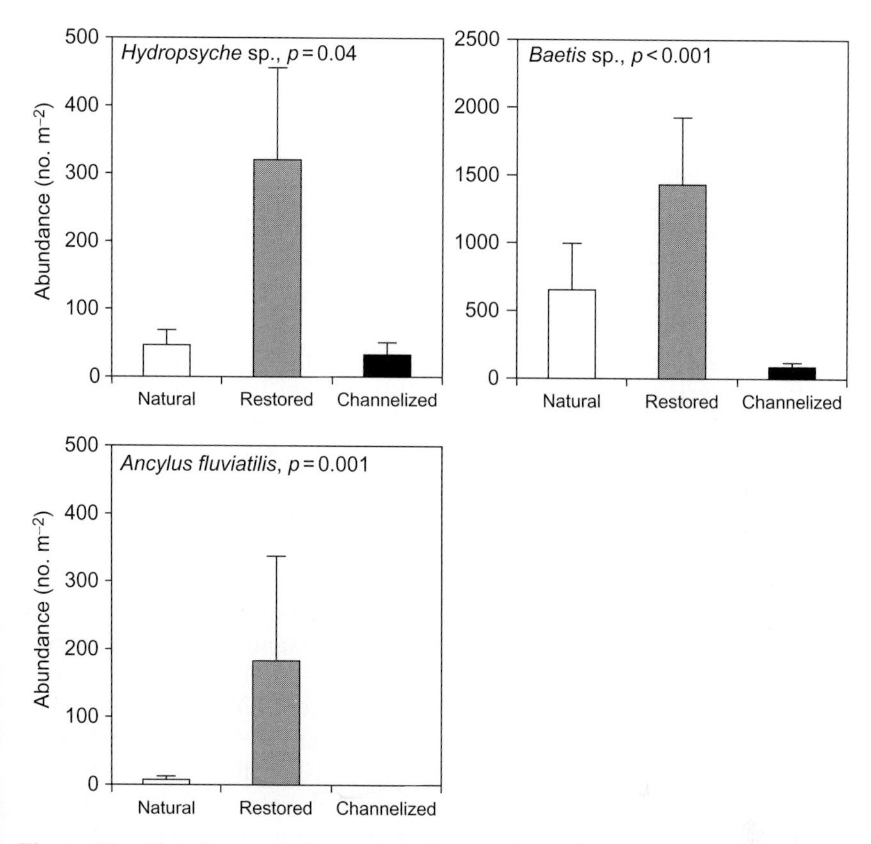


**Figure 23** Abundance of three selected macroinvertebrate taxa associated with coarse substrates. Differences in abundance were tested using one-way ANOVA on log-transformed data.

degree of success is dependent on both the context and spatio-temporal scaling of each scheme. Riparian vegetation can effectively buffer nutrients and sediments with various positive effects on the instream fauna and flora. The placement of (natural) instream habitat structures does lead to measurable effects on the instream fauna. Finally, the removal of weirs and other small barriers does restore the longitudinal connectivity and habitat diversity upstream.

The problem is, however, that many detected effects occur only in the short-term and at the local (site) scale, which raises the question of what scale is most appropriate for this kind of restoration. Temporal and spatial scales in ecology are often strongly interconnected and this property of 'ergodicity' (Olesen *et al.*, 2010; Woodward and Hildrew, 2002) is especially important in riverine landscapes from a restoration perspective, where long-term responses require large-scale manipulations. We cannot provide unequivocal evidence for the most appropriate temporal scale, but many taxa will take more than one generation to recover and our review strongly supports the conclusion that the local scale is inappropriate to achieve long-term measurable improvements. Local habitat enhancement measures are often 'swamped' by reach- or watershed-scale pressures upstream that continue to affect the treated sites. These limitations imply that the spatial scaling of restoration schemes *must* fit the scaling of degradation, that is, the scale of the pressures impacting the system, a point that has been largely ignored to date.

The same applies to hydrological and geomorphological processes, which are often neglected (e.g. Beechie *et al.*, 2010), but which constitute important characteristics that determine the level of success of restoration. Peak discharges caused by a high degree of impervious areas upstream, for instance, need to be taken into consideration *before* site-scale habitat restoration is being implemented, and they must, therefore, be mitigated or managed. Hydrological regimes need to include consideration of the frequency and magnitude of floods in order to balance short-term deleterious effects of weir and dam removal due to, for instance, increased sediment loads downstream of the former impounded section: even a cursory examination of the annual hydrograph is likely to prove invaluable here, yet, again, this is rarely incorporated into restoration schemes. Such negative effects are reported to last at least up to 5 years, or even more (Bednarek, 2001) and might be mitigated by natural flood regimes that help transport fine sediment further downstream where it can accumulate in natural sinks, such as pools: the timing and position of barrier removal are therefore critical when trying to maximize the reward:risk ratio of restoration.

All these findings emphasise that restoration measures must be designed integrated at appropriate spatial scales. Local (site scale) measures need to be accompanied by reach-scale measures further upstream to control potential confounding effects of watershed-scale pressures. Further, restoration measures need to be monitored beyond the timescale of typical experimental studies (i.e. the 3–4 years of many research grant funding schemes), in order to detect long-term intergenerational recovery, but also potential adverse effects and potential longer-term reversion to a degraded condition. The current knowledge on long-term restoration effects is scarce. Given the long history of river basin degradation in large areas of Europe, ecologists, restorationists and water managers need to remain patient, at least until some of the more recent, better-designed and integrated restoration schemes start to show the desired effects (or not, as the case may be) in the near future. Only a sufficiently frequent and long-term monitoring scheme will help provide more insight into the spatial and temporal effects of restoration. This knowledge is likely to be the key to design effective and successful ecological restoration schemes in the future, and whilst this might appear to be an expensive extravagance to many river managers or funding bodies, in the longer term it would represent only a tiny proportion of the overall project cost.

In light of the Water Framework Directive (WFD) and further European water-related legislation, we recommend that restoration schemes should be designed and developed at the water-body scale. According to the WFD, a water body constitutes the spatial entity or unit for river basin management. This can be a lake or a section of a river that belongs to the same lake or stream type, respectively, and is characterised by the same (set of) pressure(s). All European Union Member States have already divided their surface waters into water bodies, which form the basis for the development of management plans within the WFD. Although the water-body scale is a WFD-specific characteristic, the operational implementation of river basin management at the water-body scale is probably easily transferable to other regions (e.g. North America). The same or similar entities should, thus, be followed for the planning and implementation of restoration and mitigation schemes, in Europe and elsewhere. That is, restoration needs to consider wholesale water-body characteristics and possible limitations that may counteract restoration targets.

## B. Appropriate Indicators Are Required

Numerous bioindicators were developed and combined in ecological assessment systems, for example, in context of the WFD and earlier biomonitoring schemes (Birk *et al.*, 2010; Friberg *et al.*, 2011; Furse *et al.*, 2006; Lorenz *et al.*, 2004), with a focus on compositional, abundance, richness and functional measures of the plant and animal communities (Friberg *et al.*, 2011). The same indicators are often being applied to assess biological effects and recovery after restoration (e.g. Jähnig *et al.*, 2009b; Matthews *et al.*, 2010), but this reliance on these structural metrics is highly questionable for two reasons. First, such indicators were often developed to assess the overall impact of pressures at a water-body scale, so they tend to respond to a rather general degradation caused by several pressures acting in concert within a water body. In the case of agricultural land use, these stressors often act at the catchment scale, and the resulting impacts may have lasted for decades or even centuries. This phenomenon is often referred to as 'the ghost of land use past' (Harding *et al.*, 1998) and implies that its legacy might continue to limit full restoration success in the future. Such indicators of degradation are therefore very unlikely to detect changes at the (small) site or reach scale and in the short-term in response to restoration, unless land use legacy and/or other pressures are mitigated in parallel.

In addition to the (large-scale) indicators applied in routine monitoring schemes, restoration indicators need to be specifically designed to also detect short-term effects of specific restoration measures at the reach or even smaller scale. Such indicators exist (e.g. benthic macroinvertebrate metrics: Matthews *et al.*, 2010; riparian ground beetles and floodplain vegetation: Jähnig *et al.*, 2009a; Januschke *et al.*, 2009), but they are rarely being used in the context of river restoration, probably because they are not compliant with the explicit requirements of the WFD and so are probably unfamiliar to many practitioners on the ground. Jähnig *et al.* (2009a) compared hydromorphological and biological characteristics of re-braided mountain river section in Germany with unrestored control sites further upstream, and found that the floodplain vegetation richness responded particularly rapidly to restoration of the river's course. Habitat enhancement can also have positive effects on riparian ground beetle richness (Januschke *et al.*, 2009) that are adapted to the dynamic environmental conditions (e.g. frequent flood events causing erosion and deposition) typical of habitats adjacent to rivers. Both floodplain vegetation and riparian ground beetles are likely to be effective indicators of the effects of habitat enhancement in the short-term because of their high dispersal capabilities.

In order to monitor restoration schemes addressing the improvement of (dynamic) riverine processes, functional indicators are required (Feld *et al.*, 2009, 2010). Such indicators may include measures of organic matter processing, for instance, the number and density of shredders and detrital feeders (i.e. macroinvertebrates that feed on coarse and fine particulate organic matter, respectively) or algal production rates, which can provide insight into the functioning of riverine food webs (e.g. Hladyz *et al.*, 2011a,b). These and other functional aspects, such as food web dynamics, however, remain widely unconsidered in restoration monitoring, presumably because of their perceived complexity and basis in more classical theoretical ecology (Friberg *et al.*, 2011). A better, mechanistic understanding of cause–effect relationships will be required as a basis for the development of practical, functional indicators for application in standard restoration monitoring schemes in the future.

Our review revealed that in particular indicators of restoration effects on primary producers (both aquatic macrophytes and algae) are rare. Plants, however, are considered better suited to indicate nutrient enrichment than, for instance, fishes and benthic macroinvertebrates (e.g. Johnson and Hering, 2009), as the linkage between eutrophication and primary production is often clear and relatively direct, while the linkage between nutrients and fishes or benthic macroinvertebrates is rather indirect, although often evident (Woodward, 2009). Aquatic macrophytes and benthic algae clearly need to be better involved in restoration monitoring schemes, if we are to gain a more holistic understanding of whole-system restoration.

Concerted effort is also required to streamline indicator development, in particular to account for biological recovery beyond the scope of, for instance, the WFD or similar water management-related policies on other continents. Aquatic and terrestrial ecosystems are often subject to different environmental legislations and policies, ignoring the interconnections between them. In Europe, for instance, rivers, lakes, riparian buffers, floodplains and wetlands are subject to several nature conservation directives and initiatives (e.g. Habitats Directive, Natura 2000, European and global biodiversity policies). In this context, semi-aquatic or even terrestrial indicators may provide an invaluable potential as indicators for restoration effects and success. Riparian and floodplain vegetation, carabid beetles, or riparian birds may add to fishes, benthic macroinvertebrates and other aquatic indicator groups. This would also help integrate ecosystem monitoring across various types of legislation and could lead to a more effective, synergistic use of resources.

## C. Ecological Constraints can Determine Ecological Success

The lack of biological recovery following river restoration measures has been attributed to ecological constraints that may have limited or even inhibited recovery in restoration studies (e.g. Jähnig *et al.*, 2009a; Lorenz *et al.*, 2009). Such constraints probably include meta population dynamics of available source populations (e.g. Shields *et al.*, 1995b, 2006) and the dispersal capabilities of the community members to recover (e.g. Shields *et al.*, 1995b), although evidence on the underlying mechanisms is still lacking from restoration studies.

In order to assist recovery of riverine organisms, their source populations need to be large enough to release sufficient propagules to establish a new reproductive population at a restored site, while their location relative to the restored site needs to fit their dispersal capacities (e.g. through drift, compensation flight, active migration) (Brooks *et al.*, 2002; Parkyn *et al.*, 2003). For both species whose life cycle is entirely aquatic (e.g. fishes: Shields *et al.*, 2006) and species whose life cycle involves terrestrial (flying) stages (e.g. caddisflies: Smith *et al.*, 2009), the pathways between source population and restored sites are critical. Habitat conditions need to be suitable for migration and also for establishment of the species (presence of instream and riparian habitats to complete the full life cycle, absence of migration barriers, etc.).

Restoration practitioners should aim to take into account the meta population aspects of recolonisation, when planning a restoration scheme. A simple inventory of available source populations of targeted organisms prior to restoration, for instance, may help identify priority areas (where rapid recolonisation is likely), and may avoid restoration in areas (yet) too far away from existing source populations: such data are often held by regulatory bodies and are often collected as part of statutory biomonitoring schemes. Together with data on species' dispersal capabilities at different life stages, dispersal modelling is possible (e.g. Dedecker *et al.*, 2007; Smith *et al.*, 2009), which could inform restoration practitioners in advance about the probability of recovery at a candidate restoration site before proceeding.

## D. Hierarchical Pressures Require Hierarchical Restoration

Not all pressures impose the same stress level on ecosystems: some constitute chief sources of impact while others are subordinate (e.g. Palmer *et al.*, 2010; Roni *et al.*, 2008). In river ecosystems, one can distinguish between pressures that affect the water quantity and quality itself (i.e. the quality of the medium 'water') and others that act on the quality of the physical structure and geo-/hydromorphological setting (i.e. the quality of the matrix 'river/floodplain network'). For instance, if organic pollution or eutrophication occurs in a river stretch that is subject to restoration, the pollution is often of chief importance and *must* be reduced or otherwise mitigated *before* physical habitat and geomorphological processes are being restored. Several restoration studies suggested that ongoing water quality problems in the catchment upstream of a restoration were possible causes of failure as they were not addressed by (the spatial scale of) restoration (e.g. Pretty *et al.*, 2003; Roni *et al.*, 2008; see Palmer *et al.*, 2010 for a recent review). Penczak (1995), however, suggested the opposite and attributed positive impacts of riparian buffer restoration on fish community composition and biomass to the concurrent overall increase in water quality in the catchment upstream of the restored river section. In other words, a poor medium 'water' flowing in a good matrix is probably an insufficient precondition for recovery.

Conversely, if the water quality is sufficient for recovery, it is the chief geomorphological processes or physical structures that may hinder recovery (e.g. Shields *et al.*, 2008). For instance, excessive fine sediment entry from adjacent croplands upstream of a restoration were often suggested in the restoration literature to counteract physical habitat improvements (Larson *et al.*, 2001; Levell and Chang, 2008; Merz *et al.*, 2004; Moerke *et al.*, 2004). Fine sediment layers cover coarse substrates (gravel, cobbles) and limit, for instance, the availability of spawning habitats for salmonid fish (e.g. Moerke and Lamberti, 2003). Consequently, restoration schemes not only need to be spatially integrative, but they also need to account for a pressure hierarchy, that is, the hierarchy (and possible interaction) of the underlying mechanisms: one stressor (scale) may be superior to another, which implies that superior stressors have to be mitigated first (or in parallel) with (subordinate) others. Palmer *et al.* (2010) suggested the following hierarchy of measures:

protection of critical habitats in the watershed > water quality improvement > restoration of watershed processes (e.g. habitat connectivity, hydrology) > instream habitat enhancement. Our literature review, and in particular, the potentially limiting factors of recovery reported in many restoration studies, in general support this ordering, although it is important to remember these components are also interlinked and various feedbacks may be in operation. Surrounding land use, riparian conditions, instream substrate composition and water temperature can act alone and/or interact to determine the recovery of stream biota (Entrekin *et al.*, 2008).

## E. Future Research Needs

Several shortcomings in river restoration ecology and the implementation of ecological restoration have already been discussed in Sections III–V and earlier in this section. Nonetheless, practitioners and restorationists are forced to continue to restore rivers in Europe, North America and elsewhere, irrespective of these shortcomings. The European WFD (2000/60/EC), for example, requires water managers at the pan-European scale to improve the ecological quality of many river systems in the near future, a great effort that inevitably involves restoration and requires huge investments. We cannot precisely estimate the investments that will be required for river restoration in Europe, but given the amount of ca. 200 million EUR that has been spent on river restoration in Denmark alone since the early 1980s, the annual investments in Europe may easily exceed 1 billion EUR to meet the goals of the WFD. This is similar to the river restoration efforts in the U.S., which were estimated to be $\sim$1 billion USD annually (Bernhardt *et al.*, 2005).

In order not to render these huge investments a waste of time and money, restoration ecology must address several research needs and help fill the knowledge gaps to provide the data needed to design ecologically effective restoration schemes. The first dilemma of a spatial mismatch of pressures and their impacts on the one side, and of restoration on the other side requires further research in order to identify the minimum spatial extend of various kinds of restoration measures (e.g. habitat enhancement, re-meandering, removal of barriers) to be ecologically effective. A series of well-designed restoration measures integrated in large-scale restoration schemes (i.e. integrated restoration) might help generate the data basis to identify the processes and mechanisms of recovery at several spatial scales, from the (sub-)catchment to the local site scale. The example of the Danish River Skjern (see Section VII.B) showed that large-scale restoration facilitated both biological recovery and geomorphological processes that shape dynamic habitat conditions (e.g. through erosion and deposition).

The second dilemma of a temporal mismatch of restoration and monitoring might be overcome if water managers and practitioners agree to monitor restoration effects in the long-term (i.e. >7–10 years). Our review revealed that the majority of restoration studies spanned a period of 1–7 years. This period is relatively short-term and often insufficient to detect restoration effects and biological recovery. In the Kissimmee River project in Florida, the time-span required by different organism groups to recover was estimated at 3–8 years for aquatic plants, 10–12 years for benthic macroinvertebrates and 12–20 years for fish (Trexler, 1995). It is crucial therefore that restoration schemes are accompanied by tailor-made (BACI design) monitoring programmes and consider both the kind of restoration measures and the anticipated time-span required for recovery. As common indicators of degradation may turn out to be inappropriate to detect restoration effects and progress in the short term, there is also a need for indicators of restoration progress, suited to reliably track changes and their trends towards recovery (Matthews *et al.*, 2010). Such indicators may involve physical habitat measures (e.g. relation erosional:depositional habitats, flow diversity and dynamics) or biological traits (e.g. relation r:K strategists) and should detect changes rather than states.

In order to implement large-scale integrated restoration schemes and monitoring programmes, it is crucial that water managers, practitioners and restoration ecologists closely cooperate in the framework of joint projects (see also Palmer, 2009). The cooperation with practitioners facilitates the design of practicable restoration measures and the involvement of practitioners may bring additional synergistic benefits: within a larger scheme of individual restoration actions, local branches of nature protection organisations, for instance, may support individual local restoration measures, while regional organisations may help negotiate land use changes in the floodplain with landowners and landowner unions.

The research needs outlined above require additional resources and efforts. It is expected, however, that the costs for future research on ecological restoration and monitoring will make up only a small portion of the annual investments in river restoration, and more carefully designed schemes in the future could provide significant improvements to what is currently a rather *ad hoc* and often inefficient process: the restoration of our rivers is an important goal and one that clearly merits our careful attention.

## ACKNOWLEDGEMENTS

We are grateful to two referees, who provided helpful comments to improve the chapter. Special thanks go to the attendees of a workshop on conceptual modelling in November 2009, held at the Alterra Institute in Wageningen, the Netherlands, for stimulating discussions and valuable contributions to the development of conceptual models. A large part of the review and development of the conceptual framework model is a result of the project WISER (Water bodies in Europe: Integrative Systems to assess Ecological status and Recovery) funded by the European Union under the 7th Framework Programme, Theme 6 (Environment including Climate Change) (contract No. 226273), www.wiser.eu.

# APPENDIX A

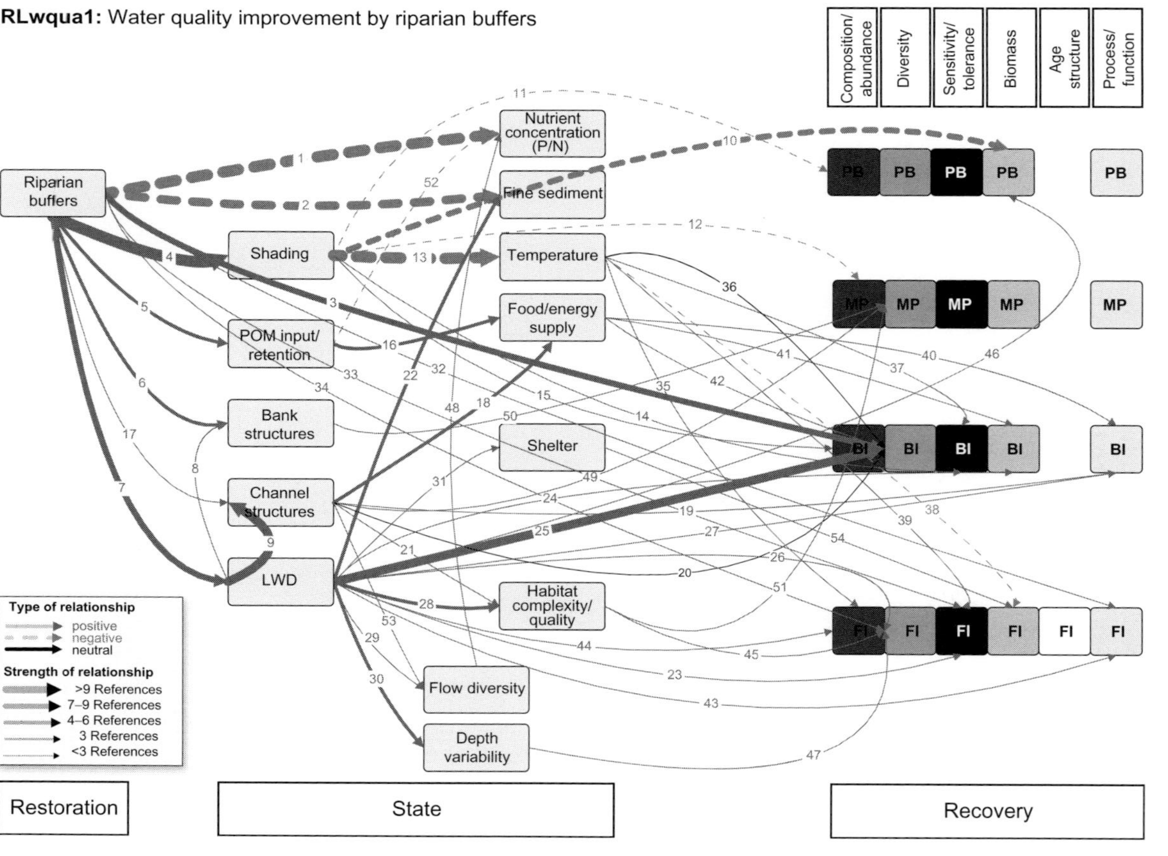


Figure A1

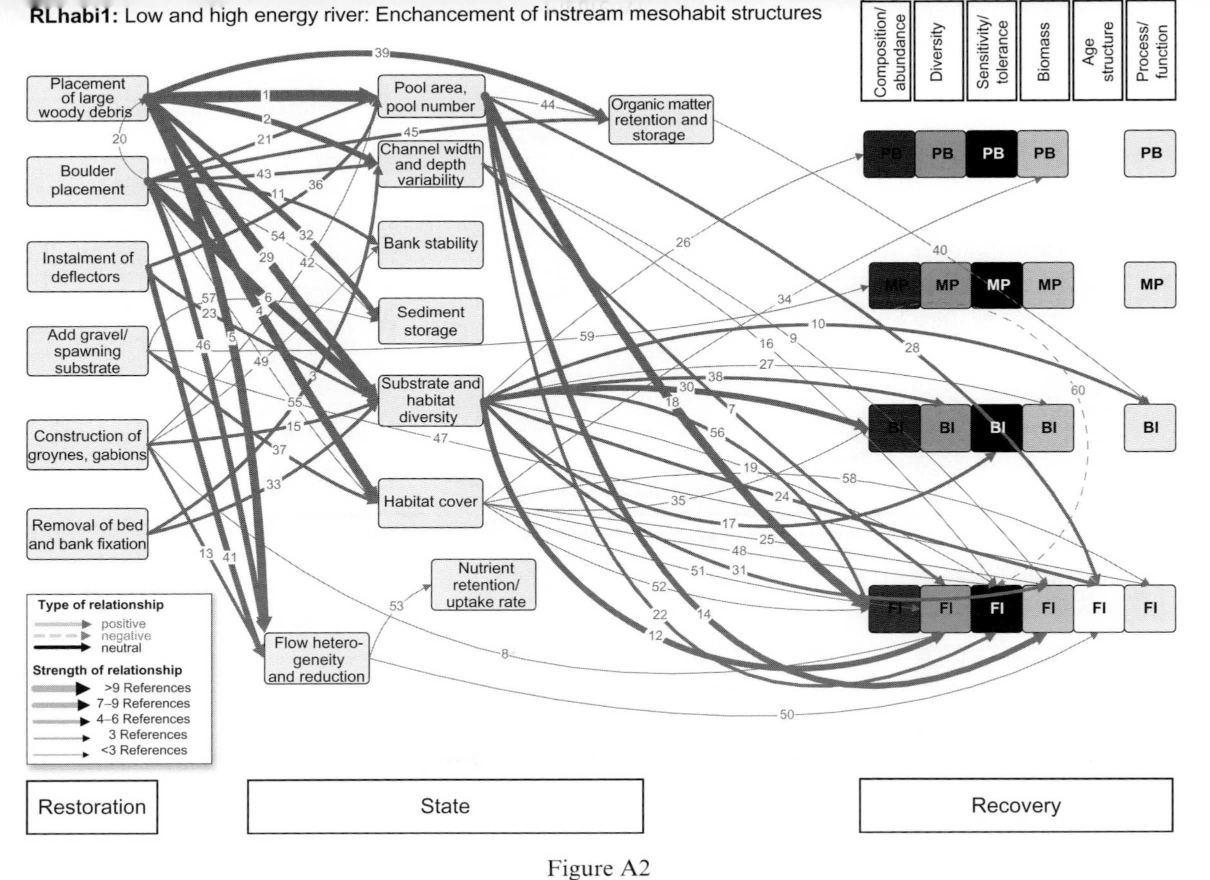
RLhabi1: Low and high energy river: Enchancement of instream mesohabit structures
Placement of large woody debris
Boulder placement
Instalment of deflectors
Add gravel/ spawning substrate
Construction of groynes, gabions
Removal of bed and bank fixation
Pool area, pool number
Channel width and depth variability
Bank stability
Sediment storage
Substrate and habitat diversity
Habitat cover
Nutrient retention/ uptake rate
Flow hetero-geneity and reduction
Organic matter retention and storage
Composition/ abundance
Diversity
Sensitivity/ tolerance
Biomass
Age structure
Process/ function
PB
MP
BI
FI
Type of relationship
positive
negative
neutral
Strength of relationship
>9 References
7–9 References
4–6 References
3 References
<3 References
Restoration
State
Recovery

Figure A2

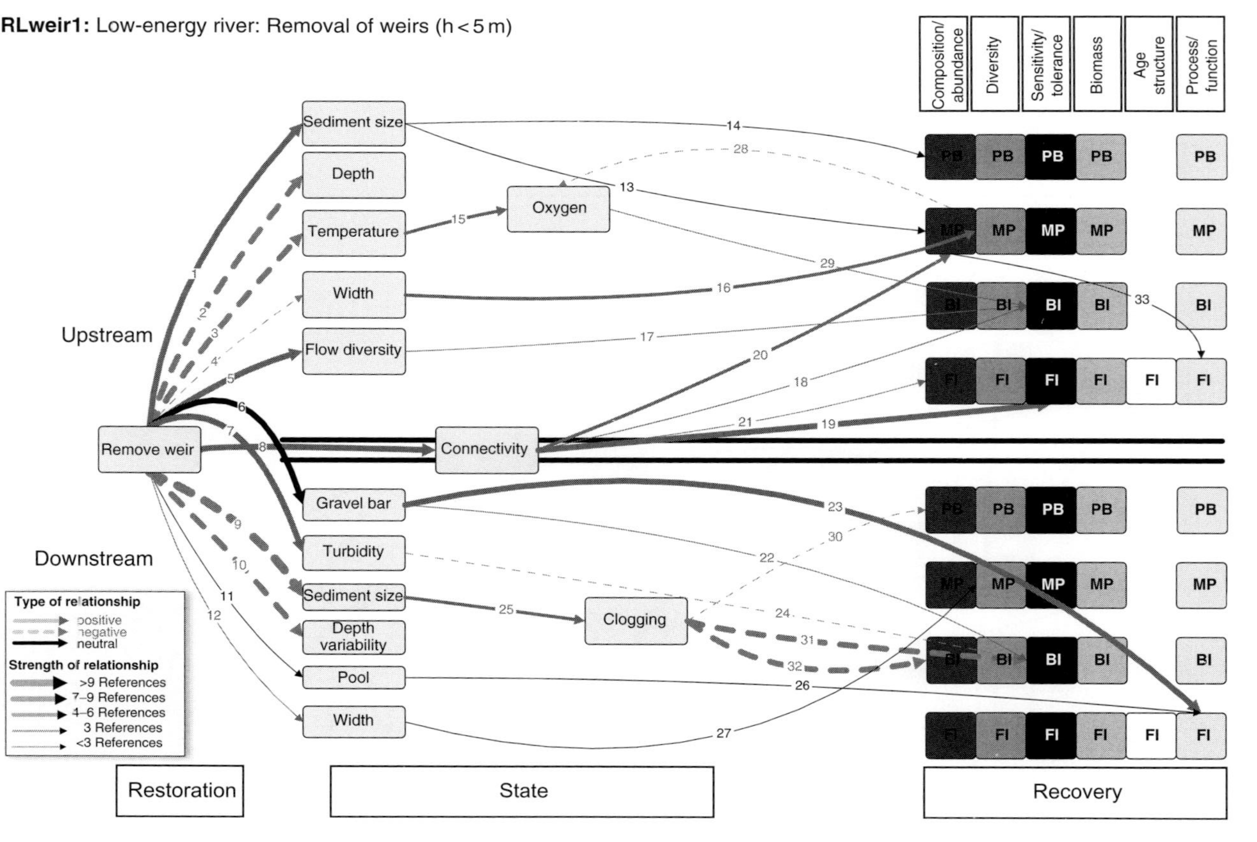
RLweir1: Low-energy river: Removal of weirs (h < 5 m)
Composition/abundance
Diversity
Sensitivity/tolerance
Biomass
Age structure
Process/function
Upstream
Downstream
Remove weir
Sediment size
Depth
Temperature
Width
Flow diversity
Oxygen
Connectivity
Gravel bar
Turbidity
Sediment size
Depth variability
Pool
Width
Clogging
PB
MP
BI
FI
Type of relationship
positive
negative
neutral
Strength of relationship
>9 References
7–9 References
4–6 References
3 References
<3 References
Restoration
State
Recovery

Figure A3

# APPENDIX B

**Table B1** Matrix of arrow (link) numbers of the conceptual model on riparian buffer management (Figure A1) and restoration studies that refer to the links.

| | Model link No. | | | | | | | | | | | | | | | | | | | | | | | | | | | | | | | | | | | | | | | | | | | | | | | | | | | | | |
|---|---|---|---|---|---|---|---|---|---|---|---|---|---|---|---|---|---|---|---|---|---|---|---|---|---|---|---|---|---|---|---|---|---|---|---|---|---|---|---|---|---|---|---|---|---|---|---|---|---|---|---|---|---|---|
| Serial No. | 1 | 2 | 3 | 4 | 5 | 6 | 7 | 8 | 9 | 10 | 11 | 12 | 13 | 14 | 15 | 16 | 17 | 18 | 19 | 20 | 21 | 22 | 23 | 24 | 25 | 26 | 27 | 28 | 29 | 30 | 31 | 32 | 33 | 34 | 35 | 36 | 37 | 38 | 39 | 40 | 41 | 42 | 43 | 44 | 45 | 46 | 47 | 48 | 49 | 50 | 51 | 52 | 53 | First author | Year |
| 1 | x | x | | | | | | | | | | | | | | | | | | | | | | | | | | | | | | | | | | | | | | | | | | | | | | | | | | | | Dosskey MG | 2001 |
| 2 | x | x | | | | | | | | | | | | | | | | | | | | | | | | | | | | | | | | | | | | | | | | | | | | | | | | | | | | Mankin KR | 2007 |
| 3 | | | | x | | | | | | x | | | | | | | | | | | | | | | | | | | | | | | | | | | | | | | | | | | | | | | | | | | | Davies-Colley RJ | 1998 |
| 4 | x | x | | | | | | | | | | | | | | | | | | | | | | | | | | | | | | | | | | | | | | | | | | | | | | | | | | | | Schultz R | 1995 |
| 5 | | | x | x | | | | | | | | | x | | | | | | | | | | | | | | | | | | | | | | | | | | | | | | | | | | | | | | | | | Broadmeadow S | 2004 |
| 6 | | | | | | | | | | | | | | | | | | | | | | | | | | | | | | | | | | | | | | x | | | | | | | | | | | | | | | | Weatherley N | 1990 |
| 7 | x | | | | | | | | | x | x | x | x | x | | | | | | | | | | | | | | | | | | | | | | | | | | | | | | | | | | | | | | | | Parkyn SM | 2003 |
| 8 | | | | | | | | | x | | | | | | | | | | | | | | | | x | | x | | | | | | | | | | | | | | | | | | | | | | | | | | | Larson M | 2001 |
| 9 | | | | | | | | | | | | | | | | | | | | | | | | | x | | | | | | | | | | | | | | | | | | | | | | | | | | | | | Gerhard M | 2000 |
| 10 | | | | | | | | | x | | | | | | | | | | | | | x | | | | | | x | | x | | | | | | | | | | | | | | | x | | x | | | | | | | Brooks AP | 2004 |
| 11 | | | | | x | | | | | | | | | | | x | | | | | | | | | | | | | | | | | | | | | | | | | x | x | | | | | | | | | | | | Wallace JB | 1997 |
| 12 | | | | | x | | | | | | | | | | | x | | | | | | | | | | | | | | | | | | | | | | | | | | x | | | | | | | | | | | | Northington RM | 2006 |
| 13 | | | | | | | | x | x | | | | | | | | | x | | | | x | | x | x | x | | x | x | x | | | | | | | | | | | | | | | | | | | | | | | | Lester RE | 2008 |
| 14 | | | | | | | | | | | | | | | | | | x | x | | x | | | | | | | | | | | | | | | | | | | | | | | | x | | | | | | | | | Muotka T | 2007 |
| 15 | | | | | | | | | | | | | | | | | | | | | | | | | x | | | | | | | | | | | x | x | | | | | | | | | | | | | | | | | Miller SW | 2010 |
| 16 | | | | | | | | | | | | | | | | | | x | | | | | | | | | | | | | | | | | | | | | | x | | | | | | | | | | | | | | Lepori F | 2005 |
| 17 | | | | x | | | x | | x | | | | x | | | | | | | | | | | | | | | | | | | | | | | | | | | | | | | | | | | | | | | | | Opperman JJ | 2004 |
| 18 | | | | | | | | | | | | | | | | | | | | x | | | | | | | | | | | | | | | | | | | | | | | | | | | | | | | | | | Harrison SS | 2004 |
| 19 | x | | | | | | | | | | | | | | | | | | | | | | | | | | | | | | | | | | | | | | | | | | | | | | | | | | | | | Sutton AJ | 2009 |
| 20 | | | | | | | | | | | | | | | | | | | | | | | | | | | | | | | | | | | | | | | | | | | | | | | | | x | x | x | | | Moustgaard-Pedersen T( | 2006 |
| 21 | | | | | | | | | | | | | | | | | | | | | | | | | | | | | | | | | | | | | | | | | | | | | | | | | | | | x | | Aldridge KT | 2009 |
| 22 | | | | | | | | | | | | | | | | | | | | | | | | | | | | | | | | | | | | | | | | | | | | | | | | x | | | | | x | Bukaveckas PA | 2007 |
| 23 | | | | | | | | | | | | | | | | | | | | | | | | | x | | | | | | | | | | | | | | | | | | | | | x | | | | | | | | Coe HJ | 2009 |
| 24 | | | | | | | | | | | | | | | | | | | | | | | | x | x | | x | | | | | | | | | | | | | | | | | | | | | | | | | | | Entrekin SA | 2009 |
| 25 | | | | | | | | | | | | | | | | | | | | | | | | | | | | | | | | | | | | | | | | | | | x | x | | | | | | | | | | Cederholm CJ | 1997 |
| 26 | | | | | | | | | x | | | | | | | | | | | | | | | | x | | | | | | | | | | | | | | | | | | | | | | | | | | | | | Hilderbrand | 1997 |
| 27 | | | x | | | | | | | | | | | | | | | | | | | | | | | | | | | | | | | x | | | | | | | | | | | | | | | | | | | | Jowett | 2009 |
| 28 | | x | x | x | | x | x | | | x | | | | | | | | | | | | | | | | | | | | | | | | | | | | | | | | | | | | | | | | | | | | Quinn J | 2009 |
| 29 | x | | | | | | | | | | | | | | | | | | | | | | | | | | | | | | | | | | | | | | | | | | | | | | | | | | | | | Kaushal SS | 2008 |
| 30 | | | | | | | x | | | | | | | | | | | | | | | | | | | | | | | | | | | | | | | | | | | | | | | | | | | | | | | McBride M | 2008 |
| 31 | | | | | | | | | | | | | | | | | | | x | | | | | | | | | | | | | | | | | | | | | | | | | | | | | | x | | | | | Muotka T | 2002 |
| 32 | | | | | | | | | | | | | | | | | | | | | | | | | | | | | | | | x | x | x | | | | | | | | | | | | | | | | | | | | Penczak T | 1995 |
| 33 | | | | | | x | | | | | | | | | | | | | | | | | | | | | | | | | | | | | | | | | | | | | | | | | | | | | | | | Shields FD | 1995b |
| 34 | | | x | x | | | | | | | | | x | | x | | | | | | | | | | | | | | | | | | | | | | | | | | | | | | | | | | | | | | | Becker A | 2009 |
| 35 | | | x | | | | | | | | | | | | | | x | | | | x | | | | | | | | | | | | | | | | | | | | | | | | | | | | | | | | | Pedraza GX | 2008 |
| 36 | x | x | | | | | | | | | | | | | | | | | | | | | | | | | | | | | | | | | | | | | | | | | | | | | | | | | | | | Correll DL | 2005 |
| 37 | | | | | | | x | | x | | | | | | | | | | | | | | | | | | | | | | | | | | | | | | | | | | | | | | | | | | | | | Warren DR | 2009 |
| 38 | | | | | | | | | | | | | | | | | | | | | | | | | | | | | | | | | | | | | x | | | | | | | | | | | | | | | | | Haidekker A | 2007 |
| 39 | | | | | | | | | | x | | | | | | | | | | | | | | | | | | | | | | | | | | | | | | | | | | | | | | | | | | | | Ghermandi A | 2009 |
| 40 | | | | | | | | x | x | | | | | | | | | | | | | x | | | | | | x | x | x | | | | | | | | | | | | | | | | | | | | | | | | Chen X | 2008 |
| 41 | | | | x | | | x | | | | | | x | | | | | | | | | | | | | | | | | | | | | | | | | | | | | | | | | | | | | | | | | Davies-Colley RJ | 2009 |
| 42 | x | x | | x | | x | | | | | | | | | | | | | | | | | | | | | | | | | | | | | | | | | | | | | | | | | | | | | | | | Parkyn SM | 2005 |
| 43 | x | x | x | x | | | | | | | | | x | | | | | | | | | | | | | | | | | | | | | | | | | | | | | | | | | | | | | | | | | Castelle AJ | 1994 |
| 44 | | | | x | | | | | | | | | x | | | | | | | | | | | | | | | | | | | | | | x | | | | x | | | | | | | | | | | | | | | Barton DR | 1985 |
| 45 | x | x | | x | | | | | | | | | x | | | | | | | | | | | | | | | | | | | | | | | | | | | | | | | | | | | | | | | | | Osborne LL | 1993 |
| 46 | x | x | | x | x | | x | | x | | | | x | | | x | | | | | | | x | | | | | | | | x | | | | | | | | | | | | | | | | | | | | | | | Wenger S | 1999 |
| 47 | x | | | | | | | | | | | | | | | | | | | | | | | | | | | | | | | | | | | | | | | | | | | | | | | | | | | | | Fennessy MS | 1997 |
| 48 | | | | x | | | | | | | | | x | | | | | | | | | | | | | | | | | | | | | | x | | | | x | | | | | | | | | | | | | | | Whitledge GW | 2006 |

**Table B2** Matrix of arrow (link) numbers of the conceptual model on instream mesohabitat enhancement (Figure A2) and restoration studies that refer to the links

| Serial no. | Model link no. | | | | | | | | | | | | | | | | | | | | | | | | | | | | | | | | | | |
|---|---|---|---|---|---|---|---|---|---|---|---|---|---|---|---|---|---|---|---|---|---|---|---|---|---|---|---|---|---|---|---|---|---|---|---|
| | 1 | 2 | 3 | 4 | 5 | 6 | 7 | 8 | 9 | 10 | 11 | 12 | 13 | 14 | 15 | 16 | 17 | 18 | 19 | 20 | 21 | 22 | 23 | 24 | 25 | 26 | 27 | 28 | 29 | 30 | 31 | 32 | 33 | 34 | 35 |
| 1 | | | | | | | | | | | | | | | | | | | | | | | x | | | | | | | | x | | | | |
| 2 | x | x | | x | | | x | | x | | | x | | x | | x | | | | | | x | | | | | | | x | | | | | | |
| 3 | x | | | x | | | | | | | x | | | x | | | | | | | | | | | x | | | | | | | | | | |
| 4 | x | | | | | | | | | | | | | | | | | | | | | | | | | | | | x | | | x | | | |
| 5 | x | | | | | | | | | | | | | x | | | | | | | | | | | | | | x | | | | | | | |
| 6 | | | | | | | | | | | | x | | | | | | | | | | | | | | | | | | x | | | x | | |
| 7 | | | | x | | | | | | | | | | | | | | | | | | | | | | | | | | | | | | x | x |
| 8 | x | | | | | | | | | | | | | | | | x | x | | | | | | | | | | | | | | | | | |
| 9 | | | | | | | | | | | | | | | | | | | | | | | | | | | x | | | x | | | | | x |
| 10 | | | | | | | | | | | | | | | | | | | | | | | | | | | | | | | | | | | |
| 11 | | | | | | | | | | | | | | | | | | | | | | | | | | | | | | | | | | | |
| 12 | | x | | | | | | | | | | | | | | | | | | | | | | | | | | | x | x | | | | | |
| 13 | | | | | | | | | | | | | | | | | | | | | | | x | | | | | | | | | | | | |
| 14 | | | | | | | | | | | | | | | | | | | | | | | | | | | | | | | | | | | |
| 15 | x | | | | | | | | | | | | | | | | | | | | | | | | | | | | | | | | | | |
| 16 | | | | | | x | | | | | | | | | | | | x | | | x | | | | | | | | | | | | | | |
| 17 | | | | | | | | | | | | | | | | | | | | | | | | | | | | | | | | | | | |
| 18 | | | | | | x | | | | | | | | | | | | | | | | | | | | | | | | | | | | | |
| 19 | x | | | | | | | | | | | | | | | | | x | | | | | | | | | | x | | | | | | | |
| 20 | | | x | | | | | x | | | | | x | | x | | | | | | | | | | | | | | | | x | | | | |
| 21 | | | x | | | | x | | | | | x | x | x | x | x | | | | | | | | x | | | | | | | | | | | |
| 22 | | | | | | x | | | | | | | | | | | | | | | | | x | | | | | | | | | | | | |
| 23 | x | | | | | | | | | | | | | | | | | | | | | | | | | | | | | | | x | | | |
| 24 | x | x | | | x | | | | | | | | | x | | | | x | | | | | | | | | | | x | | x | x | | | |
| 25 | x | | | | | | | | | | | | | | | | | | | | | | | | | | | | | | | | | | |
| 26 | | | | | | | | | | | | | | | | | | | | | | | | | | | | | | | | | | | |
| 27 | | | | | | x | | | | | | | | | | | | | | | | | | | | | | | | | | | | | |
| 28 | | | | | | | | | | | | | | | | | | | | | | | | | | | | | | | | | | | |
| 29 | | | | | | | | | | | | | | | | | | | | | | | | | | | | | | | | x | | | |
| 30 | | | | | | | | | | | | | | | | | | | | | | | | | | | | | | | | | | | |
| 31 | | | | | | | | | | | | | | | | | | | | | | | | | | | | | | | | | | | |
| 32 | | | | | | | | | | | | | | | | | | | | | | | | | | | | | | | | | | | |
| 33 | | | | | | | | | | | | | | | | | | | | | | | | | | | | | | | | | | | |
| 34 | | | | | | | | | | | | | | | | | | | | | | | | | | | | | x | | | | | | |
| 35 | x | | | | | x | | | | | | | | | | | | | | | | | | | | | | | x | | | | | | |
| 36 | x | | | | | | | | | | | | | | | | | | | | | | | | | x | x | | x | x | | | | | |
| 37 | | | x | | | | | | | | | | | | | | x | | x | | | | | | | | | | | | | | | | |
| 38 | | | | | | x | | | | | | | | | | | | | | | | | | | | | | | | | | | | | |
| 39 | | | | | | x | | | | x | | | | | | | | | | | | | | | | | | | | | | | | | |
| 40 | | | | | | | | | | | | | | | x | | | | | | | | | | | | | | | x | | | | | |
| 41 | | | | | | x | | | | | | | | | | | | | | | | | | | | | | | | x | | | | | |
| 42 | | | | | | x | | | | | | | | | | | | | | | | | | | | | | | | | | | | | |
| 43 | | | | | | | | | | | | | | | | | | | | | | | | | | | | | | | | | | | |
| 44 | | | | x | | | | | | | | | | | | | | | | | | | | | | | | | | | | | | | |
| 45 | | | | x | | | | | | | | | | | | | | | | | | | | | | | | | | | | | | | |
| 46 | | x | | | x | | | | | | | | | | | | | | | | | | | | | | | | | | | | | | |
| 47 | | | | x | | | | | | | x | | | | | | | | | | | | | | | | | | | | | | | | |
| 48 | | | | | | | | | | | | | | | | | | | | | | | | | | | | | | | | | | | |
| 49 | x | x | | x | x | | | | x | | | | | x | | | | x | | | | | | | | | | | | | | | | | |
| 50 | | | | | | | | | | | | | | | | | | | | | | | | | | | | | | | | | | | |
| 51 | x | | | | | x | | | | | | | | | | | | x | | x | x | x | | | | | | | | | | | | | |
| 52 | x | | | | | | | | | | | | | | | | | | | | | | | | | | | | | | | | | | |
| 53 | x | | | | | | | | | | | | | | | | | | | | | | | | | | | | | | | | | | |
| 54 | x | | | | | | | | | | | | | | | | | x | | | | x | | | | | | | | | | | | | |
| 55 | x | | | | | | | | | | x | | | | | | | x | | | | x | x | | | | | x | | | | | | | |
| 56 | | | | | | | | | | | | | | | | | | | | | | | | | | | | | | | | | | | |
| 57 | | | | | | | | | | | | | | | | | | | | | | x | x | | | | | x | | | | | | | |
| 58 | | | | | | | | | | | | | | | | | | | | | | | | x | | | | | x | | | x | | | |
| 59 | | | | | x | | | | | | | | | | | | | | | | | | | | | | | | | | | x | | | |
| 60 | | | | | | | | | | | | | x | | x | | | x | | | | | | | | | | x | | | | | | | |
| 61 | | | | | | | | | | | x | x | | | | | | | | | x | | | x | | | | | | | x | | | | |
| 62 | | | | | x | | | | | | | x | | | | | | | | | | | | | | | | | x | x | | x | | | |
| 63 | x | | | x | x | | | | | | | | | | | | | x | | | | | | | | | | | | | | | | | |
| 64 | | | | | x | | | | | | | | | | | | | | | | | | | | | | | | | | | | | | |
| 65 | | | | | | | | | | x | | | | | | | | | | | | | x | | | | | | | | | | | | |
| 66 | x | x | | | x | | | | | | | | | | | | | | | | | | | | | | | | | | | | | | |
| 67 | | | | | | | | | | | | | | | | | | | | | | | | | | | | | | | | | | | |
| 68 | | | | | | x | | | | | | | | | | | x | | | | | | | | | | | | | | | | | | |
| 69 | | | | | | | | | | | | | | | | | | | | | | | | | | | | | | | | | x | | |
| 70 | | | | | | | | | | | | | | | | | | | | | | | | | | | | | | | | | x | | |
| 71 | x | x | | x | x | | | | | | | | | x | | | | | | | | | | | x | | | | | | | | | | |
| 72 | x | | | | | | | | | | | x | | | | | | | | | | | | | | | | | x | | | x | | | |
| 73 | | x | | | | x | | | | x | | | | | | | x | | | | | | | | | | | | x | | | | | | |
| 74 | | | | | | | | | | | | | | | | | | | | | | | | | | | | | x | | | | | | |
| 75 | | | | | | | | | | | | | | | | | | | | | x | | | | | | | | | | | | | | |

| 36 | 37 | 38 | 39 | 40 | 41 | 42 | 43 | 44 | 45 | 46 | 47 | 48 | 49 | 50 | 51 | 52 | 53 | 54 | 55 | 56 | 57 | 58 | 59 | 60 | 61 | 62 | 63 | 64 | First author | Year |
|---|---|---|---|---|---|---|---|---|---|---|---|---|---|---|---|---|---|---|---|---|---|---|---|---|---|---|---|---|---|---|
| | | | | | | | | | | | | | | | | | | | | | | | | | | | | | Avery L | 2004 |
| | | | | | | | | | | | | | | | | | | | | | | | | | | | | | Baldigo BP | 2008 |
| | | | | | | | | | | | | | | | | | | | | | | | | | | | | | Binns NA | 2004 |
| | | | | | | | | | | | | | | | | | | | | | | | | | | | | | Brooks AP | 2006 |
| | | | | | | | | | | | | | | | | | | | | | | | | | | | | | Cederholm CJ | 1997 |
| | | | | | | | | | | | | | | | | | | | | | | | | | | | | | Chovanec A | 2002 |
| | | | | | | | | | | | | | | | | | | | | | | | | | | | | | Coe HJ | 2009 |
| x | | | | | | | | | | | | | | | | | | | | | | | | | | | | | Crispin V | 1993 |
| | x | x | | | | | | | | | | | | | | | | | | | | | | | | | | | Edwards CJ | 1984 |
| | | | x | | | | | | | | | | | | | | | | | | | | | | | | | | Entrekin SA | 2008 |
| | | | x | x | | | | | | | | | | | | | | | | | | | | | | | | | Entrekin SA | 2009 |
| | | | | | | | | | | | | | | | | | | | | | | | | | | | | | Gerhard M | 2000 |
| | | | | | x | | | | | | | | | | | | | | | | | | | | | | | | Hammond D | 2009 |
| | | | | | x | | | | | | | | | | | | | | | | | | | | | | | | Harrison SS | 2004 |
| | | | | | | | | | | | | | | | | | | | | | | | | | | | | | Hilderbrand RH | 1997 |
| | | | | | | x | | | | | | | | | | | | | | | | | | | | | | | House RA | 1996 |
| | | | | | | | | | | | | | | | | | | | | | | | | | | | | | Howson TJ | 2009 |
| | | | | | | | x | | | | | | | | | | | | | | | | | | | | | | Huusko A | 1995 |
| | | | | | | | | | | | | | | | | | | | | | | | | | | | | | Johnson SL | 2005 |
| | | x | | | | | | | | | | | | | | | | | | | | | | | | | | | Jungwirth M | 1993 |
| | | | | | | | | | | | | | | | | | | | | | | | | | | | | | Jungwirth M | 1995 |
| | | | | | x | | x | | | | | | | | | | | | | | | | | | | | | | Laasonen P | 1998 |
| | | | | | | | | | | | | | | | | | | | | | | | | | | | | | Larson MG | 2001 |
| | | | | | | | | | | | | | | | | | | | | | | | | | | | | | Lehane BM | 2002 |
| | | | | | | | | x | | | | | | | | | | | | | | | | | | | | | Lemly AD | 2000 |
| | | | x | | | | | | x | | | | | | | | | | | | | | | | | | | | Lepori F | 2005 |
| | | | | | | | x | | | x | | | | | | | | | | | | | | | | | | | Lepori F | 2005 |
| | | | | | | | | | x | | | | | | | | | | | | | | | | | | | | Lepori F | 2006 |
| | | | | | | | | | | | | | | | | | | | | | | | | | | | | | Levell AP | 2008 |
| | | | | | | | | | | | x | | | | | | | | | | | | | | | | | | Merz JE | 2004 |
| | | | | | | | | | | | | | | | | | | | | | | | x | x | | | | | Merz JE | 2008 |
| | x | | | | | | | | | | | x | | | | | | | | | | | | | | | | | Merz JE | 2004 |
| | x | | | | | | | | | | | | | | | | | | | | | | | | | | | | Merz JE | 2005 |
| | | | x | | | | | | | | | | | | | | | | | | | | | | | | | | Millington CE | 2007 |
| | | | | | | | | | | | | | | | | | | | | | | | | | | | | | Moerke AH | 2003 |
| | | | | | | | | | | | | | | | | | | | | | | | | | | | | | Moerke AH | 2004 |
| | | | | | | | | | | | | | | | | | | | | | | | | | | | | | Muhar S | 2008 |
| | | | | | | | | | x | x | | | | | | | | | | | | | | | | | | | Muotka T | 2002 |
| | | | | | | | | | x | x | | | | | | | | | | | | | | | | | | | Muotka T | 2002 |
| | | x | | | | | | | | | | | x | | | | | | | | | | | | | | | | Nakano D | 2006 |
| | | | | | | | | | x | x | | | | | | | | | | | | | | | | | | | Negishi | 2003 |
| | | | | | | | | | | x | | | | x | | | | | | | | | | | | | | | Palm D | 2010 |
| | | | | | | | | | | | x | | | | | | | | | | | | | | | | | | Palm D | 2007 |
| | | | | | | | | | | | | | | | x | | | | | | | | | | | | | | Pander J | 2010 |
| | | | x | | | | | | | | | | | | | x | | | | | | | | | | | | | Pretty JL | 2004 |
| | | | | | | | | | | | | | | | | | | | | | | | | | | | | | Pretty JL | 2003 |
| | | | | | | | | | | | | | | | | | | | | | | | | | | | | | Quinn JW | 2000 |
| | | | | | | | | | | | | | | | | | | | | | | | | | | | | | Raborn SW | 2003 |
| | | | | | | | | | | | | | | | | | | | | | | | | | | | | | Riley SC | 1995 |
| | | | | | | | | | | | | | | | | | x | | | | | | | | | | | | Roberts BJ | 2007 |
| | | | | | | | | | | | | | | | | | | | | | | | | | | | | | Roni P | 2006 |
| | | | | | | | | | | | | | | | | | | | | | | | | | | | | | Roni P | 2003 |
| | | | | | | | | | | | | | | | | | | | | | | | | | | | | | Roni P | 2001a |
| | | | | | | | | | | | | | | | | | | | | | | | | | | | | | Roni P | 2001b |
| | | | | | | | | | | | | | | | | | | | | | | | | | | | | | Saunders JW | 1962 |
| | | | | | | | | | | | | | | | | | | | | | | | | | | | | | Schmetterling DA | 1999 |
| x | | | | | | | | | | | | | | | | | | | | | | | | | | | | | Shetter DS | 1946 |
| | | | | | | | | | | | | | | | | | | | | | | | | | | | | | Shields FD | 2006 |
| | | | | | | | | | | | | | | | | | | | | | | | | | | | | | Shields FD | 2003 |
| | | | | | | x | | | | | | | | | | | | | | | | | | | | | | | Shields FD | 1995a |
| x | | | | | x | | | | | | | | | | | | | | | | | | | | | | | | Shields FD | 1993 |
| | | x | | | | | | | | | | | | | | | | | | | | | | | | | | | Shields FD | 2008 |
| | | | | | | | | | | | | | | | | | | | | | | | | | | | | | Solazzi MF | 2000 |
| | | | | | | | | | | | | | | | | | | | | | | | | | | | | | Spänhoff B | 2006 |
| x | | | | | x | | | | | | | | | | | | | | | | | | | | | | | | Tarzwell CM | 1937 |
| | | | | | | | | | | | | | | | | | | | | | | | | | | | | | Thompson | 2002 |
| | | | | | | | | | | | | | | | | | | | | | | | | | | | | | Tullos DD | 2006 |
| | | | | | | | x | | | x | | | | | | | | x | x | x | | | | | | | | | Van Zyll de Jong MC | 1997 |
| | x | | | | | | | | | | | | | | | | | | | | x | x | | | | | | | Zauner G | 2001 |
| | | | | | | | | | | | | | | | | | | | | x | | | | | | | | | Zauner G | 2003 |
| | | | x | | | | | | | | | | | | | x | | | | | | | | | | | | | Zika U | 2002 |
| | | | | | | | | | | | | | | | | | | | | | | | | | | | | | Brooks AP | 2004 |
| | x | | | | x | | x | | | x | | | | | | | | | | | | | | | | | | | Muotka T | 2007 |
| | | x | | | | | | | | | | | | | | | | | | | | | | | | | | | Miller SW | 2010 |
| | | | | | | | | | | | | | | | | | x | | | x | x | | | | | | | | Kaushal SS | 2008 |

**Table B3** Matrix of arrow (link) numbers of the conceptual model on weir removal (Figure A3) and restoration studies that refer to the links

| A | Model link No. | | | | | | | | | | | | | | | | | | | | | | | | | | | | | | | | | |
|---|---|---|---|---|---|---|---|---|---|---|---|---|---|---|---|---|---|---|---|---|---|---|---|---|---|---|---|---|---|---|---|---|---|---|
| Serial No. | 1 | 2 | 3 | 4 | 5 | 6 | 7 | 8 | 9 | 10 | 11 | 12 | 13 | 14 | 15 | 16 | 17 | 18 | 19 | 20 | 21 | 22 | 23 | 24 | 25 | 26 | 27 | 28 | 29 | 30 | 31 | 32 | 33 | First author | Year |
| 1 | | | | | | | | | | | | | | | | | | | | | | | | | | x | | | | | | | | Schlosser | 1982 |
| 2 | | | | | | | | | | | | | | | | | | | | | | | | | | x | | | | | | | | Harvey | 1991 |
| 3 | | | | | | | | | | | | | | | | | | | x | | x | | | | | | | | | | | | | Iversen | 1993 |
| 4 | | | x | | x | | | | | | | | | | x | | | | | | | | | | | | | | | | | | x | Hill | 1993 |
| 5 | x | | | | | | | | | | | | | | | | | | | | | | | | | | | | | | | | | Kanehl | 1997 |
| 6 | | | | | | | | x | | | | | | | | | | | | | | | | | | | | | | | | | | Poff | 1997 |
| 7 | | | | | | | | | | | | | x | x | | | | | | | | | | | | | | | | | | | | Baattrup-Pedersen | 1999 |
| 8 | | | | | | | | | | | | | | | | | | | | | | | x | | | | | | | | | | | Bednarek | 2001 |
| 9 | | x | x | | x | | | | | x | | | | | x | | | | | | | | | | | | | | x | | x | | | Bushaw-Newton | 2002 |
| 10 | | | x | | x | x | | x | | | | | | | | | | | x | x | x | | x | | | | | | | | | | | Gregory | 2002 |
| 11 | | x | | x | x | | | x | x | | | | | | | x | | | x | | | | | | | | | | | | | | | Hart | 2002 |
| 12 | | | | | | | | | | | | | | | | | | | | | | | | | | | | | | | | | | Pizzuto | 2002 |
| 13 | | | | x | | | | | | | | | | | | x | | | | x | | | | | | | x | | | | | | | Shafroth | 2002 |
| 14 | | | x | | | | | | | | x | | | | x | | | | | | | x | | | | x | | | | | | | | Stanley | 2002 |
| 15 | | x | x | | | | x | | | | | | | | x | | | | | | | | | | | | | x | | | | | | Chaplin | 2003 |
| 16 | x | | | | | x | x | | x | | | | | | | | | | | | | | x | | x | | | | | | | | | Hart | 2003 |
| 17 | x | x | | | | x | x | | x | x | | | | | | | | | | | | | | | | | | | | | | | | Randle | 2003 |
| 18 | | | | | | x | x | | x | x | x | x | | | | | | | | x | | | | | x | | | | | | | | | Rathburn | 2003 |
| 19 | x | | | | | | | | x | | | | | | | | | | | | | | | | x | | | | | | x | x | | Pollard | 2004 |
| 20 | | | | | | | | | | | | | | | | | | | | | | | | | | | | | | | | | | Doyle | 2005 |
| 21 | | | | | | | | x | | | | | | | | | | | x | | | | | | | | | | | | | | | Schmitt | 2005 |
| 22 | | | | | | | | | x | | | | | | | | | | | | | | x | | | | | | | x | x | x | | Thomson | 2005 |
| 23 | | | | | | | | | | | | | | | | | | | | | | | | x | | | | | | x | x | x | | Orr | 2006 |
| 24 | | | | | | | x | | x | x | | | | | | | | | | | | | | | | | | | | | | | | Cheng | 2007 |
| 25 | | | | | | | | | | | | | x | x | | | | | | | | | | | | | | | | | | | | Kuhar | 2007 |
| 26 | | | | | | | | | | | | | | | | | | | x | | | | | | | | | | | | | | | Leaniz | 2008 |
| 27 | | | | | | | | | | | | | | | | | x | x | | | | | | | | | | | | | | | | Maloney | 2008 |
| 28 | | x | | x | | | | | | x | x | x | | | | | | | | | | | | | | | | | | | | | | Burroughs | 2009 |
| 29 | | | | | | | | | x | | x | | | | | | | | | | | | | | | | | | | | | | | Ahearn | 2005 |
| 30 | | | | | | | | | | | | | | | | | | | | | | | | | | | | | | | | | | Ashley | 2006 |
| 31 | | | | x | x | | | | | | | | | | | | | | | | | | | | | | | | | | | | | Evans | 2007 |
| 32 | | | | | | | | | | | | | | | | | | | | | | | | | | | | | | | | | | Velinsky | 2006 |
| 33 | | | | | | | | | | | | | | | | | | | x | | x | | | | | | | | | | | | | Stanley | 2008 |
| 34 | | | | | | | | | x | | | | | | | | | | | | | | | | x | | | | | x | x | x | | Orr | 2008 |
| 35 | x | | | x | x | | | | x | | | x | | | | | | | | | | | | | | | | | | | | | | Rumschlag | 2007 |
| 36 | | | | | | | | | | x | | | | | | | | | | | | x | | | | | | | | | x | x | | Tzsydel | 2009 |

# REFERENCES


Acreman, M., and Dunbar, M.J. (2004). Defining environmental river flow requirements - a review. *Hydrol. Earth Syst. Sci.* **8**, 861–876.

Ahearn, D.S., and Dahlgren, R.A. (2005). Sediment and nutrient dynamics following a low-head dam removal at Murphy Creek. *California. Limnol. Oceanogr.* **50**(6), 1752–1762.

Aldridge, K.T., Brookes, J.D., and Ganf, G.G. (2009). Rehabilitation of stream ecosystem functions through the reintroduction of coarse particulate organic matter. *Restor. Ecol.* **17**, 97–106.

Allan, J.D. (1995). *Stream Ecology: Structure and Function of Running Waters.* Chapman and Hall, London, 388 pp.

Allan, J.D. (2004). Landscapes and riverscapes: The Influence of Land Use on Stream Ecosystems. *Annu. Rev. Ecol. Evol. Syst.* **35**, 257–284.

Arnold, C.L., Jr., and Gibbons, C.L. (1996). Impervious surface coverage: The emergence of a key environmental indicator. *J. Am. Plan. Assoc.* **62**(2), 243–258.

Ashley, J.T.F., Bushaw-Newton, K., Wilhelm, M., Boettner, A., Drames, G., and Velinsky, D.J. (2006). The effects of small dam removal on the distribution of sedimentary contaminants. *Environ. Monit. Assess.* **114**(1–3), 287–312.

Avery, L. (2004). A compendium of 58 trout stream habitat development evaluations in Wisconsin - 1985–2000. *Wisconsin Depart. Nat. Resour. Res. Rep.* **187**, 1–97.

Baattrup-Pedersen, A., and Rijs, T. (1999). Macrophyte diversity and composition in relation to substratum characteristics in regulated and unregulated Danish streams. *Freshwat. Biol.* **42**, 375–385.

Baillie, B.R., Garrett, L.G., and Evanson, A.W. (2008). Spatial distribution and influence of large woody debris in an old-growth forest river system, New Zealand. *For. Ecol. Manage.* **256**, 20–27.

Bakker, E.S., van Donka, E., Declerck, S.A.J., Helmsing, N.R., Hidding, B., and Nolet, B.A. (2010). Effect of macrophyte community composition and nutrient enrichment on plant biomass and algal blooms. *Basic Appl. Ecol.* **11**, 432–439.

Baldigo, B.P., Warren, D.R., Ernst, A.G., and Mulvihill, C.I. (2008). Response of fish populations to natural channel design restoration in streams of the Catskill Mountains. *New York. North Am. J. Fish. Manage.* **28**, 954–969.

Barton, D.R., Taylor, W.D., and Biette, R.M. (1985). Dimensions of Riparian Buffer Strips Required to Maintain Trout Habitat in Southern Ontario Streams. *North Am. J. Fish. Manage.* **5**, 364–378.

Becker, A., and Robson, B.J. (2009). Riverine macroinvertebrate assemblages up to 8 years after riparian restoration in a semi-rural catchment in Victoria. *Australia. Mar. Freshwat. Res.* **60**, 1309–1316.

Bednarek, A.T. (2001). Undamming rivers: A review of the ecological impacts of dam removal. *Environ. Manage.* **27**(6), 803–814.

Beechie, T.J., Sear, D.A., Olden, J.D., Pess, G.R., Buffington, J.M., Moir, H., Roni, P., and Pollock, M.M. (2010). Process-based principles for restoring river ecosystems. *Bioscience* **60**, 209–222.

Bernhardt, E.S., Palmer, M.A., Allan, J.D., Alexander, G., Barnas, K., Brooks, S., Carr, J., Clayton, S., Dahm, C., Follstad-Shah, J., Galat, D., Gloss, S., *et al.* (2005). Ecology—Synthesizing US river restoration efforts. *Science* **308**, 636–637.

Binns, N.A. (2004). Effectiveness of habitat manipulation for wild salmonids in Wyoming streams. *North Am. J. Fish. Manage.* **24**, 911–921.

Birk, S., Strackbein, J., and Hering, D. (2010). WISER methods database. Version: September 2010. Available at http://www.wiser.eu/programme-and-results/data-and-guidelines/method-database/. Accessed on 4 February 2011.

Booth, D.B., and Jackson, C.R. (1997). Urbanization of aquatic systems: Degradation thresholds, stormwater detection, and the limits of mitigation. *J. Am. Water Resour. Assoc.* **33**, 1077–1090.

Borja, A., Miles, A., Occhipinti-Ambrogi, A., and Berg, T. (2009a). Current status of macroinvertebrate methods used for assessing the quality of European marine waters: Implementing the Water Framework Directive. *Hydrobiologia* **633**, 181–196.

Borja, Á., Rodríguez, J.G., Black, K., Bodoy, A., Emblow, C., Fernandes, T.F., Forte, J., Karakassis, I., Muxika, I., Nickell, T.D., Papageorgiou, N., Pranovi, F., *et al.* (2009b). Assessing the suitability of a range of benthic indices in the evaluation of environmental impact of fin and shellfish aquaculture located in sites across Europe. *Aquaculture* **293**, 231–240.

Born, S.M., Filbert, T.L., Genskow, K.D., Hernandez-Mora, N., Keefer, M.L., and White, K.A. (1996). *The removal of small dams: An institutional analysis of the Wisconsin experience*. Cooperative Extension Report 96-1, May. Department of Urban and Regional Planning, University of Wisconsin–Madison, 34 pp.

Bradley, D.C., Clough, S., German, S., Robinson, K., and Bowles, F. (2009). Establishing the real Bourne identity—A twin track approach to assessing river habitat diversity and identifying appropriate river restoration needs. In: *Proceedings of the 10th Ann. River Rest. Centre Conference,* pp. 42–52, Nottingham.

Bradley, D.C., and Ormerod, S.J. (2002). Long-term effects of catchment liming on invertebrates in upland streams. *Freshwat. Biol.* **47**, 161–171.

Broadmeadow, S., and Nisbet, T. (2004). The effects of riparian forest management on the freshwater environment: A literature review of best management practice. *Hydrol. Earth Syst. Sci.* **8**, 286–305.

Brooks, A.P., Gehrke, P.C., Jansen, J.D., and Abbe, T.B. (2004). Experimental reintroduction of woody debris on the Williams River. *NSW: Geomorphic and ecological responses. River Res. Appl.* **20**, 513–536.

Brooks, A.P., Howell, T., Abbe, T.B., and Arthington, A.H. (2006). Confronting hysteresis: Wood based river rehabilitation in highly altered riverine landscapes of south-eastern Australia. *Geomorphology* **79**, 395–422.

Brooks, S.S., Palmer, M.A., Cardinale, B.J., Swan, C.M., and Ribblett, S. (2002). Assessing stream ecosystem rehabilitation: Limitations of community structure data. *Restor. Ecol.* **10**, 156–168.

Bukaveckas, P.A. (2007). Effects of channel restoration on water velocity, transient storage, and nutrient uptake in a channelized stream. *Environ. Sci. Technol.* **41**, 1570–1576.

Bunn, S.E., and Arthington, A.H. (2002). Basic principles and ecological consequences of altered flow regimes for aquatic biodiversity. *Environ. Manage.* **30**, 492–507.

Burroughs, B.A., Hayes, D.B., Klomp, K.D., Hansen, J.F., and Mistak, J. (2009). Effects of Stronach Dam removal on fluvial geomorphology in the Pine River, Michigan, United States. *Geomorphology* **110**, 96–107.

Bushaw-Newton, K.L., Hart, D.D., Pizzuto, J.E., Thomson, J.R., Egan, J., Ashley, J.T., Johnson, T.E., Horwitz, R.J., Keeley, M., Lawrence, J., Charles, D., Gatenby, C., *et al.* (2002). An integrative approach towards understanding ecological responses to dam removal: The Manatawny Creek Study. *J. Am. Water Resour. Assoc.* **38**, 1581–1599.

Carl, J. (2000). Habitat Surveys as a Tool to assess the benefits of Stream rehabilitation II: Fish. *Verh. Int. Verein. Theor. Angew. Limnol.* **27**, 1515–1519.

Castelle, A.J., Johnson, A.W., and Conolly, C. (1994). Wetland and stream buffer size requirements—A review. *J. Environ. Qual.* **23**, 878–882.

Cederholm, C.J., Bilby, R.E., Bisson, P.A., Bumstead, T.W., Fransen, E.R., Scarlett, W.J., and Ward, J.W. (1997). Response of juvenile coho salmon and steelhead to placement of large woody debris in a coastal Washington stream. *North Am. J. Fish. Manage.* **17**, 947–963.

Chaplin, J.J. (2003). Framework for monitoring and preliminary results after removal of Good Hope Mill Dam. In: *Dam Removal Research: Status and Prospects* (Ed. by W.L. Graf). The H. John Heinz III Center for Science, Economics and the Environment, Washington, DC.

Chen, X., Wei, X., Scherer, R., and Hogan, D. (2008). Effects of large woody debris on surface structure and aquatic habitat in forested streams, southern interior British Columbia. *Canada. River Res. Appl.* **24**, 862–875.

Cheng, F., and Granata, T. (2007). Sediment transport and channel adjustments associated with dam removal: Field observations. *Water Resour. Res.* **43**, 1–14.

Chovanec, A., Schiemer, F., Weidbacher, H., and Spolwind, R. (2002). Rehabilitation of a heavily modified river section of the Danube in Vienna (Austria): Biological assessment of landscape linkages on different scales. *Int. Rev. Hydrobiol.* **87**, 183–195, Halbgeviert.

Clews, E., and Ormerod, S.J. (2008). Improving bio-diagnostic monitoring using simple combinations of standard biotic indices. *River Res. Appl.* **24**, 1–14.

Cloern, J.E. (2001). Our evolving conceptual model of the coastal eutrophication problem. *Mar. Ecol. Prog. Ser.* **210**, 223–253.

Coe, H.J., Kiffney, P.M., Pess, G.R., Kloehn, K.K., and McHenry, M.L. (2009). Periphyton and invertebrate response to wood placement in large pacific coastal rivers. *River Res. Appl.* **25**, 1025–1035.

Correll, D.L. (2005). Principles of planning and establishment of buffer zones. *Ecol. Eng.* **24**, 433–439.

Crispin, V., House, R., and Roberts, D. (1993). Changes in instream habitat, large woody devris, and salmon habitat after restructuring of a coastal Oregon stream. *North Am. J. Fish. Manage.* **13**, 96–102.

Davies-Colley, R.J., Meleason, M.A., Hall, G.M., and Rutherford, J.C. (2009). Modelling the time course of shade, temperature, and wood recovery in streams with riparian forest restoration. *N. Z. J. Mar. Freshwat. Res.* **43**, 673–688.

Davies-Colley, R.J., and Quinn, J.M. (1998). Stream lighting in five regions of North Island, New Zealand: Control by channel size and riparian vegetation. *N. Z. J. Mar. Freshwat. Res.* **32**, 591–605.

Dedecker, A.P., van Melckebeke, K., Goethals, P.L.M., and de Pauw, N. (2007). Development of migration models for macroinvertebrates in the Zwalm river basin (Flanders, Belgium) as tools for restoration management. *Ecol. Modell.* **203**, 72–86.

de Leaniz, C.G. (2008). Weir removal in salmonid streams: Implications, challenges and practicalities. *Hydrobiologia* **609**, 83–96.

Diaz, R., and Rosenberg, R. (2008). Spreading dead zones and consequences for marine ecosystems. *Science* **321**, 926–929.

Dister, E., Gomer, D., Obrdlik, P., Petermann, P., and Schneider, E. (1990). Water mangement and ecological perspectives of the upper rhine's floodplains. *Regul. Rivers: Res. Managem.* **5**, 1–15.

Doeg, T.J., and Koehn, J.D. (1994). Effects of draining and desiltation of a small weir on downstream fish and macroinvertebrates. *Regul. Rivers: Res. Managem.* **9**, 263–277.

Dosskey, M.G. (2001). Toward quantifying water pollution abatement in response to installing buffers on crop land. *Environ. Manage.* **28**, 577–598.

Downes, B.J., Barmuta, L.A., Fairweather, P.G., Faith, D.P., Keough, M.J., Lake, P.S., Mapstine, B.D., and Quinn, G.P. (2002). *Monitoring Ecological Impacts—Concepts and Practice in Flowing Waters.* Cambridge University Press, Cambridge, England, 452 pp.

Downs, P.W. (1995). River channel classification for channel management purposes. In: *Changing River Channels* (Ed. by A. Gurnell and G. Petts), pp. 347–365. John Wiley and Sons Ltd, UK.

Doyle, M.W., Stanley, E.H., Orr, C.H., Selle, A.R., Sethi, S.A., and Harbor, J.M. (2005). Stream ecosystem response to small dam removal: Lessons from the Heartland. *Geomorphology* **71**, 227–244.

Duarte, C.M., Conley, D.J., Carstensen, J., and Sánchez-Camacho, M. (2009). Return to neverland: Shifting baselines affect eutrophication restoration targets. *Estuaries Coasts* **32**, 29–36.

Dunbar, M.J., Pedersen, M.L., Cadman, D., Extence, C.A., Waddingham, J., Chadd, R.P., and Larsen, S.E. (2010a). River discharge and local scale physical habitat influence macroinvertebrate LIFE score. *Freshwat. Biol.* **55**, 226–242.

Dunbar, M.J., Warren, M., Extence, C.A., Baker, L., Cadman, D., Mould, D., Hall, J., and Chadd, R.P. (2010b). Interaction between macroinvertebrates, discharge and physical habitat in upland rivers. *Aquat. Conserv.: Mar. Freshwat. Ecosyst.* **20**, 31–44.

Edwards, C.J., Griswold, B.L., Tubb, R.A., Weber, E.C., and Woods, C.L. (1984). Mitigating effects of artificial riffles and pools on the fauna of a channelized warmwater streams. *North Am. J. Fish. Manage.* **4**, 194–203.

EEA (European Environment Agency) (2007). *Halting the loss of biodiversity by 2010: Proposal for a first set of indicators to monitor progress in Europe.* EEA technical report 11/2007, Luxembourg, 38 pp.

Elliott, M. (2002). The role of the DPSIR approach and conceptual models in marine environmental management: An example for offshore wind power. *Mar. Pollut. Bull.* **44**, 3–7.

Entrekin, S.A., Tank, J.L., Rosi-Marshall, E.J., Hoellein, T.J., and Lamberti, G.A. (2008). Responses in organic matter accumulation and processing to an experimental wood addition in three headwater streams. *Freshwat. Biol.* **53**, 1642–1657.

Entrekin, S.A., Tank, J.L., Rosi-Marshall, E.J., Hoellein, T.J., and Lamberti, G.A. (2009). Response of secondary production by macroinvertebrates to large wood addition in three Michigan streams. *Freshwat. Biol.* **54**, 1741–1758.

Evans, J.E., Huxley, J.M., and Vincent, R.K. (2007). Upstream channel changes following dam construction and removal using a GIS/remote sensing approach. *J. Am. Water Resour. Assoc.* **43**(3), 683–697.

Feist, G.W., Webb, M.A.H., Gundersen, D.T., Foster, E.P., Schreck, C.B., Maule, A.G., and Fitzpatrick, M.S. (2005). Evidence of detrimental effects of environmental contaminants on growth and reproductive physiology of white sturgeon in impounded areas of the Columbia River. *Environ. Health Perspect.* **113**, 1675–1682.

Feld, C.K. (2004). Identification and measure of hydromorphological degradation in Central European lowland streams. *Hydrobiologia* **516**, 69–90.

Feld, C.K., and Hering, D. (2007). Community structure or function: Effects of environmental stress on benthic macroinvertebrates at different spatial scales. *Freshwat. Biol.* **52**, 1380–1399.

Feld, C.K., Martins da Silva, P., Sousa, J.P., de Bello, F., Bugter, R., Grandin, U., Hering, D., Lavorel, S., Mountford, O., Pardo, I., Pärtel, M., Römbke, J., *et al.* (2009). Indicators of biodiversity and ecosystem services: A synthesis across ecosystems and spatial scales. *Oikos* **118**, 1862–1871.

Feld, C.K., Sousa, J.P., Martins da Silva, P., and Dawson, T.P. (2010). Indicators for biodiversity and ecosystem services: Towards an improved framework for ecosystems assessment. *Biodivers. Conserv.* **19**, 2895–2919.

Fitzpatrick, F.A., Scudder, B.C., Lenz, B.N., and Sullivan, D.J. (2001). Effects of multi-scale environmental characteristics on agricultural stream biota in eastern Wisconsin. *J. Am. Water Resour. Assoc.* **37**, 1489–1507.

Folt, C.L., Chen, C.Y., Moore, M.V., and Burnaford, J. (1999). Synergism and antagonism among multiple stressors. *Limnol. Oceanogr.* **44**, 864–873.

Friberg, N. (2010a). Pressure-response relationships in stream ecology: Introduction and synthesis. *Freshwat. Biol.* **55**, 1367–1381.

Friberg, N. (2010b). Ecological consequences of river channel management. In: *Handbook of Catchment Management* (Ed. by R.C. Ferrier and A. Jenkins), pp. 77–106. Wiley-Blackwell, Chichester, UK.

Friberg, N., Hansen, H.O., and Kronvang, B. (2000). Habitat Surveys as a Tool to assess the benefits of Stream rehabilitation. II: Macroinvertebrate Communities. *Verh. Int. Verein. Theor. Angew. Limnol.* **27**, 1510–1514.

Friberg, N., Hansen, H.O., Kronvang, B., and Svendsen, L.M. (1998). Long-term, habitat-specific response of a macroinvertebrate community to river restoration. *Aquat. Conserv.: Mar. Freshwat. Ecosyst.* **8**, 87–99.

Friberg, N., Kronvang, B., Svendsen, L.M., Hansen, H.O., and Nielsen, M.B. (1994). Restoration of channellized reach of the River Gelså, Denmark. *Aquat. Conserv.: Mar. Freshwat. Ecosyst.* **4**, 289–297.

Friberg, N., Bonada, N., Bradley, D.C., Dunbar, M.J., Edwards, F.K., Grey, J., Hayes, R.B., Hildrew, A.G., Lamouroux, N., Trimmer, M., and Woodward, G. (2011). Biomonitoring of human impacts in natural ecosystems: The good, the bad and the ugly. *Adv. Ecol. Res.* **44**, 1–68.

Frissell, C.A., Liss, W.J., Warren, C.E., and Hurley, M.D. (1986). A hierarchical framework for stream habitat classification: Viewing streams in a watershed context. *Environ. Manage.* **12**, 199–214.

Furse, M., Hering, D., Moog, O., Verdonschot, P., Johnson, R.K., Brabec, K., Gritzalis, K., Buffagni, A., Pinto, P., Friberg, N., Murray-Bligh, J., Kokes, J., *et al.* (2006). The STAR project: Context, objectives and approaches. *Hydrobiologia* **566**, 3–29.

Galli, J. (1991). *Thermal Impacts Associated with Urbanization and Stormwater Management Best Management Practices.* Metropolitan Washington Council of Governments, Maryland Department of Environment, Washington, DC, 188 pp.

Gerhard, M., and Reich, M. (2000). Restoration of streams with large wood: Effects of accumulated and built-in wood on channel morphology, habitat diversity and aquatic fauna. *Int. Rev. Hydrobiol.* **85**, 123–137.

Ghermandi, A., Vandenberghe, V., Benedetti, L., Bauwens, W., and Vanrolleghem, P. (2009). Model-based assessment of shading effect by riparian vegetation on river water quality. *Ecol. Eng.* **35**, 92–104.

Gregory, S., Li, H., and Li, J. (2002). The conceptual basis for ecological responses to dam removal. *Bioscience* **52**, 713–723.

Gurnell, A.M., Gregory, K.J., and Petts, G.E. (1995). The role of coarse woody debris in forest aquatic habitats: Implications for management. *Aquat. Conserv.: Mar. Freshwat. Ecosyst.* **5**, 143–166.

Gurnell, A.M., Piégay, H., Swanson, F.J., and Gregory, S.V. (2002). Large wood and fluvial processes. *Freshwat. Biol.* **47**, 601–619.

Haidekker, A., and Hering, D. (2007). Relationship between benthic insects (Ephemeroptera, Plecoptera, Coleoptera, Trichoptera) and temperature in small and medium-sized streams in Germany: A multivariate study. *Aquat. Ecol.* **42**, 463–481.

Hammond, D., Mant, J., Janes, M., and Fellick, A. (2009). River restoration assessment of the STREAM project - Executive summary. Report by the River Restoration Centre (UK). http://www.streamlife.org.uk/actions/survey/. Accessed on 4 February 2011.

Hansen, H.O., and Baattrup-Pedersen, A. (2006). A new development: Stream restoration. In: *Running Waters* (Ed. by K. Sand-Jensen, N. Friberg and J. Murphy). National Environmental Research Institute, Silkeborg, Denmark.

Harding, J.S., Benfield, E.F., Bolstad, P.V., Helfman, G.S., and Jones, E.B.D. (1998). Stream biodiversity: The ghost of land use past. *PNAS* **95**, 14843–14847.

Harrison, S.S., Pretty, J.L., Shepherd, D., Hildrew, A.G., Smith, C., and Hey, R.D. (2004). The effect of instream rehabilitation structures on macroinvertebrates in lowland rivers. *J. Appl. Ecol.* **41**, 1140–1154.

Hart, D.D., Johnson, T.E., Bushaw-Newton, K.L., Horwitz, R.J., Bednarek, A.T., Charles, D.F., Kreeger, D.A., and Velinsky, D.J. (2002). Dam removal: Challenges and opportunities for ecological research and river restoration. *Bioscience* **52**, 669–681.

Hart, D.D., Johnson, T.E., Bushaw-Newton, K.L., Horwitz, R.J., and Pizzuto, J.E. (2003). Ecological effects of dam removal: An integrative case study and risk assessment framework for prediction. In: *Dam Removal Research: Status and Prospects* (Ed. by W.L. Graf). The H. John Heinz III Center for Science, Economics and the Environment, Washington, DC.

Harvey, B.C., and Stewart, A.J. (1991). Fish size and habitat depth relationships in headwater streams. *Oecologia* **87**, 336–342.

Henley, W., Patterson, M., Neves, R., and Lemly, A.D. (2000). Effects of sedimentation and turbidity on lotic food webs: A concise review for natural resource managers. *Rev. Fish. Sci.* **8**(2), 125–139.

Henry, C.P., and Amoros, C. (1995). Restoration ecology of Riverine Wetlands: I. A scientific base. *Environ. Manage.* **19**, 891–902.

Herbst, D.B., and Kane, J.M. (2009). Responses of aquatic macroinvertebrates to stream channel reconstruction in a degraded rangeland creek in the Sierra Nevada. *Ecol. Restor.* **27**, 76–88.

Hering, D., Moog, O., Sandin, L., and Verdonschot, P.F.M. (2004). Overview and application of the AQEM assessment system. *Hydrobiologia* **516**, 1–20.

Hilderbrand, R., Lemly, A., Dolloff, C., and Harpster, K. (1997). Effects of large woody debris placement on stream channels and benthic macroinvertebrates. *Can. J. Fish. Aquat. Sci.* **54**, 931–939.

Hill, M.J., Long, E.A., and Hardin, S. (1993). *Effects of dam removal on Dead Lake, Chipola River, Florida. Apalachi-cola River Watershed Investigations, Florida Game and Fresh Water Fish Commission*. A Wallop-Breaux Project F-39-R, 12 pp.

Hladyz, S., Tiegs, S.D., Gessner, M.O., Giller, P.S., Risnoveanu, G., Preda, E., Nistorescu, M., Schindler, M., and Woodward, G. (2010). Leaf-litter breakdown in pasture and deciduous woodland streams: A comparison among three European regions. *Freshwat. Biol.* **55**, 1916–1929.

Hladyz, S., Abjornsson, K., Giller, P.S., and Woodward, G. (2011a). Impacts of an aggressive riparian invader on community structure and ecosystem functioning in stream food webs. *J. Appl. Ecol.* **48**(2), 443–452.

Hladyz, S., Åbjörnsson, K., Chauvet, E., Dobson, M., Elosegi, A., Ferreira, V., Fleituch, T., Gessner, M.O., Giller, P.S., Gulis, V., Hutton, S.A., Lacoursière, J.O., *et al.* (2011b). Stream ecosystem functioning in an agricultural landscape: The importance of terrestrial-aquatic linkages. *Adv. Ecol. Res.* **44**, 211–276.

Horner, R.R., Booth, D.B., Azous, A., and May, C.W. (1997). Watershed determinants of ecosystem functioning. ,In: *Proc. Engineer. Found. Conf.* pp. 251–274. Snowbird, Utah, USA (4–9 August 1996).

House, R.A. (1996). An evaluation of stream restoration structures in a coastal Oregon stream, 1981–1993. *North Am. J. Fish. Manage.* **16**, 272–281.

Howson, T.J., Robson, B.J., and Mitchell, B.D. (2009). Fish assemblage response to rehabilitation of a sand-slugged lowland river. *River Res. Appl.* **25**, 1251–1267.

Huryn, A.D., Huryn, V.M.B., Arbuckle, C.J., and Tsomides, L. (2002). Catchment land-use, macroinvertebrates and detritus processing in headwater streams: Taxonomic richness versus function. *Freshwat. Biol.* **47**, 401–415.

Huusko, A., and Yrjänä, T. (1995). Évaluation de la restauration des rivières chenalisées pour le flottage du bois: étude du cas de la rivière Kutinjoki, Finlande du Nord. *Bulletin Français de la Pêche et de la Pisciculture* **337/338/339**, 407–413.

Hynes, H.B.N. (1960). *The Biology of Polluted Waters.* Liverpool University Press, Liverpool, England, 202 pp.

Iversen, T.M., Kronvang, B., Madsen, B.L., Markmann, P., and Nielsen, M.B. (1993). Re-establishment of Danish streams: Restoration and maintenance measures. *Aquat. Conserv.: Mar. Freshwat. Ecosyst.* **3**, 73–92.

Jähnig, S.C., Brunzel, S., Gacek, S., Lorenz, A.W., and Hering, D. (2009a). Effects of re-braiding measures on hydromorphology, floodplain vegetation, ground beetles and benthic invertebrates in mountain rivers. *J. Appl. Ecol.* **46**, 406–416.

Jähnig, S.C., Lorenz, A.W., and Hering, D. (2009b). Re-meandering German lowland streams: Qualitative and quantitative effects of restoration measures on hydromorphology and macroinvertebrates. *Environ. Manage.* **44**, 745–754.

Januschke, K., Sundermann, A., Antons, C., Haase, P., Lorenz, A., and Hering, D. (2009). Untersuchung und Auswertung von ausgewählten Renaturierungsbeispielen repräsentativer Fließgewässertypen der Flusseinzugsgebiete Deutschlands. [Investigation and analysis of selected restoration case studies at representative river types in Germany.]. *Deut. Rat Landespfl.* **82**, 23–39.

Jensen, F. (1995). Døgnflue- og slørvingebestanden udvikling og status I Skern Å-systemet. Naturhistorisk Museum, University of Århus, Denmark, 53 pp.

Jeppesen, E., Søndergaard, M., Jensen, J.P., Havens, K., Anneville, O., Carvalho, L., Coveney, M.F., Deneke, R., Dokulil, M., Foy, B., Gerdeaux, D., Hampton, S.E., *et al.* (2005). Lake responses to reduced nutrient loading—An analysis of contemporary long-term data from 35 case studies. *Freshwat. Biol.* **50**, 1747–1771.

Johnson, R.K., and Hering, D. (2009). Response of river inhabiting organism groups to gradients in nutrient enrichment and habitat physiography. *J. Appl. Ecol.* **46**, 175–186.

Johnson, S.L., Rodgers, J.D., Solazzi, M.F., and Nickelson, T.E. (2005). Effects of an increas ein large wood on abundance and survival of juvenile Salmonids (Oncorhynchus spp.) in an Oregon coastal stream. *Can. J. Fish. Aquat. Sci.* **62**, 412–424.

Jowett, I.G., Richardson, J., and Boubée, J.A. (2009). Effects of riparian manipulation on stream communities in small streams: Two case studies. *N. Z. J. Mar. Freshwat. Res.* **43**, 763–774.

Jungwirth, M., Moog, O., and Muhar, S. (1993). Effects of river bed restructuring on fish and benthos of a fifth order stream, Melk. *Austria. Regul. Rivers: Res. Manage.* **8**, 195–204.

Jungwirth, M., Muhar, S., and Schmutz, S. (1995). The effects of recreated instream and ecotone structures on the fish fauna of an epipotamal river. *Hydrobiologia* **303**, 195–206.

Kail, J., Hering, D., Muhar, S., Gerhard, M., and Preis, S. (2007). The use of large wood in stream restoration: Experiences from 50 projects in Germany and Austria. *J. Appl. Ecol.* **44**, 1145–1155.

Kanehl, P.D., Lyons, J., and Nelson, J.E. (1997). Changes in the habitat and fish community of the Milwaukee River, Wisconsin, following removal of the Woolen Mills Dam. *North Am. J. Fish. Manage.* **17**, 387–400.

Karageorgis, A.P., Skourtos, M.S., Kapsimalis, V., Kontogianni, A.D., Skoulikidis, N.T., Pagou, K., Nikolaidis, N.P., Drakopoulou, P., Zanou, B., Karamanos, H., Levkov, Z., and Anagnostou, C. (2005). An integrated approach to watershed management within the DPSIR framework: Axios River catchment and Thermaikos Gulf. *Reg. Environ. Change* **5**, 138–160.

Kaushal, S.S., Groffman, P.M., Mayer, P.M., Striz, E., and Gold, A.J. (2008). Effects of stream restoration on denitrificationin an urbanizing watershed. *Ecol. Appl.* **18**, 789–804.

King, E.G., and Hobbs, R.J. (2006). Identifying linkages among conceptual models of ecosystem degradation and restoration: Towards an integrative framework. *Restor. Ecol.* **14**, 369–378.

Kronvang, B., Svendsen, L.M., Ottosen, O., Nielsen, M.B., and Johannesen, L. (1998). Re-meandering of rivers: Short-term implications for sediment and nutrient transport. *Verh. Int. Ver. Theoret. Angew. Limnol.* **26**, 929–934.

Kuhar, U., Gregorc, T., Rencelj, M., Sraj-Krzic, N., and Gaberscik, A. (2007). Distribution of macrophytes and condition of the physical environment of streams flowing through agricultural landscape in north-eastern Slovenia. *Limnologica* **37**, 146–154.

Laasonen, P., Muotka, T., and Kivijärvi, I. (1998). Recovery from macroinvertebrate communities from stream habitat restoration. *Aquat. Conserv.: Mar. Freshwat. Ecosyst.* **8**, 101–113.

Lake, P.S., Bond, N., and Reich, P. (2007). Linking ecological theory with stream restoration. *Freshwat. Biol.* **52**, 597–615.

Larson, M.G., Booth, D.B., and Morley, S.A. (2001). Effectiveness of large woody debris in stream rehabilitation projects in urban basins. *Ecol. Eng.* **18**, 211–226.

Layer, K., Riede, J.O., Hildrew, A.G., and Woodward, G. (2010). Food web structure and stability in 20 streams across a wide pH gradient. *Adv. Ecol. Res.* **42**, 265–301.

Layer, K., Hildrew, A.G., Jenkins, G.B., Riede, J.O., Rossiter, S.J., Townsend, C.R., and Woodward, G. (2011). Long-term dynamics of a well-characterised food web: Four decades of acidification and Recovery in the Broadstone stream model system. *Adv. Ecol. Res.* **44**, 69–118.

Lehane, B.M., Giller, P.S., O'Halloran, J., Smith, C., and Murphy, J. (2002). Experimental provision of large woody debris in streams as a trout management technique. *Aquat. Conserv.: Mar. Freshwat. Ecosyst.* **12**, 289–311.

Lemly, A.D., and Hilderbrand, R.H. (2000). Influence of large woody debris on stream insect communities and benthic detritus. *Hydrobiologia* **421**, 179–185.

Lepori, F., Gaul, D., Palm, D., and Malmqvist, B. (2006). Food-web responses to restoration of channel heterogeneity in boreal streams. *Can. J. Fish. Aquat. Sci.* **63**, 2478–2486.

Lepori, F., Palm, D., Brännäs, E., and Malmqvist, B. (2005a). Does Restoration of Structural Heterogeneity in Streams Enhance Fish and Macroinvertebrate Diversity? *Ecol. Appl.* **15**, 2060–2071.

Lepori, F., Palm, D., and Malmqvist, B. (2005b). Effects of stream restoration on ecosystem functioning: Detritus retentiveness and decomposition. *J. Appl. Ecol.* **42**, 228–238.

Lester, R.E., and Boulton, A.J. (2008). Rehabilitating agricultural streams in Australia with wood: A review. *Environ. Manage.* **42**, 310–326.

Levell, A.P., and Chang, H. (2008). Monitoring the channel process of a stream restoration project in an urbanizing watershed: A case study of Kelley Creek, Oregon. *USA. River Res. Appl.* **182**, 169–182.

Lorenz, A., Hering, D., Feld, C.K., and Rolauffs, P. (2004). A new method for assessing the impact of hydromorphological degradation on the macroinvertebrate fauna of five German stream types. *Hydrobiologia* **516**, 107–127.

Lorenz, A.W., Jähnig, S.C., and Hering, D. (2009). Re-meandering German lowland streams: Qualitative and quantitative effects of restoration measures on hydromorphology and macroinvertebrates. *Environ. Manage.* **44**, 745–754.

MA (Millennium Ecosystem Assessment) (2005). Ecosystems and Human Well-being: Synthesis. Island Press, Washington, DC.

Maloney, K.O., Dodd, H.R., Butler, S.E., and Wahl, D.H. (2008). Changes in macroinvertebrate and fish assemblages in a medium-sized river following a breach of a low-head dam. *Freshwat. Biol.* **53**, 1055–1068.

Mankin, K.R., Ngandu, D.M., Barden, C.J., Hutchinson, S.L., and Geyer, W.A. (2007). Grass-shrub riparian buffer removal of sediment, phosphorus, and nitrogen from simulated runoff. *J. Am. Water Resour. Assoc.* **43**, 1108–1116.

Matthaei, C.D., Weller, F., Kelly, D.W., and Townsend, C.R. (2006). Impacts of fine sediment addition to tussock, pasture, dairy and deer farming streams in New Zealand. *Freshwat. Biol.* **51**, 2154–2172.

Matthews, J., Reeze, B., Feld, C.K., and Hendriks, A.J. (2010). Lessons from practice: Assessing early progress and success in river rehabilitation. *Hydrobiologia* **655**, 1–14.

McBride, M., Hession, W., and Rizzo, D. (2008). Riparian reforestation and channel change: A case study of two small tributaries to Sleepers River, northeastern Vermont, USA. *Geomorphology* **102**, 445–459.

McCoy, E.D., and Bell, S.S. (1991). Habitat structure: The evolution and diversification of a complex topic. In: *Habitat Structure: The Physical Arrangement of Objects in Space* (Ed. by S.S. Bell, E.D. McCoy and H.R. Mushinsky), pp. 3–27. Chapman and Hall, London.

Merz, J.E., and Chan, L.K.O. (2005). Effects of gravel augmentation on macroinvertebrate assemblages in a regulated California river. *River Res. Appl.* **21**, 61–74.

Merz, J.E., and Setka, J.D. (2004). Evaluation of a spawning habitat enhancement site for Chinook salmon in a regulated California river. *North Am. J. Fish. Manage.* **24**, 397–407.

Merz, J.E., Setka, J.D., Pasternack, G.B., and Wheaton, J.M. (2004). Predicting benefits of spawning-habitat rehabilitation to salmonid (*Oncorhynchus* spp.) fry production in a regulated California river. *Can. J. Fish. Aquat. Sci.* **24**, 397–407.

Merz, J.E., Smith, J.R., Workman, M.L., Setka, J.D., and Muchaey, B. (2008). Aquatic macrophyte encroachment in chinook salmon spawning beds: Lessons learned from gravel enhancement monitoring in the lower Mokelumne River. *California. North Am. J. Fish. Manage.* **28**, 1568–1577.

Miller, S.W., Budy, P., and Schmidt, J.C. (2010). Quantifying macroinvertebrate responses to in-stream habitat restoration: Applications of meta-analysis to river restoration. *Restor. Ecol.* **18**, 8–19.

Millington, C.E., and Sear, D.A. (2007). Impacts of river restoration on small-wood dynamics in a low-gradient headwater stream. *Earth* **1218**, 1204–1218.

Moerke, A., Gerard, K., Latimore, J., Hellenthal, R., and Lamberti, G. (2004). Restoration of an Indiana, USA, stream: Bridging the gap between basic and applied lotic ecology. *J. North Am. Benthol. Soc.* **23**, 647–660.

Moerke, A.H., and Lamberti, G.A. (2003). Responses in fish community structure to restoration of Two Indiana streams. *North Am. J. Fish. Manage.* **23**, 748–759.

Moerke, A.H., and Lamberti, G.A. (2004). Restoring stream ecosystems: Lessons from a midwestern state. *Restor. Ecol.* **12**, 327–334.

Moss, B., Hering, D., Green, A.J., Aidoud, A., Becares, E., Beklioglu, M., Bennion, H., Boix, D., Brucet, S., Carvalho, L., Clement, B., Davidson, T., *et al.* (2009). Climate change and the future of freshwater biodiversity in Europe: A primer for policy-makers. *Freshwat. Rev.* **2**, 103–130.

Moustgaard-Pedersen, T.C., Baattrup-Pedersen, A., and Madsen, T.V. (2006). Effects of stream restoration and management on plant communities in lowland streams. *Freshwat. Biol.* **51**, 161–179.

Muhar, S., Jungwirth, M., Unfer, G., Wiesner, C., Poppe, M., Schmutz, S., Hohensinner, S., and Habersack, H. (2008). Restoring riverine landscapes at the Drau River: Successes and deficits in the context of ecological integrity. *Dev. Earth Surf. Proc.* **11**, 779–803.

Mulder, C., Boit, A., Bonkowski, M., De Ruiter, P.C., Mancinelli, G., Van der Heijden, M.G.A., van Wijnen, H.J., Vonk, J.A., and Rutgers, M. (2011). A below-ground perspective on Dutch agroecosystems: How soil organisms interact to support ecosystem services. *Adv. Ecol. Res.* **44**, 277–358.

Muotka, T., and Laasonen, P. (2002). Ecosystem recovery in restored headwater streams: The role of enhanced leaf retention. *J. Appl. Ecol.* **39**, 145–156.

Muotka, T., Paavola, R., Haapala, A., Novikmec, M., and Laasonen, P. (2002). Long-term recovery of stream habitat structure and benthic invertebrate communities from in-stream restoration. *Biol. Conserv.* **105**, 243–253.

Muotka, T., and Syrjänen, J. (2007). Changes in habitat structure, benthic invertebrate diversity, trout populations and ecosystem processes in restored forest streams: A boreal perspective. *Freshwat. Biol.* **52**, 724–737.

Naiman, R.J., and Decamps, H. (1997). The ecology of interfaces—Riparian zones. *Annu. Rev. Ecol. Syst.* **28**, 621–658.

Nakano, D., and Nakamura, F. (2006). Responses of macroinvertebrate communities to river restoration in a channelized segment of the Shibetsu River. *Northern Japan. River Res. Appl.* **22**, 681–689.

Negishi, J.N., and Richardson, J.S. (2003). Responses of organic matter and macroinvertebrates to placements of boulder clusters in a small stream of southwestern British Columbia. *Canada. Can. J. Fish. Aquat. Sci.* **60**, 247–258.

Niemi, G.J., DeVore, P., Detenbeck, N., Yount, J.D., Lima, A., Pastor, J., and Naiman, R.J. (1990). Overview of case studies on recovery of aquatic systems from disturbance. *Environ. Manage.* **14**, 571–588.

Nilsson, C., Lepori, F., Malmqvist, B., Törnlund, E., Hjerdt, N., Helfield, J.M., Palm, D., Östergren, J., Jansson, R., Brännäs, E., and Lundqvist, H. (2005). Forecasting environmental responses to restoration of rivers used as Log floatways: An interdisciplinary challenge. *Ecosystems* **8**, 779–800.

Northington, R.M., and Hershey, A.E. (2006). Effects of stream restoration and wastewater treatment plant effluent on fish communities in urban streams. *Freshwat. Biol.* **51**, 1959–1973.

Olesen, J.M., Dupont, Y.L., O'Gorman, E.J., Ings, T.C., Layer, K., Melián, C.J., Troejelsgaard, K., Pichler, D.E., Rasmussen, C., and Woodward, G. (2010). From Broadstone to Zackenberg: Space, time and hierarchies in ecological networks. *Advances Ecol. Research* **42**, 1–71.

Opperman, J.J., and Merenlender, A.M. (2004). The Effectiveness of Riparian Restoration for Improving Instream Fish Habitat in Four Hardwood-Dominated California Streams. *North Am. J. Fish. Manage.* **24**, 822–834.

Multiple Stressors in Freshwater Ecosystems. Ormerod, S.J., Dobson, M., Hildrew, A.H. and Townsend, C. (Eds.), (2010) *Freshwat. Biol.* **55**, 1–269.

Orr, C.H., Kroiss, S.J., Rogers, K.L., and Stanley, E.H. (2008). Downstream benthic responses to small dam removal in a coldwater stream. *River Res. Appl.* **24**(6), 804–822.

Orr, C.H., Rogers, K.L., and Stanley, E.H. (2006). Channel morphology and P uptake following removal of a small dam. *J. North Am. Benthol. Soc.* **25**, 556–568.

Osborne, L.L., and Kovacic, D.A. (1993). Riparian vegetated buffer strips in water-quality restoration and stream management. *Freshwat. Biol.* **29**, 243–258.

Palm, D., Brännäs, E., Lepori, F., Nilsson, K., and Stridsman, S. (2007). The influence of spwaning habitat restoration on juvenile brown trout (Salmo trutta) density. *Can. J. Fish. Aquat. Sci.* **64**, 509–515.

Palm, D., Lepori, F., and Brännas, E. (2010). Influence of habitat restoration on post-emergence displacement of brown trout (Salmo trutta L.): A case study in a northern Swedish stream. *River Res. Appl.* **26**, 742–750.

Palmer, M.A. (2009). Invited Odum essay: Reforming Watershed Restoration: Science in Need of Application and Applications in Need of Science. *Estuaries Coasts* **32**, 1–17.

Palmer, M., Allan, J.D., Meyer, J., and Bernhardt, E.S. (2007). River restoration in the twenty-first century: Data and experiential knowledge to inform future efforts. *Restor. Ecol.* **15**, 472–481.

Palmer, M.A., Ambrose, R.F., and Poff, N.L. (1997). Ecological theory and community. *Restor. Ecol.* **5**(4), 291–300.

Palmer, M.A., Bernhardt, E.S., Allan, J.D., Lake, P.S., Alexander, G., Brooks, S., Carr, J., Clayton, S., Dahm, C.N., Follstad Shah, J., Galat, D.L., Loss, S.G., *et al.* (2005). Standards for ecologically successful river restoration. *J. Appl. Ecol.* **42**, 208–217.

Pander, J., and Geist, J. (2010). Seasonal and spatial bank habitat use by fish in highly altered rivers—A comparison of four different restoration measures. *Ecol. Freshwat. Fish* **19**, 127–138.

Palmer, M.A., Menninger, H., and Bernhardt, E.S. (2010). River restoration, habitat heterogeneity and biodiversity: A failure of theory or practice? *Freshwat. Biol.* **55** (Suppl. 1), 205–222.

Parkyn, S., Davies-Colley, R., Cooper, A., and Stroud, M. (2005). Predictions of stream nutrient and sediment yield changes following restoration of forested riparian buffers. *Ecol. Eng.* **24**, 551–558.

Parkyn, S.M., Davies-Colley, R.J., Halliday, N.J., Costley, K.J., and Croker, G.F. (2003). Planted riparian buffer zones in New Zealand: Do they live up to expectations? *Restor. Ecol.* **11**, 436–447.

Paul, M.J., and Meyer, J.L. (2001). Streams in the urban landscape. *Annu. Rev. Ecol. Syst.* **32**, 333–365.

Pedersen, M.L., Friberg, N., Skriver, J., Baattrup-Pedersen, A., and Larsen, S.E. (2007). Restoration of Skjern River and its valley—Short-term effects on river habitats, macrophytes and macroinvertebrates. *Ecol. Eng.* **30**, 145–156.

Pedraza, G.X., Giraldo, L.P., and Chará, J.D. (2008). Effect of restoration of riparian corridors on the biotic and abiotic characteristics of streams in cattle ranching areas of La Vieja river catchment in Colombia [Efecto de la restauración de corredores ribereños sobre características bióticas y abióticas]. *Zootecnia Tropical* **26**, 179–182.

Penczak, T. (1995). Effects of removal and regeneration of bankside vegetation on fish population dynamics in the Warta River, Poland. *Hydrobiologia* **303**, 207–210.

Perkins, D.M., Reiss, J., Yvon-Durocher, G., and Woodward, G. (2010). Global change and food webs in running waters. *Hydrobiologia* **657**, 181–198.

Pizzuto, J. (2002). Effects of dam removal on river form and process. *Bioscience* **52**, 683–691.

Poff, N.L. (1997). Landscape filters and species traits: Towards mechanistic understanding and prediction in stream ecology. *J. North Am. Benthol. Soc.* **16**(2), 391–409.

Pollard, D.A., and Hannan, J.C. (1994). The ecological effects of structural flood mitigation works on fish habitats and fish communities in the lower Clarence River system of south-eastern Australia. *Estuaries* **17**, 427–461.

Pollard, A.I., and Reed, T. (2004). Benthic invertebrate assemblage change following dam removal in a Wisconsin stream. *Hydrobiologia* **513**, 51–58.

Pretty, J.L., and Dobson, M. (2004). The response of macroinvertebrates to artificially enhanced detritus levels in plantation streams. *Hydrol. Earth Syst. Sci.* **8**, 550–559.

Pretty, J.L., Harrison, S.S.C., Shepherd, D.J., Smith, C., Hildrew, A.G., and Hey, R.D. (2003). River rehabilitation and fish populations: Assessing the benefit of instream structures. *J. Appl. Ecol.* 51–265.

Pusch, M., Andersen, H.E., Bäthe, J., Behrendt, H., Fischer, H., Friberg, N., Gancarczyk, C., Hoffmann, C.C., Hachot, J., Kronvang, B., Nowacki, F., Pedersen, M.L., *et al.* (2009). Rivers of the central European highlands and plains—Skjern River. In: *Rivers of Europe* (Ed. by K. Tockner, C. Robinson and U. Uehlinger), pp. 552–556. Academic Press, Heidlberg.

Quinn, J., Croker, G.F., Smith, B.J., and Bellingham, M.A. (2009). Integrated catchment management effects on flow, habitat, instream vegetation and macroinvertebrates in Waikato, New Zealand, hill-country streams. *N. Z. J. Mar. Freshwat. Res.* **43**, 775–802.

Quinn, J.W., and Kwak, T.J. (2000). Use of rehabilitated habitat by Brown Trout and Rainbow Trout in an Ozark Tailwater River. *North Am. J. Fish. Manage.* **20**, 737–751.

Raborn, S.W., and Schramm, H.L. (2003). Fish assemblage response to recent mitigation of a channelized warmwater stream. *River Res. Appl.* **19**, 289–301.

Rambusch, S.H.G. (1900). *Studier over Ringkjøbing Fjord.* Det Nordiske Forlag, Copenhagen, Denmark.

Randle, T.J. (2003). Dam removal and sediment management. In: *Dam Removal Research: Status and Prospects* (Ed. by T.J. Randle and W.L. Graf), pp. 81–104. The Heinz Center for Science, Economics, and the Environment, Washington, DC.

Rathburn, S.L., and Wohl, E.E. (2003). Sedimentation hazards downstream from reservoirs. In: *Dam Removal Research: Status and Prospects* (Ed. by W.L. Graf), pp. 105–118. The Heinz Center for Science, Economics, and the Environment, Washington, DC.

Reice, S.R. (1985). Experimental disturbance and the maintenance of species diversity in a stream community. *Oecologia* **67**, 191–195.

van Riel, M.C., van der Velde, G., Rajagopal, S., Marguillier, S., Dehairs, F., and bij de Vaate, A. (2006). Trophic relationships in the Rhine food web during invasions and after establishment of the Ponto-Caspian invader Dikerogammarus villosus. *Hydrobiologia* **565**, 39–58.

Riley, S.C., and Fausch, K.D. (1995). Trout population response to habitat enhancement in six northern Colorado streams. *Can. J. Fish. Aquat. Sci.* **52**, 34–53.

Roberts, B.J., Mulholland, P.J., and Houser, J.N. (2007). Effects of upland disturbance and instream restoration on hydrodynamics and ammonium uptake in headwater streams. *J. North. Am. Benthol. Soc.* **26**, 38–53.

Rolauffs, P., Stubauer, I., Zahrádková, S., Brabec, K., and Moog, O. (2004). Integration of the saprobic system into the European Union Water Framework Directive. *Hydrobiologia* **516**, 285–298.

Roni, P. (2003). Responses of benthic fishes and Giant Salamanders to placement of large woody debris in small Pacific Northwest streams. *North Am. J. Fish. Manage.* **23**, 1087–1097.

Roni, P., Bennett, T., Morley, S., Pess, G.R., and Hanson, K. (2006). Rehabilitation of bedrock stream channels: The effects of boulder weir placement on aquatic habitat and biota. *River Res. Appl.* **980**, 967–980.

Roni, P., Hanson, K., and Beechie, T. (2008). Global review of the physical and biological effectiveness of stream habitat rehabilitation techniques. *North Am. J. Fish. Manage.* **28**, 856–890.

Roni, P., and Quinn, T.P. (2001a). Density and size of juvenile salmonids in response to placement of large woody debris in western Oregon and Washington streams. *Can. J. Fish. Aquat. Sci.* **58**, 282–292.

Roni, P., and Quinn, T.P. (2001b). Effects of wood placement on movements of trout and juvenile coho salmon in natural and artificial stream channels. *Trans. Am. Fish. Soc.* **130**, 675–685.

Rosgen, D. (1994). A classification of natural rivers. *Catena* **22**, 169–199.

Roy, A.H., Rosemond, A.D., Pauls, M.J., Leigh, D.S., and Wallace, J.B. (2003). Stream macroinvertebrate response to catchment urbanisation (Georgia, USA). *Freshwat. Biol.* **48**, 329–346.

Rumschlag, J.H., and Peck, J.A. (2007). Short-term sediment and morphologic response of the middle Cuyahoga River to the removal of the Munroe Falls Dam, Summit County, Ohio. *Great Lakes Res.* **33 (SI2)**, 142–153.

Sand-Jensen, K., Friberg, N. and Murphy, J. (Eds.) (2006) *Running Waters. Historical Development and Restoration of Lowland Danish Streams*. National Environmental Research Institute, Denmark. 159 pp. (http://www2.dmu.dk/1_viden/2_Publikationer/3_Ovrige/rapporter/RW_web.pdf; checked on 28 Feb 2011).

Sand-Jensen, K., Mebus, J., and Madsen, T.V. (1996). Macrophytes affect current and sediment. *Vand og Jord* **5**, 191–195.

Saunders, J.W., and Smith, M.W. (1962). Physical alteration of stream habitat to improve brook trout production. *Trans. Amer. Fish. Soc.* **9**, 185–188.

Schlosser, I.J. (1982). Fish community structure and function along 2 habitat gradients in a headwater stream. *Ecol. Monogr.* **52**, 395–414.

Schmetterling, D.A., and Pierce, R.W. (1999). Success of instream habitat structures after a 50-year flood in Gold Creek. *Montana. Restor. Ecol.* **7**, 369–375.

Schmitt, F. (2005). *Impact écologique de l'effacement des barrages dans le Grand-Est*. Conseil supérieur de la pêche, Marly, France, 17 pp.

Schulz, R. (2004). Field studies on exposure, effects, and risk mitigation of aquatic nonpoint-source insecticide pollution: A review. *J. Environ. Qual.* **33**, 419–448.

Schultz, R., Colletti, J., Isenhart, T., Simpkins, W., Mize, C., and Thompson, M. (1995). Design and placement of a multi-species riparian buffer strip system. *Agrofor. Syst.* **29**, 201–226.

Shafroth, P.B., Friedman, J.M., Auble, G.T., Scott, M.L., and Braatne, J.H. (2002). Potential responses of riparian vegetation to dam removal. *Bioscience* **52**, 703–712.

Shetter, D.S., Clark, O.H., and Hazzard, A.S. (1946). The effect of deflectors in a section of a Michigan trout stream. *Trans. Amer. Fish. Soc.* **76**, 248–278.

Shields, F. (2003). Stream corridor restoration research: A long and winding road. *Ecol. Eng.* **20**, 441–454.

Shields, F.D., Bowie, A.J., and Cooper, C.M. (1995a). Control of streambank erosion due to bed degradation with vegetation and structure. *Water Resour. Bull.* **31**, 475–489.

Shields, F.D., Cooper, C.M., and Knigth, S.S. (1993). Initial habitat response to incised channel rehabilitation. *Aquat. Conserv.: Mar. Freshwat. Ecosyst.* **3**, 93–103.

Shields, F.D., Knight, S.S., and Cooper, C.M. (1995b). Incised stream physical habitat restoration with stone weirs. *Regu. Rivers: Res. Managem.* **10**, 181–198.

Shields, F.D., Jr., Knight, S.S., Morin, N., and Blank, J. (2003). Response of fishes and aquatic habitats to sand-bed stream restoration using large woody debris. *Hydrobiologia* **494**, 251–257.

Shields, F.D., Knight, S.S., and Stofleth, J.M. (2006). Large wood addition for aquatic habitat rehabilitation in an incised, sand-bed stream, Little Topashaw Creek. *Mississippi. River Res. Appl.* **22**, 803–817.

Shields, F.D., Pezeshki, S.R., Wilson, G.V., Wu, W., and Dabney, S.M. (2008). Rehabilitation of an incised stream using plant materials: The dominance of geomorphic processes. *Ecol. Soc.* **13**, 54.

Shuman, J.R. (1995). Environmental considerations for assessing dam removal alternatives for river restoration. *Reg. Rivers: Res. Manage.* **11**, 249–261.

Skriver, J., and Nielsen, H.T. (2006). Streams and their future inhabitants. In: *Running Waters* (Ed. by K. Sand-Jensen, N. Friberg and J. Murphy), pp. 115–122. Historical development and restoration of lowland Danish streams, National Environmental Research Institute, Denmark.

Smith, E.P. (1993). Impact assessment using the Before-After-Control-Impact (BACI) model: Concerns and comments. *Can. J. Fish. Aquat. Sci.* **50**, 627–637.

Smith, R.F., Alexander, L.C., and Lamp, W.O. (2009). Dispersal by terrestrial stages of stream insects in urban watersheds: A synthesis of current knowledge. *J. North Am. Benth. Soc.* **28**, 1022–1037.

Solazzi, M.F., Nickelson, T.E., Johnson, S.L., and Rodgers, J.D. (2000). Effects of increasing winter rearing habitat on abundance of salmonids in two coastal Oregon streams. *Can. J. Fish. Aquat. Sci.* **57**, 906–914.

Spänhoff, B., Riss, W., Jägel, P., Dakkak, N., and Meyer, E.I. (2006). Effects of an experimental enrichment of instream habitat heterogeneity on the stream bed morphology and chironomid community of a straightened section in a sandy lowland stream. *Environ. Manage.* **37**, 247–257.

Stanford, J.A., Ward, J.V., Liss, W.J., Frissell, C.A., Williams, R.N., Lichatowich, J.A., and Coutant, C.C. (1996). A general protocol for restoration of regulated rivers. *Reg. Rivers: Res. Manage.* **12**, 391–413.

Stanley, E.H., Catalano, M.J., Mercado-Silva, N., and Orra, C.H. (2007). Effects of dam removal on brook trout in a Wisconsin stream. *River Res. Appl.* **23**(7), 792–798.

Stanley, E.H., and Doyle, M.W. (2003). Trading off: The ecological effects of dam removal. *Front. Ecol. Environ.* **1**, 15–22.

Stanley, E.H., Luebke, M.A., Doyle, M.W., and Marshall, D.W. (2002). Short-term changes in channel form and macro invertebrate communities following low-head dam removal. *J. North Am. Benthol. Soc.* **21**, 172–187.

Stanners, D., Bosch, P., Dom, A., *et al.* (2007). Frameworks for environmental assessment and indicators at the EEA. In: *Sustainability Indicators. A Scientific Assessment* (Ed. by T. Hak, B. Moldan and A.L. Dahl), pp. 127–144. Island Press, Washington.

Stauffer, J., Goldstein, R., and Newman, R. (2000). Relationship of wooded riparian zones and runoff potential to fish community composition in agricultural streams. *Can. J. Fish. Aquat. Sci.* **57**, 307–316.

Steedman, R.J. (1988). Modification and assessment of an index of biotic integrity to quantify stream quality in southern Ontario. *Can. J. Fish. Aquat. Sci.* **45**, 492–501.

Sutton, A.J., Fisher, T.R., and Gustafson, A.B. (2009). Effects of restored stream buffers on water quality in non-tidal streams in the Choptank River basin. *Water Air Soil Poll.* **208**, 101–118.

Svendsen, L.M., and Hansen, H.O. (1997). *Skjern Aa. Sammenfatning af den eksisterende viden om de fysiske, kemiske og biologiske forhold i den nedre del af Skjern Aa.* Miljø-og Energiministeriet, Danmarks Miljøundersøgelser, Denmark, 198 pp.

Sweeney, B.W., Bott, T.L., Jackson, J.K., Kaplan, L.A., Newbold, J.D., Standley, L.J., Hession, W.C., and Horwitz, R.J. (2004). Riparian deforestation, stream narrowing, and loss of stream ecosystem services. *PNAS* **101**, 14132–14137.

Tarzwell, C.M. (1937). Experimental evidence on the value of trout stream improvement in Michigan. *Amer. Fisheries Soc.* **66**, 177–187.

Tews, J., Brose, U., Grimm, V., Tielbörger, K., Wichmann, M.C., Schwager, M., and Jelsch, F. (2004). Animal species diversity is driven by habitat heterogeneity/diversity: The importance of keystone structures. *J. Biogeogr.* **31**, 79–92.

Thompson, D.M. (2002). Long-Term Effect of Instream Habitat-Improvement Structures on Channel Morphology Along the Blackledge and Salmon Rivers, Connecticut. *USA. Environ. Manage.* **29**, 250–265.

Thomson, J.R., Hart, D.D., Charles, D.F., Nightingale, T.L., and Winter, D.M. (2005). Effects of removal of a small dam on downstream macroinvertebrate and algal assemblages in a Pennsylvania stream. *J. North Am. Benthol. Soc.* **24**, 192–207.

Tilman, D., Kilham, S.S., and Kilham, P. (1982). Phytoplankton Community Ecology: The Role of Limiting Nutrients. *Ann. Rev. Ecol. Syst.* **13**, 349–372.

Tockner, K., Robinson, C.T. and Uehlinger, U. (Eds.) (2009) *Rivers of Europe,* Academic Press, Heidelberg, 700 pp.

Trexler, J.C. (1995). Restoration of the Kissimmee river—A conceptual model of past and present fish communities and its consequences for evaluating restoration success. *Rest. Ecol.* **3**, 195–210.

Tszydel, M., Grzybkowska, M., and Kruk, A. (2009). Influence of dam removal on trichopteran assemblages in the lowland Drzewiczka River, Poland. *Hydrobiologia* **630**, 75–89.

Tullos, D.D., Penrose, D.L., and Jennings, G.D. (2006). Development and application of a bioindicator for benthic habitat enhancement in the North Carolina Piedmont. *Environ. Eng.* **27**, 228–241.

Underwood, A.J. (1994). On beyond BACI: Sampling designs that might reliably detect environmental disturbances. *Ecol. Appl.* **4**, 3–15.

U.S. Senate (1972). Federal Water Pollution Control Act. http://epw.senate.gov/water.pdf. Accessed on 4 February 2011.

Van der Wal, D., Pye, K., and Neal, A. (2002). Long-term morphological change in the Ribble Estuary, northwest England. *Mar. Geol.* **189**(3–4), 249–266.

Vaughan, I.P., Diamond, M., Gurnell, A.M., Hall, K.A., Jenkins, A., Milner, N.J., Naylor, L.A., Sear, D.A., Woodward, G., and Ormerod, S.J. (2009). Integrating ecology with hydromorphology: A priority for river science and management. *Aquat. Conserv.: Mar. Freshwat. Ecosyst.* **19**, 113–125.

Velinsky, D.J., Bushaw-Newton, K.L., Kreeger, D.A., and Johnson, T.E. (2006). Effects of small dam removal on stream chemistry in southeastern Pennsylvania. *J. North Am. Benthol. Soc.* **25**(3), 569–582.

Vitousek, P.M., Mooney, H.A., Lubchenco, J., and Melillo, J.M. (1997). Human Domination of Earth's Ecosystems. *Science* **277**, 494–499.

Vörösmarty, C.J., McIntyre, P.B., Gessner, M.O., Dudgeon, D., Prusevich, A., Green, P., Glidden, S., Bunn, S.E., Sullivan, C.A., Liermann, C.R., and Davies, P.M. (2010). Global threats to human water security and river biodiversity. *Nature* **467**, 555–561.

Wallace, J.B. (1997). Multiple trophic levels of a forest stream linked to terrestrial litter inputs. *Science* **277**, 102–104.

Walter, R.C., and Merritts, D.J. (2008). Natural streams and the legacy of water-powered mills. *Science* **319**, 299–304.

Wang, L., Lyons, J., Kanehl, P., and Gatti, R. (1997). Influences of watershed land use on habitat quality and biotic integrity in Wisconsin streams. *Fisheries* **22**(6), 6–12.

Warren, D.R., Kraft, C.E., Keeton, W.S., Nunery, J.S., and Likens, G.E. (2009). Dynamics of wood recruitment in streams of the northeastern US. *For. Ecol. Manage.* **258**, 804–813.

Weatherley, N.S., and Ormerod, S.J. (1990). Forests and the temperature of upland streams in Wales: A modelling exploration of the biological effects. *Freshwat. Biol.* **24**, 109–122.

Wenger, S. (1999). *A review of the scientific literature on riparian buffer width, extent and vegetation*. Review for the Institute of Ecology, University of Georgia, Athens, USA, 59 pp.

Whitledge, G.W., Rabeni, C.F., Annis, G., and Sowa, S.P. (2006). Riparian shading and groundwater enhance growth potential for smallmouth bass in ozark streams. *Ecol. Appl.* **16**, 1461–1473.

Wood, P.J., Agnew, M., and Petts, G. (2000). Flow variations and macroinvertebrate community responses in a small groundwater-dominated stream in south east England. *Hydrol. Process.* **14**, 3133–3147.

Wood, P.J., and Armintage, P.D. (1997). Biological effects of fine sediment in the lotic environment. *Environ. Manage.* **21**, 203–217.

Woodward, G. (2009). Biodiversity, ecosystem functioning and food webs in fresh-waters: Assembling the jigsaw puzzle. *Freshwat. Biol.* **54**, 2171–2187.

Woodward, G., and Hildrew, A.G. (2002). Food web structure in riverine landscapes. *Freshwat. Biol.* **47**, 777–798.

Woodward, G., Benstead, J.P., Beveridge, O.S., Blanchard, J., Brey, T., Brown, L., Cross, W.F., Friberg, N., Ings, T.C., Jacob, U., Jennings, S., Ledger, M.E., *et al.* (2010a). Ecological Networks in a Changing Climate. *Adv. Ecol. Res.* **42**, 72–138.

Woodward, G., Friberg, N., and Hildrew, A.G. (2010b). Science and non-science in the biomonitoring and conservation of fresh waters. In: *Freshwater Ecosystems and Aquaculture Research* (Ed. by F. de Carlo and A. Bassano). 978-1-60741-707-1. Nova Science Publishing, USA.

Woodward, G., Perkins, D.M., and Brown, L.E. (2010c). Climate change in freshwater ecosystems: Impacts across multiple levels of organisation. *Phil. Trans. Royal Soc. B* **365**, 2093–2106.

Woodward, J., and Foster, I. (1997). Erosion and suspended sediment transfer in river catchments: Environmental controls, processes and problems. *Geography* **82**, 353–376.

Woolsey, S., Capelli, F., Gonser, T.O., Hoehn, E., Schweizer, S., Tiegs, S.D., Tockner, K., and Weber, C. (2007). A strategy to assess river restoration success. *Freshwat. Biol.* **52**, 752–769.

Yoder, C.O., Miltner, R.J., and White, D. (1999). Assessing the status of aquatic life designated uses in urban and suburban watersheds. In: *Proceedings of the National Conference for Retrofit Opportunities for Water Resource Protection in the Urban Environment,* pp. 16–28. EPA Report EPA/625/R-99/002.

Zauner, G. (2003). *Fischökologische Evaluierung der Biotopprojekte Ybbser-Scheibe und Diedersdorfer Haufen - Studie im Auftrag der Wasserstraßendirektion.* Ezb - TB Zauner, Engelhartszell, Austria, 70 pp.

Zauner, G., Pinka, P., and Moog, O. (2001). *Pilotstudie oberes Donautal - Gewässerökologische Evaluierung neugeschaffener Schotterstrukturen im Stauwurzelbereich des Kraftwerks Aschach.* Bundesministerium für Verkehr, Innovation und Technologie, Wien, 132 pp.

ZEC (2000). Directive 2000/60/EC of the European Parliament and of the Council of 23 October 2000 establishing a framework for Community action in the field of water policy. *Official Journal of the European Communities*, (Vol. L 327, p. 72), Brussels.

Zelinka, M., and Marvan, U.P. (1961). Zur Präzisierung der biologischen Klassifizierung der Reinheit fließender Gewässer. *Arch. Hydrobiol.* **57**, 389–407.

Zika, U., and Peter, A. (2002). The introduction of woody debris into a channelized stream: Effect on trout populations and habitat. *River Res. Appl.* **18**, 355–366.

Zyll, V., de Jong, M.C., Cowx, I.G., and Scruton, D.A. (1997). An evaluation of instream habitat restoration techniques on salmonid populations in a Newfoundland stream. *Reg. Rivers: Res. Manage.* **13**, 603–614.

# Stream Ecosystem Functioning in an Agricultural Landscape: The Importance of Terrestrial–Aquatic Linkages

SALLY HLADYZ, KAJSA ÅBJÖRNSSON, ERIC CHAUVET, MICHAEL DOBSON, ARTURO ELOSEGI, VERÓNICA FERREIRA, TADEUSZ FLEITUCH, MARK O. GESSNER, PAUL S. GILLER, VLADISLAV GULIŞ, STEPHEN A. HUTTON, JEAN O. LACOURSIÈRE, SYLVAIN LAMOTHE, ANTOINE LECERF, BJÖRN MALMQVIST, BRENDAN G. MCKIE, MARIUS NISTORESCU, ELENA PREDA, MIIRA P. RIIPINEN, GETA RÎŞNOVEANU, MARKUS SCHINDLER, SCOTT D. TIEGS, LENA B.-M. VOUGHT AND GUY WOODWARD



ADVANCES IN ECOLOGICAL RESEARCH VOL. 44 0065-2504/11 $35.00

DOI: 10.1016/B978-0-12-374794-5.00004-3

## SUMMARY

The loss of native riparian vegetation and its replacement with non-native species or grazing land for agriculture is a worldwide phenomenon, but one that is prevalent in Europe, reflecting the heavily-modified nature of the continent's landscape. The consequences of these riparian alterations for freshwater ecosystems remain largely unknown, largely because bioassessment has traditionally focused on the impacts of organic pollution on community structure. We addressed the need for a broader perspective, which encompasses changes at the catchment scale, by comparing ecosystem processes in woodland reference sites with those with altered riparian zones. We assessed a range of riparian modifications, including clearance for pasture and replacement of woodland with a range of low diversity plantations, in 100 streams to obtain a continental-scale perspective of the major types of alterations across Europe. Subsequently, we focused on pasture streams, as an especially prevalent widespread riparian alteration, by characterising their structural (e.g. invertebrate and fish communities) and functional (e.g. litter decomposition, algal production, herbivory) attributes in a country (Ireland) dominated by this type of landscape modification, via field and laboratory experiments. We found that microbes became increasingly important as agents of decomposition relative to macrofauna (invertebrates) in impacted sites in general and in pasture streams in particular. Resource quality of grass litter (e.g., carbon : nutrient ratios, lignin and cellulose content) was a key driver of decomposition rates in pasture streams. These systems also relied more heavily on autochthonous algal production than was the case in woodland streams, which were more detrital based. These findings suggest that these pasture streams might be fundamentally different from their native, ancestral woodland state, with a shift towards greater reliance on autochthonous-based processes. This could have a destabilizing effect on the dynamics of the food web relative to the slower, detrital-based pathways that dominate in woodland streams.

# I. INTRODUCTION

## A. Impacts of Agriculture on European Streams: Pollution, River Engineering and Clearance of Riparian Zones

Europe's landscape has been altered profoundly by human activity for millennia. Since the Industrial Revolution (ca. 1840), and particularly after the Second World War (1945), agricultural practices have intensified across the continent, and now operate at truly industrial scales (Feld *et al.*, 2011; Friberg *et al.*, 2011; Mulder *et al.*, 2011). The consequences of this are manifested in the continent's freshwaters which, as typically small and relatively isolated water bodies in a largely agricultural landscape, are particularly sensitive to the land use in their catchments. Europe's streams and rivers receive runoff loaded with agrochemicals, pesticides, sewage and other pollutants (Feld *et al.*, 2011; Friberg *et al.*, 2011) and they have also been channelised, deepened and straightened for land drainage, flood management and the transport of humans and goods (e.g. timber floating). In recent years, increasing attempts have been made to restore water quality through enhanced pollution legislation (e.g. EU Habitats Directive and EU Water Framework Directive—WFD), more enlightened catchment management regimes, and even river restoration schemes (Feld *et al.*, 2011; Friberg *et al.*, 2011). In addition to this primarily post-industrialisation intensification of agriculture, many catchments have been stripped of their native deciduous woodland; a process that has been ongoing since the Neolithic. Of these three main types of perturbation, the most is known about the ecological impacts on running waters of pollutants (mostly in the form of organic pollution), less about the effects of river engineering (but see Feld *et al.*, 2011; Harrison *et al.*, 2004), and least about the consequences of altering the riparian vegetation (but see Hladyz *et al.* 2010; Riipinen *et al.*, 2010): it is this latter anthropogenic impact that forms the focus of this chapter.

Most biomonitoring and bioassessment in running waters to date has focussed on measuring the impacts of organic pollution on community structure (Bonada *et al.*, 2006; Woodward, 2009), but the need to measure the impacts of a wider range stressors and to consider the functional responses of ecosystems is now being recognised (Gessner and Chauvet, 2002), even if it is yet to be formally incorporated into environmental legislation, such as the EU Water Framework Directive (Friberg *et al.*, 2011). Bioassessment of 'ecological status' for freshwaters in the WFD is predominately based on structural indicators (e.g. measures of diversity, or indices based on sensitive taxa). This implies that the WFD assessment of 'ecological status' assumes that the biological structure of an ecosystem directly relates to the functioning of an ecosystem, which is not necessarily the case (Friberg *et al.*, 2011; Gessner and Chauvet, 2002). A more integrated structural–functional approach, which also considers the influence of the surrounding terrestrial environment, is needed if we are to move beyond primarily descriptive biomonitoring towards developing a deeper understanding that can help to predict, and ultimately mitigate future change in what is already a heavily modified landscape (Hladyz *et al.*, 2011; Moss 2008; Perkins *et al.* 2010a,b; Woodward, 2009). Recent studies have started to address the current lack of knowledge by investigating how environmental perturbations affect the decomposition rate of terrestrial leaf litter in streams (Hladyz *et al.*, 2010; Huryn *et al.*, 2002; McKie and Malmqvist, 2009; Riipinen *et al.*, 2010; Young and Collier, 2009). Allochthonous leaf litter is the dominant basal resource in many stream food webs (Cummins *et al.*, 1989; Wallace *et al.*, 1999; Woodward *et al.*, 2005) and its decomposition represents a key component of ecosystem functioning.

Woodland has been cleared to provide pasture for livestock grazing across Europe for millennia, long before the advent of artificial fertilisers, agrochemicals or river engineering: as such, this is arguably the earliest form of large-scale alteration of land use, although its impacts on terrestrial–aquatic riparian zones and the attendant implications for stream food webs remain largely unknown (but see Hagen *et al.*, 2010; Hladyz *et al.*, 2010, 2011; Young *et al.*, 1994). The fact that this dramatic land-use change started so early in human history might explain the otherwise surprising lack of studies into its ecological impacts in running waters—it is now so ubiquitous that it is widely viewed as simply being an integral part of the familiar European landscape, even to the extent that many of our moorland, heathland, fell, pastures and meadows are regarded as having high conservation status, despite not being natural climax communities (Friberg *et al.*, 2011). In Ireland, where much of the current study is based, this process of removing large tracts of the native vegetation began as early as 5400 years ago, with extensive and widespread forest clearance starting about 600 years ago (Dodson and Bradshaw, 1987). Agricultural or semi-natural pasture, moorland and heathland now dominate many European, and most Irish, catchments, and this is also increasingly true of many other parts of the world (Fujisaka *et al.*, 1996; Menninger and Palmer, 2007; Reid *et al.*, 2008). The vegetation of riparian zones has also been altered in many other ways, but typically to a lesser extent, than conversion to pasture, including, for instance, the introduction or invasion of exotic plants used for forestry (e.g. conifer plantations, Riipinen *et al.*, 2010; eucalyptus plantations, Ferreira *et al.*, 2006), the creation of monospecific forests for forestry (e.g. Lecerf *et al.*, 2005), forest clear-cutting (McKie and Malmqvist, 2009), biofuel production, ornamental purposes, or the provision of shelter for livestock (e.g. *Rhododendron* invasion Hladyz *et al.*, 2011).

These riparian alterations have the potential to trigger significant changes in stream ecosystem functioning (Figure 1). For instance, Hladyz *et al.* (2011) examined the influence of three vegetation types on community structure and

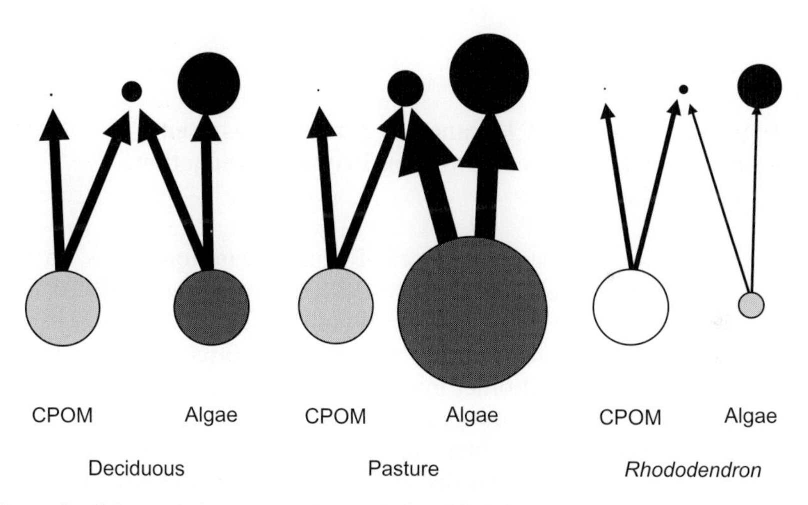


**Figure 1** Schematic representations of simplified food webs linking the primary consumers to the dominant basal resources in three vegetation types. Among the basal resources, the diameter of the circles denoting algae is scaled linearly to primary production per degree day per unit area of stream bed, and the diameter of the CPOM circles scales linearly to the standing biomass per unit area of stream bed. Among the primary consumers, grazers (e.g. *Baetis* spp.), generalists (e.g. *Gammarus* spp.) and shredders (e.g. Limnephilidae caddis larvae), the diameter of each circle scales linearly to the standing biomass per unit area of stream bed. The intensity of the colour of each circle denotes tissue C:N (black = 0–5; dark grey = 5–10; mid-grey = 10–100; white = 100–1000 [molar ratios]). The arrows denote energy flux from resources to consumers: the width of each arrow indicates whether the flux is degraded in terms of quality (thin arrows) or quality and magnitude (thinnest arrows) or if it is enhanced in quality and magnitude (thick arrows), relative to conditions in the deciduous woodland 'reference' streams. Redrawn after Hladyz *et al.* (2011).

three key ecosystem processes (decomposition, primary production and herbivory) in nine Irish streams bordered by three characteristic vegetation types, deciduous woodland, pasture or *Rhododendron ponticum* L. (an aggressive riparian invader). Community structure and ecosystem processes differed among vegetation types, with autochthonous pathways being relatively more important in the pasture streams than in the woodland reference streams (Figure 1). However, in *Rhododendron*-invaded streams overall ecosystem functioning was compromised because both allochthonous and autochthonous inputs were impaired, that is, *Rhododendron*'s poor quality litter and densely shaded canopy suppressed decomposition rates and algal production, and the availability of resources to consumer assemblages (Figure 1). In general, the consequences of riparian vegetation changes, for stream ecosystems are therefore potentially profound, yet still largely unknown, as few studies have investigated these processes in 'pasture' streams or other altered riparian zones relative to those bordered by native vegetation (but see Hladyz *et al.*, 2010, 2011; Riipinen *et al.*, 2010).

## B. Impacts of Riparian Clearance on Stream Ecosystem Functioning: Detrital Decomposition, Primary Production and Consumption Rates

Detritus dominates the basal resources of many stream food webs, particularly in the upper reaches of river networks (Cummins *et al.*, 1989; Wallace *et al.*, 1999; Webster and Benfield, 1986; Woodward *et al.*, 2005). It is derived mostly from allochthonous subsidies of riparian leaf litter, which are broken down to produce $CO_2$ and other inorganic compounds, dissolved and fine-particulate organic matter, and consumer biomass (Gessner *et al.*, 1999). The principal biological agents of litter decomposition are detritivorous invertebrate 'shredders' and microbial decomposers (bacteria and aquatic hyphomycete fungi; Hieber and Gessner, 2002), and decomposition rates are mediated via the combined influences of resource quality, temperature, and consumer abundance (Figure 2; Boyero *et al.* 2011; Gessner *et al.*, 2010; Hladyz *et al.*, 2009; Reiss *et al.*, 2010). Shredders often account for the majority of leaf mass loss, at least in the temperate woodland streams that have been most intensively studied (Hieber and Gessner, 2002; Hladyz *et al.*, 2009; Irons *et al.*, 1994); their abundance and activity are determined by the quality, quantity and timing of litter inputs. There are, however, suggestions that microbial decomposers can be more important in pasture streams because invertebrate shredders may be scarce or absent, even though total decomposition (i.e. consumption by invertebrates + microbes) rates may be similar to those in woodland streams (Hladyz *et al.*, 2010).

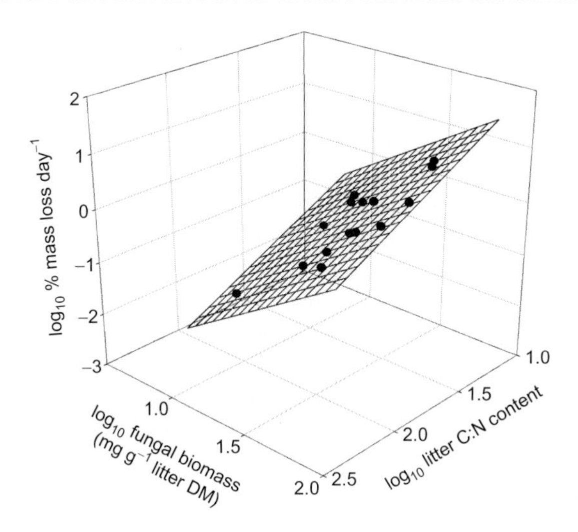


**Figure 2** Within-stream constraints on leaf-litter decomposition rates. 91% of variance in litter decomposition rates within a single woodland stream as a function of resource quality for invertebrate shredders, as measured by the degree of microbial conditioning and C:N content of litter—91% of the variance is accounted for by these two variables alone. Redrawn after Hladyz *et al.* (2009).

In theory, because pasture streams lack a dense, overhanging canopy and significant leaf litter inputs (Campbell *et al.*, 1992; Reid *et al.*, 2008), algal production could underpin a greater proportion of secondary production, via autochthonous-based pathways in food webs, relative to the role of detritus in woodland streams (Figure 1; Delong and Brusven, 1998; Hladyz *et al.*, 2011). In this way, pasture systems could retain the potential to process leaf litter (Gessner *et al.*, 1998), even though that ability might not normally be expressed (Hladyz *et al.*, 2011). Although leaf litter *per se* may be scarce, these systems can receive appreciable terrestrial inputs of grass litter as an alternative detrital resource (Hladyz *et al.*, 2011; Leberfinger and Bohman, 2010; Menninger and Palmer, 2007). Surprisingly, little is known about how grass litter is processed in pasture streams (but see Menninger and Palmer, 2007; Young *et al.*, 1994), but there are indications that it is used primarily by microbes rather than invertebrate consumers (Dangles *et al.*, 2011; Hladyz *et al.*, 2010; Niyogi *et al.*, 2003). This suggests a fundamental shift in the driving agents of decomposition in particular and overall ecosystem functioning in general, in terms of reliance on autochthonous versus allochthonous pathways, at the base of the food web (Hladyz *et al.*, 2010). In other words, the structure and dynamics of these systems might be very different from their woodland counterparts.

One of the principal aims in this chapter was to compare litter decomposition rates in a set of 100 European streams, half of which were reference sites bordered by native vegetation, and the other half had altered riparian vegetation. These were selected to represent the major types of riparian alteration, on a continental-scale, with the 10 regions representing the major European ecoregions as defined in the WFD: the alterations investigated included riparian zones that had been cleared for pasture (in Ireland, Romania and Switzerland) or forestry (N. Sweden), and a range of low diversity plantations (e.g. monospecific beech forests in France; eucalypt plantations in Portugal and Spain). We then sought to characterise community structure and ecosystem functioning in a set of nine pasture streams in Ireland, to gain a better understanding of how these systems operate in their own right, using a combination of survey and experimental approaches (Table 1; Figure 3).

**Table 1** Outline of the tiered approach to the study, ranging from an extensive pan-European field bioassay experiment (Tier I) through to increasingly more controlled field-based (Tiers II–III) and laboratory studies (Tier IV)

| | Tier I | Tier II | Tier III | Tier IVa | Tier IVb |
|---|---|---|---|---|---|
| Number of study sites | 100 | 9 | 1 | n/a | n/a |
| Water chemistry characterised/controlled | ✓/× | ✓/× | ✓/✓ | ✓/✓ | ✓/✓ |
| Oak litter decomposition | ✓ | ✓ | ✓ | × | ✓ |
| Grass litter decomposition | × | ✓ | ✓ | ✓ | ✓ |
| Grass litter quality controlled | × | × | ✓ | ✓ | ✓ |
| 1° consumer assemblage composition characterised/controlled | ×/× | ✓/× | ✓/✓ | ✓/✓ | ✓/✓ |
| 1° consumer abundance characterised/controlled | ×/× | ✓/× | ✓/✓ | ✓/✓ | ✓/✓ |
| 1° consumer—basal resource stable isotope signatures characterised | × | ✓ | × | × | × |
| 2° consumer assemblage composition characterised/controlled | ×/× | ✓/× | ✓/✓ | ×/× | ×/× |

Tier I: RIVFUNCTION pan-European study (including 15 pasture streams—5 in Ireland); Tier II: Irish multiple sites field experiments; Tier III: Irish single site (Dripsey River) field experiments; Tier IVa: Irish laboratory mesocosm experiment (multiple-choice feeding trial); Tier IVb: Irish laboratory microcosm experiment (single-choice feeding trial). See Section II for full details and Figure 3 for a schematic depiction of the connections between the drivers and responses for Tiers II–IV.

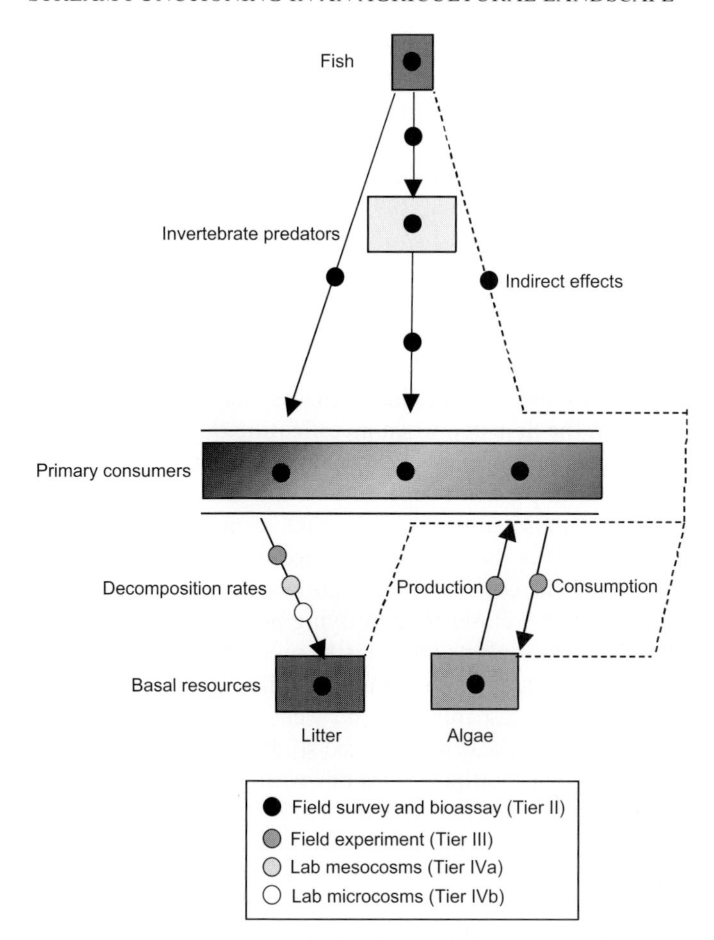


**Figure 3** Generalised schematic representation of the putative main drivers of community structure and ecosystem functioning in the intensive studies carried out in the Irish pasture streams (Tiers II–IV of the study design—see Table 1 for details).

## C. The Potential for Indirect Food Web Effects to Influence Stream Ecosystem Functioning

The vast majority of ecosystem process studies in running waters have focussed on trophic interactions at the base of the food web (i.e. litter decomposition by detritivores and decomposers). However, ecosystem processes can also be affected indirectly via top-down effects exerted by the higher trophic levels, including, for instance, predatory fishes. Focussing on basal resources and their primary consumers, therefore, provides only a partial picture of the potential biological drivers (Power, 1990; Woodward *et al.*, 2008). Indirect food web effects could also interact with the effects of riparian clearance. For instance, fishes can induce trophic cascades by altering the density and/or behaviour of primary consumers (Nakano *et al.*, 1999), which in turn can affect the use of algae and detritus by consumers at lower trophic levels (Abrams, 1995; Holomuzki and Stevenson, 1992; Oberndorfer *et al.*, 1984; Peacor and Werner, 1997; Power, 1990; Short and Holomuzki, 1992; Turner, 1997; Woodward *et al.*, 2008).

Food web structure and ecosystem functioning are clearly inextricably linked and hence need to be considered in parallel to obtain a complete picture of how these higher levels of organisation operate (Gessner *et al.*, 2010). This interconnection underpinned our reasons for considering the wider community food web in the intensive studies in the Irish systems. Surveys are often the first step when investigating these higher level patterns and processes since appropriate experimental approaches tend to be logistically unfeasible to implement, especially as the drivers and responses often operate at larger spatiotemporal scales that cannot be manipulated easily (e.g. reaches, years; Woodward, 2009). By decreasing in scale and simplifying complexity from field conditions to tightly controlled laboratory experiments, one might move closer to identifying potential drivers, which are far harder to resolve in the field as they may be modulated by a complex array of variables (e.g. water chemistry, community differences).

To date most of the research on food webs and ecosystem processes has focussed on small spatial scales in isolated studies of single, or a few, systems, rather than employing multiple-scales of investigation in replicated systems across environmental gradients (Woodward, 2009; but see Brown *et al.*, 2011; Layer *et al.*, 2010b; Ledger *et al.*, 2011). We attempted to integrate field and laboratory experiments with empirical survey data in a standardised manner across multiple sites to explore the connexions between structure and function and to explore potential links between different organisational levels.

## D. Linking Ecosystem Structure and Functioning Across Multiple Levels of Organisation via Experimental and Empirical Approaches

In the first part of this chapter, we compared a range of different anthropogenic alterations to riparian zones, including replacement of natural woodland with pasture and plantations, in a continental-scale study of 100 streams across multiple ecoregions as part of the pan-European RIVFUNCTION research project (e.g. Hladyz *et al.*, 2010; Riipinen *et al.*, 2010). This involved conducting field-based bioassays of leaf litter decomposition rates in response to a range of different riparian alterations (Table 2), with the specific type of impact representing a dominant perturbation in each of the nine countries.

**Table 2** Geographical co-ordinates and physicochemistry of the 100 European reference and impacted streams, mean (min/max) given per region

| European region | Stream types | Longitude °E/W | Latitude °N | Temp. (°C) | TON ($\mu g\ L^{-1}$) | SRP ($\mu g\ L^{-1}$) | pH | Conduc. ($\mu S\ cms^{-1}$) |
|---|---|---|---|---|---|---|---|---|
| Ireland[a] | Reference | −9.56/−9.39 | 53.90/53.96 | 5.7–6.0 | 121(55/220) | 4(4/18) | 7.32(6.99/7.64) | 166(114/211) |
| | Impact (pasture) | −9.57/−9.31 | 53.92/54.00 | 5.1–5.7 | 42(19/61) | 6(3/18) | 7.36(7.27/7.48) | 124(87/150) |
| Switzerland[a] | Reference | 7.93/9.65 | 53.90/48.08 | 4.9–8.4 | 1870(669/2636) | 5(6/8) | | 449(323/517) |
| | Impact (pasture) | 8.03/9.57 | 47.18/47.83 | 6.3–8.0 | 1800(560/3261) | 5(5/7) | | 394(287/502) |
| Romania[a] | Reference | 25.03/25.32 | 63.88/44.92 | 3.7–6.3 | 341(117/679) | 5(1/10) | 7.69(7.30/7.91) | 249(218/320) |
| | Impact (pasture) | 25.10/25.73 | 44.68/45.09 | 2.5–6.7 | 1230(163/2074) | 5(1/24) | 7.81(7.49/8.05) | 398(218/670) |
| N. Sweden[b] | Reference | 19.86/20.38 | 44.86/64.39 | 0.3–0.9 | 39(9/129) | 7(2/9) | 6.49(5.65/7.03) | 54(39/85) |
| | Impact (forest clearfell) | 19.88/20.60 | 63.94/63.34 | 0.3–0.5 | 23(7/57) | 9(4/16) | 6.32(5.81/6.84) | 39(30/47) |
| S. Sweden | Reference | 13.07/14.07 | 55.73/56.04 | 1.8–4.8 | 3386(1247/7760) | 21(12/32) | 7.44(7.25/7.86) | 268(149/519) |
| | Monospecific forest (Beech) | 13.12/14.08 | 55.71/56.07 | 1.8–4.9 | 3120(1184/7747) | 18(9/39) | 7.27(6.85/7.64) | 213(96/478) |
| France[c] | Reference | 2.09/2.41 | 43.41/43.49 | 6.4–8.4 | 1384(1064/1804) | 3(2/3) | 6.41(6.32/6.51) | 43(33/62) |
| | Monospecific forest (Beech) | 2.21/2.44 | 43.41/43.49 | 7.7–9.1 | 954(396/2239) | 3(2/5) | 6.44(5.71/7.26) | 82(27/210) |
| Great Britain[d] | Reference | −1.95/−1.61 | 53.42/54.40 | 3.9–5.7 | 1042(294/2680) | 4(3/7) | 5.67(4.33/6.79) | 83(75/169) |

(continued)

**Table 2** (*continued*)

| European region | Stream types | Longitude °E/W | Latitude °N | Temp. (°C) | TON ($\mu$g L$^{-1}$) | SRP ($\mu$g L$^{-1}$) | pH | Conduc. ($\mu$S cms$^{-1}$) |
|---|---|---|---|---|---|---|---|---|
| | Conifer forest | −1.84/−1.75 | 53.41/53.44 | 3.9–5.7 | 606(456/790) | 4(3/4) | 6.24(5.00/7.10) | 80(53/96) |
| Poland[d] | Reference | 20.07/20.63 | 49.47/49.61 | 1.5–2.7 | 792(521/1288) | 39(33/53) | 8.06(7.77/8.28) | 187(167/208) |
| | Conifer forest | 20.02/20.53 | 49.42/49.60 | 1.1–3.0 | 548(340/914) | 49(39/67) | 8.10(7.73/8.26) | 184(167/196) |
| Portugal[e] | Reference | −8.30/−8.21 | 40.06/40.51 | 8.3–12.5 | 230(110/356) | 3(3/4) | 6.47(6.29/6.72) | 31(23/41) |
| | Eucalyptus plantation | −8.37/−8.22 | 40.45/40.52 | 11.3–13.0 | 250 (137/388) | 3(2/5) | 6.49(6.18/6.68) | 48(34/81) |
| Spain[e,f] | Reference | −3.33/−3.23 | 43.21/43.32 | 7.8–10.9 | 861(245/1560) | 6(3/9) | 7.00(6.44/7.55) | 92(68/141) |
| | Eucalyptus plantation | −3.33/−3.23 | 43.17/43.35 | 8.7–10.6 | 5734(428/767) | 9(8/10) | 6.81(6.07/7.19) | 116(67/214) |

Superscript denotes previous published data comparing reference and impacted streams: [a]Hladyz *et al.* (2010), [b]McKie and Malmqvist (2009), [c]Lecerf *et al.* (2005), [d]Riipinen *et al.* (2009, 2010), [e]Ferreira *et al.* (2006) and [f]Elosegi *et al.* (2006).

This 100-stream dataset included 50 reference (native woodland) and 50 impacted sites, of which 15 were pasture streams in three countries, Ireland, Switzerland and Romania (Hladyz *et al.*, 2010). Leaf litter decomposition rates in all 100 streams were measured in coarse and fine-mesh bags, as proxy measures of the relative importance of invertebrates and microbes, respectively, as the agents of decomposition (after Hladyz *et al.*, 2010).

In the following year, we focussed on a set of nine pasture streams in Ireland, three of which were also included in the earlier pan-European project, in a more intensive case study. Here, we took a more in-depth and holistic view of overall ecosystem functioning and the potential for indirect food web effects to influence process rates, in addition to characterising community assemblages directly, rather than using proxy measures. These field experiments were complemented with a range of laboratory trials, in which we attempted to identify the key drivers of litter decomposition. We have subdivided the study into four tiers of approach (Table 1), which follow a logical progression from the most extensive but least detailed field experiment in 100 European streams (Tier I: RIVFUNCTION study), to a more intensive field-based study in nine Irish streams (Tier II), followed by better controlled field experiments in a single focal stream in Ireland (Tier III) and finally to highly controlled laboratory experiments (Tier IV).

In the Irish studies, we expanded the range of both responses and predictors considered in the pan-European study. We quantified three key ecosystem processes (decomposition, primary production and herbivory) and community structure (invertebrate and fish population abundance and biomass) in nine pasture streams across a gradient of intensity of agricultural land use using a suite of field and laboratory experiments. Resource quality, in terms of nutrient status (C:N:P ratios) and physical toughness (% lignin and % cellulose content), of grass litter was characterised from each stream, to assess its influence on process rates. We also compared decomposition rates among streams in parallel using a single, standardised source of oak litter (*Quercus robur* L.), as in the pan-European field experiment, as this species constitutes a widespread and significant component of the detrital pool in many woodland 'reference' streams. The other major energy input to the food webs, autochthonous primary production, was assayed using algal colonisation tiles, and top-down effects of herbivores were gauged by excluding crawling grazers from half of these (after Hladyz *et al.*, 2011; Lamberti and Resh, 1983). Finally, we quantified the abundance and biomass of invertebrate and fish populations, which we used to make inferences about the potential for indirect food web effects to influence process rates at the base of the web (e.g. Woodward *et al.*, 2008). In summary, our key objectives were as follows:

1. Compare leaf litter decomposition rates in reference sites versus those with altered riparian zones using a standardised pan-European bioassay experiment in 100 streams. We predicted that, overall, impacted streams would be impaired and therefore have slower decomposition rates than reference streams.
2. Characterise ecosystem functioning (litter decomposition rates, algal production and herbivory) and community attributes (abundance and species composition of fish and invertebrate assemblages) of pasture streams. We predicted the more nutrient-enriched streams would exhibit faster process rates at the base of the food web and that this would be reflected by increased consumer abundance.
3. Determine relationships between resource quality and decomposition rates of grass and leaf litter. We predicted that grass litter provides, in general, a poorer quality resource than leaf litter, but that in addition its quality will vary considerably across sites, with a general increase associated with agricultural improvement (e.g. anthropogenic fertilisation of pasture catchments and riparian zones).
4. Assess the potential for subtle indirect food web effects to be manifested, to test the hypothesis that top-down effects (e.g. herbivory) would also increase with agricultural intensification (as with bottom-up effects), because detrital subsidies should support more consumers per unit biomass with increasing litter quality (e.g. carbon:nutrient ratios in grasses). Additionally, predatory fish were predicted to influence the abundance of organisms at lower trophic levels and, hence, to affect process rates indirectly, via predation on invertebrate grazers and detritivores.

# II. METHODS

## A. Tier I. RIVFUNCTION Field Experiment: Impacts of Riparian Alterations on Decomposition Rates in 100 European Streams

Ten research teams from nine countries (France, Great Britain, Ireland, Poland, Portugal, Romania, Spain, Northern Sweden, Southern Sweden and Switzerland) carried out a single co-ordinated field experiment to assess the effects of altered riparian vegetation on leaf litter decomposition rates in 100 streams across Europe (Table 2; Figure 4). Stream characteristics (other than riparian vegetation) were standardised as far as possible between impacted and reference sites, both within and among regions, in order to isolate the effects of alterations to the riparian zone. Dissolved nutrient concentrations within each country reflected regional baselines that were relatively free of agricultural

**Figure 4** Tier I: Location of the 100 stream sites used in the pan-European RIVFUNCTION study (see Section II).

runoff and sewage effluents, and all streams were $<5$ m wide, $<50$ cm deep at winter baseflow, 1st–4th order, with a stony substrate and reference sites were all bordered with native woodland (after Hladyz *et al.*, 2010; Riipinen *et al.*, 2010). Water samples filtered over Whatman GF/F glass fibre filters (average pore size 0.7 $\mu$m) were analysed in the laboratory for total oxidised nitrogen ($NO_3^- + NO_2^-$) and soluble reactive phosphorus ($SRP \sim PO_4^{3-}$). Conductivity, pH and stream temperature were measured in the field.

Decomposition rates of alder (*Alnus glutinosa* (L.) Gaertn.) and oak (*Q. robur* L.) leaf litter, sourced locally in each country, were measured in a single large-scale trial conducted during autumn/winter 2002/2003. Mesh bags, each containing $5.00 \pm 0.25$ g of air-dried leaves, were deployed in 10 streams per region (Hladyz *et al.* 2010; Riipinen *et al.* 2010). Mesh apertures of 10 ('coarse mesh' hereafter) or 0.5 mm ('fine mesh') were used to permit or prevent invertebrate colonisation, respectively. In total, over 2400 leaf bags were exposed (6 replicates × 2 mesh sizes × 2 leaf species × 10 streams × 10 regions), which were retrieved when additional coarse-mesh bags (sampled repeatedly at reference sites; data not shown here—see Hladyz *et al.*, 2010 for details) had lost approx. 50% of their initial mass ($T_{50}$) to standardise for the degree of decomposition, rather than exposure time, among regions and leaf species (after Hladyz *et al.*, 2010; Riipinen *et al.*, 2010). The retrieved leaf litter was frozen at −20 °C, and subsequently oven-dried at 105 °C, with a subsample combusted at 550 °C to calculate ash-free dry mass (AFDM).

Litter decomposition rates were expressed as the exponential decay rate coefficient, $k$, in the model $(m_t/m_0) = e^{-kt}$, where $m_0$ is the initial AFDM and $m_t$ is AFDM at time $t$ (Boulton and Boon, 1991). Both types of rate coefficient were calculated for total decomposition in the coarse-mesh bags ($k_{total}$) and for microbial decomposition in fine-mesh bags ($k_{microbial}$). Invertebrate-mediated decomposition was calculated as the difference between percent mass remaining in coarse-mesh and fine-mesh bags, and this was also then converted to a decomposition coefficient ($k_{invert}$; after McKie *et al.*, 2006). Finally, a dimensionless metric was calculated as the ratio of invertebrate-mediated decomposition coefficient to microbial decomposition coefficient (i.e. $k_{invert}/k_{microbial}$). To correct for potential temperature differences among streams and regions, $t$ was also expressed in terms of thermal sums (degree days).

### B. Tier II. Irish Field Experiments and Surveys: Decomposition, Algal Production and Herbivory Rates and Community Structure in Nine Pasture Streams

The same source of air-dried oak litter that was used in the pan-European field experiment was employed in the subsequent Irish studies described below (Tiers II–IV, inc.), to provide a standardised detrital resource base for comparative purposes. In addition, senescent grass litter was collected from the banks of each stream and used in the Irish trials to examine the effects of local differences in resource quality. Litter was weighed to 5.00 ± 0.25 g per ‘coarse’ or ‘fine’ mesh bag, as described for the RIVFUNCTION trial. Each grass litter pack was tied into a cylindrical bundle using two cable ties, to mimic natural benthic ‘litter-packs’.

The extensive field experiments and surveys that comprised the Tier II study in nine Irish streams were carried out in winter (November–December) 2003 and spring (April–May) 2004, to coincide with the seasonal peaks in leaf litter inputs and algal production, respectively. Streams were standardised as far as possible to fall within the same criteria for physical characteristics used in the RIVFUNCTION study: that is, all nine streams were 1st–2nd order, $<5$ m wide, with stony substrata. The streams were chosen to span a gradient of agricultural activity, from unimproved rough pasture (in the six northern sites in Co. Mayo) to improved pasture. All catchments were dominated by pasture ($>90\%$ areal coverage), as is typical of much of Ireland. Three of the less agriculturally productive sites in Co. Mayo (Srahrevagh, Goulaun and Yellow) were also used previously in the Tier I RIVFUNCTION experiments. The three additional Clare Island sites, also in Co. Mayo, are located in a very low density agricultural area (approx. 7.8 people km$^{-2}$) of high conservation value (Guiry *et al.*, 2007) and, like the three nearby mainland Mayo sites, are rough pasture as is typical of much of the north of the country. The three sites in Cork were more productive, improved pasture with higher densities of livestock, which is typical of the southern counties: that is, overall, the nine sites represented a general gradient of increasing agricultural intensity (Clare Island $\rightarrow$ Mayo $\rightarrow$ Cork) (Table 3).

Conductivity and pH were measured in the field and filtered (0.7 $\mu$m pore size) water samples were analysed in the laboratory for a suite of chemical variables (Table 3): all nine streams were pH$>7$, with similar temperature ranges, and all were below the median values of SRP for Europe (Figure 5) and not showing any obvious signs of organic pollution. Stream temperatures were measured continuously throughout the experiment using data loggers (ACR Systems Inc., BC Canada).

Nine replicate leaf bags were used per stream in a nested, randomised block design (i.e. 2 mesh types $\times$ 2 litter types [oak/local grass] $\times$ 9 replicates $\times$ 3 streams $\times$ 3 regions $\times$ 2 seasons $=$ 648 litter bags). After 28 days exposure the bags were collected and frozen at $-20°$ C. After thawing, invertebrates were separated from the litter in the laboratory (using a 500-$\mu$m sieve), preserved in 70% ethanol, assigned to functional feeding groups (after Cummins and Klug, 1979) and counted. The remaining litter was processed and decomposition rates calculated as described for the RIVFUNCTION study above.

Resource quality of the different litter types was assessed by measuring initial C:N:P ratios and lignin and cellulose content of oven-dried litter (Table 4). Samples were ground into a fine powder (Culatti DFH48 mills, 1 mm screen), and carbon and nitrogen content was determined using a Perkin Elmer Series II CHNS/O analyser. Phosphorus was determined spectrophotometrically at 700 nm, after mixed acid digestion (15 min at 325 °C; Allen 1989). C:N:P ratios were expressed as molar ratios, whereas lignin and cellulose concentrations were determined gravimetrically (after Gessner, 2005).

**Table 3** Geographical co-ordinates and physicochemistry of the nine Irish pasture streams over two seasons (winter 2003/spring 2004)

| Region | Stream | Stream code | Longitude °W | Latitude °N | Temp. (°C) | TON ($\mu g\ L^{-1}$) | $NH_4$ ($\mu g\ N\ L^{-1}$) | SRP ($\mu g\ L^{-1}$) | pH | Conduc. ($\mu S\ cms^{-1}$) |
|---|---|---|---|---|---|---|---|---|---|---|
| Clare Island | Bunnamohaun | Bu | −10.04 | 54.80 | 6.5/10.9 | 181/218 | 15/8 | 4/4 | 8.11/8.05 | 471/579 |
| | Owenmore | Ow | −10.01 | 53.79 | 6.6/10.5 | 103/73 | 11/21 | 1/5 | 8.06/7.92 | 245/343 |
| | Dorree | Do | −9.98 | 53.81 | 6.8/11.8 | 33/99 | 33/120 | 2/4 | 8.02/7.54 | 220/332 |
| Mayo | Yellow | Ye | −9.55 | 53.93 | 6.6/9.3 | 104/24 | 11/0 | 4/2 | 7.82/7.21 | 76/91 |
| | Goulaun | Go | −9.58 | 54.00 | 6.9/9.7 | 39/8 | 19/0 | 4/<1 | 7.94/7.05 | 80/85 |
| | Srahrevagh | Ro | −9.56 | 53.98 | 6.2/9.3 | 68/40 | 30/4 | 16/8 | 8.17/8.01 | 113/183 |
| Cork | Dripsey | Dr | −8.76 | 51.97 | 8.5/9.1 | 5921/4743 | 8/10 | 27/16 | 7.31/7.46 | 221/225 |
| | Morning star | Ms | −8.37 | 52.40 | 7.8/9.5 | 2407/1749 | 16/15 | 28/32 | 8.35/8.21 | 481/463 |
| | Aherlow | Ah | −8.30 | 52.37 | 7.1/9.1 | 2214/1126 | 18/75 | 31/47 | 8.39/8.90 | 316/288 |

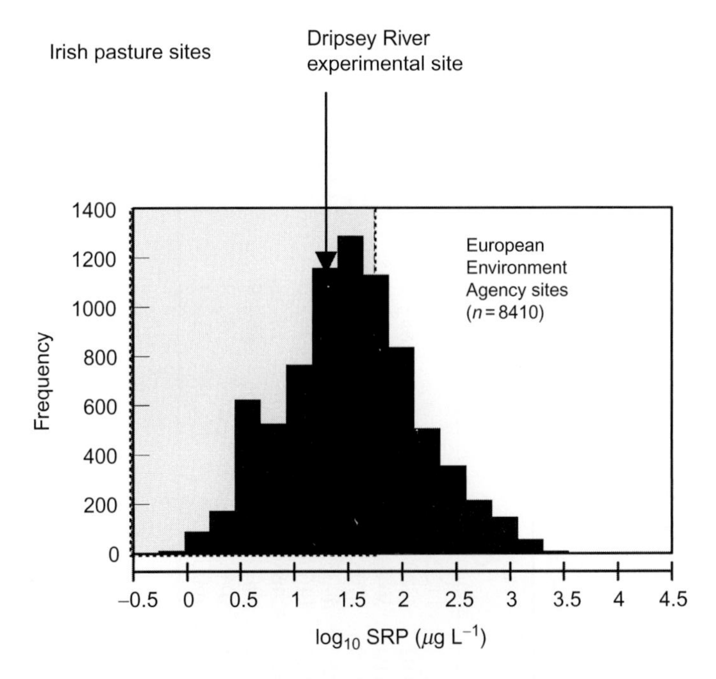


**Figure 5** Europe-wide SRP concentrations in 8410 running waters (EEA database), with the range included in the nine Irish pasture streams study (Tier II) delimited by the shaded box bounded within the dashed lines and the position of the Dripsey River focal experimental site (Tier III) highlighted with an arrow.

Algal production and herbivory rates were measured over 28 days exposure of colonisation tiles (10 cm × 10 cm unglazed quarry tiles) attached in pairs to house bricks. The vertical edges of one tile per brick were coated with petroleum jelly to deter crawling gazers, and the other was left as a control, to obtain a proxy measure of grazing pressure (after Hladyz *et al.*, 2011). Algal samples were frozen within 1 h of collection and chlorophyll *a* concentration was determined subsequently using standard spectrophotometric techniques, after overnight ethanol extraction (Jespersen and Christoffersen, 1987).

Nine Surber samples (25 cm × 25 cm quadrat, 500-μm mesh size) were taken per stream per season to quantify benthic invertebrate abundance, and processed as per the litter bags. Invertebrate body mass data were obtained by applying published length–mass regression to linear dimensions (e.g. head capsule width, body length), after Layer *et al.* (2010a,b) measured under a dissecting microscope. A 50-m reach in each of the nine sites was electrofished during the spring sampling period (this could not be done in the winter because these sites were subject to legal protection at this time), using

**Table 4** Measures of resource quality for grass and oak litter (mean ± S.E.) used in the Irish studies (Tiers II–IV as described in Section II)

| Region | Stream code | Litter | %C | %N | %P | C:P | C:N | N:P | % lignin | % cellulose |
|---|---|---|---|---|---|---|---|---|---|---|
| Cork | | Oak | 49.00 | 2.24 | 0.11 | 1122 | 26 | 44 | 40.5 ± 0.7 | 31.1 ± 1.0 |
| Clare Island | Bu | Grass | 40.52 ± 4.39 | 1.04 ± 0.11 | 0.04 ± 0.02 | 4389 ± 2917 | 47 ± 10 | 85 ± 45 | 8.7 ± 0.6 | 36.0 ± 1.3 |
| | Do | | 41.68 ± 1.90 | 1.22 ± 0.28 | 0.05 ± 0.04 | 5048 ± 3923 | 43 ± 12 | 101 ± 65 | 11.8 ± 0.4 | 42.7 ± 3.5 |
| | Ow | | 42.84 ± 2.57 | 1.36 ± 0.22 | 0.06 ± 0.00 | 1706 ± 131 | 37 ± 4 | 47 ± 8 | 12.2 ± 2.2 | 35.7 ± 1.5 |
| Mayo | Go | | 45.50 ± 0.16 | 1.13 ± 0.05 | 0.03 ± 0.02 | 4735 ± 2497 | 47 ± 2 | 99 ± 49 | 7.8 ± 0.6 | 36.8 ± 0.6 |
| | Ro | | 44.40 ± 0.35 | 1.29 ± 0.24 | 0.08 ± 0.02 | 1460 ± 284 | 42 ± 8 | 35 ± 0 | 6.9 ± 0.4 | 31.8 ± 0.9 |
| | Ye | | 45.82 ± 0.04 | 1.25 ± 0.00 | 0.05 ± 0.01 | 2468 ± 349 | 43 ± 0 | 58 ± 8 | 11.5 ± 0.8 | 34.1 ± 1.0 |
| Cork | Ah | | 45.17 ± 0.30 | 1.32 ± 0.17 | 0.07 ± 0.02 | 1906 ± 615 | 41 ± 6 | 46 ± 9 | 7.1 ± 0.6 | 35.5 ± 2.0 |
| | Dr | | 44.21 ± 0.04 | 1.69 ± 0.34 | 0.11 ± 0.01 | 1048 ± 87 | 32 ± 6 | 34 ± 4 | 3.2 ± 0.2 | 20.5 ± 0.9 |
| | Ms | | 43.97 ± 1.72 | 1.65 ± 0.05 | 0.12 ± 0.02 | 960 ± 173 | 31 ± 2 | 31 ± 3 | 3.7 ± 0.4 | 25.4 ± 0.2 |

three depletion runs between stop-nets (after Bohlin *et al.*, 1989). All captured fishes were identified to species, measured (fork length and body mass) and released.

Stable isotope analyses (SIA) of primary consumers and basal resources within each food web were used to assess consumer–resource interactions among regions and between seasons. Five replicates of CPOM, epilithic biofilm and the dominant invertebrate taxa from the main functional feeding groups were collected from each stream during winter and spring sampling for isotopic determination. For instance, the freshwater shrimp *Gammarus duebeni* (Liljeborg) was used as a representative generalist ('shredder-grazer') able to exploit both algae and detrital food chains (cf. Hladyz *et al.*, 2011), and the mayflies Heptageniidae and *Baetis* spp. represented 'grazers' (after Cummins and Klug, 1979). SIA samples were frozen within 1 h of collection. The SIA samples were combusted in a PDZ Europa ANCA-GSL preparation model and passed through a PDZ Europa 20–20 stable isotope analyser (PDZ Europa Ltd., Sandbach, UK). Pee Dee Belemnite was used as the standard ratio for C and atmospheric N for N. All isotope values are given in per mille, and standard $\delta$ notation is used to describe the relative difference in isotope ratio between the samples and known standards.

### C. Tier III. Intensive Experimental Study of Grass Litter Decomposition Within a Single Field Site

In addition to the extensive field experiments carried out in the nine Irish streams, a complementary intensive field trial was carried out in a single site (the Dripsey River), in which we used all nine grass types from the different streams during winter (February–March) 2005 (Table 4). In this experiment, we used coarse and fine-mesh bags in a randomised block design, with five replicates per treatment, which were deployed simultaneously for 28 days to quantify decomposition rates whilst controlling for potential site-specific differences in community composition and physicochemical parameters. The Dripsey River was selected because it was the most nutrient-rich site within the Irish study, and hence had the largest range of potential consumer species that could respond to differences in resource quality, and because it lay closest to the mean, median and modal values for European streams (Figure 5). Consumer abundance and identity were thus controlled for, unlike in the multiple sites study, insofar as each litter bag was exposed to an identical regional pool of potential colonists, so any differences in decomposition rates could be ascribed to differential local-scale responses among the consumers to litter quality: that is, shredders were free to choose which litter type to exploit (cf. Hladyz *et al.*, 2009).

## D. Tier IV. Laboratory Experiments: Resource Quality and Decomposition Rates

To complement the field experiments and to identify the drivers of decomposition rates under more tightly controlled conditions, we carried out two laboratory experiments that examined the feeding preferences of two detritivore consumers, the caddis larvae *Halesus radiatus* Curtis (Order Trichoptera) and the freshwater shrimp *G. duebeni* Liljeborg (Order Amphipoda), common taxa in Irish pasture streams (e.g. Hladyz *et al.*, 2011). The taxa were offered either a single resource (single-choice) or a choice of the nine types of grass litter resources (multiple-choice) used in the Tier II and III studies, respectively, and we hypothesised that the shredders would feed preferentially on the better quality resources. In general, the more intensively farmed Cork grass litters were characterised by higher %P and %N content, whereas the rough pasture grasses were characterised by higher % lignin and % cellulose. These experiments were designed to complement the previous studies, by focussing on specific drivers and responses: that is, we controlled for consumer species identity and abundance and also the physicochemical environment (temperature, light regime, water chemistry) in order to isolate the effects of resource type *per se*.

The first (Tier IVa) of these laboratory experiments measured consumer preference of grass litter (multiple-choice trials) using 60 L mesocosms. In this study, 3 g ($\pm$0.25 g) of air-dried grass litter from all nine sites was preconditioned for 2 weeks in Dripsey River water and subsequently added to arenas containing different consumer assemblages (monocultures which consisted of either 90 *Gammarus* individuals or 90 *Halesus* individuals and a microbial-only control treatment) in a C–T room at 10 °C. Five replicate arenas were used per consumer treatment and in each arena a paired control of each grass litter type was used to assess microbial-only decomposition (grass litter enclosed in 500-$\mu$m fine-mesh bag versus the equivalent litter in a coarse-mesh bag). The experiment ran for 2 weeks after the initial preconditioning period, over the same period as the Dripsey River field experiment (Tier III). The mesocosms were examined daily for dead animals, which were removed and replaced. Survival rates were between 89% and 93% for *Gammarus* and *Halesus*, respectively. At the end of the trial, invertebrates were separated from the litter, and litter was dried to constant mass at 105 °C and decomposition rates, $-k$, were calculated for each replicate.

The second (Tier IVb) laboratory trial was carried out using standardised consumer assemblages composed of 10 individuals of either the cased caddis *Halesus* or the freshwater shrimp *Gammarus*, in which we measured decomposition rates of grass litter and oak litter in 1 L microcosms (after McKie *et al.*, 2008; single-choice trials). Grass or oak litter were air-dried to constant mass, weighed to 3 g ($\pm$0.25 g), and then one of each of the 10 resource types (nine grasses, or oak) was conditioned in streamwater collected from the Dripsey River in each microcosm for 2 weeks, prior to the introduction of the invertebrates (five replicates of each). The entire study was conducted in a C–T room at 10 °C under a 12-h light:dark regime for 2 weeks after the initial conditioning period. Microcosms were examined daily for dead animals, which were removed and replaced. Survival rates over the duration of the trial were 88% and 91% for *Gammarus* and *Halesus*, respectively. At the end of the trial, invertebrates were removed and the remaining litter was oven-dried to constant mass at 105 °C and a decomposition rate, $k$, was calculated for each microcosm.

### 1. Statistical Analyses

Two analyses were performed on the RIVFUNCTION Tier I dataset: first, we examined differences in decomposition coefficients and ratios between impact and reference streams among regions. We then tested for differences in decomposition coefficients and ratios between pasture streams and other riparian impacts. Linear mixed effects models (LMEM) were used to account (i) for the hierarchical nature of the experimental design, with litter bags nested within individual streams and streams nested within pairs and (ii) for the incorporation of both fixed and random effects in the design. The following variables were fitted as fixed effects in the analyses: region, impact, leaf species and mesh type. We treated region as a fixed effect in the following analyses: regions were chosen *a priori* to include the different types of the major riparian alteration impacts commonplace across Europe, and they also represented the main ecoregions defined within the WFD. Streams and stream pairs were fitted as random effects. Since our experimental design was unbalanced because of the loss of some litter bags during field exposure, we used the restricted maximum likelihood method (REML) to estimate error terms (after Hladyz *et al.*, 2011). The optimal model structure was determined using the hypothesis testing approach (e.g. likelihood ratio test) following the mixed effects model selection procedure outlined in Zuur *et al.* (2009). Pairwise comparisons on main fixed effects were performed using Bonferroni *post hoc* tests. Decomposition coefficients and ratios were $\log_{10}$ transformed to normalise the data. If homogeneity was violated separate LMEMs analyses were carried out. The statistical analyses were performed with PASW Statistics Version 18.0 (SPSS Inc., Chicago, IL).

In the Irish Tier II study, decomposition rates, algal production and grazing rates and invertebrate abundances in the extensive field trial were analysed using LMEMs and also partial least squares (PLS) regression (after Eriksson *et al.*, 1999). In the former, region and season were fitted as fixed effects in all analyses. We treated region as a fixed effect in these analyses as we were interested in examining differences among the regions which represented a general gradient of increasing agricultural intensity (Clare

Island → Mayo → Cork). In addition the decomposition rates analysis included litter type and mesh type and the algal production analysis included grazing treatment. Sample-units and streams were fitted as random effects. REML was used to estimate error terms due to our unbalanced designs as a result of losses of a few litter-bags or tiles in the field. Data were $\log_{10}$-transformed (or $\log_{10}x+1$ where zero counts were present) to meet the assumptions of the tests, where appropriate.

Relationships between ecosystem process rates and water chemistry variables and/or resource quality and/or biotic variables were examined using PLS regression. PLS extracts components from a set of variables which, as in principal components analysis, are orthogonal and so eliminates multi-colinearity. In addition, PLS maximises the explained covariance between the variables. The constructed components are used to create a model for the response variable with the relative importance of the predictor variables ranked with variable importance on the projection (VIP) values (Eriksson *et al.*, 1999). The VIP values reflect the importance of terms in the model both with respect to $y$ and with respect to $x$ (the projection). VIP is normalised and the average squared value is 1, so terms in the model with a VIP > 1 are important. PLS analyses were conducted using SIMCA-P (version 11.5; Umetrics AB, Umeå, Sweden). Additional components were extracted until the increase in explained variance fell below 10%.

Multi-variate analyses of water chemistry, resource quality and community structure among regions was analysed using PRIMER 6 (v 6.1.13) and PERMANOVA (v.1.0.3) (PRIMER-E Ltd., Plymouth, UK) (after Anderson *et al.*, 2008). PERMANOVA is more robust than MANOVA since calculations of $P$-values are via permutations, thus avoiding the assumption of data normality. To visualise patterns evident from PERMANOVA, we used metric multi-dimensional scaling, principal coordinate analysis (PCO), which is an unconstrained ordination method. In order to highlight patterns in the PCO, we used vectors based on Spearman correlations (greater than 0.5 to target variables with high correlations) which highlights the overall increasing or decreasing relationships of individual variables across the plot ignoring all other variables.

Carbon isotope signatures of biofilm and terrestrial detritus were overlapping in many of our streams or were outside the limits of the consumer isotope signatures, so we were unable to use mixing models to discern their relative contributions to diet. Instead of using mixing models the $\delta^{13}C$ signatures for consumers were related to values of biofilm and CPOM via PLS regression analysis with both biofilm and CPOM used as predictor variables on consumer signatures.

The data from the Tier III field experiment study was analysed with PLS regression. In the laboratory studies for the multi-choice treatments (Tier IVa), we evaluated food preferences (using untransformed decomposition rates, $-k$)

**Table 5** pan-European RIVFUNCTION field experiment (Tier I): Linear mixed effects model results of comparisons of standard ($k_d$) and temperature-normalised ($k_{dd}$) decomposition rate coefficients of litter in coarse-mesh and fine-mesh bags among regions and impacts

| | | | *F*-ratio | |
|---|---|---|---|---|
| Comparison | $df_N$ | $df_D$ | $\log_{10} k_d$ | $\log_{10} k_{dd}$ |
| Region | 9 | 39.98/39.56 | 10.30*** | 29.36*** |
| Impact | 1 | 39.64/39.17 | $0.053^{ns}$ | $0.012^{ns}$ |
| Mesh | 1 | 251.67/251.81 | 776.58*** | 748.29*** |
| Leaf | 1 | 252.59/252.77 | 805.21*** | 587.31*** |
| Impact × mesh | 1 | 251.67/251.81 | 10.64** | 10.48** |
| Region × impact | 9 | 39.52/39.05 | $1.22^{ns}$ | 2.50* |
| Impact × leaf | 1 | 252.31/252.48 | $0.823^{ns}$ | $0.611^{ns}$ |
| Region × mesh | 9 | 251.67/251.81 | 12.51*** | 11.97*** |
| Mesh × leaf | 1 | 251.67/251.81 | $3.89^{ns}$ | $3.84^{ns}$ |
| Region × leaf | 9 | 252.44/252.61 | 14.24*** | 13.23*** |
| Region × impact × mesh | 9 | 251.67/251.81 | 3.12** | 3.04** |
| Region × mesh × leaf | 9 | 251.67/251.81 | 2.55** | 2.45* |

Non-significant interactions and parameters omitted. *$P<0.05$; **$P<0.01$; ***$P<0.001$; ns: $P>0.05$.

using Friedman's test, which is based on ranks (after Canhoto *et al.*, 2005) and also analysed decomposition rates using PLS regression. In the single-choice treatment (Tier IVb), where consumers were presented with only a single resource type, decomposition rates ($-k$) were analysed using PLS regression.

# III. RESULTS

## A. Tier I. RIVFUNCTION Field Experiment: Impacts of Riparian Alterations on Decomposition Rates in 100 European Streams

Decomposition rates differed among regions, mesh types and leaf types, with faster decomposition in coarse mesh and for alder litter (Table 5; Figure 6). There was no main effect of riparian alterations *per se*, but there were significant two-way and three-way interactions, indicating that these impacts were contingent upon other factors. For instance, the impact × mesh interaction revealed that decomposition rates were similar for fine mesh between treatments whereas for coarse-mesh rates were generally faster in reference streams than in impacted streams. After correction for temperature effects all significant effects remained and regional differences actually increased,

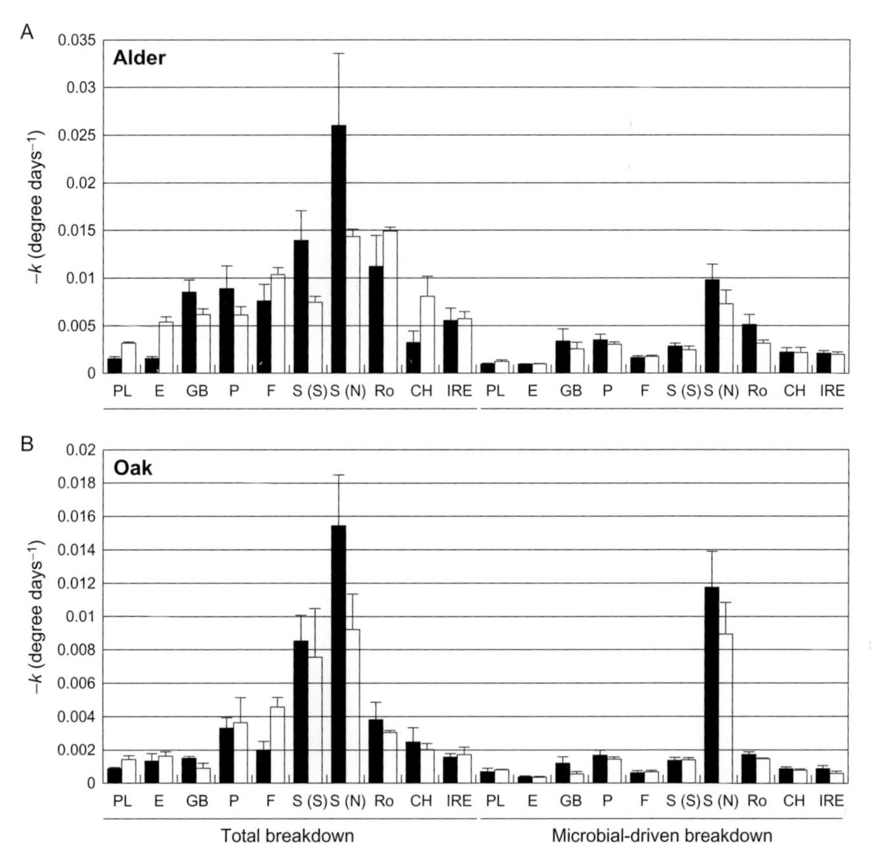


**Figure 6** Tier I: Impacts of anthropogenic alterations to riparian zones on alder (A) and oak (B) litter decomposition rates in 100 European streams across 10 regions in nine countries, grouped by type of alteration. Portugal (PL) and Spain (E) impacted sites were invaded by *Eucalyptus* spp.; Great Britain (GB) and Poland (P) were impacted by conifer plantations; France (F) and southern Sweden (S (S)) were impacted by replacement of native forest with beech (*Fagus sylvatica*) plantations and four sites had their native riparian woodland cleared for forestry (northern Sweden (S (N))) or pasture (Romania (Ro), Switzerland (CH) and Ireland (IRE)). Shaded bars represent impacted streams, white bars denote reference streams (means per region $\pm$ S.E.).

suggesting that these were not simply due to variation in latitudinal or altitudinal thermal regimes. The region $\times$ impact interaction further highlighted the regional contingency, which might reflect differences in the type of impact and/or consumer communities (Table 5). Invertebrate:microbial decomposition rates differed among regions and were lower in impacted versus reference streams, although this outcome was also dependent upon region (Table A1).

The subsequent analysis between pasture land use and other riparian impacts revealed faster decomposition rates in pasture streams than in streams

**Table 6** pan-European RIVFUNCTION field experiment (Tier I): Linear mixed effects model results of comparisons of standard ($k_d$) and temperature-normalised ($k_{dd}$) decomposition rate coefficients of litter in coarse-mesh and fine-mesh bags between pasture and other riparian impacts

| | | | *F*-ratio | |
|---|---|---|---|---|
| Comparison | $df_N$ | $df_D$ | $\log_{10} k_d$ | $\log_{10} k_{dd}$ |
| Impact | 1 | 48.16/± | 4.49* | |
| Mesh | 1 | 145.03/145.99 | 177.32*** | 177.55*** |
| Leaf | 1 | 145.79/146.32 | 208.31*** | 164.67*** |

±, non-significant interactions and parameters omitted. *$P<0.05$; ***$P<0.001$.

subjected to other types of riparian impact (Table 6). Decomposition was also fastest for alder litter and coarse-mesh bags. After correcting for temperature effects, the differences between pasture streams and streams subjected to other riparian impacts became non-significant (Table 6). Invertebrate:microbial decomposition rates were lower in pasture streams compared with streams subjected to other riparian impacts (LMEM $F\ df_{N1}\ df_{D48.43}=6.87$, $P<0.05$), but there was no differences between oak and alder litter.

## B. Tier II. Irish Field Experiments and Surveys: Decomposition, Algal Production and Herbivory Rates and Community Structure in Nine Pasture Streams

### *1. Tier II.i. Chemical Characteristics of Regions*

Physicochemical parameters (SRP, TON, $NH_4$, pH and conductivity) differed among streams, revealing a general gradient of agricultural intensity: the first axis of a PCO explained 59.5% of the chemical data and was associated with SRP, TON, pH. The second axis explained a further 18.7% and was associated with TON and $NH_4$ (Figure 7A).

### *2. Tier II.ii. Ecosystem Functioning: Detrital Pathways*

The spring grass resource quality measures (%C, %N, %P, % lignin and % cellulose) differed among streams (Figure 7B). This separation was evident on the first axis of an associated PCO which explained a total of 91.8% of the variation in the resource quality matrix. In general, the more intensively farmed

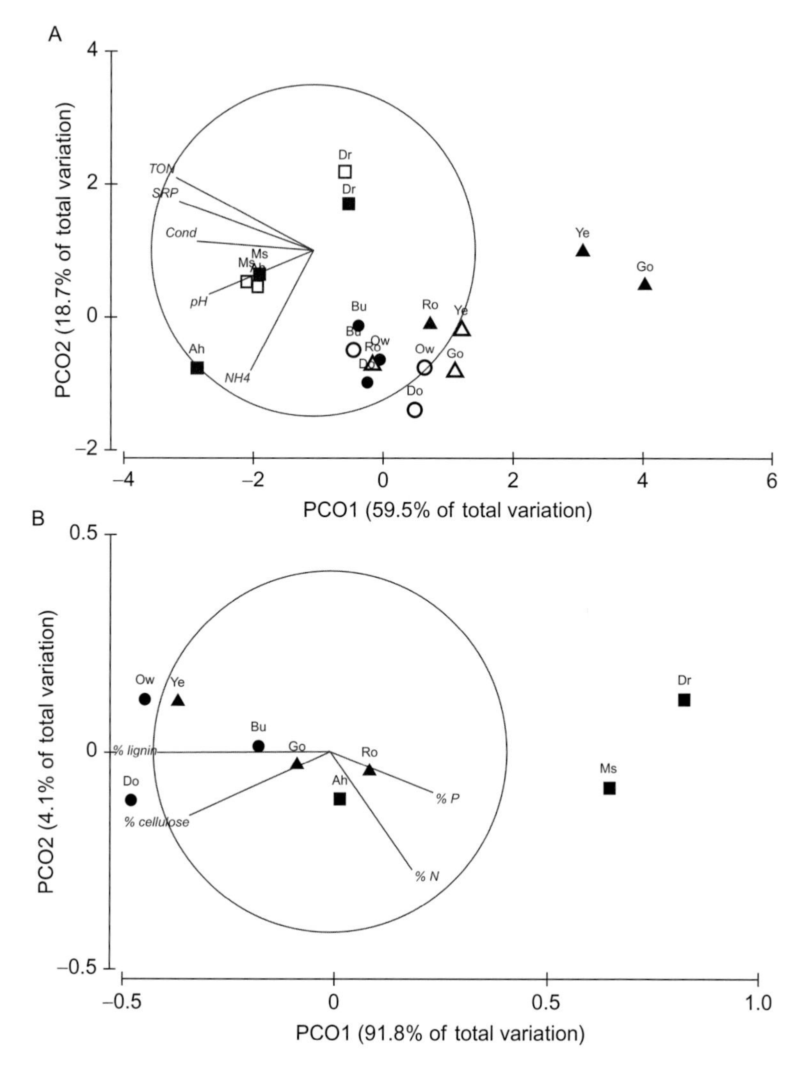


**Figure 7** Tier II: PCO of physicochemical variables among Irish pasture streams for two seasons (A) and PCO of grass resource quality variables among Irish pasture streams in spring (B). Open symbols denote winter and closed denote spring. Squares denote Cork streams, triangles denote Mayo streams and circles denote Clare Island streams, respectively. Vector overlay denotes Spearman correlations, showing vectors longer than 0.5 in length (see Section II). Labels denote stream names; Bu, Bunnamohaun; Ow, Owenmore; Do, Dorree; Ye, Yellow; Go, Goulaun; Ro, Srahrevagh; Dr, Dripsey; Ms, Morning Star; Ah, Aherlow.

**Table 7** Irish multiple sites field experiments (Tier II): Linear mixed effects model results of comparisons of standard ($k_d$) and temperature-normalised ($k_{dd}$) decomposition rate coefficients of litter in coarse-mesh and fine-mesh bags among regions and between seasons

| | | | | *F*-ratio | |
|---|---|---|---|---|---|
| Comparison | Mesh size | $df_N$ | $df_D$ | $\log_{10} k_d$ | $\log_{10} k_{dd}$ |
| Region | C | 2 | 5.99 | 3.44$^{ns}$ | 2.72$^{ns}$ |
| Leaf | C | 1 | 298.00/296.00 | 33.30*** | 33.71*** |
| Season | C | 1 | 298.06/296.06 | 152.24*** | 81.75*** |
| Region × leaf | C | 2 | 298.00/296.00 | 5.30** | 5.40** |
| Region × season | C | 2 | ±/296.06 | | 3.43* |
| Leaf | F | 1 | 282.33/282.32 | 260.73*** | 261.12*** |
| Season | F | 1 | 282.40/282.39 | 77.75*** | 34.99*** |

±, non-significant interactions and parameters omitted. *$P<0.05$; **$P<0.01$; ***$P<0.001$; ns: $P>0.05$.

Cork sites were characterised by higher %P and %N content, whereas the rough pasture grasses were characterised by higher % lignin and % cellulose.

In general, litter decomposition in coarse and fine-mesh bags was fastest in spring and for grass litter (Table 7; Figure 8). For coarse-mesh bags, there was no main effect of region, though there was a significant interaction with region × litter type: oak decomposition in Cork streams was 3.7× and 2.4× faster than in Mayo and Clare Island streams, respectively; whereas for grass litter these regional differences were about 30% less marked. Microbial-only decomposition did not differ among regions. After correcting for temperature effects, seasonal differences were reduced but still significant for both coarse and fine-mesh bags (Table 7). The ratio of $k_{invert}/k_{microbial}$ was highest in spring and on oak litter, with no main effect of region (LMEM Season, $F$ $df_{N1}$, $df_{D25}=5.02$, $P<0.05$, Leaf, $F$ $df_{N1}$, $df_{D25}=52.15$, $P<0.001$; Figure 8). PLS regression revealed that in winter grass decomposition in coarse mesh was driven by litter quality (e.g. C:P, %P) and shredder abundance; whereas, in spring, abiotic variables (e.g. pH, SRP, TON) were also significant in addition to the previously mentioned drivers in winter (Table 8; Figures 9 and 10A). Similar patterns occurred with oak litter decomposition in coarse-mesh bags, with shredders being important drivers in winter and the same abiotic variables (e.g. pH, SRP, TON) primarily associated in spring (Table 8; Figures 9 and 10A). For fine mesh, a combination of abiotic and resource quality attributes were important for grass litter decomposition in both seasons (e.g. temperature, pH, N:P). For oak litter, SRP and temperature in winter and TON in the spring, respectively, were important for decomposition rates (Table 8). Correcting for temperature effects in general

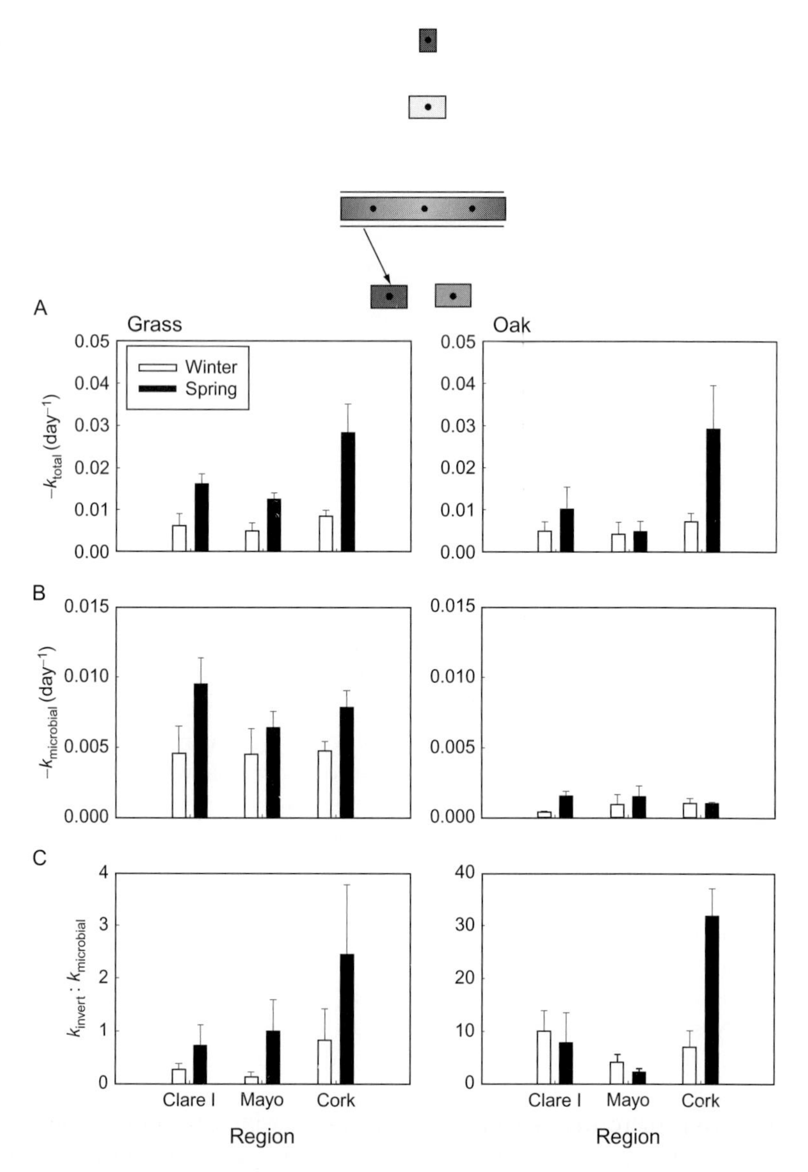


**Figure 8** Tier II. Grass and oak litter decomposition among Irish pasture streams in winter and spring: (A) total decomposition rates (coarse mesh), (B) microbial decomposition rates (fine mesh), (C) ratio of invertebrate-mediated to microbial decomposition rates. Individual values were averaged using streams as replicates and data presented show mean values of streams ± 1 S.E.

**Table 8** Summary table of partial least-squares (PLS) regression output for litter decomposition rates ($-k_d$) from Irish large-scale field (Tier II) trials, Irish single site field experiment (Tier III) and laboratory trials (IVa,b) (see Tables A2, A3, A10–A12 for VIP, slope for all variables and $k_{dd}$ results when applicable)

| Tier | Response | Season | Variables[a] | $R^2Y$ |
|---|---|---|---|---|
| II | Grass total | Winter | C:P, %P, N:P, B.ShrA, %Gra, %N, C:N, Temp., %Pre | 0.93 |
| | | Spring | pH, SRP, Grass Shr $g^{-1}$, TON, N:P, C:P, Temp., %P | 0.77 |
| | Oak total | Winter | Oak Shr $g^{-1}$, B.ShrA | 0.81 |
| | | Spring | pH, SRP, Oak Shr $g^{-1}$, TON | 0.83 |
| | Grass microbial | Winter | Temp, N:P, %P, C:P | 0.85 |
| | | Spring | pH, Temp., %C, TON, N:P, conduc. | 0.87 |
| | Oak microbial | Winter | SRP, Temp., conduc. | 0.72 |
| | | Spring | TON | 0.33 |
| III | Grass total | | Lig, cell, Shr $g^{-1}$, N:P | 0.87 |
| | Grass microbial | | Lig, cell, N:P | 0.84 |
| IVa | *Halesus* | | Lig, cell, N:P, %P | 0.72 |
| | *Gammarus* | | Lig, cell, N:P | 0.77 |
| | Microbes | | Lig, cell, N:P | 0.84 |
| IVb | *Halesus* | | Lig, N:P, cell, C:P, %P | 0.52 |
| | *Gammarus* | | Lig, N:P,%P, cell, C:P | 0.72 |
| | Microbes | | Lig, %P, N:P, cell, C:P | 0.79 |

[a]Predictor variables: *biotic attributes*: fish biomass (spring only; FishB), fish abundance (spring only; FishA), invertebrate biomass (spring only; InvB), Grass shredders $g^{-1}$ (Grass only), Oak shredders $g^{-1}$ (Oak only), benthos shredder abundance (B.ShrA), % invertebrate predators in benthos (%Pre), % invertebrate grazers in benthos (%Gra), % invertebrate shredders in benthos (%Shr); *abiotic attributes*: stream temperature (Temp.), TON, SRP, pH, conductivity (conduc.), $NH_4$. *Resource quality attributes* (Grass only): %P, %C, %N, C:P, C:N, N:P, % lignin content (lig), % cellulose content (cell).

did not markedly increase the amount of explained variance in decomposition rates (Tables A2 and A3).

Shredder abundance $g^{-1}$ litter was highest in spring (LMEM $F$ $df_{N1}$ $df_{D146.69}=19.93$, $P<0.001$). There was no main effect of region (LMEM $F$ $df_{N2}$ $df_{D6.00}=2.21$, $P>0.05$) or litter type (LMEM $F$ $df_{N1}$ $df_{D151.68}=0.198$, $P>0.05$) but there were significant two-way interactions with region × leaf (LMEM $F$ $df_{N2}$ $df_{D151.63}=6.38$, $P<0.01$) and region × season (LMEM $F$ $df_{N2}$ $df_{D146.67}=11.95$, $P<0.001$): Cork streams had higher abundances on oak litter than Clare Island and Mayo streams; whereas, grass litter abundances were similar among regions. Cork streams had higher abundances in spring than Clare Island and Mayo streams; whereas, in winter, abundances were similar among regions, as revealed by a region × season interaction.

Invertebrate assemblages $g^{-1}$ litter varied among sites, seasons and litter types (Table A4; Figure A1). The shredder guild in the Mayo streams were

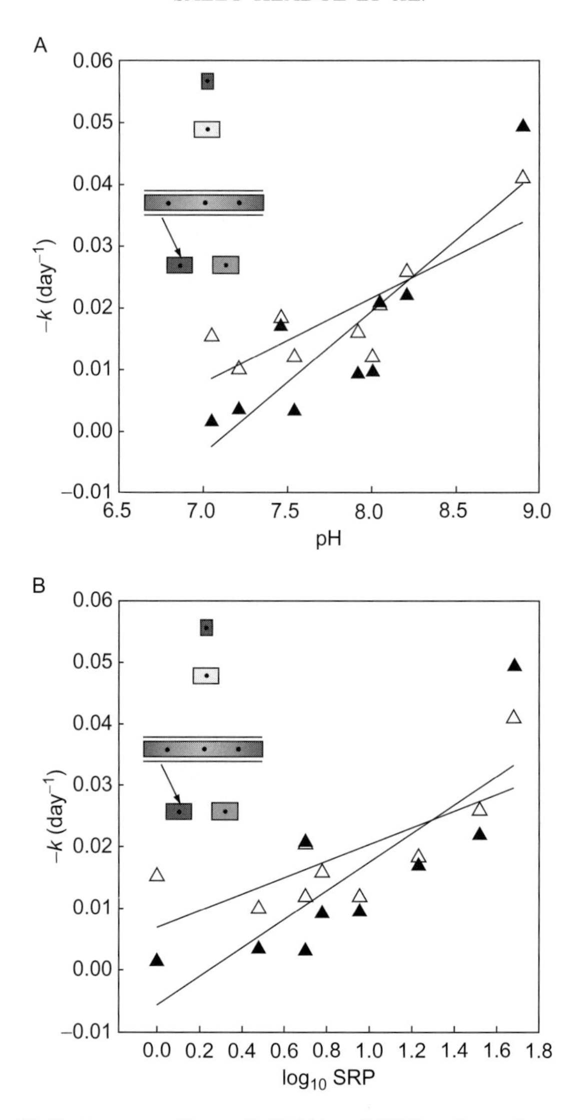


**Figure 9** Tier II: Bottom-up effects of pH (A) and SRP on litter decomposition rates in spring ($-k_{d\text{total}}$) in Irish pasture streams (B). Figures denote linear relationships of daily litter decomposition and two measures of stream water chemistry (pH and SRP) for the nine Irish pasture streams (closed triangle denotes oak litter decomposition rates; SRP, $r^2=0.65$, $P<0.01$; pH, $r^2=0.77$, $P<0.01$; open triangle denotes grass litter decomposition rates; SRP, $r^2=0.54$, $P<0.05$; pH, $r^2=0.66$, $P<0.01$).

characterised by mostly stonefly species (e.g. from the genera *Leuctra*, *Protonemura* and *Amphinemura*) whereas Cork streams were distinguished by a range of shredder consumers (e.g. Limnephilidae caddis larvae, *Gammarus* spp.

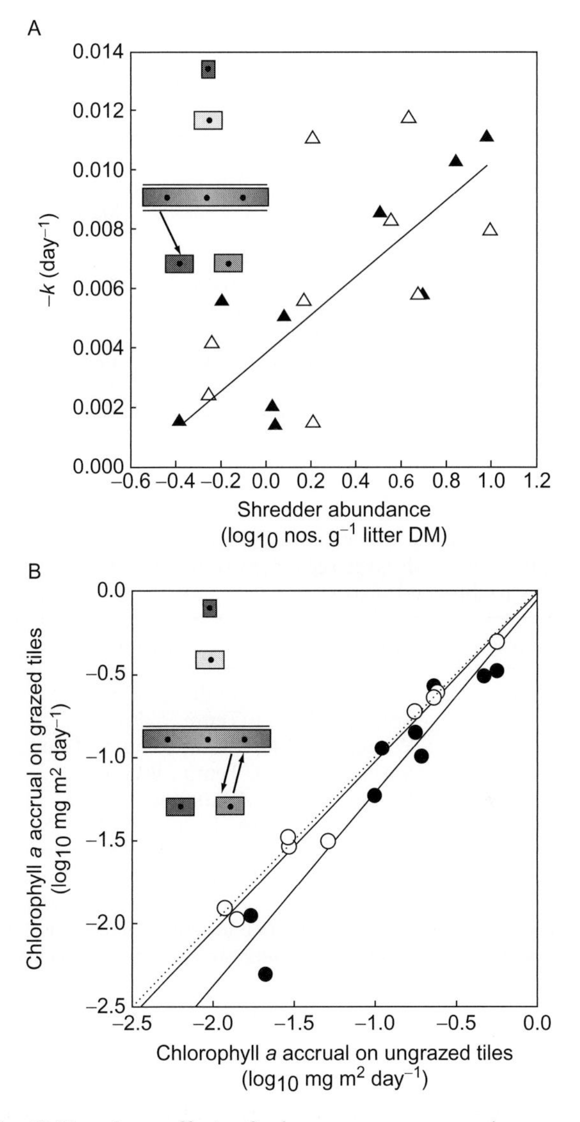


**Figure 10** Tier II: Top-down effects of primary consumers on decomposition rates in winter (A) (closed triangle; Oak $r^2 = 0.71$, $P < 0.01$, open triangle; Grass non-significant as a bivariate relationship in winter, as depicted here, although it is a significant predictor in PLS regression during spring; Table 8) and direct top-down effects of primary consumers on chlorophyll production (B): distance below the 1:1 line reveals the strength of herbivory. Daily chlorophyll *a* production in nine streams for two seasons. Open circles denote winter values, closed circles denote spring values.

**Table 9** Irish multiple sites field experiments (Tier II): Linear mixed effects model results of comparisons of standard ($A_d$) and temperature-normalised ($A_{dd}$) algal production ($\log_{10}$ chlorophyll *a* mg $m^{-2}$) on grazed and ungrazed tiles among regions and between seasons

| | | | *F*-ratio | |
|---|---|---|---|---|
| Comparison | $df_N$ | $df_D$ | $\log_{10} A_d$ | $\log_{10} A_{dd}$ |
| Region | 2 | 6.04/6.04 | 15.36** | 18.07** |
| Treatment | 1 | 150.00/150.00 | 38.93*** | 38.93*** |
| Season | 1 | 144.12/142.13 | 8.50** | $0.17^{ns}$ |
| Region × treatment | 2 | 150.00/150.00 | 3.80* | 3.80* |
| Treatment × season | 1 | 150.00/150.00 | 13.50*** | 13.50*** |
| Region × season | 2 | ±/142.13 | ± | 4.95** |

±, non-significant interactions and parameters omitted. *$P<0.05$; **$P<0.01$; ***$P<0.001$; ns: $P>0.05$.

**Table 10** Summary of table of partial least-squares (PLS) regression output for algal tiles and ratio from Irish large-scale field trial (Tier II) ($A_d$) (see Table A5 for VIP, slope for all variables and $A_{dd}$)

| Tier | Response | Season | Variables[a] | $R^2Y$ |
|---|---|---|---|---|
| II | Grazed tiles | Winter | %Pre, conduc., %Gra | 0.67 |
| | | Spring | %Gra, Temp., %Pre | 0.78 |
| | Ungrazed tiles | Winter | %Pre, %Gra, conduc., Temp. | 0.65 |
| | | Spring | %Gra, Temp., B.GraA | 0.87 |
| | Ungrazed: Grazed | Winter | B.GraA, conduc. | 0.93 |
| | | Spring | B.GraA, FishA, conduc., $NH_4$, TON | 0.85 |

[a]Predictor variables: *biotic attributes*: fish biomass (spring only; FishB), fish abundance (spring only; FishA), invertebrate biomass (spring only; InvB), benthos grazer abundance (B.GraA), % invertebrate predators in benthos (%Pre), % invertebrate grazers in benthos (%Gra), % invertebrate shredders in benthos (%Shr); *abiotic attributes*: stream temperature (Temp.), TON, SRP, pH, conductivity (conduc.), $NH_4$.

amphipods), with Clare Island streams being intermediate between the two (Figure A2).

### 3. *Tier II.iii. Ecosystem Functioning: Algal Pathways and Herbivory*

Algal production increased across the gradient of agricultural intensity (Clare Island < Mayo < Cork) and was highest on ungrazed tile surfaces and in spring (Table 9; Figure 10B). A significant season × grazing treatment interaction revealed that the proportion of production consumed was lower in

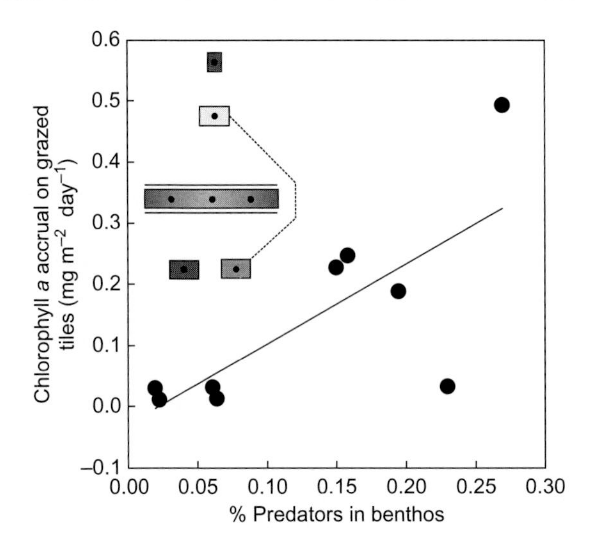


**Figure 11** Tier II: Potential for indirect food web effects on ecosystem processes—algal production on grazed tiles versus invertebrate predator relative abundance in winter ($r^2 = 0.54$, $P < 0.05$).

spring (70.3%) than in winter (93.2%). After correcting for temperature effects, however, the main effect of season was no longer significant, although its interactions were. There was also a region × grazing treatment interaction, with Mayo (92.1%) streams having markedly higher proportions of production consumed compared with Cork (71.6%), with both regions having a high supply rate, and Clare Island streams, which had both low supply and consumption rates (59.7%). PLS regressions indicated the importance of the invertebrate community in relation to algal accrual on tiles, with % grazers in the benthos in spring (negative association) and % invertebrate predators (positive association) in the winter being significant predictors (Table 10; Figure 11; Table A5), with stream conductivity and temperature being significant abiotic drivers (Table 10; Table A5). Important variables predicting the ratio of algal accrual on ungrazed:grazed tiles were grazer and fish abundance, conductivity, $NH_4$, and TON (Table 10; Table A5).

### *4. Tier II.iv. Community Structure and Ecosystem Functioning: Effects of Consumer Assemblages*

The composition and absolute abundances of the invertebrate assemblages in the benthos differed among sites, as revealed on the PCO (Table A6; Figure A3). Benthic abundances of shredders were highest in spring (LMEM $F$ $df_{N1}$ $df_{D150} = 14.26$, $P < 0.001$), and no main effect of region was evident (LMEM $F$ $df_{N2}$ $df_{D6} = 1.34$, $P > 0.05$). There was, however, a significant

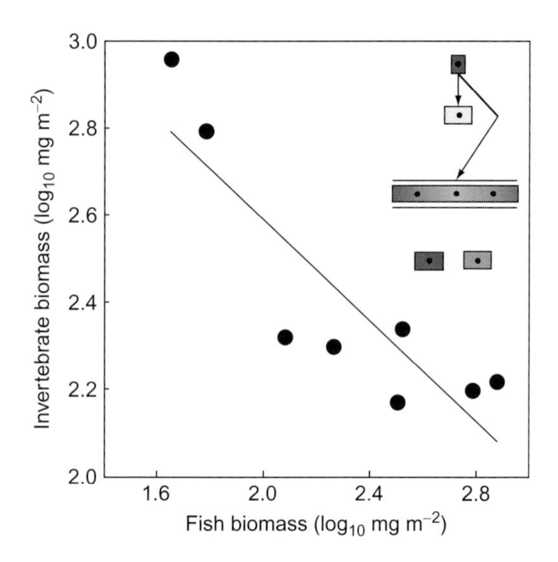


**Figure 12** Tier II: Potential top-down food web effects: invertebrate biomass versus fish biomass ($r^2=0.77$, $P<0.01$).

season × region interaction (LMEM $F$ $\mathrm{df_{N2}}$ $\mathrm{df_{D150}}=5.76$, $P<0.01$): Cork streams had higher abundances than the other sites in spring, whereas abundances in winter were more similar among regions, and the same was true for grazers (LMEM $F$ $\mathrm{df_{N2}}$ $\mathrm{df_{D150}}=5.26$, $P<0.01$). There was, however, no main effect for region (LMEM $F$ $\mathrm{df_{N2}}$ $\mathrm{df_{D6}}=0.83$, $P>0.05$) or season for grazers (LMEM $F$ $\mathrm{df_{N1}}$ $\mathrm{df_{D150}}=0.87$, $P>0.05$). Invertebrate biomass declined as fish biomass increased, suggesting the potential for top-down control of the former by the latter (Figure 12; Table A7) and invertebrate-mediated decomposition for oak litter was also reduced as fish biomass increased (negative association; Figure 13A; Table A8). Fish abundance was also an important predictor of herbivory on the algal tiles: invertebrate grazing pressure declined as fish biomass increased (Table 10; Figure 13B).

The mass–abundance scaling plots of the consumer communities revealed that the general direction of energy flux in the food web was smaller, more abundant taxa to larger, rare and less diverse consumers. The number of nodes increased across the gradient of agricultural intensification, particularly among the higher trophic levels (i.e. the predatory fishes), which were more prevalent in the Cork sites and least so in the rough pasture streams on Clare Island, one of which was the only fishless stream (Figure 14).

The stable isotope data were highly variable among streams (Figure 15). PLS regression confirmed that the grazing mayflies *Baetis* spp. and Heptageniidae fed mostly on algal biofilm (Table A9). In contrast, one of the dominant consumers, *Gammarus* spp., appeared to exploit both biofilm and

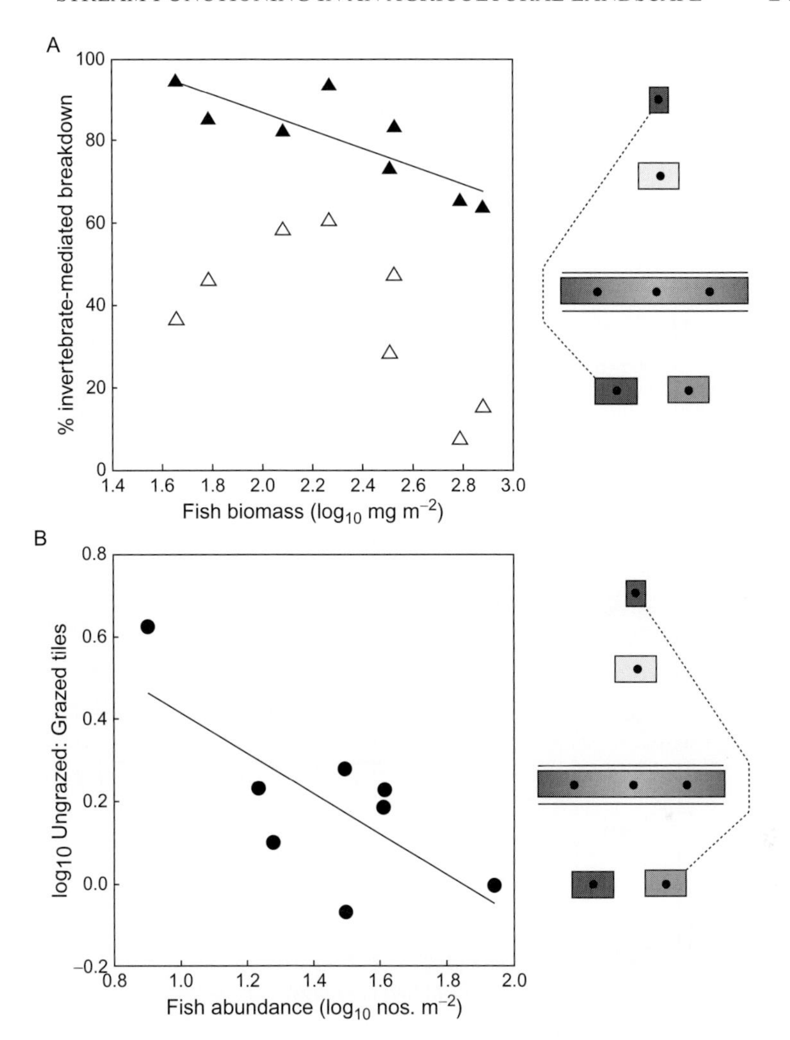


**Figure 13** Tier II: Potential cascading indirect (e.g. mediated via fish predation on invertebrate predators) top-down effects: (A) detritivory versus fish biomass (closed symbols denote oak decomposition rates, $r^2 = 0.68$, $P < 0.05$; open symbols denote grass decomposition rates, non-significant as a bivariate relationship, as depicted here, although it is a significant predictor in PLS regression; Table A8) and (B) invertebrate algal grazing versus fish abundance ($r^2 = 0.52$, $P < 0.04$).

CPOM (cf. Hladyz *et al.*, 2011) as revealed by the close match to the 1:1 line (assuming no carbon fractionation; after Nyström *et al.*, 2003), with the latter being more important in the winter and the former in the spring, tracking changes in resource availability (Figure 15; Table A9).

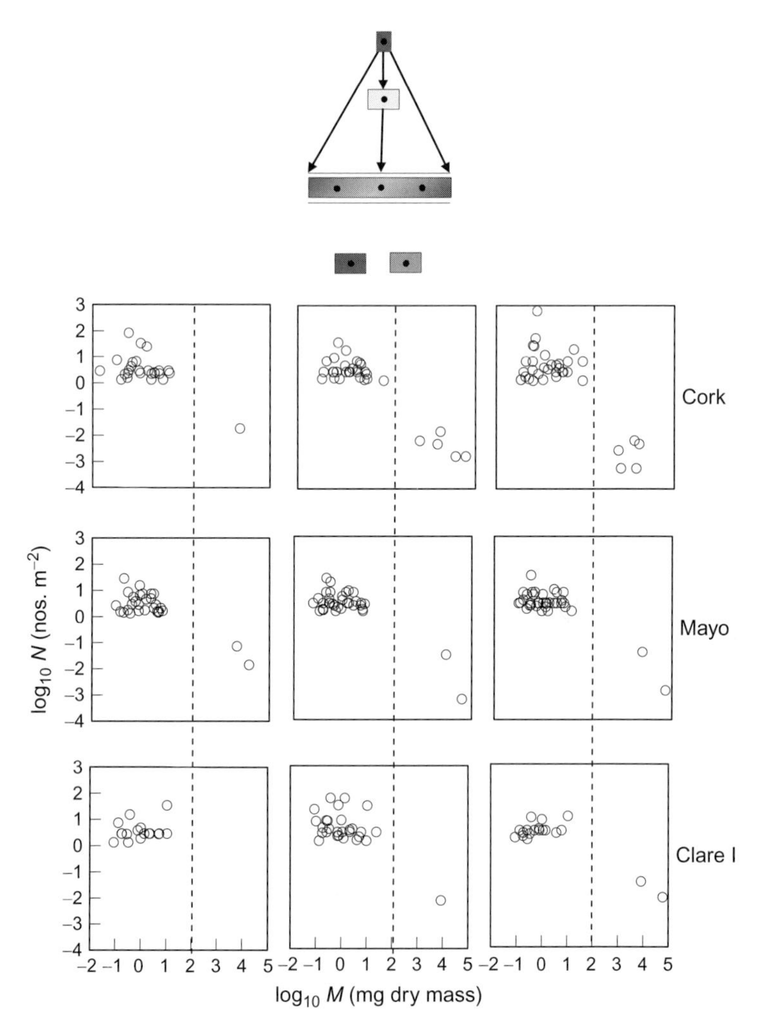


**Figure 14** Tier II: Food web structure in the nine food webs, showing mass–abundance scaling relationships among fish and invertebrate consumers. The fish assemblage in each panel occurs to the right of the dashed line.

## C. Tier III. Intensive Experimental Study of Grass Litter Decomposition Within a Single Field Site

In the intensive field trial where all nine grass litter types were placed within the Dripsey River, PLS regression analysis identified litter quality (in this instance % lignin content, % cellulose content and N:P) and shredder

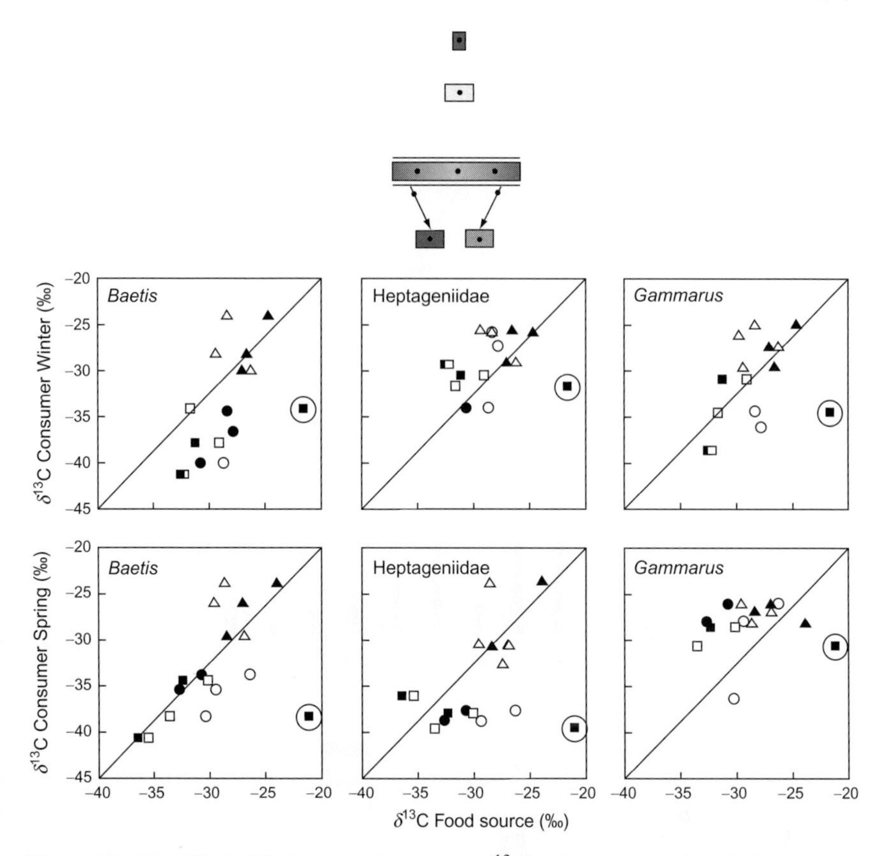


**Figure 15** Tier II: Stable isotope signatures ($\delta^{13}$C) of consumers in relation to two potential basal resources: epilithic biofilm (solid symbols) and CPOM (open symbols) among regions and between seasons. Squares represent Cork streams, triangles Mayo streams and circles Clare Island streams. Outlier circled in plots and excluded from analyses. Identical values on the $x$ and $y$ axes would lie on the diagonal 1:1 line (assuming no fractionation) (see Section III).

abundance as significant predictors of decomposition rates (Figure 16; Table 8; Table A10), as in the Tier II study in multiple sites.

## D. Tier IV. Laboratory Experiments: Resource Quality and Decomposition Rates

In the Tier IVa multiple-choice feeding trial decomposition rates were driven by resource quality, and declined with increasing % lignin content (Figure 17). *Gammarus* preferred higher quality grasses (Friedman's test $P < 0.001$; pairwise comparisons Ms, Ah, Dr > Do; Dr, Ms > Ye) and similar

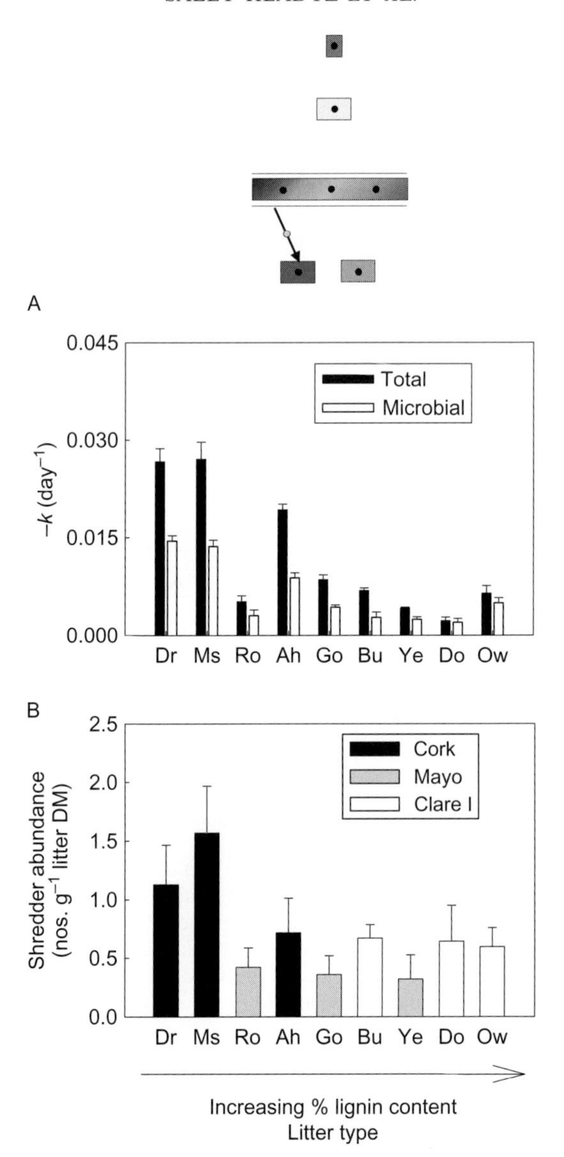


**Figure 16** Tier III: Dripsey River field trial of decomposition rates of all nine grass litter types in a single stream (A) and shredder abundance in litter bags (B). Data displayed are mean values ± S.E. Litter is ranked from lowest to highest in % lignin content.

patterns were evident for *Halesus* (Friedman's test $P < 0.001$; pairwise comparisons Ms, Ah, Dr > Do; Dr, Ms > Ye; Ro > Do). In the final experiment (Tier IVb), where only a single resource was given (single-choice), *Halesus*

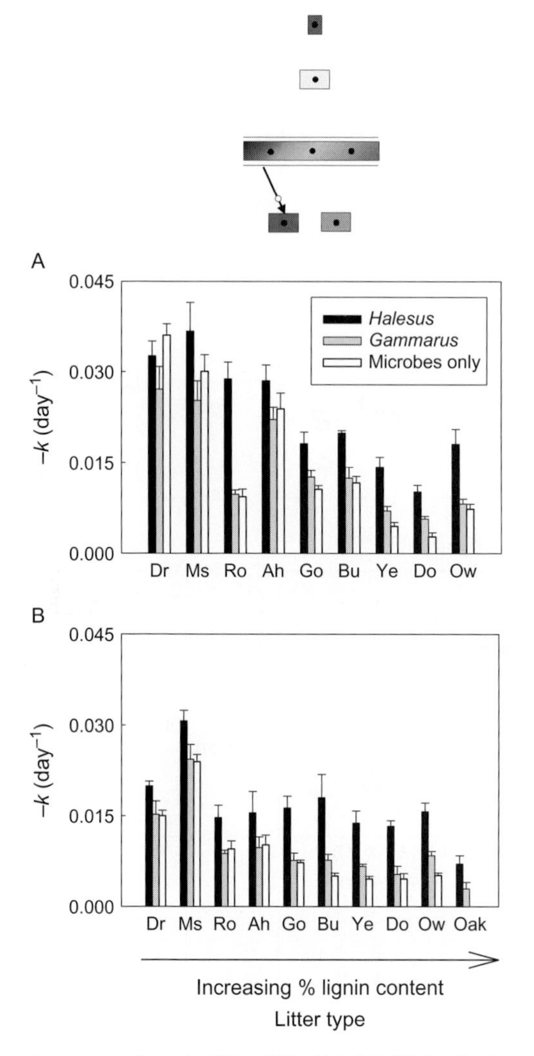


**Figure 17** Laboratory experiments (Tier IV): (A): Multiple-choice grass litter preference trials with 90 shredders consumers (*Halesus* or *Gammarus*) and microbial-only decomposition and (B) single-choice grass and oak litter decomposition trials with 10 shredder (*Halesus* or *Gammarus*) consumers and microbial-only decomposition. Litter is ranked from lowest to highest in % lignin content.

and *Gammarus* decomposition rates were also highly correlated suggesting that they were feeding on litter types in a comparable manner ($r^2 = 0.92$, $P < 0.001$). PLS regressions from both laboratory experiments confirmed the importance of % lignin content to decomposition rates, other measures of

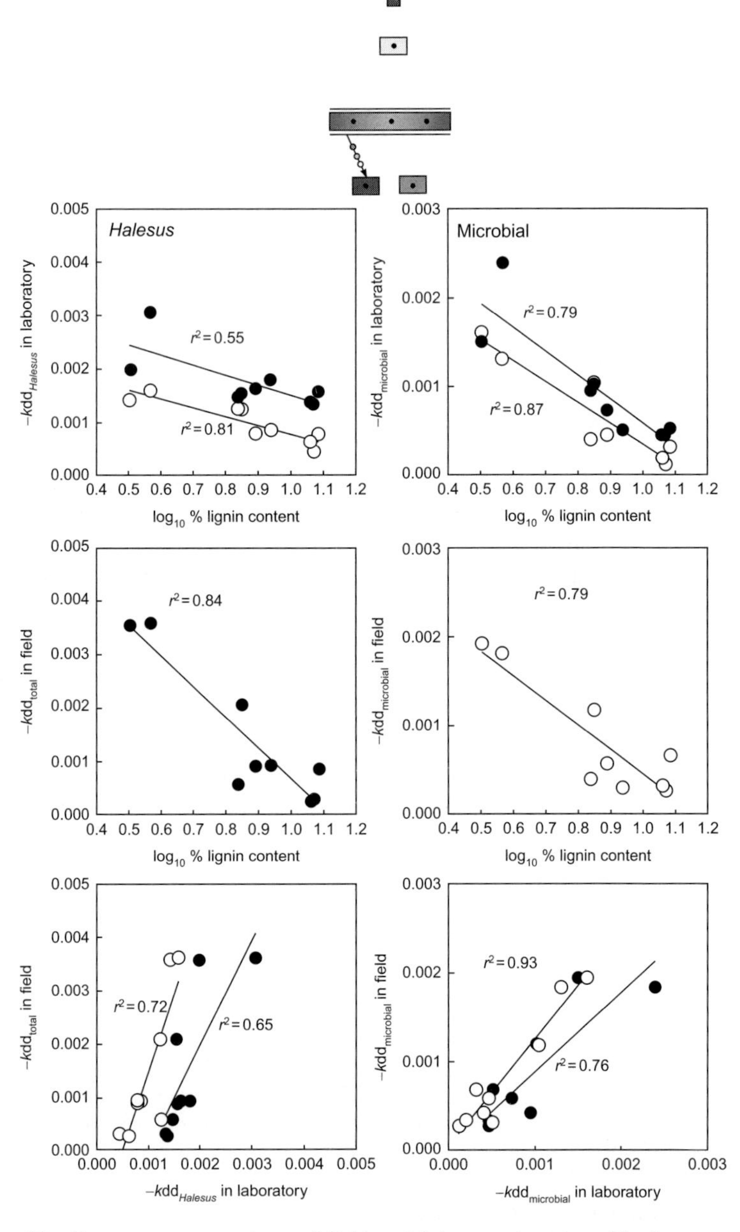


**Figure 18** Summary comparison of field and lab experiments, with decomposition rates expressed per degree day to normalise for temperature differences among sites and trials (*Halesus* and microbial only). Note the high degree of congruence across the different experimental tiers, from field experiments (Tier III) to laboratory mesocosms (Tier IVa, open circles) to laboratory microcosms (Tier IVb, closed circles). All regressions were significant ($P < 0.05$).

quality that were also identified included % cellulose content, %P and N:P and C:P ratios (Table 8; Tables A11 and A12).

In summary, our pan-European RIVFUNCTION experiment revealed that in many (but not all) instances, total litter decomposition rates were faster in reference streams, but microbial-mediated decomposition was similar in absolute terms: that is, the relative importance of microbes tended to be higher in impacted sites, and especially so in pasture streams. In our more intensive studies on pasture systems in Ireland, experiments and surveys revealed that resource quality was a fundamental driver of decomposition rates, but that its effect was modulated by other drivers, particularly as complexity and realism increased from laboratory microcosms to field conditions. In the latter, there were a range of additional underlying drivers which influenced process rates, including water chemistry variables (e.g. SRP, pH) and local differences in community abundance and composition. When we focussed on one stream (Tier III) as a model system to control for differences in water chemistry and community composition and abundance *per se*, the importance of resource quality (e.g. % lignin content and nutrient content) was more marked than in the Tier II studies, as was also the case in the simplest, most tightly controlled Tier IV laboratory experiments (Figure 18). Figure 18 demonstrates this high degree of congruence, using linear regression, across the different experimental tiers, for grass litter decomposition in response to resource quality, from field experiments to laboratory experiments with *Halesus* and microbes. This congruence in results across these different scales and levels of organisation suggested that similar fundamental drivers were operating and that their effects were relatively direct, given that similar responses were observed in simple single-species microcosms and under less controlled field conditions.

# IV. DISCUSSION

We found marked differences between ecosystem process rates in native woodland (i.e. reference) streams and those with altered riparian vegetation, although the magnitude and direction of these differences varied with the type of impact across Europe. Overall, there was no consistent main effect of altered riparian zones *per se*, since some impacts increased overall decomposition rates (McKie and Malmqvist, 2009), whereas others had no effect or decreased rates. Effects on total decomposition rates were contingent upon the type of impact and/or the particular European region of study, although the general effects of riparian alterations were broadly consistent among regions where similar types of impact (i.e. pasture vs. woodland in Switzerland, Ireland and Romania) were compared (cf. Hladyz *et al.*, 2010; Riipinen *et al.*, 2010). This is perhaps not surprising, given that riparian alterations could potentially cover a vast range of scenarios, from improving to impairing the quality and magnitude of litter inputs, suggesting that a blanket effect of riparian degradation is perhaps a somewhat naïve and unrealistic expectation. Although this large-scale field study was focussed on decomposition rates, inferences could be made about the roles of different consumer types by proxy, based on the mesh aperture used, as in other recent studies (e.g. Dangles *et al.*, 2011; Hladyz *et al.*, 2009, 2010; Riipinen *et al.*, 2010). This revealed more consistent effects of riparian alterations, especially via the mesh type × impact interaction, with a general increase in the importance of microbes relative to invertebrates as important agents of decomposition in impacted streams, especially in pasture compared to woodland streams. Although invertebrates were still important, and in many cases still *the* most important agents, this response suggests that the invertebrate detritivores were more sensitive to perturbations than the microbial decomposers.

The apparent decline in the relative importance of invertebrates as agents of decomposition in pasture streams has been suggested previously (Bird and Kaushik, 1992; Danger and Robson, 2004; Huryn *et al.*, 2002) and may be because grass litter is generally a poorer quality resource than leaf litter, being less accessible or palatable to shredders. A recent laboratory study using two shredder species (the caddis larvae *Limnephilus bipunctatus* and stonefly *Nemoura* sp.) from open-canopy streams revealed that senescent grass, despite being the most abundant food source year round, was relatively unfavoured when compared with leaf litter of terrestrial trees and shrubs, fresh grass litter, moss and benthic algae (Leberfinger and Bohman, 2010). Another potential reason for the decline in the role of invertebrates is that food webs in pasture streams rely more on inputs from autochthonous algal production than on allochthonous terrestrial detritus (Delong and Brusven, 1998; England and Rosemond, 2004; Hladyz *et al.*, 2010, 2011; McCutchan and Lewis, 2002; but see Leberfinger *et al.*, 2011). Therefore, not only is the detrital resource base impaired in pasture streams, but there is also greater availability of higher quality algal food as a more palatable alternative than terrestrial leaf litter.

The results of the intensive studies we conducted in Ireland are among the first attempts to characterise structural and functional attributes in a standardised manner across multiple sites via the integration of experiments with empirical survey data. One key point that emerged from this study was the high level of congruence in results obtained at different spatial scales in the field and the laboratory, which suggested that similar mechanisms were operating and that key system properties could be replicated under controlled conditions (e.g. Figure 18). It was apparent, for instance, that essentially the same drivers (i.e. measures of resource quality) of decomposition rates in the laboratory experiments were also evident in the field, despite the vast increase in the complexity of the systems under study.

### *1. Ecosystem Functioning: Bottom-Up Effects and Direct Consumer–Resource Interactions*

The laboratory and field experiments in Ireland all revealed strong effects of resource quality on decomposition rates. There were also significant influences of consumer identity in response to resource quality, with oak litter being processed primarily by invertebrates, and with grass litter, particularly the higher quality litter, being processed to a greater extent by microbial activity. The fact that oak is itself not a particularly favoured resource relative to other types of leaf litter (Hladyz *et al.*, 2009) further highlights the apparent poor quality of grass as a food source.

In general, the results obtained in the laboratory and the intensive field trial in the Dripsey River mirrored those from the extensive trials across all nine streams. For instance, the sites with higher stream nutrient concentrations and better quality grass litter were associated with enhanced litter decomposition rates and algal production, and higher invertebrate abundance, which, when taken together, are suggestive of bottom-up driven systems. These positive associations between stream nutrient concentrations and decomposition rates are also in line with our initial predictions based on previous studies (cf. Niyogi *et al.*, 2003; Pearson and Connolly, 2000; Robinson and Gessner, 2000). Rates of microbial decomposition, however, showed no clear and consistent response among the nine streams: thus, differences in total decomposition are more likely to be attributable to the shredder guild, as suggested by the strong positive correlation between their abundance and decomposition rates.

It is intriguing that resource quality was a powerful predictor of microbial decomposition in the laboratory experiments. The fact that the same pattern was not so evident in the field is certainly suggestive of the role of shredders as key modulators of process rates: some shredders turned out to be particularly reliant on microbial conditioning to improve litter palatability, but others (e.g. *Halesus* in the Tier IV experiments) appeared to be far less dependent (Figure 17). Initial resource quality prior to microbial conditioning, as measured, for instance, by concentrations of leaf nutrients (Gessner 1991; Petersen and Cummins, 1974; Quinn *et al.* 2000; Suberkropp *et al.*, 1976; Webster and Benfield, 1986) and physical toughness (Gessner and Chauvet, 1994; Melillo *et al.*, 1982; Royer and Minshall, 2001) of litter also determines decomposition rates, and indeed the higher quality grasses broke down especially rapidly in both the laboratory and field trials, with shredders consuming more of the better quality resources per unit time.

In other controlled laboratory studies species richness, consumer abundance and body size have been positively correlated with decomposition rates (Jonsson *et al.*, 2001; McKie *et al.*, 2008). The more nutrient-enriched sites in

Ireland were indeed characterised by an abundant and diverse guild of large shredders, including caddis and *Gammarus* species, whereas the Mayo sites had slower decomposition rates and a distinct shredder guild dominated by small stoneflies (Figure A2). It is often difficult to disentangle the effects of size and species identity *per se*, as they tend to be confounded in experimental studies, but some recent research indicates that the former, rather than the latter, is the key driver in freshwater assemblages (Reiss *et al.*, 2009, 2010). In the context of the current study, this suggests that similar process rates might be maintained in taxonomically different assemblages, so long as consumer sizes are comparable (cf. Figure 14).

Despite differences in the relative importance of microbes versus invertebrates, total decomposition rates for grass and oak litter were often similar. Shearer and Webster (1985) and Metwalli and Shearer (1989) found that the composition of fungal assemblages in forested streams differed from those in sparsely vegetated areas. Gulis (2001) found in a study of 92 streams that grass blades had different fungal assemblages from those supported by leaf litter, and inferred that pasture sites might contain aquatic fungal species adapted to processing grass litter. In our study, the invertebrate-mediated: microbial decomposition ratio was higher for oak than for grass litter, even though the former had far higher lignin content, which is a strong predictor of decomposition rates (Gessner and Chauvet, 1994; Hladyz *et al.*, 2009). However, oak litter had lower C:N and N:P ratios than most of the grass litters, which implies that invertebrates respond to more than one aspect of resource quality (Hladyz *et al.*, 2009). Further, the grass litter used in our experiments, like most grasses, probably had a higher silica content (e.g. Lanning and Eleuterius, 1987) than the oak litter, and this could also have contributed to lower rates of invertebrate-mediated decomposition and shredder colonisation. Similarly, a study examining decomposition of herbs and grasses in an open-canopy stream found lower colonisation of grasses by shredders compared to herb leaf litter and inferred that this reflected the lower resource quality of grass litter (Menninger and Palmer, 2007).

Taken together, our results indicate that pasture streams have an impaired detrital base, in terms of the rate at which energy per unit mass of resource is transferred to higher trophic levels. This appears to be a result not only of reduced litter inputs but also, in accordance with a few previous studies have also suggested that grass litter might indeed be a relatively poor food resource (Leberfinger and Bohman, 2010; Niyogi *et al.*, 2003; Young *et al.*, 1994). Nonetheless, even though pasture streams receive limited leaf litter inputs, the ability of these ecosystems to process this resource was clearly retained by the microbes and some shredders that consumed grass litter. Similar observations have been made in alpine streams above the tree line (Gessner *et al.*, 1998; Robinson *et al.*, 1998). This could confer a degree of inbuilt resilience upon the maintenance of overall ecosystem functioning, whereby the potential for the food web to switch to a detrital resource base is retained should conditions change (e.g. via successional reversion of pasture to woodland). Despite the lack of significant leaf litter inputs, standing crops of detritus can still be large in pasture streams, although they are composed primarily of grasses, macrophytes and herbs (e.g. Hladyz *et al.*, 2011; Leberfinger and Bohman, 2010; Menninger and Palmer, 2007). Our results suggest that pasture streams can support shredder biomass, particularly in winter, as suggested by the apparent switching by *Gammarus* from reliance on detritus (e.g. CPOM) in the winter, to increased biofilm consumption in the spring, as resource availability changed (Figure 15). Several other studies have shown that detrital resources can be of importance in open-canopy streams, and potentially more so than algal resources for particular consumers (e.g. shredding caddis larvae; Hladyz *et al.*, 2011; Leberfinger *et al.*, 2011).

Seasonal effects on ecosystem functioning were also important, with litter decomposition, algal production and grazing pressure all being higher in spring than in winter. These differences were often still apparent, albeit less so, after temperature correction of process rates, suggesting that seasonal differences were due to more than simply different thermal regimes. Shredder abundance was higher in the spring, and in combination with higher temperature rates most likely accounted for the elevated decomposition rates. In contrast, grazer abundance did not change seasonally, suggesting that per capita algal consumption was higher in spring during its peak in primary production (Cox, 1990; Francoeur *et al.*, 1999; Minshall, 1978). Overall the enhanced process rates observed in spring generally reflected a combination of increased temperatures, consumer abundance, per capita consumption and availability of algae.

### *2. Higher-Level Drivers: The Potential for Trophic Cascades and Indirect Food Web Effects*

Most studies into ecosystem functioning have tended to focus on the lower trophic levels, namely primary producers, detritus and primary consumers, rather than the higher, predatory levels (Woodward, 2009). This is a potential shortcoming because predators can exert powerful indirect effects on primary production and herbivory (e.g. Power, 1990; Scheffer, 1998) and there is increasing evidence that both fish and invertebrate predators can reduce leaf litter decomposition rates by influencing the abundance and/or activity of shredders (e.g. Oberndorfer *et al.*, 1984; Woodward *et al.*, 2008). Alterations to riparian vegetation could therefore modulate these indirect predator impacts, as suggested by the presence of 'apparent trophic cascades' in streams where inputs of terrestrial detritus had been curtailed (Nakano *et al.*, 1999).

The combined effects of reduced detrital resource quality and increased algal availability could, in theory, result in less dynamically stable food webs in pasture streams, due to a reduction in the damping effect of 'slow' detritus-based food chains relative to the 'faster' algal-based food chains (Rooney *et al.*, 2006). This could increase the possibility of trophic cascades arising because interaction strengths should increase, at least at the base of the web, as the system becomes less donor-controlled (Woodward, 2009; Woodward and Hildrew 2002). The increase in the diversity and abundance of predatory fish across the gradient of agricultural intensity and nutrient enrichment in the Irish streams also suggests that the food webs may be less dynamically stable than those in the less productive agricultural regions, since this is likely to further increase interaction strengths within the food web (cf. Layer *et al.*, 2010b, 2011).

Very few studies have assessed both community structure and the contribution of detrital and algal food chains to overall ecosystem functioning in pasture streams, which is surprising given the ubiquity of these systems in both Europe and elsewhere (Fujisaka *et al.*, 1996; Huryn *et al.*, 2002; Young *et al.*, 1994), and these suggestions still remain to be tested more rigorously via a full characterisation of the trophic networks of these types of streams. Nonetheless, the correlational data presented here indicate the potential for decreased food web stability in these systems relative to their native woodland counterparts, and also as they become more enriched.

Further manipulative experiments are required to identify causal mechanisms behind the potential impact of predators on basal processes (e.g. litter decomposition, algal consumption). Nonetheless, we can still make some inferences about their possible importance in pasture streams based on literature, correlations in our data, and the structure of the communities we observed. The potential for fish to exert indirect effects on the base of the food web (e.g. via suppressed rates of litter decomposition and herbivory when they are abundant; Figure 13) are hinted at in correlations in the data: for instance, the negative correlation between fish and invertebrate biomass is opposite to the positive relationship that would be expected if bottom-up effects took precedence. Top-down effects could thus account for at least some of the increased variability in the data and the weakening of resource quality constraints in the field versus the lab, even though the pattern and direction of responses was consistent. This generally high congruence between the results from the field and laboratory studies suggests that, overall, decomposition rates were most likely driven primarily by short, direct pathways within the food web (i.e. consumption by primary consumers). Indirect effects (e.g. via trophic cascades), although potentially more prevalent than in woodland streams, were likely to be of secondary importance. To explore these suggestions further, it would be instructive to perform large-scale and long-term experimental manipulations of multiple trophic levels and ecosystem processes under field conditions. The results of such experiments could provide critical information to understand and predict the structure and dynamics of stream ecosystems in general.

### *3. Caveats, Conclusions, and Future Directions*

Although the replacement of forests by human-modified vegetation is a worldwide phenomenon, our understanding of the community and ecosystem-level consequences of the loss of riparian woodland is clearly far from complete (Dodds, 1997; Hladyz *et al.*, 2010). Several lines of evidence presented here, however, suggest that streams with altered riprarian zones might, in general, be functionally fundamentally different from those in their ancestral woodland (i.e. 'reference') state, as revealed, for instance, by the significant interaction between mesh type x vegetation types. In particular, in many instances microbes played an increasingly important role as agents of decomposition, compared with the predominance of invertebrate detritivores in woodland sites, even though total decomposition rates were similar. This could have wider implications for the transfer of energy and nutrients in perturbed systems, since the capacity of microbes to correct for impaired detritivore activity might ultimately be somewhat limited (i.e. shredders are far larger consumers and many can macerate even very tough litter very effectively). In addition, microbial-mediated decomposition might operate via different routes of carbon and nutrient cycling, which might, for instance, benefit other types of invertebrate consumers (e.g. collector-gatherers vs. shredders). Further studies are clearly needed to assess how widespread this phenomenon of (relative) increased microbial decomposition is on a global scale, how consistent it is across different types of riparian modification (including those arising from the wide range of exotic plants that are invading riparian zones across the world). It is also important to develop a better understanding of the mechanisms that underpin the observed responses, and particularly to unravel the relative roles of different microbial and invertebrate consumers experimentally (Reiss *et al.*, 2010).

Differences between streams that flowed through woodland versus those that flowed through altered riparian zones were especially evident in systems dominated by pasture. Firstly, within pasture streams, algal pathways were relatively more important, due to a combination of increased light availability and poor detrital food quality, as grass litter was clearly not a favoured food among the invertebrate primary consumers (cf. Hladyz *et al.*, 2011).

Further investigations are required to quantify the extent to which increased algal contributions might compensate for, or even potentially exceed, those of allochthonous carbon, especially as this could have wider implications for carbon cycling and food web stability in freshwaters (Woodward *et al.*, 2010). Secondly, resource quality of grass litter exerted strong effects on consumption rates of both invertebrates and microbes, even though it was generally of poorer quality than the leaf litter that dominates native woodland streams (cf. Hladyz *et al.*, 2009). Thirdly, microbial decomposers become increasingly predominant relative to the invertebrate detritivores that are the principal agents of litter decomposition in woodland streams (Cummins *et al.*, 1989; Hladyz *et al.*, 2011; Wallace *et al.*, 1999). Similar responses have been reported from a recent study in 30 Swiss, Romanian, and Irish streams (Hladyz *et al.*, 2010), in an Australian pasture stream (Danger and Robson, 2004), 12 New Zealand streams in which fungal biomass was strongly correlated with grass litter decomposition rates (Niyogi *et al.*, 2003), and also in 24 high-altitude grassland streams in Ecuador (Dangles *et al.*, 2011). This suggests that the responses observed and described here might be global phenomena. Further studies in pasture streams would also clearly benefit from a closer focus on the microbial component of the food web, and it could be instructive to apply some of the new molecular and metagenomic approaches and next-generation sequencing techniques to identify the drivers behind these processes *in situ*, in terms of the identity, abundance and, potentially, the activity of both bacteria and fungi (Bärlocher *et al.*, 2009; Purdy *et al.*, 2010).

Much of our understanding of ecosystem processes and food web dynamics in running waters is derived from studies of (a few) wooded headwater streams (e.g. Cummins *et al.*, 1989; Fisher *et al.*, 1982; Gregory *et al.*, 1991; Wallace *et al.*, 1999; Woodward *et al.*, 2005). Far less attention has been paid to those in more obviously human-modified environments, where most studies have focussed on biomonitoring of the invertebrate assemblage (i.e. the nodes in part of the food web) in response to organic pollution. Other stressors, including the alteration of riparian vegetation have been largely overlooked in terms of their impacts on the higher levels of organisation and the links between community structure and ecosystem functioning (Hladyz *et al.*, 2010, 2011; Leberfinger and Bohman, 2010; Menninger and Palmer, 2007). It is also becoming increasingly clear that alterations to riparian zones and their allochthonous inputs to freshwater food webs could have important, but still largely unknown, synergies with the impacts of climate change in the near future (Boyero *et al.*, 2011; Perkins *et al.*, 2010a,b; Woodward *et al.*, 2010). It is of vital importance to understand the patterns and processes that operate in these altered streams if we are to improve stream ecosystem management, restoration and/or rehabilitation schemes, as well as to gain insight into the ecology of these systems in their own right, and we hope that the current study goes some way towards redressing this imbalance.

## ACKNOWLEDGEMENTS

We would like to thank the E. U. for funding the RIVFUNCTION research project, which was supported by the European Commission under the Fifth Framework Programme and the Swiss State Secretariat for Research and Education. We also wish to thank Manuel A. S. Graça, Jesús Pozo, Helen Cariss and Christian K. Dang and all the field and laboratory assistants who made invaluable contributions to the RIVFUNCTION study. Finally, we would like to thank two anonymous reviewers, whose comments greatly improved the chapter.

# APPENDIX

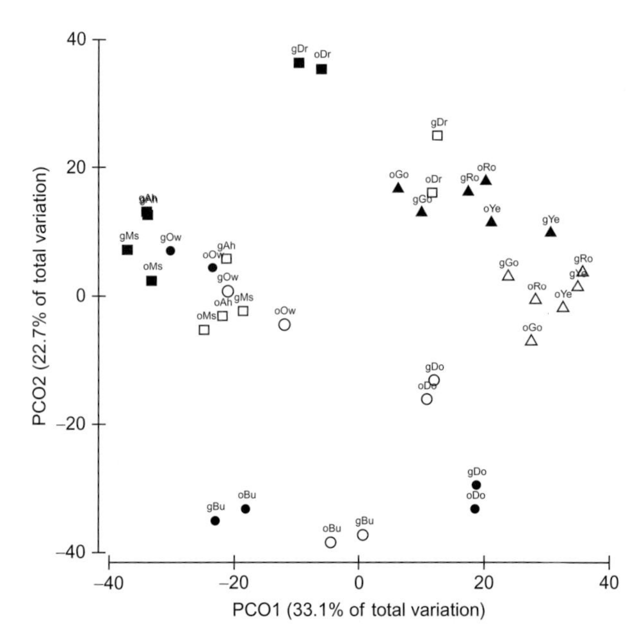


**Figure A1** Tier II: Two factor PCO plot of the centroids of the macroinvertebrate assemblages in mesh bags on the basis of the adjusted Bray–Curtis dissimilarity measure, showing the factors of region (squares denote Cork streams, triangles denote Mayo streams and circles denote Clare Island streams, respectively) and season (closed symbols denote spring samples and open symbols denote winter samples, respectively). ‘g’ denotes grass litter bags and ‘o’ denotes oak litter bags, respectively. Labels denote stream names (see Table 3 for legend).

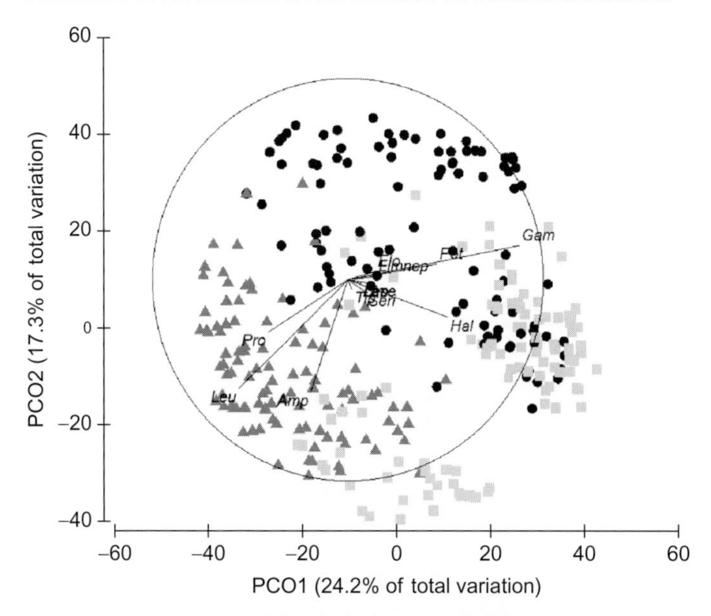


**Figure A2** Tier II: One factor PCO plot of the macroinvertebrate assemblages in mesh bags on the basis of the adjusted Bray–Curtis dissimilarity measure, showing the factor region (squares denote Cork streams, triangles denote Mayo streams and circles denote Clare Island streams, respectively). Vector overlay denotes Spearman correlations, displaying the shredder guild. Gam, *Gammarus* spp., Ase, *Asellus aquaticus* (L.), Pot, *Potamophylax* spp., Hal, *Halesus* spp., Lim, *Limnephilus* spp., Limnep, Limnephilidae indet., Seri, *Sericostoma personatum* (Kirby and Spence), Lep, *Lepidostoma* spp., Elo, *Elodes* spp. larvae, Pro, *Protonemura* spp., Amp, *Amphinemura sulcicollis* (Stephens), Leu, *Leuctra* spp., Tip, *Tipula* spp.

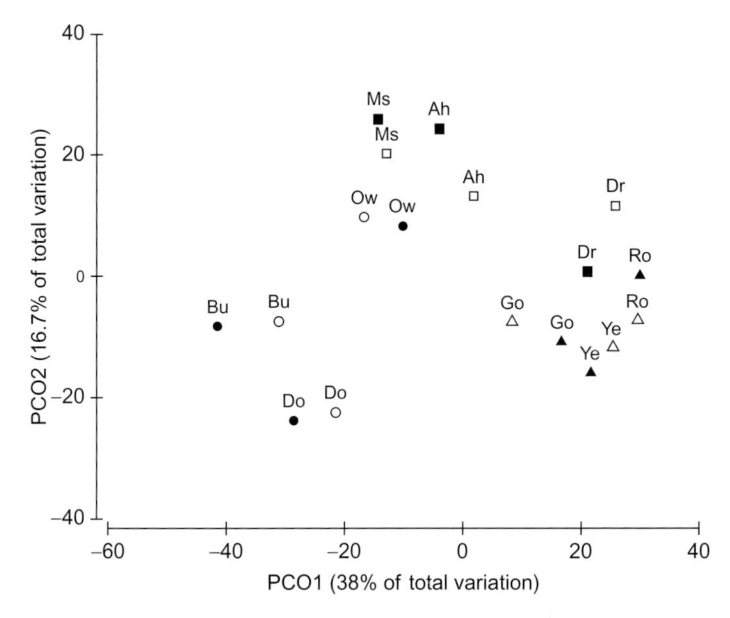


**Figure A3** Tier II: Two factor PCO plot of the centroids of the macroinvertebrate assemblages in stream benthos on the basis of the adjusted Bray–Curtis dissimilarity measure, showing the factors of region (squares denote Cork streams, triangles denote Mayo streams and circles denote Clare Island streams, respectively) and season (closed symbols denote spring samples and open symbols denote winter samples, respectively). Labels denote stream names (see Table 3 for legend).

**Table A1** pan-European RIVFUNCTION field experiment (Tier I): Linear mixed effects model results of comparisons of the ratio of invertebrate-mediated decomposition rates and microbial decomposition rates ($\log_{10} k_{invert}$:$k_{microbial}$) among European regions and impact

| Comparison | Leaf | $df_N$ | $df_D$ | *F*-ratio |
|---|---|---|---|---|
| Region | A | 9 | 40.90 | 3.13** |
| Impact | A | 1 | 39.45 | 14.59*** |
| Region × impact | A | 9 | 39.33 | 4.15** |
| Region | O | 9 | 85.00 | 7.37*** |
| Impact | O | 1 | 84.98 | 4.82* |

Non-significant interactions and parameters omitted. $^{*}P<0.05$, $^{**}P<0.01$, $^{***}P<0.001$.

**Table A2** Tier II: Partial least-squares (PLS) regression output for litter decomposition rates ($-k$) in coarse-mesh bags from Irish large-scale field trial

| $-k$ | Season | Variable[a] | VIP | Slope | Components | $R^2Y$ |
|---|---|---|---|---|---|---|
| Grass total | Winter | C:P/B.ShrA | 1.26/1.25 | −0.26/−0.71 | 3/4 | 0.93/0.96 |
| | | %P/N:P | 1.25/1.17 | 0.25/−0.49 | | |
| | | N:P/%Gra | 1.21/1.16 | −0.29/−0.36 | | |
| | | B.ShrA/%Pre | 1.16/1.14 | −0.33/−0.32 | | |
| | | %Gra/Grass Shr $g^{-1}$ | 1.14/1.14 | −0.18/0.55 | | |
| | | %N/C:P | 1.13/1.12 | 0.11/−0.34 | | |
| | | C:N/%P | 1.13/1.12 | −0.12/0.34 | | |
| | | Temp./%Shr | 1.1/1.00 | −0.39/0.002 | | |
| | | %Pre/SRP | 1.07/1.00 | −0.17/−0.38 | | |
| | | Constant | | 1.81/1.84 | | |
| Grass total | Spring | pH | 1.82/1.79 | 0.3/0.32 | 2/2 | 0.77/0.76 |
| | | SRP | 1.44/1.45 | 0.2/0.23 | | |
| | | Grass Shr $g^{-1}$ | 1.21/1.23 | 0.19/0.22 | | |
| | | TON/%C | 1.18/1.16 | 0.15/0.19 | | |
| | | N:P/TON | 1.17/1.16 | 0.06/0.17 | | |
| | | C:P | 1.15/1.12 | 0.11/0.07 | | |
| | | Temp./N:P | 1.04/1.12 | −0.18/0.06 | | |
| | | %P/ | 1.04/ | 0.04/ | | |
| | | Constant | | 1.98/1.80 | | |
| Oak total | Winter | Oak Shr $g^{-1}$ | 1.62/1.60 | 0.41/0.45 | 2/2 | 0.81/0.73 |
| | | B.ShrA | 1.26/1.17 | 0.32/0.32 | | |
| | | Constant | | 1.54/1.51 | | |
| Oak total | Spring | pH | 1.51/1.43 | 0.32/0.36 | 2/2 | 0.83/0.86 |
| | | SRP | 1.33/1.28 | 0.23/0.28 | | |
| | | Oak Shr $g^{-1}$ | 1.14/1.13 | 0.22/0.27 | | |
| | | TON | 1.11/1.04 | 0.17/0.19 | | |
| | | Constant | | 1.01/0.97 | | |

$-k_d$ and $k_{dd}$ are separated by '/'. Variables are listed with their regression slopes in descending VIP (variable importance to the projection) index order. Slope coefficients are not independent (unlike MLR), as the variables may be collinear. The VIP values reflect the importance of terms in the model both with respect to $y$ and with respect to $x$ (the projection). VIP is normalised and the average squared value is 1, so terms in the model with a VIP > 1 are important (other variables are not shown). $R^2Y$ is the % of the variation of $y$ explained by the model.

[a]Predictor variables: *biotic attributes*: fish biomass (spring only; FishB), fish abundance (spring only; FishA), invertebrate biomass (spring only; InvB), Grass shredders $g^{-1}$ (Grass only), Oak shredders $g^{-1}$ (Oak only), benthos shredder abundance (B.ShrA), % invertebrate predators in benthos (%Pre), % invertebrate grazers in benthos (%Gra), % invertebrate shredders in benthos (%Shr); *abiotic attributes*: stream temperature (Temp.), TON, SRP, pH, conductivity (conduc.), $NH_4$. *Resource quality attributes* (Grass only): %P, %C, %N, C:P, C:N, N:P, % lignin content (lig), % cellulose content (cell). Temperature was excluded from DD analyses.

**Table A3** Tier II: Partial least-squares (PLS) regression output for litter decomposition rates ($-k$) in fine-mesh bags from Irish large-scale field trial

| $-k$ | Season | Variable[a] | VIP | Slope | Components | $R^2Y$ |
|---|---|---|---|---|---|---|
| Grass microbial | Winter | Temp./N:P | 1.52/1.23 | −0.37/−0.52 | 2/3 | 0.85/0.83 |
| | | N:P/SRP | 1.34/1.20 | −0.33/−0.49 | | |
| | | %P/pH | 1.2/1.18 | 0.26/0.28 | | |
| | | C:P/TON | 1.19/1.14 | −0.25/−0.41 | | |
| | | /%C | /1.10 | /0.16 | | |
| | | /%P | /1.04 | /0.36 | | |
| | | /C:P | /1.03 | /−0.36 | | |
| | | Constant | | 1.98/1.82 | | |
| Grass microbial | Spring | pH | 1.32/1.78 | −0.82/−0.97 | 4/4 | 0.87/0.83 |
| | | Temp./%C | 1.29/1.32 | 0.36/−0.45 | | |
| | | %C/TON | 1.2/1.04 | −0.32/0.83 | | |
| | | TON/SRP | 1.11/1.03 | 0.76/0.13 | | |
| | | N:P/ | 1.02/ | 0.36/ | | |
| | | Conduc./ | 1/ | −0.14/ | | |
| | | Constant | | 3.13/3.67 | | |
| Oak microbial | Winter | SRP | 1.32/1.82 | 0.98/0.53 | 3/1 | 0.72/0.37 |
| | | Temp./ | 1.15/ | −0.92/ | | |
| | | Conduc./ | 1.06/ | 0.15/ | | |
| | | Constant | | 1.14/1.08 | | |
| Oak microbial | Spring | TON/pH | 1.7/1.18 | −0.58/−1.10 | 2/4 | 0.33/0.81 |
| | | /TON | /1.07 | /−2.47 | | |
| | | Constant | | 1.31/1.27 | | |

$-k_d$ and $k_{dd}$ are separated by '/'.

[a]Predictor variables: *abiotic attributes*: stream temperature (Temp.), TON, SRP, pH, conductivity (conduc.), $NH_4$. *Resource quality attributes* (Grass only): %P, %C, %N, C:P, C:N, N:P, % lignin content (lig), % cellulose content (cell). Temperature was excluded from DD analyses.

**Table A4** Irish multiple sites field experiments (Tier II): Permutational MANOVA (PERMANOVA) of 74 invertebrate species abundance variables in mesh bags (√transformed nos. $g^{-1}$ litter DM), based on the adjusted Bray–Curtis dissimilarity measure

| Comparison | $df_N$ | $df_D$ | Pseudo $F$-ratio |
|---|---|---|---|
| Region | 2 | 6 | 2.97* |
| Leaf | 1 | 6.01 | 3.79* |
| Season | 1 | 6 | 5.02** |
| Region × leaf | 2 | 6 | $1.09^{ns}$ |
| Region × season | 2 | 6 | 2.73* |
| Leaf × season | 1 | 6 | 4.49*** |

Three-way interactions were non-significant and therefore omitted. *$P<0.05$; **$P<0.01$; ***$P<0.001$; ns: $P>0.05$.

**Table A5** Tier II: Partial least-squares (PLS) regression output for algal tiles and ungrazed/grazed ratio from Irish large-scale field trial

| Algal response | Season | Variable[a] | VIP | Slope | Components | $R^2Y$ |
|---|---|---|---|---|---|---|
| Grazed tiles | Winter | %Pre | 1.34/1.37 | 0.28/0.30 | 2/2 | 0.67/0.68 |
| | | Conduc. | 1.27/1.34 | −0.41/−0.47 | | |
| | | %Gra | 1.22/1.24 | −0.17/−0.18 | | |
| | | Constant | | 0.86/0.94 | | |
| | Spring | %Gra | 1.86/1.70 | −1.16/−0.98 | 3/3 | 0.78/0.75 |
| | | Temp./%Pre | 1.29/1.15 | −0.07/−0.07 | | |
| | | %Pre/$NH_4$ | 1.2/1.02 | −0.3/−0.04 | | |
| | | Constant | | 1.19/1.18 | | |
| Ungrazed tiles | Winter | %Pre | 1.28/1.36 | 0.26/0.29 | 2/2 | 0.65/0.63 |
| | | %Gra | 1.17/1.24 | −0.15/−0.17 | | |
| | | Conduc. | 1.14/1.21 | −0.38/−0.42 | | |
| | | Temp./ | 1.05/ | 0.24/ | | |
| | | Constant | | 0.83/0.91 | | |
| | Spring | %Gra | 1.82/1.84 | −1.21/−1.15 | 3/3 | 0.87/0.87 |
| | | Temp./FishB | 1.23/1.09 | 0.01/−0.33 | | |
| | | B.GraA | 1.15/1.09 | −0.24/−0.22 | | |
| | | Constant | | 1.1/1.1 | | |
| Ungrazed: Grazed | Winter | B.GraA | 1.75 | 0.59 | 2 | 0.93 |
| | | Conduc. | 1.24 | 0.17 | | |
| | | Constant | | 4.5 | | |
| | Spring | B.GraA | 1.21 | 0.61 | 3 | 0.85 |
| | | FishA | 1.19 | −0.55 | | |
| | | Conduc. | 1.14 | −0.51 | | |
| | | $NH_4$ | 1.08 | 0.47 | | |
| | | TON | 1.05 | −0.46 | | |
| | | Constant | | 1.78 | | |

$A_d$ and $A_{dd}$ are separated by '/'.

[a]Predictor variables: *biotic attributes*: fish biomass (spring only; FishB), fish abundance (spring only; FishA), invertebrate biomass (spring only; InvB), benthos grazer abundance (B.GraA), % invertebrate predators in benthos (%Pre), % invertebrate grazers in benthos (%Gra), % invertebrate shredders in benthos (%Shr); *abiotic attributes*: stream temperature (Temp.), TON, SRP, pH, conductivity (conduc.), $NH_4$. Temperature was excluded from DD analyses.

**Table A6** Permutational MANOVA (PERMANOVA) of 73 invertebrate species abundance variables in stream benthos ($log_{10}$ nos. 0.0625 $m^{-2}$), based on the adjusted Bray–Curtis dissimilarity measure

| Comparison | $df_N$ | $df_D$ | Pseudo *F*-ratio |
|---|---|---|---|
| Region | 2 | 6 | 3.35** |
| Season | 1 | 6 | 3.26* |
| Region × season | 2 | 6 | $0.107^{ns}$ |

*$P<0.05$; **$P<0.01$; ns: $P>0.05$.

**Table A7** Tier II: Partial least-squares (PLS) regression output for potential indirect food web effects on invertebrate biomass from Irish large-scale field trial

| Response | Variable[a] | VIP | Slope | Components | $R^2Y$ |
|---|---|---|---|---|---|
| Invertebrate biomass | FishB | 1.60 | −0.29 | 3 | 0.96 |
| | B.GraA | 1.54 | 0.24 | | |
| | B.ShrA | 1.49 | 0.28 | | |
| | FishA | 1.32 | −0.15 | | |
| | %Pre | 1.06 | −0.21 | | |
| | SRP | 1.06 | −0.14 | | |
| | Conduc. | 1.02 | 0.06 | | |
| | Grass Shr $g^{-1}$ | 1.01 | 0.09 | | |
| | Constant | | 8.17 | | |

[a]Predictor variables: *biotic attributes*: fish biomass (FishB), fish abundance (FishA), Grass shredders $g^{-1}$, Oak shredders $g^{-1}$, benthos shredder abundance (B.ShrA), benthos grazer abundance (B.GraA), % invertebrate predators in benthos (%Pre), % invertebrate grazers in benthos (%Gra), % invertebrate shredders in benthos (%Shr); *abiotic attributes*: stream temperature (Temp.), TON, SRP, pH, conductivity (conduc.), $NH_4$, mg chlorophyll *a* accrual on ungrazed tiles $day^{-1}$ ($A_dU$), mg chlorophyll *a* accrual on grazed tiles $day^{-1}$ ($A_dG$). *Resource quality attributes* (Grass only): %P, %C, %N, C:P, C:N, N:P, % lignin content (lig), % cellulose content (cell).

**Table A8** Tier II: Partial least-squares (PLS) regression output for potential indirect food web effects on % invertebrate-mediated decomposition from Irish large-scale field trial

| Litter type | Variable[a] | VIP | Slope | Components | $R^2Y$ |
|---|---|---|---|---|---|
| Grass | $NH_4$ | 1.53 | 0.31 | 3 | 0.80 |
| | FishA | 1.38 | −0.32 | | |
| | Grass Shr $g^{-1}$ | 1.28 | 0.11 | | |
| | Cell | 1.24 | 0.08 | | |
| | FishB | 1.21 | −0.18 | | |
| | Temp. | 1.12 | −0.04 | | |
| | pH | 1.11 | 0.24 | | |
| | TON | 1.07 | 0.18 | | |
| | Lig | 1.05 | −0.15 | | |
| | Constant | | 1.94 | | |
| Oak | FishB | 1.30 | −0.30 | 3 | 0.90 |
| | Oak Shr $g^{-1}$ | 1.26 | 0.22 | | |
| | %Gra | 1.21 | −0.62 | | |
| | B.GraA | 1.17 | −0.04 | | |
| | InvB | 1.08 | 0.32 | | |
| | B.ShrA | 1.06 | 0.08 | | |
| | pH | 1.03 | 0.33 | | |
| | Constant | | 6.87 | | |

[a]Predictor variables: *biotic attributes*: fish biomass (FishB), fish abundance (FishA), Invertebrate biomass (InvB), Grass shredders $g^{-1}$, Oak shredders $g^{-1}$, benthos shredder abundance

(B.ShrA), benthos grazer abundance (B.GraA), % invertebrate predators in benthos (%Pre), % invertebrate grazers in benthos (%Gra), % invertebrate shredders in benthos (%Shr); *abiotic attributes*: stream temperature (Temp.), TON, SRP, pH, conductivity (conduc.), $NH_4$, mg chlorophyll *a* accrual on ungrazed tiles $day^{-1}$ ($A_dU$), mg chlorophyll *a* accrual on grazed tiles $day^{-1}$ ($A_dG$). *Resource quality attributes* (Grass only): %P, %C, %N, C:P, C:N, N:P, % lignin content (lig), % cellulose content (cell).

**Table A9** Tier II: Partial least-squares (PLS) regression output for consumer carbon isotope signatures in relation to potential food sources in winter and spring

| Consumer | Season | Variable | VIP | Slope | $R^2Y$ |
|---|---|---|---|---|---|
| *Baetis* | Winter | CPOM | 0.58 | 0.25 | 0.58 |
| | | Biofilm | 1.29 | 0.55 | |
| | | Constant | | −5.99 | |
| | Spring | CPOM | 0.81 | 0.35 | 0.72 |
| | | Biofilm | 1.16 | 0.51 | |
| | | Constant | | −5.79 | |
| Heptageniidae | Winter | CPOM | 0.92 | 0.33 | 0.50 |
| | | Biofilm | 1.08 | 0.39 | |
| | | Constant | | −6.99 | |
| | Spring | CPOM | 0.54 | 0.24 | 0.58 |
| | | Biofilm | 1.31 | 0.58 | |
| | | Constant | | −6.23 | |
| *Gammarus* | Winter | CPOM | 0.50 | 0.19 | 0.40 |
| | | Biofilm | 1.32 | 0.49 | |
| | | Constant | | −9.83 | |
| | Spring | CPOM | 1.35 | 0.47 | 0.27 |
| | | Biofilm | 0.41 | 0.14 | |
| | | Constant | | −8.56 | |

All models contained one component.

**Table A10** Tier III: Partial least-squares (PLS) regression output for litter decomposition rates ($-k$ $day^{-1}$) from Irish single site (Dripsey River) field experiment

| Type of decomposition | Variable[a] | VIP | Slope | Components | $R^2Y$ |
|---|---|---|---|---|---|
| Grass total | Lig | 1.33 | −0.31 | 2 | 0.87 |
| | Cell | 1.21 | −0.29 | | |
| | Shr $g^{-1}$ | 1.15 | 0.23 | | |
| | N:P | 1.01 | −0.11 | | |
| | Constant | | 1.21 | | |
| Grass microbial | Lig | 1.34 | −0.33 | 2 | 0.84 |
| | Cell | 1.31 | −0.34 | | |
| | N:P | 1.07 | −0.18 | | |
| | Constant | | 1.29 | | |

[a]Predictor variables used in this model included: Shredders $g^{-1}$ (Coarse-mesh only, Shr $g^{-1}$), % P, %C, %N, C:P, C:N, N:P, % lignin content (lig), % cellulose content (cell).

**Table A11** Tier IVa: Partial least-squares (PLS) regression output for litter decomposition rates ($-k$ day$^{-1}$) from multiple-choice mesocosm grass litter trials

| Consumer | Variable[a] | VIP | Slope | Components | $R^2Y$ |
|---|---|---|---|---|---|
| *Halesus* | Lig | 1.36 | −0.18 | 1 | 0.72 |
| | Cell | 1.21 | −0.16 | | |
| | N:P | 1.14 | −0.15 | | |
| | %P | 1.10 | 0.15 | | |
| | Constant | | 2.57 | | |
| *Gammarus* | Lig | 1.47 | −0.37 | 2 | 0.77 |
| | Cell | 1.26 | −0.31 | | |
| | N:P | 1.02 | −0.14 | | |
| | Constant | | 1.76 | | |
| Microbes | Lig | 1.43 | −0.37 | 2 | 0.84 |
| | Cell | 1.31 | −0.35 | | |
| | N:P | 1.04 | −0.16 | | |
| | Constant | | 1.27 | | |

[a]Predictor variables used in this model included: Grass litter quality: %P, %C, %N, C:P, C:N, N:P, % lignin content (lig), % cellulose content (cell).

**Table A12** Tier IVb: Partial least-squares (PLS) regression output for litter decomposition rates ($-k$ day$^{-1}$) from single-choice microcosm grass litter trials

| Consumer | Variable[a] | VIP | Slope | $R^2Y$ |
|---|---|---|---|---|
| *Halesus* | Lig | 1.29 | −0.15 | 0.52 |
| | N:P | 1.18 | −0.13 | |
| | Cell | 1.12 | −0.13 | |
| | C:P | 1.1 | −0.13 | |
| | %P | 1.09 | 0.12 | |
| | Constant | | 3.29 | |
| *Gammarus* | Lig | 1.23 | −0.16 | 0.72 |
| | N:P | 1.18 | −0.16 | |
| | %P | 1.14 | 0.15 | |
| | Cell | 1.12 | −0.15 | |
| | C:P | 1.08 | −0.14 | |
| | Constant | | 1.83 | |
| Microbes | Lig | 1.26 | −0.18 | 0.79 |
| | %P | 1.14 | 0.16 | |
| | N:P | 1.12 | −0.16 | |
| | Cell | 1.09 | −0.15 | |
| | C:P | 1.03 | −0.14 | |
| | Constant | | 1.48 | |

All models consisted of component.

[a]Predictor variables used in this model included: Grass litter quality: %P, %C, %N, C:P, C:N, N: P, % lignin content (lig), % cellulose content (cell).

# REFERENCES

Abrams, P.A. (1995). Implications of dynamically variable traits for identifying, classifying and measuring direct and indirect effects in ecological communities. *Am. Nat.* **146**, 112–134.

Anderson, M.J., Gorley, R.N., and Clarke, K.R. (2008). *PERMANOVA + for PRIMER: Guide to Software and Statistical Methods*. PRIMER-E, Plymouth, UK.

Bärlocher, F., Charette, N., Letourneau, A., Nikolcheva, L.G., and Sridhar, K.R. (2009). Sequencing DNA extracted from single conidia of aquatic hyphomycetes. *Fungal Ecol.* **3**, 115–121.

Bird, G.A., and Kaushik, N.K. (1992). Invertebrate colonisation and processing of maple leaf litter in a forested and an agricultural reach of a stream. *Hydrobiologia* **234**, 65–77.

Bohlin, T., Hamrin, S., Heggberget, T., Rasmussen, G., and Saltveit, S. (1989). Electrofishing—Theory and practice with special emphasis on salmonids. *Hydrobiologia* **173**, 9–43.

Bonada, N., Prat, N., Resh, V.H., and Statzner, B. (2006). Developments in aquatic insect biomonitoring: A comparative analysis of recent approaches. *Annu. Rev. Entomol.* **51**, 495–523.

Boulton, A.J., and Boon, P.I. (1991). A review of methodology used to measure leaf litter decomposition in lotic environments: Time to turn over an old leaf. *Aust. J. Mar. Freshw. Res.* **42**, 1–43.

Boyero, L., Pearson, R.G., Gessner, M.O., Barmuta, L.A., Ferreira, V., Graça, M.A.S., Dudgeon, D., Boulton, A.J., Callisto, M., Chauvet, E., Helson, J.E., Bruder, A., *et al.* (2011). A global experiment suggests climate warming will not accelerate litter decomposition in streams but might reduce carbon sequestration. *Ecol. Lett.* **14**, 289–294.

Brown, L.E., Edwards, F.K., Milner, A.M., Woodward, G., and Ledger, M.E. (2011). Food web complexity and allometric–scaling relationships in stream mesocosms: Implications for experimentation. *J. Anim. Ecol.* doi: 10.1111/j.1365-2656.2011.01814.x.

Campbell, I.C., James, K.R., Hart, B.T., and Devereaux, A. (1992). Allochthonous coarse particulate organic material in forest and pasture reaches of 2 south-eastern Australian streams. 1. Litter accession. *Freshw. Biol.* **27**, 341–352.

Canhoto, C., Graça, M.A.S., and Bärlocher, F. (2005). Feeding preferences. In: *Methods to Study Litter Decomposition: A Practical Guide* (Ed. by M.A.S. Graça, F. Bärlocher and M.O. Gessner), pp. 297–302. Springer, Dordrecht, The Netherlands.

Cox, E.J. (1990). Studies on the algae of a small softwater stream. 2. Algal standing crop (measured by Chlorophyll a) on soft and hard substrata. *Arch. Hydrobiol.* **83**, 553–566.

Cummins, K.W., and Klug, M.J. (1979). Feeding ecology of stream invertebrates. *Annu. Rev. Ecol. Syst.* **10**, 147–172.

Cummins, K.W., Wilzbach, M.A., Gates, D.M., Perry, J.B., and Taliaferro, W.B. (1989). Shredders and riparian vegetation: Leaf litter that falls into streams influences communities of stream invertebrates. *Bioscience* **39**, 24–30.

Danger, A.R., and Robson, B.J. (2004). The effects of land use on leaf-litter processing by macroinvertebrates in an Australian temperate coastal stream. *Aquat. Sci.* **66**, 296–304.

Dangles, O., Crespo-Peréz, V., Andino, P., Espinosa, R., Calvez, R., and Jacobsen, D. (2011). Predicting richness effects on ecosystem function in natural communities: Insights from high elevation streams. *Ecology* **92**, 733–734.

Delong, M.D., and Brusven, M.A. (1998). Macroinvertebrate community structure along the longitudinal gradient of an agriculturally impacted stream. *Environ. Manage.* **22**, 445–457.

Dodds, W.K. (1997). Distribution of runoff and rivers related to vegetative characteristics, latitude, and slope: A global perspective. *J. N. Am. Benthol. Soc.* **16**, 162–168.

Dodson, J.R., and Bradshaw, R.H.W. (1987). A history of vegetation and fire 6600 B.P. to present county Sligo Western Ireland. *Boreas (Oslo)* **16**, 113–123.

Elosegi, A., Basaguren, A., and Pozo, J. (2006). A functional approach to the ecology of Atlantic Basque streams. *Limnetica* **25**, 123–134.

England, L.E., and Rosemond, A.D. (2004). Small reductions in forest cover weaken terrestrial–aquatic linkages in headwater streams. *Freshw. Biol.* **49**, 721–734.

Eriksson, L., Johansson, E., Kettaneh-Wold, N., and Wold, S. (1999). *Introduction to Multi- and Megavariate Data Analysis Using Projection Methods (PCA and PLS)*. Umetrics AB, Umeå.

Feld, C.K., Birk, S., Bradley, D.C., Hering, D., Kail, J., Marzin, A., Melcher, A., Nemitz, D., Pedersen, M.L., Pletterbauer, F., Pont, D., Verdonschot, P.F.M., *et al.* (2011). From natural to degraded and back again: Is river restoration working? *Adv. Ecol. Res.* **44**, 119–210.

Ferreira, V., Elosegi, A., Gulis, V., Pozo, J., and Graça, M.A.S. (2006). Eucalyptus plantations affect fungal communities associated with leaf-litter decomposition in Iberian streams. *Arch. Hydrobiol.* **166**, 467–490.

Fisher, S.B., Gray, L.J., Grimm, N.B., and Busch, D.E. (1982). Temporal succession in a desert stream ecosystem following flash flooding. *Ecol. Monogr.* **52**, 93–110.

Francoeur, S.N., Biggs, B.J.F., Smith, R.A., and Lowe, R.L. (1999). Nutrient limitation of algal biomass accrual in streams: Seasonal patterns and a comparison of methods. *J. N. Am. Benthol. Soc.* **18**, 242–260.

Friberg, N., Bonada, N., Bradley, D.C., Dunbar, M.J., Edwards, F.K., Grey, J., Hayes, R.B., Hildrew, A.G., Lamouroux, N., Trimmer, M., and Woodward, G. (2011). Biomonitoring of human impacts in natural ecosystems: The good, the bad and the ugly. *Adv. Ecol. Res.* **44**, 1–68.

Fujisaka, S., Bell, W., Thomas, N., Hurtado, L., and Crawford, E. (1996). Slash-and-burn agriculture, conversion to pasture, and deforestation in two Brazilian Amazon colonies. *Agric. Ecosyst. Environ.* **59**, 115–130.

Gessner, M.O. (1991). Differences in processing dynamics of fresh and dried leaf litter in a stream ecosystem. *Freshw. Biol.* **26**, 387–398.

Gessner, M.O. (2005). Proximate lignin and cellulose. In: *Methods to Study Litter Decomposition: A Practical Guide* (Ed. by M.A.S. Graça, F. Bärlocher and M.O. Gessner), pp. 61–66. Springer, Dordrecht, The Netherlands.

Gessner, M.O., and Chauvet, E. (1994). Importance of stream microfungi in controlling breakdown rates of leaf litter. *Ecology* **75**, 1807–1817.

Gessner, M.O., and Chauvet, E. (2002). A case for using litter breakdown to assess functional stream integrity. *Ecol. Appl.* **12**, 498–510.

Gessner, M.O., Robinson, C.T., and Ward, J.V. (1998). Leaf breakdown in streams of an alpine glacial floodplain: Dynamics of fungi and nutrients. *J. N. Am. Benthol. Soc.* **17**, 403–419.

Gessner, M.O., Dobson, M., and Chauvet, E. (1999). A perspective on leaf litter breakdown in streams. *Oikos* **85**, 377–384.

Gessner, M.O., Swan, C.M., Dang, C.K., McKie, B.G., Bardgett, R.D., Wall, D.H., and Hättenschwiler, S. (2010). Diversity meets decomposition. *Trends Ecol. Evol.* **25**, 372–380.

Gregory, S.V., Swanson, F.J., McKee, W.A., and Cummins, K.W. (1991). An ecosystem perspective of riparian zones. *Bioscience* **41**, 540–551.

Guiry, M.D., John, D.M., Rindi, F. and McCarthy, T.K. (Eds.) (2007). *New Survey of Clare Island. The Freshwater and Terrestrial Algae*, Vol. 6. 978-1-904890-31-7. Royal Irish Academy, Dublin.

Gulis, V. (2001). Are there any substrate preferences in aquatic hyphomycetes? *Mycol. Res.* **105**, 1088–1093.

Hagen, E.M., McTammany, M.E., Webster, J.R., and Benfield, E.F. (2010). Shifts in allochthonous input and autochthonous production in streams along an agricultural land-use gradient. *Hydrobiologia* **655**, 61–77.

Harrison, S.S.C., Pretty, J.L., Shepherd, D., Hildrew, A.G., Smith, C., and Hey, R.D. (2004). The effect of instream rehabilitation structures on macroinvertebrates in lowland rivers. *J. Appl. Ecol.* **41**, 1140–1154.

Hieber, M., and Gessner, M.O. (2002). Contribution of stream detrivores, fungi, and bacteria to leaf breakdown based on biomass estimates. *Ecology* **83**, 1026–1038.

Hladyz, S., Gessner, M.O., Giller, P.S., Pozo, J., and Woodward, G. (2009). Resource quality and stoichiometric constraints on stream ecosystem functioning. *Freshw. Biol.* **54**, 957–970.

Hladyz, S., Tiegs, S.D., Gessner, M.O., Giller, P.S., Rîşnoveanu, G., Preda, E., Nistorescu, M., Schindler, M., and Woodward, G. (2010). Leaf-litter breakdown in pasture and deciduous woodland streams: A comparison among three European regions. *Freshw. Biol.* **55**, 1916–1929.

Hladyz, S., Åbjörnsson, K., Giller, P.S., and Woodward, G. (2011). Impacts of an aggressive riparian invader on community structure and ecosystem functioning in stream food webs. *J. Appl. Ecol.* **48**, 443–452.

Holomuzki, J.R., and Stevenson, R.J. (1992). Role of predatory fish in community dynamics of an ephemeral stream. *Can. J. Fish. Aquat. Sci.* **46**, 2322–2330.

Huryn, A.D., Butz Huryn, V.M., Arbuckle, C.J., and Tsomides, L. (2002). Catchment land-use, macroinvertebrates and detritus processing in headwater streams: Taxonomic richness versus function. *Freshw. Biol.* **47**, 401–415.

Irons, J.G., Oswood, M.W., Stout, J.R., and Pringle, C.M. (1994). Latitudinal patterns in leaf litter breakdown: Is temperature really important? *Freshw. Biol.* **32**, 401–411.

Jespersen, A.M., and Christoffersen, K. (1987). Measurements of chlorophyll a from phytoplankton using ethanol as extraction solvent. *Arch. Hydrobiol.* **109**, 445–454.

Jonsson, M., Malmqvist, B., and Hoffsten, P. (2001). Leaf litter breakdown rates in boreal streams: Does shredder species richness matter? *Freshw. Biol.* **46**, 161–171.

Lamberti, G.A., and Resh, V.H. (1983). Stream periphyton and insect herbivores: An experimental study of grazing by a caddisfly population. *Ecology* **64**, 1124–1135.

Lanning, F.C., and Eleuterius, L.N. (1987). Silica and ash in native plants of the central and southeastern regions of the United States. *Ann. Bot.* **60**, 361–375.

Layer, K., Hildrew, A.G., Monteith, D., and Woodward, G. (2010a). Long-term variation in the littoral food web of an acidified mountain lake. *Glob. Change Biol.* **16**, 3133–3143.

Layer, K., Riede, J.O., Hildrew, A.G., and Woodward, G. (2010b). Food web structure and stability in 20 streams across a wide pH gradient. *Adv. Ecol. Res.* **42**, 265–299.

Layer, K., Hildrew, A.G., Jenkins, G.B., Riede, J.O., Rossiter, S.J., Townsend, C.R., and Woodward, G. (2011). Long-term dynamics of a well-characterised food web: Four decades of acidification and recovery in the broadstone stream model system. *Adv. Ecol. Res.* **44**, 69–118.

Leberfinger, K., and Bohman, I. (2010). Grass, mosses, algae or leaves? Food preference among shredders from open-canopy streams. *Aquat. Ecol.* **44**, 195–203.

Leberfinger, K., Bohman, I., and Herrmann, J. (2011). The importance of terrestrial resource subsidies for shredders in open-canopy streams revealed by stable isotope analysis. *Freshw. Biol.* **56**, 470–480.

Lecerf, A., Dobson, M., Dang, C.K., and Chauvet, E. (2005). Riparian plant species loss alters trophic dynamics in detritus-based stream ecosystems. *Oecologia* **146**, 432–442.

Ledger, M.E., Edward, F.K., Brown, L.E., Milner, A.M., and Woodward, G. (2011). Impact of simulated rought on ecosystem biomass production: An experimental test in stream mesocosms. *Glob. Change Biol.* doi: 10.1111/j.1365-2486.2011.02420.x.

McCutchan, J.H., and Lewis, W.M. (2002). Relative importance of carbon sources for macroinvertebrates in a Rocky Mountain stream. *Limnol. Oceanogr.* **47**, 742–752.

McKie, B.G., and Malmqvist, B. (2009). Assessing ecosystem functioning in streams affected by forest management: Increased leaf decomposition occurs without changes to the composition of benthic assemblages. *Freshw. Biol.* **54**, 2086–2100.

McKie, B.G., Petrin, Z., and Malmqvist, B. (2006). Mitigation or disturbance? Effects of liming on macroinvertebrate assemblage structure and leaf-litter decomposition in the humic streams of northern Sweden. *J. Appl. Ecol.* **43**, 780–791.

McKie, B.G., Woodward, G., Hladyz, S., Nistorescu, M., Preda, E., Popescu, C., Giller, P.S., and Malmqvist, B. (2008). Ecosystem functioning in stream assemblages from different regions: Contrasting responses to variation in detritivore richness, evenness and density. *J. Anim. Ecol.* **77**, 495–504.

Melillo, J.M., Aber, J.D., and Muratore, J.F. (1982). Nitrogen and lignin control of hardwood leaf litter decomposition dynamics. *Ecology* **63**, 621–626.

Menninger, H.L., and Palmer, M.A. (2007). Herbs and grasses as an allochthonous resource in open-canopy headwater streams. *Freshw. Biol.* **52**, 1689–1699.

Metwalli, A.A., and Shearer, C.A. (1989). Aquatic hyphomycete communities in clear-cut and wooded areas of an Illinois stream. *Trans. Ill. Acad. Sci.* **82**, 5–16.

Minshall, G.W. (1978). Autotrophy in stream ecosystems. *Bioscience* **28**, 767–771.

Moss, B. (2008). The Water Framework Directive: Total environment or political compromise? *Sci. Total Environ.* **400**, 32–41.

Mulder, C., Boit, A., Bonkowski, M., De Ruiter, P.C., Mancinelli, G., Van der Heijden, M.G.A., van Wijnen, H.J., Vonk, J.A., and Rutgers, M. (2011). A belowground perspective on Dutch agroecosystems: How soil organisms interact to support ecosystem services. *Adv. Ecol. Res.* **44**, 277–358.

Nakano, S., Miyasaka, H., and Kuhara, N. (1999). Terrestrial–aquatic linkages: Riparian arthropod inputs alter trophic cascades in a stream food web. *Ecology* **80**, 2435–2441.

Niyogi, D.K., Simon, K.S., and Townsend, C.R. (2003). Breakdown of tussock grass in streams along a gradient of agricultural development in New Zealand. *Freshw. Biol.* **45**, 1698–1708.

Nyström, P., McIntosh, A.R., and Winterbourn, M.J. (2003). Top-down and bottom-up processes in grassland and forested streams. *Oecologia* **136**, 596–608.

Oberndorfer, R.Y., McArthur, J.V., Barnes, J.R., and Dixon, J. (1984). The effect of invertebrate predators on leaf litter processing in an alpine stream. *Ecology* **65**, 1325–1331.

Peacor, S.D., and Werner, E.E. (1997). Trait-mediated indirect interactions in a simple aquatic food web. *Ecology* **78**, 1146–1156.

Pearson, R.G., and Connolly, N.M. (2000). Nutrient enhancement, food quality and community dynamics in a tropical rainforest stream. *Freshw. Biol.* **43**, 31–42.

Perkins, D.M., Reiss, J., Yvon-Durocher, G., and Woodward, G. (2010a). Global change and food webs in running waters. *Hydrobiologia* **657**, 181–198.

Perkins, D.M., McKie, B.G., Malmqvist, B., Gilmour, S.G., Reiss, J., and Woodward, G. (2010b). Environmental warming and biodiversity–ecosystem functioning in freshwater microcosms: Partitioning the effects of species identity, richness and metabolism. *Adv. Ecol. Res.* **43**, 177–209.

Petersen, R.C., and Cummins, K.W. (1974). Leaf processing in a woodland stream. *Freshw. Biol.* **4**, 343–368.

Power, M.E. (1990). Effects of fish in river food webs. *Science* **250**, 811–814.

Purdy, K.J., Hurd, P.J., Moya-Laraño, J., Trimmer, M., Oakely, B.B., and Woodward, G. (2010). Systems biology for ecology: From molecules to ecosystems. *Adv. Ecol. Res.* **43**, 87–149.

Quinn, J.M., Burrell, G.P., and Parkyn, S.M. (2000). Influences of leaf toughness and nitrogen content on in-stream processing and nutrient uptake by litter in a Waikato, New Zealand, pasture stream and streamside channels. *N.Z. J. Mar. Freshw.* **34**, 253–271.

Reid, D.J., Lake, P.S., Quinn, G.P., and Reich, P. (2008). Association of reduced riparian vegetation cover in agricultural landscapes with coarse detritus dynamics in lowland streams. *Mar. Freshw. Res.* **59**, 998–1014.

Reiss, J., Bridle, J.R., Montoya, J.M., and Woodward, G. (2009). Emerging horizons in biodiversity and ecosystem functioning research. *Trends Ecol. Evol.* **24**, 505–514.

Reiss, J., Bailey, R.A., Cássio, F., Woodward, G., and Pascoal, C. (2010). Assessing the contribution of micro-organisms and macrofauna to biodiversity–ecosystem functioning relationships in freshwater microcosms. *Adv. Ecol. Res.* **43**, 151–176.

Riipinen, M.P., Davy-Bowker, J., and Dobson, M. (2009). Comparison of structural and functional stream assessment methods to detect changes in riparian vegetation and water pH. *Freshw. Biol.* **54**, 2127–2138.

Riipinen, M.P., Fleituch, T., Hladyz, S., Woodward, G., Giller, P.S., and Dobson, M. (2010). Invertebrate community structure and ecosystem functioning in European conifer plantation streams. *Freshw. Biol.* **55**, 346–359.

Robinson, C.T., and Gessner, M.O. (2000). Nutrient addition accelerates leaf breakdown in an alpine springbrook. *Oecologia* **122**, 258–263.

Robinson, C.T., Gessner, M.O., and Ward, J.V. (1998). Leaf breakdown and associated macroinvertebrates in alpine glacial streams. *Freshw. Biol.* **40**, 215–228.

Rooney, N., McCann, K.S., Gellner, G., and Moore, J.C. (2006). Structural asymmetry and the stability of diverse food webs. *Nature* **442**, 265–269.

Royer, T.V., and Minshall, G.W. (2001). Effects of nutrient enrichment and leaf quality on the breakdown of leaves in a hardwater stream. *Freshw. Biol.* **46**, 603–610.

Scheffer, M. (1998). *Ecology of Shallow Lakes*. Chapman and Hall, London.

Shearer, C.A., and Webster, J. (1985). Aquatic hyphomycete community structure in the river Teign. I. Longitudinal distribution patterns. *Trans. Brit. Mycol. Soc.* **84**, 489–501.

Short, T., and Holomuzki, J. (1992). Indirect effects of fish on foraging behaviour and leaf processing by the isopod *Lirceus fontinalis*. *Freshw. Biol.* **27**, 91–97.
Suberkropp, K., Godshalk, G.L., and Klug, M.J. (1976). Changes in the chemical composition of leaves during processing in a woodland stream. *Ecology* **57**, 720–727.
Turner, A.M. (1997). Contrasting short-term and long-term effects of predation risk on consumer habitat use and resources. *Behav. Ecol.* **8**, 120–125.
Wallace, J.B., Eggert, S.L., Meyer, J.L., and Webster, J.R. (1999). Effects of resource limitation on a detrital-based ecosystem. *Ecol. Monogr.* **69**, 409–442.
Webster, J.R., and Benfield, E.F. (1986). Vascular plant breakdown in freshwater ecosystems. *Annu. Rev. Ecol. Syst.* **17**, 567–594.
Woodward, G. (2009). Biodiversity, ecosystem functioning and food webs in fresh waters: Assembling the jigsaw puzzle. *Freshw. Biol.* **54**, 2171–2187.
Woodward, G., and Hildrew, A.G. (2002). Body-size determinants of niche overlap and intraguild predation within a complex food web. *J. Anim. Ecol.* **71**, 1063–1074.
Woodward, G., Speirs, D.C., and Hildrew, A.G. (2005). Quantification and temporal resolution of a complex size-structured food web. *Adv. Ecol. Res.* **36**, 85–135.
Woodward, G., Papantoniou, G., Edwards, F., and Lauridsen, R.B. (2008). Trophic trickles and cascades in a complex food web: Impacts of a keystone predator on stream community structure and ecosystem processes. *Oikos* **117**, 683–692.
Woodward, G., Benstead, J.P., Beveridge, O.S., Blanchard, J., Brey, T., Brown, L., Cross, W.F., Friberg, N., Ings, T.C., Jacob, U., Jennings, S., Ledger, M.E., *et al.* (2010). Ecological networks in a changing climate. *Adv. Ecol. Res.* **42**, 72–138.
Young, R.G., and Collier, K.J. (2009). Contrasting responses to catchment modification among a range of functional and structural indicators of river ecosystem health. *Freshw. Biol.* **54**, 2155–2170.
Young, R.G., Huryn, A.D., and Townsend, C.R. (1994). Effects of agricultural-development on processing of tussock leaf-litter in high country New Zealand streams. *Freshw. Biol.* **32**, 413–427.
Zuur, A.F., Ieno, E.N., Walker, N.J., Saveliev, A.A., and Smith, G.M. (2009). *Mixed Effects Models and Extensions in Ecology with R, Statistics for Biology and Health*. Springer, New York.

# A Belowground Perspective on Dutch Agroecosystems: How Soil Organisms Interact to Support Ecosystem Services

CHRISTIAN MULDER, ALICE BOIT, MICHAEL BONKOWSKI, PETER C. DE RUITER, GIORGIO MANCINELLI, MARCEL G.A. VAN DER HEIJDEN, HARM J. VAN WIJNEN, J. ARIE VONK AND MICHIEL RUTGERS



ADVANCES IN ECOLOGICAL RESEARCH VOL. 44

0065-2504/11 $35.00
DOI: 10.1016/B978-0-12-374794-5.00005-5

## SUMMARY

1. New patterns and trends in land use are becoming increasingly evident in Europe's heavily modified landscape and else whereas sustainable agriculture and nature restoration are developed as viable long-term alternatives to intensively farmed arable land. The success of these changes depends on how soil biodiversity and processes respond to changes in management. To improve our understanding of the community structure and ecosystem functioning of the soil biota, we analyzed abiotic variables across 200 sites, and biological variables across 170 sites in The Netherlands, one of the most intensively farmed countries. The data were derived from the Dutch Soil Quality Network (DSQN), a long-term monitoring framework designed to obtain ecological insight into soil types (*STs*) and ecosystem types (*ETs*).
2. At the outset we describe *ST*s and biota, and we estimate the contribution of various groups to the provision of ecosystem services. We focused on interactive effects of soil properties on community patterns and ecosystem functioning using food web models. Ecologists analyze soil food webs by means of mechanistic and statistical modelling, linking network structure to energy flow and elemental dynamics commonly based on allometric scaling.
3. We also explored how predatory and metabolic processes are constrained by body size, diet and metabolic type, and how these constraints govern the interactions within and between trophic groups. In particular, we focused on how elemental fluxes determine the strengths of ecological interactions, and the resulting ecosystem services, in terms of sustenance of soil fertility.
4. We discuss data mining, food web visualizations, and an appropriate categorical way to capture subtle interrelationships within the DSQN dataset. Sampled metazoans were used to provide an overview of belowground processes and influences of land use. Unlike most studies to date we used data from the entire size spectrum, across 15 orders of magnitude, using body size as a *continuous trait* crucial for understanding ecological services.
5. Multimodality in the frequency distributions of body size represents a *performance filter* that acts as a buffer to environmental change. Large

differences in the body-size distributions across *ET*s and *ST*s were evident. Most observed trends support the hypothesis that the direct influence of ecological stoichiometry on the soil biota as an independent predictor (e.g. in the form of nutrient to carbon ratios), and consequently on the allometric scaling, is more dominant than either *ET* or *ST*. This provides opportunities to develop a mechanistic and physiologically oriented model for the distribution of species' body sizes, where responses of invertebrates can be predicted.
6. Our results highlight the different roles that organisms play in a number of key ecosystem services. Such a trait-based research has unique strengths in its rigorous formulation of fundamental scaling rules, as well as in its verifiability by empirical data. Nonetheless, it still has weaknesses that remain to be addressed, like the consequences of intraspecific size variation, the high degree of omnivory, and a possibly inaccurate assignment to trophic groups.
7. Studying the extent to which nutrient levels influence multitrophic interactions and how different land-use regimes affect soil biodiversity is clearly a fruitful area for future research to develop predictive models for soil ecosystem services under different management regimes. No similar efforts have been attempted previously for soil food webs, and our dataset has the potential to test and further verify its usefulness at an unprecedented space scale.

Let calculation sift his claims with faith and circumspection

(Goethe, XII-237)

# I. INTRODUCTION

We live in an increasingly human-modified planet, with all the negative implications associated with biodiversity loss and the impairment of ecosystem functioning (Vitousek *et al.*, 1997), and this is perhaps nowhere more evident than in Europe, where the legacy of the Agricultural and Industrial Revolutions is firmly imprinted on the landscape (Feld *et al.*, 2011; Friberg *et al.*, 2011; Hladyz *et al.*, 2011; Layer *et al.*, 2011). Throughout the world, different strategic action plans, discussion platforms, and research fundings have contributed to the astonishing progress in recording, digitizing and cataloguing biodiversity, which has added urgency as we experience the 6th Great Extinction, which is happening at an ever-accelerating rate. Public resources, such as Barcode of Life (molecular sequence), Tree of Life (cladistics and phylogenetic relationships), BHL (taxonomic literature), Zoobank (prototype) and IPNI (taxa and synonyms for animals and plants, respectively), MorphBank (morphospecies' images), GBIF (biogeography and museum specimens), Encylopedia of Life (natural history and evolution), and IUCN Red List (conservation biology), are just a few of the myriad examples in current use (http://blogs.plos.org/plos/2010/10/aggregating-tagging-and-connecting-biodiversity-studies/), and highlight the rapid advances made in this branch of ecology in recent years.

Unfortunately, the progress in applied ecology generally remains slower than the progress in fundamental ecology, in both terrestrial and aquatic systems (Friberg *et al.*, 2011). Between 1985 and 1992, for instance, the Dutch Program on "Soil Ecology of Arable Farming Systems" (see Section II.A), a co-operative research effort between applied research institutes, provided some of the first results to links specific ecosystem processes and population dynamics of soil organisms (Brussaard, 1994, and references therein). This program mirrors three other similarly multidisciplinary programs, also directed towards the development of sustainable farming, namely: the "Ecology of Arable Land" program in Sweden (Andrén, 1988), the shortgrass prairie program of the "Central Plains Experimental Range" (CPER) in Colorado, USA, and the comparison between different arable practices at the Experimental Horseshoe Bend sites in Athens, Georgia, USA (Moore *et al.*, 2004, and references therein).

The Dutch program focused particularly on the negative effects of inorganic fertilizers, pesticides, and tillage in modern agricultural practices on soil biodiversity. These programs were the first to characterise ecosystem services provided by soil biodiversity, and they also highlighted how the precise study of soil organisms and their communities is constrained by technical challenges associated with identifying these microscopic organisms, measuring their abundance and dynamics, and identifying the traits that potentially drive ecosystem functioning.

A full characterization of the structure and dynamics of the soil biota and how it provides essential ecosystem goods and services would be in harmony with the aims of Biodiversity Programme of UNEP World Conservation Monitoring Centre and with the Clearing House Mechanism (CHM) of the Convention on Biological Diversity (CBD), and the Biological Inventory Systems Component of the Man and the Biosphere (MAB) Programme. Recently, we have seen a rapid increase in the amount of original contributions on ecosystem functioning and ecosystem services, as shown in Figure 1. This mirrors the increasing socioeconomic and political focus on "The Ecosystem" as a whole, a trend reflected in a series of attempts to incorporate sustainable management technique as a concept relevant for stakeholders and policy-decision makers (Feld *et al.*, 2011; Friberg *et al.*, 2011; Sutherland *et al.*, 2006).

In the 1970s and 1980s, applied research projects started to focus on "sustainable agriculture", with the "Ecology of Arable Land" in Sweden

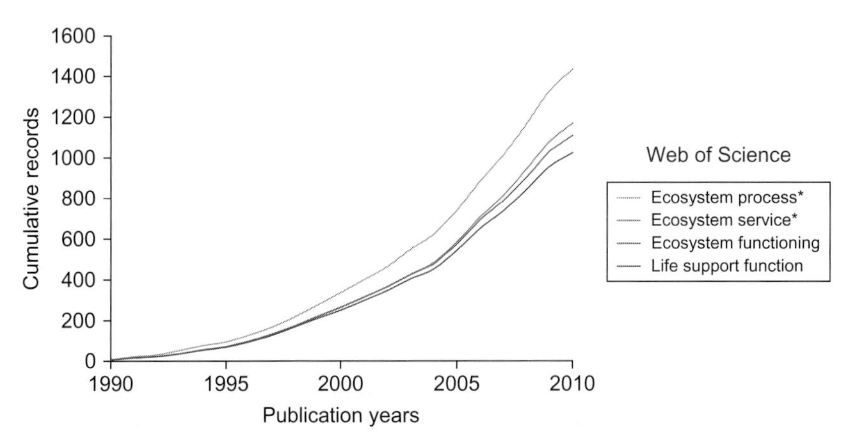


**Figure 1** Published contributions reflect an increasing focus on environmental conditions. During the past decades an increasing number of scientific papers focused on ecosystem processes and services (the literature keys are shown in the legend). Boolean analysis conducted in September 2010 within the Web of Science subject areas "Ecology", "Soil Science", "Environmental Sciences", "Zoology", "Biodiversity Conservation", "Environmental Studies", "Microbiology", "Toxicology", "Agronomy", "Agriculture, multidisciplinary", "Entomology", "Biology", "Multidisciplinary Sciences", "Forestry", or "Geosciences, multidisciplinary" with the additional topics "soil food web*" or "network*".

(Hansson and Fogelfors, 1998) and the aforementioned "Soil Ecology of Arable Farming Systems" (Brussaard, 1994) being among the most well known of these. And it was in The Netherlands, where most progresses in nematological taxonomy and molecular ecology have being achieved. Soil-community-DNA samples contain a huge proportion of nematode DNA and several pilot studies for DNA-based methods (DNA array system) have been conducted successfully (e.g. Holterman *et al.*, 2006, 2008).

In 1994, the Dutch government ratified the UNCED Biodiversity Convention of Rio de Janeiro (1992), and as a follow-up it formulated a Strategic Plan Biodiversity, in which soils biodiversity, and the functioning soil organisms perform were addressed. One of the main goals in the Dutch 1990s program was to compare the effects of conventional farming on soil biota, focusing on how so-called integrated farming system, with reduced levels of fertilizers, pesticides and tillage, might affect arable fields under rotation (Lebbink *et al.*, 1994).

These early studies were not highly resolved taxonomically, but compartmentalised the biota into broad taxonomic and dietary groups (cf. Moore *et al.*, 1988). For instance, the total biomasses of soil organisms belonging to 17 functional guilds (see Section IV), which included bacteria, fungi, protozoa, nematodes, micro-arthropods and oligochaetes. Based on coarse biomass estimates, and using literature data on growth/death rates and energy conversion efficiencies, the role of the soil food web in carbon and nutrient turnover was subsequently modelled mathematically for the first time (Bloem *et al.*, 1994; Bouwman and Zwart, 1994; De Ruiter *et al.*, 1993, 1994; Didden *et al.*, 1994; Zwart *et al.*, 1994). In one of these early pioneering studies, De Ruiter *et al.* (1993) calculated that the contribution of amoebae and nematodes to overall N mineralization in winter wheat was 18% and 5%, respectively. Their subsequent deletion from the food web model resulted in reduced N mineralization for the microfauna. In a subsequent study comparing these Polder soils with shortgrass steppe in the United States, Moore (1994) recognized common responses of soil biota to agricultural disturbances, allowing *de facto* the use of parameters previously used only in American models (Hunt *et al.*, 1987). Land-use changes became frequent during recent decades, and many more shifts are expected in The Netherlands in the near future, including transitions between farming regimes. Therefore, these results were challenging for the Dutch government, because in 1997 the government decided to investigate further the belowground ecological domain. Hence, the running Dutch Soil Quality Network (DSQN) was enlarged by a more complete biological survey in an attempt to disentangle processes and ecosystem functioning at national level.

Studying the long-term effect of nutrients on plant species diversity is of particular interest to both farmers and nature conservation managers among the main end-user communities, as well as being of fundamental importance to general ecological science. N enrichment is more intensively studied than the effects of P, K and Mg (but see De Deyn *et al.*, 2004; Wassen *et al.*, 2005), and is typically associated with decreasing biodiversity and increasing primary productivity (essay on N-deposition by Bobbink *et al.*, 1998; and Baer *et al.*, 2004; Clark *et al.*, 2007; Foster and Gross, 1998; Huberty *et al.*, 1998; Stevens *et al.*, 2004, 2006; Wardle, 2002). Under some circumstances, the opposite holds, as in the experiment in which Van der Wal *et al.* (2009a) showed that the diversity of soil biota increased by N, NPK, or PK addition, in contrast to vascular plants, whose diversity increased mainly in response to liming. Postma-Blaauw *et al.* (2010) compared agricultural systems with two extreme management regimes located close to each other: one was a conventional patch of grassland managed for more than 50 years, and the other an arable field with crop rotation that was converted from grassland more than 20 years earlier. In 2000, part of the long-term grassland was ploughed and six plots were sown in the long-term arable field. The numerical abundances and total biomasses of ciliates, omnivore nematodes, and predatory mites were highly responsive to these land-use changes (Postma-Blaauw *et al.*, 2010).

In another study, Verbruggen *et al.* (2010) performed large-scale comparisons of mycorrhizal fungal communities in agricultural soils in The Netherlands.

Focussing on sandy soils, they observed the highest richness of arbuscular mycorrhizal fungi in grassland, followed by organically managed fields, with the lowest diversity in conventionally managed fields (Verbruggen *et al.*, 2010). It is important to note that some of the latter had higher arbuscular mycorrhizal fungal species richness than some organically managed fields, showing that it is not always possible to make firm generalizations.

The aim of this chapter is to evaluate the relative effects of elemental variables and their ratios (i.e. C:N:P) and the importance of land use and soil properties on the structure and functioning of detrital soil food webs, highlighting the consequences for ecosystem services and providing an overview of the effects of agricultural management on the soil biota. First, we will introduce the investigated soil and ecosystem types within the DSQN, focusing on agroecosystems on sandy soils, and provide an overview of soil organisms groups assessed. Subsequently, we discuss how these organisms are organized in soil food webs and which—and to what extent—ecosystem services are influenced by them. We end with an analysis of mechanistic models derived from biological and ecological stoichiometry and discuss the sensitivity to disturbance of larger-sized organisms and their intraspecific size variation into food webs. Overall, we aim to highlight the significance of soil organisms for a number of key ecosystem services, including the provision of nutrients to plants, stimulating of plant productivity, and the stabilization and formation of soil.

# II. SOIL BIOTA

## A. Soil Types and the Dutch Soil Quality Network

From a geological perspective, The Netherlands are very homogeneous, with the most fertile part associated with the alluvial plains of the rivers Rhine, Meuse, Ems and Scheldt, resulting in 1,747,850 ha sandy soils and 1,606,490 ha well and fine graded soils. The major soil reference groups are Podzol, 50.1%, Cambisol and Fluvisol, 35.7%, Histosol and Gleysol, 10.6%, Luvisol, 1.6%, and Calcaric Regosol, 2% (Figure 2). These reference groups are classified according to soil characteristics, properties, horizons and profiles. Sieving soil samples, we can easily discriminate between the coarse grained soils (sandy soils *sensu stricto*) and the fine grained soils (loamy soils *sensu lato*). Coarse grained soils contain twice as much air than fine grained soils do, whereas fine grained (clay) soils contain twice as much water than coarse grained soils do: these opposite trends have important implications for all the living soil organisms. The fine grained fraction contains chemically and morphologically distinct components (silt grains vs. clay particles). To recover the clay material from our topsoil samples, we used

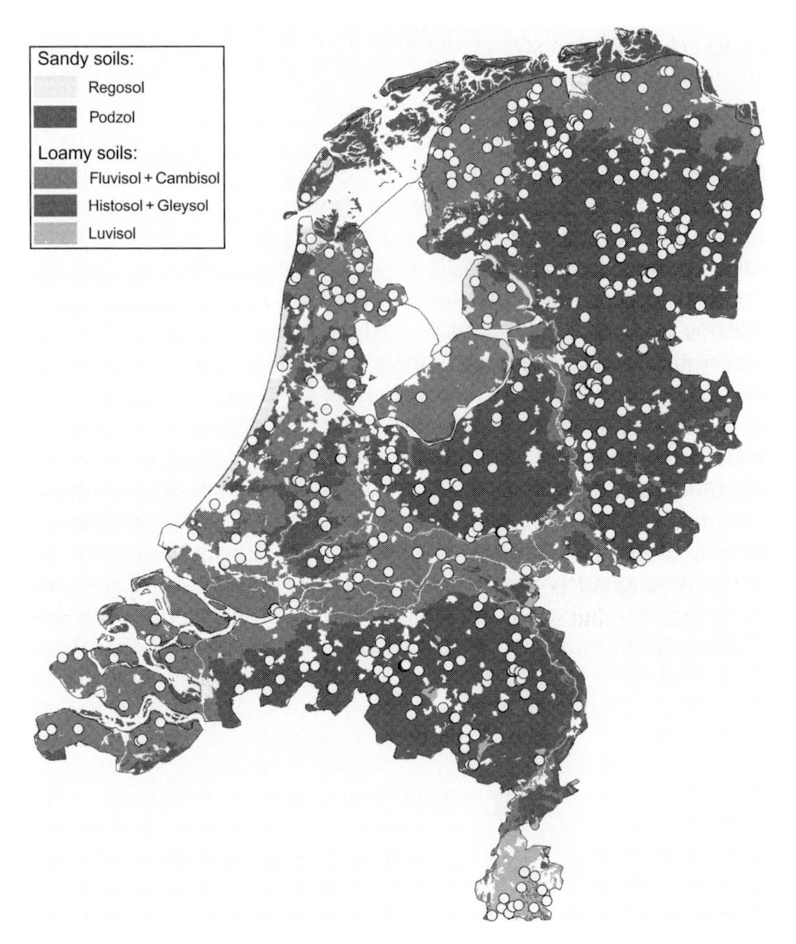


**Figure 2** Dominant soil reference groups in The Netherlands with investigated locations. The colors are those used for the FAO Soil Group Classification and the Soil Atlas of Europe (details in http://eusoils.jrc.ec.europa.eu/projects/soil_atlas/index.html). Urbanized areas were left blank. (For interpretation of the references to color in this figure legend, the reader is referred to the Web version of this chapter.)

the Stokes' law to detect the *lutum* fraction of the soil (i.e. the colloid particles with a size less than 2 $\mu$m). This unified classification for soil types (*ST*s) finally results in sandy and loamy soils, where the latter can be split further into clay (river or marine), silt loam (Löss) or organic soil (peat). The youngest *ST*s in The Netherlands (cf. ESBN, 2005) are alluvial clays, that is, Dystric Cambisols and Gleyic Fluvisols, coastal sand dunes, that is, Regosols, and (rejuvenated) loamy soils, that is, Haplic Luvisols (Figure 2).

This soil classification strongly reflects the elemental conditions of the soil system, since the elemental composition fluctuates according to the alluvial contribution of the four rivers and to land use. Taking the DSQN as a whole, the soil organic matter (SOM) fluctuates between 2% and 60% according to *ST* (Van Wezel *et al.*, 2005). Between *ST*s, the soil acidity ranges from 3.0 to 8.0 pH units (5 orders of magnitude of $H^+$ concentrations). Within *STs* the variation generally is less than 2 orders of magnitude, except for coastal ($Ca^{2+}$-enriched) sand dunes, where the scatter of soil acidity is much lower than in other inland areas, and river clay where variation is over 3 orders of magnitude (Mulder *et al.*, 2005a; Rutgers *et al.*, 2008; Van Wezel *et al.*, 2005). *ST*s are correlated with ecosystem types (*ET*s) that reflect land-use history. *ET* examples are organic or conventional farms, forests, and abandoned grasslands. Any relationship between *ET*s (for details see their classification in Section III.C) and a given soil property has to be analysed for each *ST*. If the variation in a soil property *i* (for instance, soil pH) is low within a given *ST* when compared with the variation in property *i* over all *ET*s, it indicates that soil property *i* is relatively little influenced by *ET*. If the variation in soil property *j* between *ET*s within a given *ST* is high and comparable to the variation over all areas, this might indicate that soil property *j* is correlated with (and/or influenced by) *ET*, as in the case of the example shown in Figure 3. In addition, we focused on

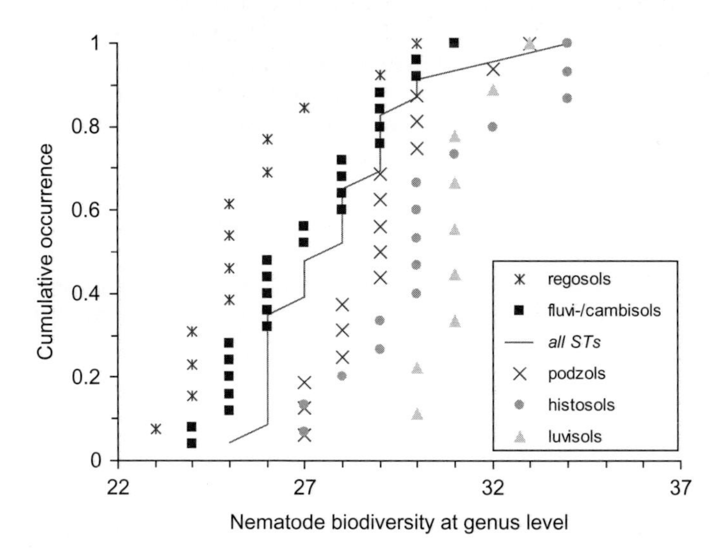


**Figure 3** Example of the variation in a specific soil property, here the nematode diversity, and the different *ST*s. Nematode scores on the vertical axis (0 = 100% absence, 1 = 100% presence). Variation in soil properties for all *ST*s together (shown trendline) and for each *ST* separately. Different symbols indicate to what extent *ST*s may deviate from other *ST*s in respect to the trend for a specific soil property computed for all *ST*s together.

systematic differences in the soil properties *i*, ..., *j* between "natural" (mostly abandoned) *ET*s and rural, actively cultivated *ET*s.

In 1993, the national DSQN monitoring program was started. The monitoring network is based on 300 sites in a stratified grid design and represents 75% of the total surface area of The Netherlands (Figure 2). The monitoring follows a 6-year cycle, 5 years field sampling, laboratory work and modelling and 1 year project evaluation by stakeholder and policy-decision makers (1993–1997, such that the first DSQN cycle was evaluated in 1998, 1999–2003, and in 2004 the evaluation of the second DSQN cycle, and 2005–2009, with the final evaluation of the third DSQN cycle in 2010).

Besides the more commonly measured abiotic variables (such as SOM, pH, *lutum* fraction, C:N:P ratios), heavy metals, polycyclic aromatic hydrocarbons and (organophosphorus) pesticides were also analyzed: the total amount of these extensive chemical measurements averaged 1140 records per year (Table A1).

## B. Soil Organisms: Types, Biology, Biogeography and Sampling Strategy

In 1994, Jones and co-workers defined "the role that many organisms play in the creation, modification and maintenance of habitats" as ecosystem engineering. Several species from different kingdoms contribute actively to this essential life-supporting process (Cuddington *et al.*, 2007), and the soil system contains numerous ecosystem engineers that "modulate the availability of resources to other species" (Jones *et al.*, 1994), with the large earthworms being the dominant physical engineer.

In 1997, a more complete biological monitoring component was embedded within DSQN that also considered these properties of both species and systems, largely in response to the Treaty on Biodiversity of Rio, which focused on sustainable development and use of biodiversity (WCED, 1987; UNCED, 1992). Since 1998, a consortium of RIVM, Alterra, Louis Bolk Institute, Wageningen University, BLGG, TNO, and Grontmij conducted nationwide investigations to develop a Biological Indicator for Soil Quality, in line with this new research agenda.

SOM incorporates dead organic matter (SOM, i.e. detritus inclusive plant roots) and fresh organic matter (FOM), which is the (elemental) contribution of all organisms. Between 0 and 20 cm depth in the soil, up to five kingdoms are represented: Monera, Protista, Fungi, Animalia and Plantae (Whittaker, 1975). This topsoil layer was sampled within DSQN, manually excluding the roots and plant remains. Of the remaining four kingdoms, Animalia and Monera received the most attention within DSQN (Table 1). Metazoans in the soil were divided into: (1) Microfauna (mostly decomposers that inhabit

**Table 1** Estimates of global (left column), local (middle) and national (right column) soil biodiversity

| Taxon | Global diversity | Sourhope site (UK) | Dutch Soil Inventory |
|---|---|---|---|
| *Phylum monera* | | | |
| Eubacteria | 6000 | 100 | One phylotype |
| *Phylum protista* | | | |
| Protozoa | 1050 | 365 | Data paucity[a] |
| *Phylum fungi* | | | |
| Arbuscular mycorrhizae | 150 | 24 | >42[b]—but in DSQN all fungi were kept as one morphospecies |
| Other fungi | 56,210[c] | 57[d] | |
| *Phylum animalia* | | | |
| Nematoda | 3000 | 129–143 | 169[d] |
| Acarina | 45,000 | 32 | 146[d] |
| Collembola | 20,000 | 12 | 37[d] |
| Enchytraeidae | 143 | 14 | 10[d] |
| Lumbricidae | 1200 | 5 | 9[d] |

Modified, with permission, from Fitter *et al.* (2005) and merged with the online inventory of Mulder *et al.* (2008). Fungal diversity from Hawksworth *et al.* (1995).
[a]Most data from fluvisols only.
[b]Verbruggen *et al.* (2010) on podzols.
[c]A. H. Fitter, personal communication.
[d]Counted as genera.

water films); (2) Mesofauna ( predators and microbial (fungal) grazers that inhabit air-filled pore spaces and use existing pore structures without altering pore structure); and (3) Macrofauna (animals that create their own spaces through burrowing activities, thus actively changing the pore structure of the soil). Monera and Fungi (cells and hyphae) were treated within the DSQN as two trophic species (Yodzis, 1982) and an inventory of all the measured microbial ecophysiological traits is shown in Table A2. Protista (see Section II.B.3) were investigated in too few DSQN locations and for too few seasons to give a comprehensive dataset for modelling their numerical abundance in space and time in The Netherlands. Below, we will briefly describe the major groups of soil organisms investigated within the DSQN (functional groups, sampling methods and detected patterns, that is, biogeography).

### 1. *Animalia*

**a. Macrofauna and Mesofauna.** *Oligochaetes* consist of lumbricids (earthworms) and potworms (enchytraeids). Earthworms are true detritivores, ingesting soil, microorganisms, and plant residuals and are well-known "ecosystem engineers". They are the largest soil invertebrates in the DSQN and the longest-lived, with a lifespan of about 8 years (Mulder *et al.*, 2007). Enchytraeids are smaller and in contrast to earthworms, little is known about their feeding strategy due to their different diets during their juvenile and adult stages.

*Sampling* was carried out in two different ways. Separate samples were collected from six discrete plots of 15 $m^2$ evenly spread across the site and located by GPS. In each plot, three subsamples were taken for the analysis of enchytraeid and earthworm communities. For earthworms, the standard approach of block sampling ($20 \times 20 \times 20$ cm) and hand-sorting was used rather than the core sampling of 15 cm depth for the much smaller enchytraeids. In the latter case, soil was sampled by (randomly chosen) Ø 5.8 cm cores. Each sample was crumbled into a sieve hung in a bowl filled with water to the edge, and kept at 10–15 °C. In few cases, the extraction process was slightly accelerated by heat (carefully using 60 W bulbs above the soil samples) to avoid the loss of juveniles which would otherwise remain in the soil (Jänsch *et al.*, 2005). Each worm was measured, and weighed individually. The complete taxonomic inventory with specific data entries is shown in Table A3.

*Biogeography*. In less acidic, alluvial soils, such as river clay soils, lumbricids form the major part of the soil fauna, with up to 1000 individuals $m^{-2}$ (comparable to 10 millions $ha^{-1}$). In our 200 grasslands on sandy soils (Section III.C), lumbricids reach an average dry biomass of 61.6 kg $ha^{-1}$ $\pm$49.4 SD (17.4–159.6 min–max), in contrast to pine forests where lumbricids are scarce (their dry biomass in Podzols is no more than 5 kg $ha^{-1}$). Lumbricids in arable fields on sand occupy an intermediate position, with an average dry biomass of 21.4 kg $ha^{-1}$. The densities of enchytraeids are less related to *ST* than those of lumbricids (Didden, 1993), although their body size (and consequently, their body mass) is inversely correlated with soil pH. Enchytraeids are mostly between 500 and 50,000 $m^{-2}$, with dry biomass values (average $\pm$ SD) of $4.5 \pm 4.4$, $5.8 \pm 6.0$ and $6.1 \pm 8.3$ kg $ha^{-1}$ in sandy soils and in marine and river clay, respectively (DSQN data).

*Micro-arthropods* are about 1 mm long, are mostly either Acarina ("Mites") or Collembola ("Springtails") and can be assigned to feeding guilds on the basis of their carbohydrase activity (Ponge, 2000; Siepel, 1994). Ecoenzymatic activity is used to determine their "Last Supper" and in the case of larger-sized adult mites, belonging to taxa such as *Isotoma*, *Isotomurus* and *Sminthurus*, one specimen was enough for carbohydrases, but in most cases up to 20 adult individuals per taxon were necessary (Jagers op Akkerhuis *et al.*, 2008; Mulder *et al.*, 2009). Ecoenzymatic activity represents the type of feeding guild, since enzyme activity depends on the food components consumed prior to sampling. Three carbohydrases were measured for the DSQN: cellulose, chitinase, and trehalase. Resulting feeding guilds range from rare bacterivore mites like *Histiostoma* and some *Eupelops* (Mulder *et al.*, 2005a, their Appendix 1), microphytophages, macrophytophages and panphytophages arthropods, up to omnivore and predatory mites (De Ruiter *et al.*, 1993; Moore *et al.*, 1988; Mulder *et al.*, 2008).

*Sampling*. Four cores (Ø 5.8 × 15 cm) were randomly selected within the 15 $m^2$ plots and kept separate until behavioural extraction (Siepel and Van de Bund, 1988), achieved by placing each sample on six discs in a Tullgren funnel (Tullgren, 1917). Temperature in the upper part of this funnel was set at 30 °C and kept at 5 °C in the lower part for a period of 7 days (sandy soils) or 15 days (loamy soils). Living organisms moved downwards to escape the heat, dropped through the funnel and were collected in a bottle with 70% ethanol. For each sample, 70 arthropods were counted and identified with a light microscope at 200–1000× via a gel-based subsampling method (Jagers op Akkerhuis *et al.*, 2008; Mulder *et al.*, 2009; Rutgers *et al.*, 2009). The complete inventory of mites and collembolans is shown in Table A4. For other taxa, here Diplurans, Pauropods, Proturans and Symphyla, please refer to the online inventory of Mulder *et al.* (2008).

*Biogeography*. Mites reach densities of $10^4$–$10^5$ $m^{-2}$ and collembolans reach densities of $10^3$–$10^5$ $m^{-2}$. Diversity and abundance of micro-arthropods show opposite trends, as very often in highly species-rich sites like forests, most species that are rare occur only once. In anthrosols (low diversity, high density) the dry biomass averages 2.6 kg $ha^{-1}$ (0.2–12.8 min–max), and in podzols (high diversity, low density) the dry biomass averages (2.1 kg $ha^{-1}$), remaining comparable. Only in fluvisols, the biomass decreases (DSQN data).

**b. Microfauna**. *Nematodes* are possibly the best investigated taxa in the soil, probably reflecting the fact that they are diverse, common, widespread (they occur in extreme environments like Antarctica as well) and, from an agricultural point of view, potentially dangerous pests. Besides plant-parasitic nematodes, these vermiform invertebrates can be bacterivores, fungivores, omnivores and predators (Yeates, 2010; Yeates *et al.*, 1993), occupying most nonbasal trophic levels in any detrital soil food web (see Section IV). They are small, with a body size of about 500 $\mu$m (a minimal length of 0.3 mm, up to a maximal length of ~1 mm; DSQN $\alpha = 0.05$), and have in comparison to other worms a very short lifespan (cf. Van Voorhies *et al.*, 2005), which makes them useful soil quality indicators. Further, the feeding strategy (bacterivores, fungivores, omnivores) of nematodes reflects their food resources. Some studies on trophic interactions between rhabditids and protozoa (Sohlenius, 1968) might suggest a more important role of protozoa as prey for microbivore nematodes than commonly thought. Moreover, the constant re-mobilization of essential elements for plant growth during microbivory by protozoa and nematodes (Bonkowski, 2004) makes nematodes "ecosystem engineers".

*Sampling*. Nematodes were sampled on each occasion since 1993. For any location one bulk sample was mixed from the soil randomly collected in 320 cores ( $2.3 \times 10$ cm) from across the study site. The bulk of approximately 500 g was kept in glass containers and stored at 4 °C prior to extraction from $\sim$100 g of soil was using the Oostenbrink method (Oostenbrink, 1960). Total numbers of living nematodes were counted in two subsamples (representing 10% of the total). After counting, samples were preserved by hot formaldehyde. All individuals within two clean 10 ml water suspensions were screened and 150 randomly chosen specimens per sample were identified under a light microscope. Identification was mostly carried out to genus level, except for the *dauerlarvae*. The adult:juvenile ratio was not taken into account due to a high sensitivity to environmental factors like temperature (after Ritz and Trudgill, 1999).

*Biogeography*. Overall, nematode diversity is rather low in clay soils, intermediate in managed sandy soils, and high in treeless nature. The total contribution of nematodes is highly responsive to pH: in nature, the total dry biomass of free-living soil nematodes per kg soil averages $3.3 \pm 1.1$ mg in abandoned mesic grasslands and $0.5 \pm 0.3$ mg in acidic heathlands, with densities of $\sim$50,000 and 20,000 nematodes, respectively. Non-parasitic nematodes are considered in The Netherlands as the best environmental indicators belowground and are the most widely investigated soil invertebrates (Table A5).

## 2. *Monera*

*Bacteria* are usually smaller than 2 $\mu$m, with densities of billions cells per gram soil. The bacterial population is dominant in the short-term resource competition (like in metabolizing carbohydrates) and—together with soil fungi—catabolize FOM (litter) and mineralize SOM. Microbial activity in soils can be determined either by measuring the short-term bacterial growth rate (the incorporation of $^3$H-thymidine into bacterial DNA simultaneously with the incorporation of $^{14}$C-leucine into proteins; Bloem and Breure, 2003) or by measuring the long-term specific respiratory rate (microbial respiration rate per unit of microbial biomass under standardized conditions and without substrate addition). Basal soil respiration by microorganisms can be quantified by measuring $CO_2$ production—the preferred method—and $O_2$ consumption (Mulder *et al.*, 2005b).

*Sampling* within DSQN occurred from 1999 onwards. Soil samples were taken at random for each site and bulked together to form one composite sample. From this composite sample, two replicate subsamples were analysed by fluorescent staining for the average body size (length and width) and the total number of bacterial cells. Bacterial cell volume was converted to dry biomass using the biovolume-to-carbon conversion factor of $3.2 \times 10^{-13}$ g C $\mu$m$^{-3}$ (Bloem *et al.*, 1995; Mulder *et al.*, 2005b; Van Veen and Paul, 1979): assuming a bacterial carbon content of 50% (Herbert, 1976), their dry biomass is equal to 100–1000 kg ha$^{-1}$ (DSQN data).

*Biogeography*. According to different *ST*s (Figure 4), bacterial biomass and growth rates were very different between clay and sand, with high bacterial growth rate and biomass in grasslands on clay and low rate and biomass in grasslands on sand (Bloem and Breure, 2003). The bacterial biomass in grasslands on sand fluctuates markedly in nutrient-rich agroecosystems, such as organic farms (Mulder *et al.*, 2003a). Bacterial cells and fungal hyphae react in very different ways to environmental conditions: for instance, a significant impact of soil pH on the fungi-to-bacteria ratio has been shown for grasslands, forests and arable systems (Högberg *et al.*, 2007; Mulder *et al.*, 2005a; Rousk *et al.*, 2009, respectively). Fungal spores are

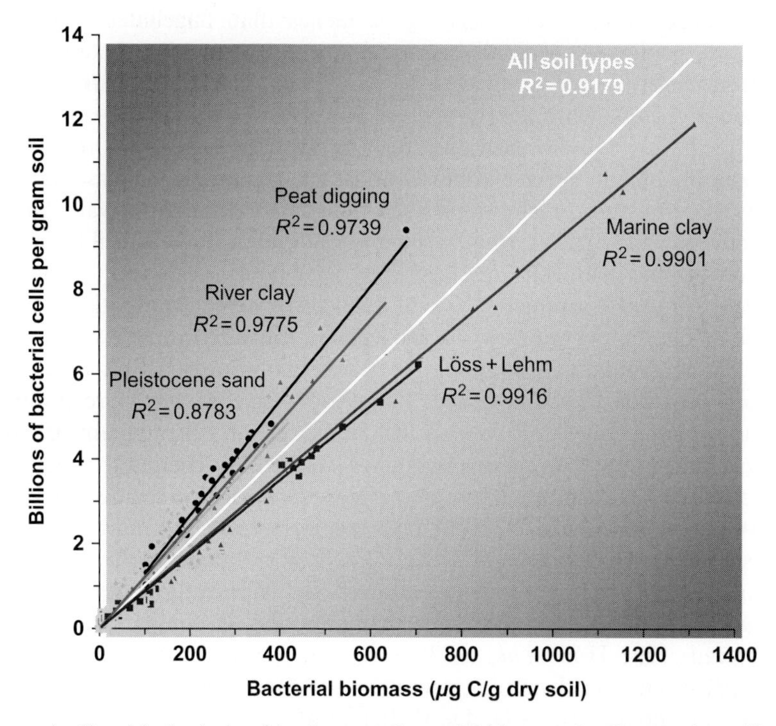


**Figure 4** Empirical relationships between bacterial biomass ($\mu$gC per g dry soil) and numerical abundance of bacterial cells (billions per g dry soil). Total biomass is numerical abundance times average cell mass, thus linear regressions for different *ST*s depict the trait variation of the body-mass average of bacterial cells as function of *ST*.

more acid tolerant than bacterial cells, and in acidic podzols these microbial shifts lead to an increased abundance of fungal-feeding arthropods and a decreased abundance of bacterial-feeding nematodes (Mulder *et al.*, 2003a, 2005a, their Figure 4 and Appendix 1).

### 3. Protista

*Protozoa* are single-celled, microscopic (2–200 μm) eukaryotic organisms. In contrast to metazoans, protists do not represent a single evolutionary lineage. One of the most recent classifications discriminated the enormous diversity of protists into six supergroups, each comprising diverse phyla (Adl *et al.*, 2005). For example, based on their molecular phylogeny, only ciliates are closely related to one another, whereas amoebae and flagellates have repeatedly evolved from very distant protistan phyla. Due to these taxonomic inconsistencies, it is often more appropriate to classify protozoa from a functional point of view: for example, protozoa with amoeboid life style will occupy more closely related ecological niches than flagellates or ciliates. Therefore, ecologists generally compile the most common heterotrophic protozoa in soil according to their general morphology into four principal functional groups: (1) naked amoebae, (2) testate amoebae, (3) flagellates and (4) ciliates (e.g. Bonkowski, 2004; Bouwman and Zwart, 1994). All protozoa in soil can survive adverse, and even extreme environmental conditions as cysts, and together with nematodes, they occupy the water microfilm between soil particles, that is, the microfaunal part of the soil food web.

Heterotrophic protozoa are considered the major consumers of bacterial production in soil, forming the base of the heterotrophic eukaryotic food web that channels the energy flow via bacteria to higher trophic levels (i.e. the bacterial energy channel). It should be mentioned, however, that the trophic niche of protozoa in soil is broad: fungivorous soil protozoa are probably common (Ekelund, 1998; Petz *et al.*, 1985), as well as obligate plant parasites and non-pathogenic endophytes (Bulman *et al.*, 2001; Müller and Döring, 2009; Neuhauser and Kirchmair, 2009); also algivorous protozoa regularly occur in the litter layer (Patterson, 1996). A high proportion of soil protozoa are predators of other protozoa, and some even prey on nematodes (Yeates and Foissner, 1995). Due to a bias in culture-based enumeration methods, only the function of heterotrophic bacterivore protists is considered here (De Ruiter *et al.*, 1993; Hunt *et al.*, 1987).

*Sampling*. Despite their high numbers and fast turnover in arable fields (Bouwman and Zwart, 1994; De Ruiter *et al.*, 1993), we still have only a vague idea of the identity of soil protozoans. There is a great discrepancy between laboratory studies showing a significant functional role of soil protozoa for nutrient cycling and on plant growth (e.g. Reiss *et al.*, 2010), and field studies where the numbers and even the identity of the most dominant soil taxa are unknown.

Most protozoa cannot be directly extracted from soil and each group requires specific observation conditions, each encumbered with logistical drawbacks (see reviews by Ekelund and Rønn, 1994; Foissner, 1987, 1997). Generally, the approaches fall into two broad categories: direct counting methods and dilution–cultivation techniques. Direct counting methods for active protozoa in soil are extremely laborious, feasible only for small sample sizes, and not all protozoan groups can be counted (Couteaux and Palka, 1988; Luftenegger *et al.*, 1988), although the direct counting of testate amoebae remains popular in vegetation science and (palaeo)ecology (Mulder and Janssen, 1999; Payne and Mitchell, 2009). Since some protozoan cells divide within hours (like bacterial cells), direct counting may reflect only a snapshot of protozoan activity.

On the other hand, common dilution–cultivation techniques (i.e. the most probable number "MPN"-method; Darbyshire *et al.*, 1974; Ekelund *et al.*, 2001) quantify only the "potential" abundance or protozoa in a field sample. The reason is that at most times a substantial part of the protozoa in soil is inactive, surviving adverse conditions in resting stages (cysts), which may remain viable for decades (Moon-Van der Staay *et al.*, 2006). The cysts that hatch during the incubation period prior to counting may result in an overestimation of the active protozoan population. On the contrary, if mainly cultivation-resistant species are favoured, these methods may greatly underestimate the potentially active protozoan population. In fact, progress made over the last decade has confirmed that there is a wide range of protozoan diversity and activity which remains mostly unexplored (Ekelund, 1998; Foissner, 2006). There have been significant improvements in soil DNA extraction, sequencing techniques, and the development of dedicated DNA databases coupled to powerful *in silico* data compilation and analysis capabilities (Purdy *et al.*, 2010). For example, novel high-throughput molecular diagnostic tools including massively parallel sequencing have identified slime moulds (Mycetozoa) as being the most abundant protist group in a sandy grassland soil, followed by Cercozoa, Plasmodiophorida and Alveolata; and also the presence of Lobosea, Acanthamoebidae, Heterolobosea, Euglenozoa and Stramenopiles was confirmed, although relatively few 18S rRNA sequences are currently known for protists (Urich *et al.*, 2008). At the same time, there is growing evidence that the role of soil fauna and their interactions with soil microbes is essential to ecosystem functioning (Bardgett, 2005; Wardle, 2002).

*Biogeography*. Rarely have all protozoan groups been considered in a single investigation. Finlay and co-workers recorded 365 protozoan species at a grassland site in Scotland with average numbers (per gram soil dry weight) of 46,400 flagellates, 17,700 naked amoebae, 11,000 testate amoebae and 4300 ciliates (Finlay *et al.*, 2000; corrigendum). In a comparison of

Dutch farming systems, the average biomass and annual production rates of bacterivore protozoa under winter wheat were estimated as 16 kg C $ha^{-1}$ and 105 kg C $ha^{-1}$ $yr^{-1}$, respectively (0–25 cm soil depth), compared to 0.33 kg C $ha^{-1}$ and 11.6 kg C $ha^{-1}$ $yr^{-1}$ for bacterivore and omnivore nematodes together (Bouwman and Zwart, 1994). Protozoa are usually small, but abundant, and their (biomass) turnover in soil is comparable to that of nematodes. Field studies have shown that protozoan biomass equals or exceeds that of microfauna and mesofauna together (Foissner, 1987; Reiss *et al.*, 2010; Schaefer and Schauermann, 1990). This leads to the paradoxical situation that, under certain circumstances, the biomass of soil fauna is made up of two main pools, the protozoa with very fast generation turnover and the earthworm "ecosystem engineers" with generation times of up to several years (see Section II.B.1.a). Although the exact contribution of protozoa to the soil energy flux is still not known, the high production rates of bacterivore protozoa can only be maintained if a substantial part of the bacterial biomass is consumed (Clarholm, 1985; Frey *et al.*, 2001; Jones *et al.*, 2009).

### 4. *Fungi*

*Fungi* form dense hyphal networks in the soil, reaching total lengths of 10–1000 km $kg^{-1}$ soil. While individual hyphae are small and thin (e.g. fungal hyphae usually have diameters of 2–10 $\mu$m), mycelial networks can be extremely large. For instance, Smith *et al.* (1992) observed that fungi are among the largest living organisms: mycelial networks formed by one individual of the fungus *Armillaria bulbosa* occupied 15 ha and weighed over 10,000 kg. Fungi provide a number of ecosystem services crucial for ecosystem functioning, as they secrete a range of extracellular enzymes that facilitate decomposition and make nutrients available for primary production (Dighton, 2003). Examples are rock dissolution, soil formation, nutrient acquisition, and nutrient cycling (Dighton, 2003; Van der Heijden *et al.*, 2008; Van der Wal *et al.*, 2006). The content of dark brown pigments in the mycelium is often high (Van der Wal *et al.*, 2009b), and pigmented hyphae have been found to be a highly preferred (much more palatable) food resource for soil micro-arthropods (Schneider and Maraun, 2005).

*Sampling* within DSQN has taken place on each occasion since 2002. Asexual states (anamorphs) are the form in which fungi are most often encountered in the soil. The increased popularity of an identification of fungal anamorphs on the basis of form, shape and pigmentation, which allow their assignment to one of the Saccardo's spore groups of mitosporic fungi, rarely ascomycetes (Mulder *et al.*, 2003b), allows the choice for a morphological investigation of the entire mycelium as well (Table A2). Hyphae were viewed with fluorescent staining and direct microscopy and the lengths of the hyphal branches were estimated by the line intercept method at 2500×. The mycelium volume $V$ ($\mu m^3$) was estimated from $V = \pi(w/2)^2 \times b$, where $w$ is the average hyphal width and $b$ the average hyphal branch length. We kept the mycelium as unity, without separation between saprotrophic and ectomycorrhizal fungi. As recognized by Rayner (1991) and Smith *et al.* (1992), thalli consist of an intricate network of anastamosing hyphae embedded in the substratum. If we assume a hyphal width of 2.5 $\mu$m and the biovolume-to-carbon factor of Bakken and Olsen (1983), the resulting superorganism reaches biomasses of 1–500 kg C $ha^{-1}$.

*Biogeography*. As with the Monera (Section II.B.2), fungi mostly show habitat–response relationships that are opposite to those of bacteria. The total lengths of the mycelium, and thus the fungal biomass as a whole, are strongly correlated with pH, soil texture, organic matter, and moisture (Oberholzer and Höper, 2000). Merging Swiss and German metadata (Höper, 1999; Oberholzer *et al.*, 1999, respectively), the multiple regression:

$$\log[C_{\rm mic}] = 1.1427 + 0.093 \times {\rm pH} + 0.329 \times \log[C_{\rm org}] + 0.311 \times \log[lutum\%] + 0.0005 \times [{\rm Rainfall\ mm/year}]$$

($\pm$0.12SE, $R^2 = 0.79$) predicts the total $C$ produced by microorganisms ($C_{\rm mic}$ = bacterial + fungal and protozoan carbon content) living in a soil with the given properties soil acidity (pH), carbon content of the organic matter ($C_{\rm org}$), percentage of soil particles smaller than 2 $\mu$m (lutum) and local rainfall (Oberholzer and Höper, 2000). We applied their model, calibrated with Lower Saxony soils and Dutch agroecosystems (Mulder *et al.*, 2005b), within the ArcMap module (ArcInfo 8). Figure 5 shows the extent to what all the three microbial groups, bacteria, fungi and protozoa, are expected to share their dependence on soil abiotics: $C_{\rm mic}$, in fact, can originate from either the microflora or from protozoa. In contrast to podzols, where most $C_{\rm mic}$ remains of bacterial origin (Mulder *et al.*, 2005b), in fluvisols a huge part of $C_{\rm mic}$ seems to be of protozoan origin (De Ruiter *et al.*, 1993, 1994). For fungi, the strongest predictor seems to be soil pH: (increasing) pH predicts 41% of the (decreasing) total fungal biomass (Mulder *et al.*, 2003a, 2005a, 2009).

# III. COMPARATIVE ECOSYSTEM ECOLOGY

## A. Building a Comprehensive Database

Many key journals are increasingly willing to introduce data-archiving policies (Whitlock *et al.*, 2010) and the need for more open data-sharing has been addressed in several recent papers (Queenborough *et al.*, 2010; Whittaker, 2010a,b; and references therein). In general, many molecular studies make

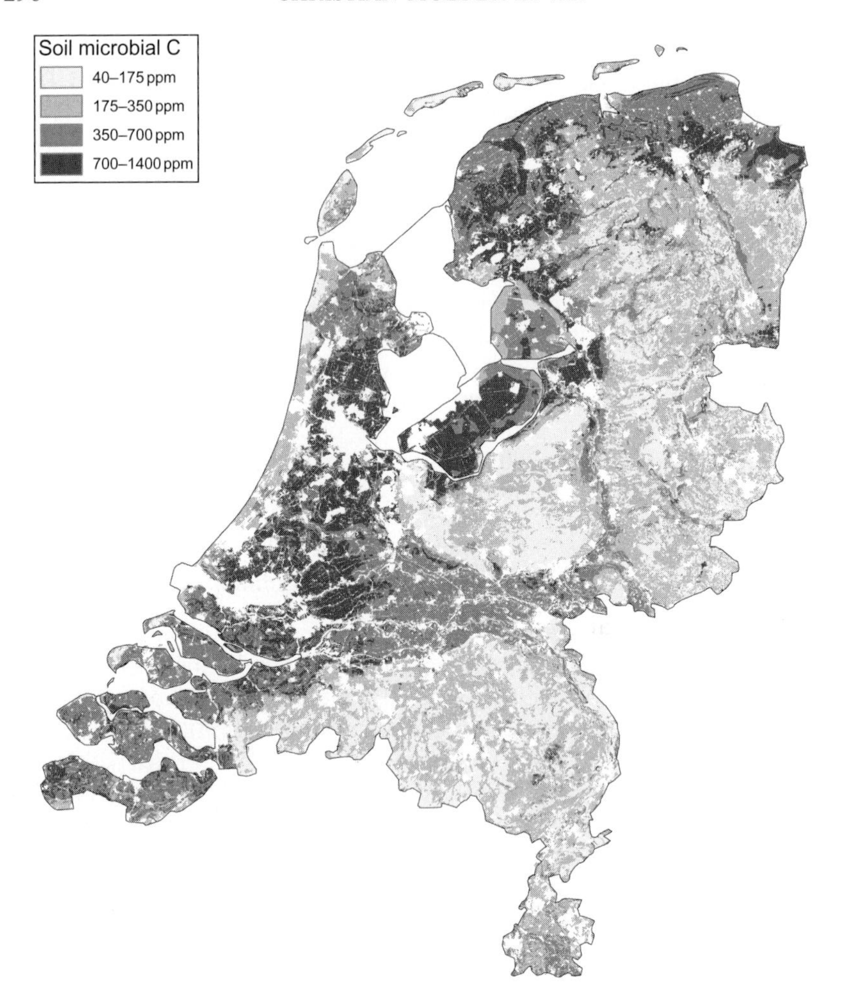


**Figure 5** Estimated total microbial carbon ($C_{mic}$) in The Netherlands according to the model of Oberholzer and Höper (2000) for Switzerland and Germany. As the GIS-data frame is a collection of layers (each layer represents a distinct set of soil properties in a particular *ST*), we calculated $C_{mic}$ for the upper 30 cm horizon from a composite polygonal layer. The contours for the soil pH, $C_{org}$ and clay attributes were derived from the Soil Map of The Netherlands 1:50,000. Each polygonal layer was then converted into a grid with a $100 \times 100$ m scale. Dataset on the mean annual rainfall from the Royal Netherlands Meteorological Institute (KNMI), same resolution. Measurements of pH in different databases were often performed in different solutions. For the soil pH layer, there was an adjustment, if necessary, to report the pH in $CaCl_2$ solution ($pH = 1.057 \times pH_{KCl} + 0.123 = 0.976 \times pH_{water} - 0.427$).

their databases publicly available, in contrast to most ecological studies, although biogeographers and macroecologists are notable exceptions, and several comprehensive floral and vegetational databases are online or in progress (e.g. Kühn *et al.*, 2004, and www.imcg.net/gpd/gpd.htm, respectively) (Box 1).

Ecologists conducting a meta-analysis therefore often face a bottleneck as they attempt to gather literature. Some meta-analyses on biodiversity, ecosystem functioning or ecosystem services, for instance, have remarkably few studies of soil biota included in their metadata (Balvanera *et al.*, 2006; De Bello *et al.*, 2010; Waide *et al.*, 1999, respectively), in contrast to the authoritative and far more exhaustive analysis by Srivastava and Vellend (2005).

Although much has been published, most empirical data remain inaccessible, representing a potentially vast but largely unexploited resource, especially within the so-called grey literature (Friberg *et al.*, 2011; Mulder, 2011). And many ecologists may have studies they decided not to publish (yet) or which were rejected by journals. This can contribute to a publication bias: the disproportionate amount of studies with statistically significant effects is a potentially confounding effect within any meta-analysis (Friberg *et al.*, 2011). This implies that it is not the use of methodological options like "the criteria for inclusion should be reasonable and scientifically defensible" (Gurevitch and Hedges, 2001) that are critical for a valid selection (Gillman and Wright, 2010), but it is the *kind* of published data. Although unbiased, the open publication of raw monitoring data may be dangerous because too much field expertise is required for further data mining and appropriate modelling. For instance, Queenborough *et al.* (2010) correctly suggested that README reports are compulsory in order to analyze the data properly (Box 1).

International inventories like the European Inventory of Existing Commercial Chemical Substances (http://ecb.jrc.ec.europa.eu/esis), as national inventories such as the RIVM e-toxBase (http://www.e-toxbase.com) were funded by public money for integrative studies. In Continental Europe, monitoring surveys funded by public money often involve grey literature in the form of reports (Schouten *et al.*, 1997, 1999) that are only occasionally in English, making a possible misanalysis of open-access data more likely. This returns to the issue of the extent of high-quality documentation for accurate interpretation and reuse of data (Mulder, 2011; Queenborough *et al.*, 2010; Whittaker, 2010a,b) and contributes to explain the choice to publish only complete, accurate appendices in peer-reviewed ecological journals, in contrast to data papers in repositories, to avoid *HARK*ing (*H*ypothesizing *A*fter the *R*esults are *K*nown *sensu* Kerr, 1998).

In our case, predictions for biogeochemical cycling, agricultural sustainability, and soil food web stability require many empirical properties as model inputs or for parameterization (cf. Holmes *et al.*, 2004). An extremely careful choice of the correct classification labels—implying a wide consensus on both the dataset as the use of ontological terms (STs, ecosystem types, taxonomy, species' traits such as body size, response traits such as bacterial ecophysiology, and many more)—was mandatory. Our DSQN sites have multiple measurements, and therefore we had to choose between a powerful interactive decision tree and a relational database scheme. The latter scheme (Figure 6) provides a flexible way to capture most of the important subtleties of multi-complex data.

## B. Ecosystem Services

The Millennium Ecosystem Assessment (2005) distinguishes broad categories for ecosystem services. Passive services (non-cultural services *sensu* MA, 2005) are grouped together in (1) "Provisioning Services" those that describe the material output from ecosystems, including food, water, fibre, wood, biofuels, plant oils and many other resources, and maintaining genetic diversity; (2) "Regulating Services" those that ecosystems provide by acting as regulators, for instance during carbon sequestration, erosion prevention, animal pollination, and pest regulation; (3) "Supporting Services" those that

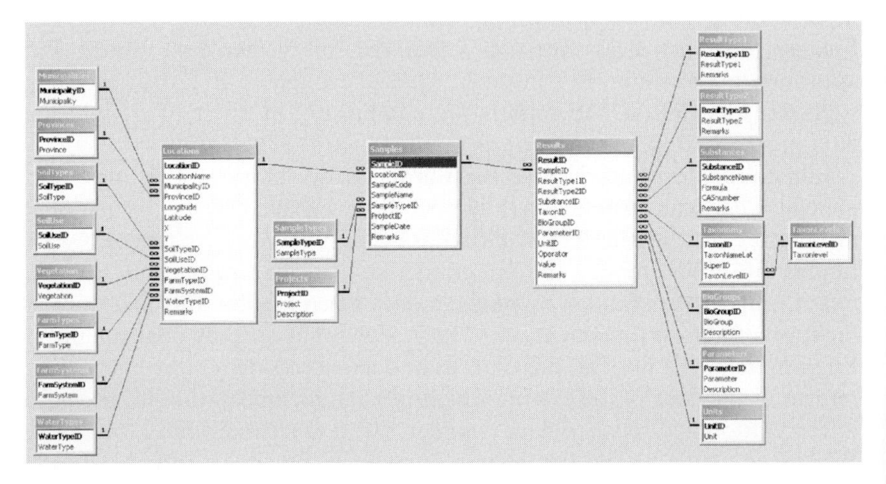


**Figure 6** Detailed representation of DSQN entities and of ontology extensions. These examples for simple derived units (single soil type under a certain vegetation at a specific time) and complex derived units composed of more locations (for instance, all locations with a given soil type or a particular land use) enable queries for all corresponding "semantic units" and "higherThan units" (*sensu* Madin *et al.*, 2007). All DSQN observations are linked to soil properties (for instance, all locations within a certain clay content or soil acidity range).

capture ecosystem processes that underpin all other services (e.g. providing habitat, soil formation, and nutrient cycling). Many of these different ecosystem services are interconnected.

In soil biota, most services are related to ecosystem processes performed by specific groups of soil organisms; for instance, earthworms, collembola, fungi and bacteria influence plant productivity, including food production, etc. After a decline in supporting services such as nutrient cycling, net primary productivity and soil formation (Table 2), ecosystems undergo substantial degeneration (Peltzer *et al.*, 2010).

The co-existence of different energetic patterns within and among the soil biota is in agreement with the complementary action of "energy transfer agents" (like the enchytraeids) versus "habitat engineer agents" (like the earthworms). Moreover, the same soil biota can provide multiple services (Bennett *et al.*, 2009; TEEB, 2010) and it is evident that the type of ecosystem also matters. Therefore, the real extent to what human well-being affects ecosystem services is still unclear and remains to be investigated more rigorously in the future (Carpenter *et al.*, 2009; Costanza *et al.*, 1997a; Daily and Matson, 2008). Any accounting for ecosystem services is also case-sensitive (Mäler *et al.*, 2008) because:

- Ecosystem dynamics varies from case to case.
- Ecosystem services vary from case to case.
- Stock definition varies from case to case.
- Institutions vary with implication for evaluation.

In Figure 7, we aimed to summarize the first two points (ecosystem dynamics and ecosystem services) for a temperate soil food web in The Netherlands.

In our study, intriguingly contrasting trends were evident for the mesofauna forest and grassland soils: a gradient of micro-arthropod diversity appeared in Liiri *et al.* (2002) unrelated to the (birch) primary production and nutrient uptake, although a comparable gradient of micro-arthropod diversity appeared in Van Eekeren *et al.* (2010) strongly (and inversely) correlated to the (grass) primary production and nutrient uptake. Still, there is little evidence to suggest that within a soil food web an increase of resource consumption due to the introduction of predators can induce trophic cascades with consequences at ecosystem level (Laakso and Setälä, 1999; Wardle, 1999), although mathematical simulations on a short timescale suggest an adaptation to the consumed resources (Rossberg *et al.*, 2010). Top-down effects due to (introduced) large predators are commonly thought to alter the food Web structure and dynamics profoundly (Díaz *et al.*, 2007; Layer *et al.*, 2011; Maron *et al.*, 2006; Wardle and Bardgett, 2004). Body size is thus a linchpin (Kaspari, 2005; LaBarbera, 1989) that shapes processes

**Table 2** Groups of soil organisms ranked according to their role in different ecosystem services, weighted for the two main soil types

| | Sandy soils (podzol) | Clays (cambisol and fluvisol) |
|---|---|---|
| *Provisioning services* | | |
| Food | Enchytraeidae (1), Lumbricidae (2), Collembola (3), Fungi (4), Bacteria (5) | Lumbricidae (1), Enchytraeidae (2), Protozoa (3), Bacteria (4), Collembola (5) |
| Fibre | Lumbricidae (1), Enchytraeidae (2), Bacteria (3), Fungi (4) | Lumbricidae (1), Enchytraeidae (2), Protozoa (3), Bacteria (4) |
| Genetic resouces (all) | Bacteria (1), Protozoa (2), Nematoda (3), Acarina(4), Collembola (5), Enchytraeidae (6), Lumbricidae (7), Fungi (8) | Bacteria (1), Protozoa (2), Nematoda (3), Acarina(4), Collembola (5), Enchytraeidae (6), Lumbricidae (7) |
| Biochemicals | Fungi (1), Bacteria (2) | Bacteria (1), Fungi (2) |
| Natural antibiotica | Fungi | Fungi |
| Freshwater | Nematoda (1), Collembola (2) | Enchytraeidae (1), Bacteria (2) |
| *Regulating services* | | |
| Climate regulation | Fungi (1), Enchytraeidae (2), Collembola (3), Lumbricidae (4), Protozoa (5), Nematoda (6) | Enchytraeidae (1), Lumbricidae (2), Bacteria (3), Protozoa (4), Collembola (5), Nematoda (6) |
| Water regulation | Lumbricidae (1), Enchytraeidae (2), Fungi (3) | Lumbricidae (1), Enchytraeidae (2), Bacteria (3) |
| Erosion regulation | Lumbricidae (1), Enchytraeidae (2), Fungi (3) | Enchytraeidae (1), Lumbricidae (2) |
| Water purification | Lumbricidae (1), Bacteria (2), Fungi (3), Enchytraeidae (4) | Enchytraeidae (1), Bacteria (2), Lumbricidae (3) |
| Disease regulation | Enchytraeidae (1), Bacteria (2), Nematoda (3) | Enchytraeidae (1), Lumbricidae (2), Bacteria (3) |
| Pest regulation | Enchytraeidae (1), Fungi (2), Nematoda (3) | Enchytraeidae |
| Natural hazard regulation | Lumbricidae (1), Enchytraeidae, (2), Fungi (3) | Lumbricidae (1), Bacteria (2), Enchytraeidae (3), Protozoa (4) |
| *Supporting services* | | |
| Nutrient cycling | Fungi (1), Lumbricidae (2), Protozoa (3), Enchytraeidae (4), Bacteria (5), Collembola (6) | Enchytraeidae (1), Collembola (2), Fungi (3), Protozoa (4), Lumbricidae (5), Bacteria (6), Nematoda (7), Acarina (8) |

| | | |
|---|---|---|
| Primary production | Fungi (1), Nematoda (2), Lumbricidae (3), Protozoa (4), Bacteria (5), Enchytraeidae (6), Collembola (7), Acarina (8) | Nematoda (1), Enchytraeidae (2), Fungi (3), Protozoa (4), Bacteria (5), Collembola (6), Lumbricidae (7), Acarina (8) |
| Soil formation | Lumbricidae (1), Enchytraeidae (2), Fungi (3), Bacteria (4) | Enchytraeidae (1), Collembola (2), Bacteria (3), Fungi (4), Lumbricidae (5) |

Many definitions of ecosystem services are still controversial and we only aim to *predict which organisms are more likely to play a relevant function*, and how their function (if relevant) might change in different *ST*s. The most basic assumption is that all these ecosystem services and all these soil organisms are comparable to each other, that is, they have the same value (in the case of ecosystem services) or perform each service in a comparable way (in the case of soil organisms). In coarse soils, such as podzol, fungi and enchytraeidae (both ranked 3.7th of our 8 groups of soil organisms) are on average the most important agents in these 16 ecosystem services, followed by lumbricidae (#4.0), bacteria (#5.4), nematoda (#6.2), collembola (#6.7), protozoa (#7.0) and acarina (#7.8); in finely graded soils, like cambisol and fluvisol, enchytraeidae are by far the most important ecosystem agents (ranked 2.6th of 8), followed by bacteria (#4.0), lumbricidae (#4.3), protozoa (#6.1), fungi (#6.4), collembola (#6.6), nematoda (#7.1) and acarina (#7.8). Oligochaetes seem to dominate ecosystem functioning, together with fungi in sand and with bacteria in loam or clay. Mites were always ranked as least relevant group of soil organisms.

**REGULATING ECOSYSTEM SERVICES**

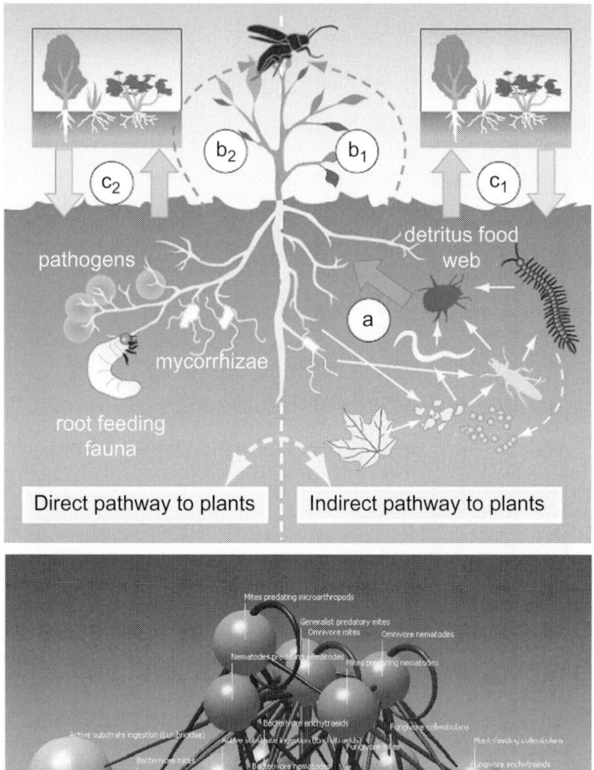


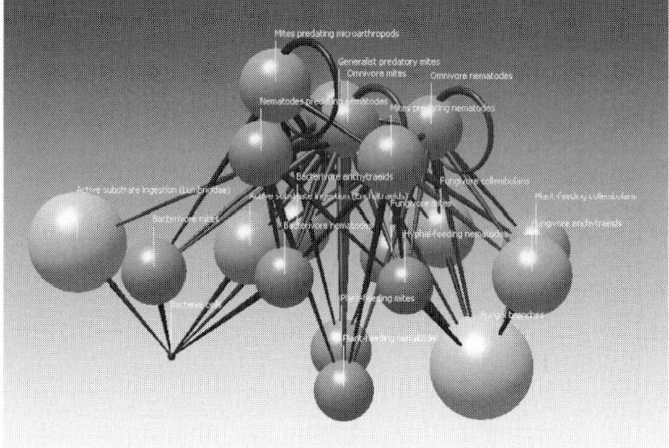

**SUPPORTING ECOSYSTEM SERVICES**

**Figure 7** Aboveground biota (green food webs and ecosystem services) are influenced by soil biota (brown food webs) in direct and indirect ways. (Upper panel, left compartment) Soil organisms exert direct effects on vascular plants by consumption, forming antagonistic or mutualistic relationships with plants. (Upper panel, right compartment) Multitrophic interactions in the brown food web (detritus, thin white arrows) stimulate nutrient turnover (thick red arrow), nutrient acquisition (a), plant growth and indirectly influence aboveground herbivores and mutualistic networks like those of plant-pollinators (red broken arrow) ($b_1$). These direct interactions influence both the performances of the plants themselves, as that of the herbivores like arthropods ($b_2$), and potentially their predators. Further, the soil food web can control the development of the resulting vegetation both directly ($c_2$) as indirectly ($c_1$),

across multiple levels of biological organisation, from individual metabolism to community dynamics, ecosystem functioning and evolution (Figure 7).

Burrowing (large-sized) organisms are also well known for the ecosystem services they provide: most of these active services are seen as life-support functions of the ecosystem in general, and the soil system in particular (Brussaard *et al.*, 1997; Schouten *et al.*, 1997, 1999). Our aim was to obtain a representative simplification of the soil system: ecosystem processes belonging to ecosystem's supporting, regulating and provisioning functions, including the responsible groups of soil organisms, are briefly summarised in Table 2. Many ecosystem services fundamental to life are provided by small soil organisms, although no consensus seems to exist as to what the practical indicators of these might be. To construct Table 2, a starting set was provided by DSQN data (Appendix) and the empirical data from outside the DSQN to which the co-authors had access. The ranking of the different groups of organisms according to their roles in two *ST*s and 16 services was realized by multicriteria analysis, taking only the absolute extent of the effects (as derived from correlation) into account, without differentiation between the positive and the negative effects.

Environmental scientists and managers are therefore faced with the difficult task of assessing ecosystem services to provide the very-basic information on the soil "black box" (Fitter, 2005) for management strategy and policy. There are different benefits and services to price. Land products can be quantified relatively directly, whereas the economical evaluation of *services* is more abstruse and therefore challenging, but still feasible. For instance, we may easily calculate the costs of water purification in the coastal dunes or the benefits of pollination service by popular insects like bees and butterflies. However, many ecosystem services of the soil are difficult to quantify as costs–benefits (although some valuable examples appeared recently, e.g. Van Eekeren *et al.*, 2010).

Therefore most approaches are focused on bioindicators, such as the fungi-to-bacteria ratio, which is supposed to provide a gauge of ecological soil quality, and the soil faunal biodiversity is also seen as a good proxy for resilience. In contrast to the soil α-diversity, which can be roughly assessed even at continental scales (Jeffery *et al.*, 2010), a quantification of the

---

and these plant community changes can in turn influence soil biota through changes in the soil C:N:P stoichiometry. (Lower panel) Multitrophic interactions within the brown food web are size-driven and process-driven. Realized faunal prey–predator (resource–consumer) links between functional boxes with average body-mass values (log-scaled, as observed in all 170 DSQN locations) are plotted at the bottom with the program NETWORK 3D. [Upper panel (green food webs) from Wardle *et al.* (2004). Reprinted with permission from AAAS.]. (For interpretation of the references to color in this figure legend, the reader is referred to the Web version of this chapter.)

composition of all these faunal taxonomic groups, completed with population densities, species' body-mass averages, total biomass and (response) traits for food web modeling, is hardly feasible on large scales and requires intensive sampling campaigns in both time and space. Still, disentangling soil biota and understanding ecosystem functioning are necessary prerequisites to fully evaluating ecosystems and ecosystem services, especially within the context of "Ecosystems must be managed within the limits of their functioning" (UNEP/CBD, 2000).

This *functionality* resembles sustainability, where "a sustainable system is one which survives or persists" (Costanza and Pattern, 1995), although this definition can be applied only *after the fact* (Costanza *et al.*, 1997b, p. 96). Such an assessment problem can be avoided if we use the ecological concept of *stability*. According to Woodward *et al.* (2005a), some of the main definitions are:

- Ability and rate of recovery (resilience);
- Resistance (degree of change in structure);
- Variability (internally or externally driven).

Ecosystem services can be seen as a currency driven and supported by the multiple responses of organisms to environmental stress over time (short-to-long-term perturbations) and space (small-to-large-scale perturbations). The responses of soil biota to perturbations can be characterised, simulated and modeled with food webs. Unmanaged "reference" sites might imply local stability, whereas most rural management is interventionist to some degree and thus implies (long-term) perturbations: this latter case is probably now the typical state of most European ecosystems. To quantify human-induced impact(s) at different levels of organisation (efficient use of non-renewable resources, maintaining soil fertility, etc.) a careful definition of the investigated agroecosystems is a necessary prerequisite.

## C. Ecosystem Types

The study sites exhibited marked differences in their aboveground management regimes both among and within *ET*s: managed grasslands, for instance, differ in terms of livestock density (with consequent grazing pressure), fertilizer inputs of N and P (organic or mineral), pesticide applications (Mulder *et al.*, 2003a), ploughing or compaction with heavy equipment (with the consequent effects on soil bulk density, water holding capacity, aeration, etc.). Eleven years (1999–2009) of biological soil monitoring within the II and III DSQN cycles (see Section II.A) have provided data from a total of 285 locations, spread across a broad range of *ET* and *ST* (i.e. including all categories described in Rutgers *et al.*, 2009). We have documented, for example, all the 9 *ET*s on the "Pleistocene sand" *ST*, giving a total of 200 sites for which all three soil macronutrients (the total contents of C, N, P) were measured from 1999 onwards. Total soil N and P content was measured directly in all sites; total soil C content was measured directly in 72.6% of the sites, and derived from SOM (C total $= 0.5508 \times$ SOM $- 0.036$; $R^2 = 0.8387$) in the remainder (see Table A1 for the complete DSQN abiotics). LGN6 1:50,000 Maps of SOM, *lutum*, and pH were used to define the *ET*s (Hazeu *et al.*, 2010). The considered *ET*s were classified as either rejuvenated (#0), managed (#1–5, from low to intensive regime), abandoned (#6), or natural (#7, 8):

(#0) 26 Arable fields, no livestock, divided into 9 organically managed fields and 17 conventionally managed fields. These rejuvenated anthrosols were almost bare soils throughout the sampling period. In the organic fields, sampled between February and March, atoms of carbon, nitrogen, and phosphorus are present on the average in the ratio of 75:4:1 [focusing on the changing balances between nutrients and phosphorus with an increasing P contribution, the soil molar C:P and N:P ratios range from 144 (least P) to 30 (most P) and from 8 (least P) to 2 (most P), respectively]. In the conventional fields, sampled in the same period, atoms of C, N and P are present on the average in the C:N:P ratio 302:12:1 (soil molar C:P and N:P ratios range from 582 to 120 and from 24 to 4, respectively).

(#1) 18 Certified organic farms (mixed biodynamic or bioorganic farms), using compost/farmyard manure and no biocides. In these soils, sampled between April and June, atoms of C, N and P are present on the average in the C:N:P ratio 131:8:1 (soil molar C:P and N:P ratios range from 243 to 68 and from 14 to 4, respectively).

(#2) 50 Conventional farms, using mineral fertilisers and to a lesser extent, farmyard manure. In these soils, sampled between March and June, atoms of C, N and P are present on the average in the ratio of 155:9:1 (soil molar C:P and N:P ratios range from 277 to 61 and from 14 to 3, respectively).

(#3) 28 Semi-intensive farms, using both organic and mineral fertilisers. In these soils, sampled between March and June, atoms of C, N and P are present on the average in the ratio of 114:7:1 (soil molar C:P and N:P ratios range from 269 to 51 and from 13 to 3, respectively).

(#4) 28 Intensive farms, using both organic and mineral fertilisers. In these soils, sampled between April and June, atoms of C, N and P are present on the average in the ratio of 95:6:1 (soil molar C:P and N:P ratios range from 209 to 50 and from 9 to 3, respectively).

(#5) 5 Pastures, no biocides. In these soils, sampled between April and May, atoms of C, N and P are present on the average in the ratio of 126:7:1 (soil molar C:P and N:P ratios range from 219 to 40 and from 11 to 5, respectively).

(#6) 10 Abandoned, mature grasslands. In these semi-natural soils, entirely covered by grass and herbs and sampled between April and June, atoms of C, N and P are present on the average in the ratio of 238:12:1 (soil molar C:P and N:P ratios range from 584 to 118 and from 23 to 5, respectively).

Land-use history plays a key role in the correct identification of an *ET*. The categorization into arable fields and/or grasslands (under any management regime), for instance, largely depends upon the kind of soil and the grass species occurring. The possibility of having an economically valuable organic farm in sandy areas depends mostly on existing good plant–soil interactions. Precipitation contributes to the evolution of a soil profile by improving degradation of organic matter and by allowing the mobility of bacterivores and the abundance of bacterial cells in moisture films.

The definition of "nature" is even more complicated due to the patchy structure in most densely populated and highly cultivated regions (Kristensen, 2001). We used the definitions provided by Schulze *et al.* (2002).

(#7) 10 Heathlands, *Molinietalia* and *Callunetea*. In these soils, sampled between April and June, atoms of C, N and P are present on the average in the molar ratios of 792:19:0.7.

(#8) 25 Forests: a minimum area $> 500$ m$^2$, a canopy $> 10\%$ and trees at least 2 m high, divided into 14 mixed woodlands and 11 Scots Pine plantations. In these mixed woodlands, sampled between April and June, atoms of C, N and P are present on the average in the extreme molar ratios of 1067:39:0.7. In the pine forests, sampled between June and (mostly) October, atoms of C, N and P are present on the average in the molar ratios of 521:22:0.7.

The extent to which *ET*s reflect the concentration of soil nutrients is shown in Figure 8. Empirical data from the DSQN database were used to derive proxies for the total C, N and P contents in the Dutch topsoil. Proxies were derived for each combination of land use and ST (Table 3). Some combinations of *ET* and *ST* were not available in the DSQN database, such as where particular land uses are not feasible on certain STs: for example, arable fields on alluvial clay. Then, proxies derived from robust empirical correlations were used instead, besides in one case, where a categorical average was used (Table 3); this allowed us to calculate the maps of total C, N and P.

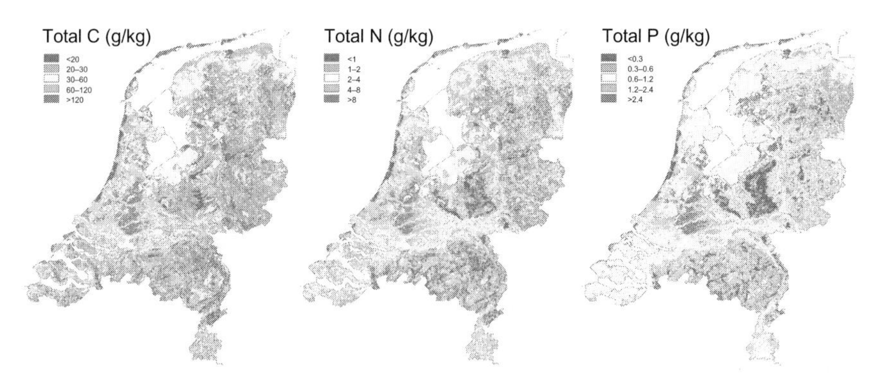


**Figure 8** Total carbon, nitrogen and phosphorus in the upper 30 cm of the soil profile of The Netherlands. All contours derived from the Soil Map of The Netherlands 1:50,000. Polygonal layers were converted at 1 × 1 km scale and merged with at 500 × 500 m scale. (For interpretation of the references to color in this figure legend, the reader is referred to the Web version of this chapter.)

# IV. SOIL FOOD WEBS

Food web models provide a way to analyze the dynamics of the constituent populations with soil trophic networks and a means of linking changes in the soil community structure to particular ecosystem services and processes. Moreover, since dynamic food webs represent transfer rates of energy, organic matter and nutrients, most such models provide a way to connect the dynamics of populations to the dynamics in ecological pathways within key nutrient cycles (Pignatti, 1994). By incorporating the habitat–response relationships for the occurrence of (soil) organisms, these models can assess the effects of (changes in) abiotic and management conditions on ecosystem functioning. The results also have the potential to provide valuable new insight into cross-cutting issues, such as the Global Taxonomy Initiative (http://www.cbd.int/gti/), where the final goal is the development of a complete information system for animals, plants, fungi and bacteria.

It was within such a framework that Wall *et al.* (2001, p. 115) claimed the *highest* priority for the development of "a strategy to increase the number of systematists working with below-surface invertebrates, as there are presently very few of these specialists globally". According to Fitter *et al.* (2005), it is not unlikely that the huge below-surface biodiversity is related to the extreme physical heterogeneity of soils, which forces us to use trophic "simplifications" of complex systems, such as food webs.

There are several possibilities to view (and compute) food webs (Table 4). According to Ulanowicz (1995), any ecosystem can be reduced to a Lindeman's tropho-dynamic sequence of discrete levels (Lindeman, 1942) to which species can be apportioned in proportion to their feeding strategy.

**Table 3** Summary of proxies (best model from multiple linear regressions) for the total soil carbon, nitrogen and phosphorus concentrations shown in Figure 8

| | Total C content (g/kg) | $R^2$ | Total N content (g/kg) | $R^2$ | Total P content (g/kg) | $R^2$ |
|---|---|---|---|---|---|---|
| Grasslands on sand (#1–5) | $5.37 \times SOM^{0.98}$ | 0.8388 | $0.29 \times SOM + 0.61$ | 0.7201 | $0.58 \times lutum^{0.22}$ | 0.1584 |
| Arable fields on sand (#0) | $6.79 \times SOM - 11.25$ | 0.9496 | $0.27 \times SOM - 0.18$ | 0.9291 | 0.46 (Mean estimate) | |
| Treeless sandy soils (#6–7) | $5.08 \times SOM + 2.92$ | 0.8434 | $0.74 \times SOM^{0.62}$ | 0.6116 | $0.02 \times lutum + 0.48$ | 0.3902 |
| Woodlands on sand (#8) | $5.76 \times SOM^{0.98}$ | 0.9045 | $0.23 \times SOM - 0.0028$ | 0.8630 | $0.02 \times pH \times lutum - 0.05$ | 0.4515 |
| Grasslands on Löss | $11.11 \times SOM^{0.59}$ | 0.5340 | $0.42 \times SOM + 0.20$ | 0.9149 | $0.29 \times SOM \times lutum^{0.24}$ | 0.4342 |
| Grasslands on river clay | $0.07 \times SOM \times lutum + 22.27$ | 0.8701 | $0.008 \times SOM \times lutum + 1.77$ | 0.8678 | $0.02 \times lutum + 0.55$ | 0.6638 |
| Arable fields on river clay | $5.08 \times SOM + 2.92$ | 0.8434 | $0.31 \times SOM + 0.39$ | 0.7570 | $-0.23 \times pH + 2.44$ | 0.3297 |
| Grasslands on marine clay | $4.77 \times SOM + 5.99$ | 0.6948 | $0.36 \times SOM + 0.82$ | 0.5036 | $0.29 \times lutum^{0.3431}$ | 0.5358 |
| Arable fields on marine clay | $0.15 \times SOM \times lutum + 12.61$ | 0.1435 | $0.01 \times SOM \times lutum + 0.69$ | 0.2514 | $0.002 \times pH \times lutum + 0.45$ | 0.4032 |
| Grasslands on peat | $5.08 \times SOM + 2.92$ | 0.8434 | $0.33 \times SOM + 0.19$ | 0.9755 | $0.10 \times lutum + 0.37$ | 0.8140 |

Linear pedo-transfer functions generated upon values reported in The Netherlands for sampling locations. Good agreements ($R^2$) were observed with DSQN and LGN6 data. Soil pH of oven-dried topsoil samples as measured in 1 M KCl. The numbers #0, #1–5, #6–7 and #8 refer to the ETs on podzol as described in Section III.C.

Alternatively, it is also possible to address changes in the power-law fashions of body-size distributions, and to assess the extent to what larger (although less abundant) species might influence network structure and dynamics. Computation possibilities provided by food web modeling therefore provide new tools to assess and evaluate ecosystem services (Table 4), and as such offer novel ways to approach biomonitoring from a new, more theoretical, perspective (Friberg *et al.*, 2011).

## A. Detrital Soil Food Webs

Almost two decades ago, De Ruiter *et al.* (1993, 1994) introduced the detrital soil food web model of Hunt *et al.* (1987) in The Netherlands (Section II.A). This model starts from observed equilibrium population abundances, making the calculation of energy flows and trophic interactions between compartments possible (Figure 9, upper panel, left). The original model was proposed as single-compartmented static model (Cohen *et al.*, 1990; Emmerson *et al.*, 2005; Fitter *et al.*, 2005; Hunt *et al.*, 1987). However, although the original parameters were kept as defined by Hunt and co-workers, it was possible to rearrange this compartmented static model into a dynamic Lotka–Volterra system when at equilibrium (De Ruiter *et al.*, 1995; Neutel *et al.*, 2002). In many Dutch environments, especially those on clay soils, the results of this approach at the level of overall mineralization fluxes are realistic (De Ruiter *et al.*, 1993, 1994). A possible drawback is the input variables, as the entire model is based on yearly averages and fluctuations due to environmental conditions are not taken into account. Especially in *ET*s on sandy soils, the highly variable fresh carbon litter input is strictly related to the type of vegetation and the kind of management regime and fluctuations of the average C:N:P ratio can be very high (Section III.C).

Therefore, as shown by Dekker *et al.* (2005), great care is necessary to avoid the overestimation of C (and N) mineralization due to the overwhelming size of the biomass fluxes associated with fast-growing and rapidly reproducing soil organisms (like bacteria and protozoa), in striking contrast to most other belowground consumers and predators, which are likely to have relatively insignificant direct effects on the computed fluxes. In particular, the dichotomy between the bacterial energy channel (where nematodes and protozoa have the greatest potential to enhance the fluxes) and the fungal energy channel (where arthropods have the greatest potential to affect it) must be parameterized to compare actual effects of human pressure on the numerical abundances of soil organisms (from perturbation to secondary extinction). This rearranged model is thus more suitable for calculating the energy fluxes and the total C and N mineralization in soil systems.

## B. Allometric Scaling of Food Webs

The detrital food web model analyzes "functional groups", where each functional group incorporates all the species that share the same feeding strategy and physiology. However, a food web can also be plotted with a much higher taxonomic (and hence trophic) level of resolution (Figure 9,

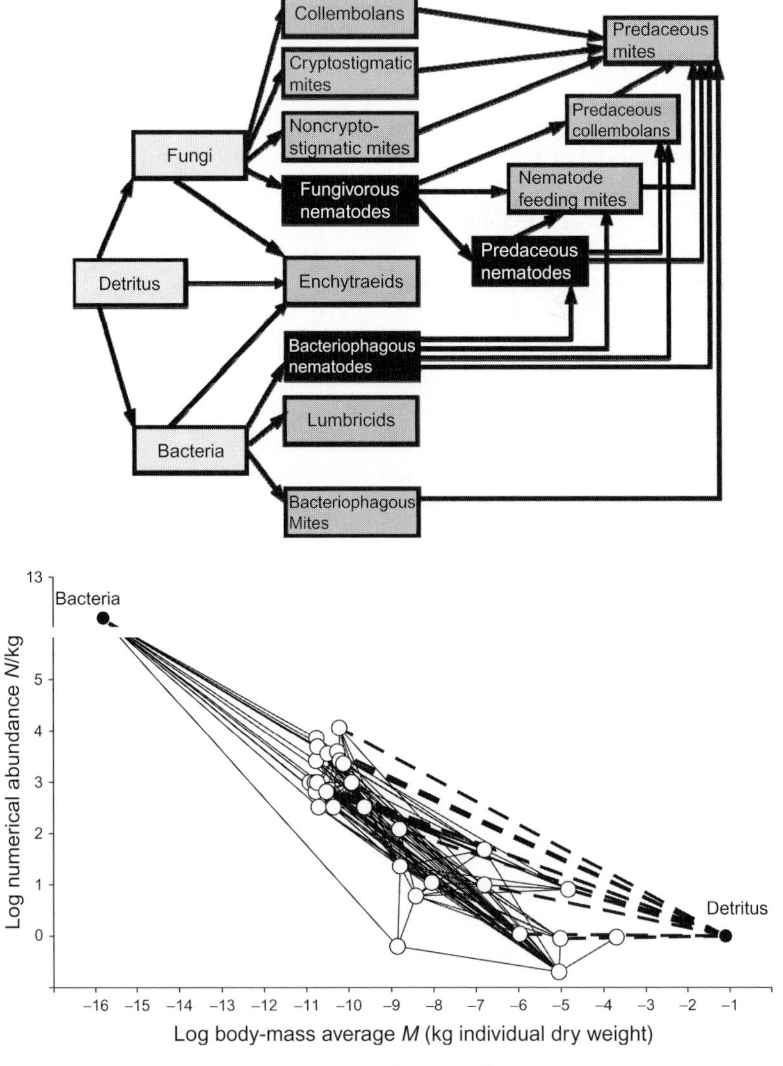


**Figure 9** (Continued)

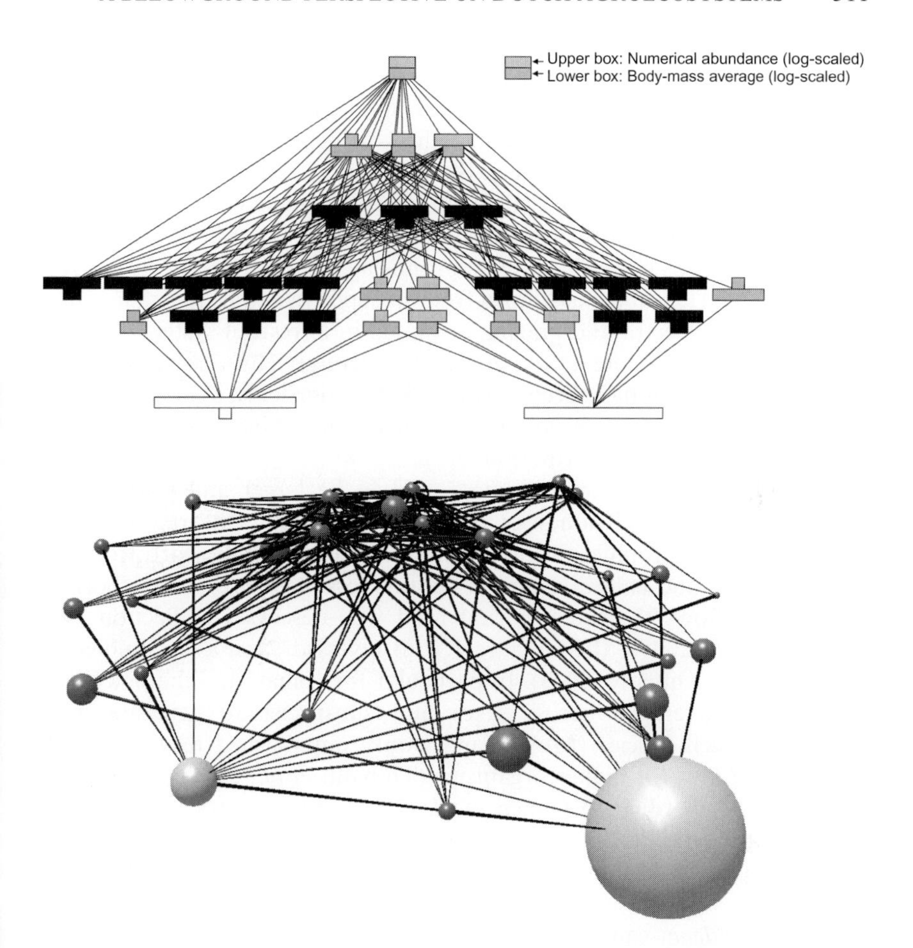


**Figure 9** Four ways to view the same data. Organic farm data, available as download from Mulder *et al.* (2005b): soil detritus, bacterial cells and non-parasitic invertebrates were sampled during June 1999. Counterclockwise: classical food web in compartments (De Ruiter *et al.*, 1993; Hunt *et al.*, 1987); mass–abundance relationship with individual trophic links (apart from the resources, the taxonomical resolution is at the genus level); constellation diagram showing the multitrophic structure of the entire soil food web, each node sized according to its biomass; and disaggregated network showing the energetic setup in terms of body mass and numerical abundance for the bacterial pathway (left) and the detrital pathway (right). The constellation diagram (bottom panel, right) was plotted with the Microsoft Research software NETWORK 3D (Williams, 2008; Yoon *et al.*, 2004, 2005); the two largest nodes in the constellation diagram represent the two main resources, the bacterial biomass (left) and the detrital resource (right).

bottom panel, left), using, instead of biomass, the mean body mass and abundance of each trophic element (≈species). Such a combination of species' body-mass data with numerical abundance data has been widely explored and has been used among others to depict aquatic communities and to show their exploitation (e.g. Blanchard *et al.*, 2009; Reiss *et al.*, 2010; Woodward *et al.*, 2010a,b). Historically, three main concepts might be recognized in allometry: the "universal" three-quarters power law, the "questioned" two-thirds power law, and a promising environmental-driven variation of allometric scaling, as known from size spectra and fisheries (Jennings, 2005) and size spectra and grazing or liming in terrestrial systems (Mulder and Elser, 2009; Mulder *et al.*, 2008, respectively). Thus, several density–scaling relationships have been postulated for ecosystems, among others the minimum size of a given population at a given time, their population energy use, and their biomass distribution.

Although higher order consumers tend to be larger and more mobile than their resource (Rooney *et al.*, 2008), body-mass averages are not always linked to trophic position. (Notable exceptions are earthworms in soil food webs and baleen whales in oceanic food webs.) Figure 9 (bottom-panel, left) shows the same detrital food web described in Section IV.A (Mulder *et al.*, 2005b, their Figure 1 with reversed axes, as recommended by Brown and Gillooly, 2003). In particular for nematodes, the body-mass values were averaged across sexes and subspecies to generate mean masses for each species (Mulder and Vonk, unpublished results), an approach used also for higher organisms (Thibault *et al.*, 2011), but see Section VII.B. According to Cohen *et al.* (2003), an isometric slope (i.e. an individual trophic link from a resource to a consumer with a slope equal to $-1$) implies that a consumer's total biomass equals its resource's total biomass.

Such *equivalency* can be applied to entire food webs only if all the taxa are equally distributed in the $n, \bar{m}$ space, because in real communities many different functional groups tend to clump together, the predators showing short trophic links and the bacterivores showing long trophic links (Figure 9, bottom panel, left), with a median quite close to the food web diagonal (+5%). Taking the bacterial-feeding nematodes into account, a slope more (or less) negative than $-1$ indicates that one of these microbial grazers has greater (or smaller) biomass, respectively, than the bacterial resource itself, assuming that the consumer is below and to the right of the resource (Figure 9, bottom panel, left). The opposite holds for fungal grazers and detritivores, assuming that these consumers are above and to the left of their specific resources. Therefore, the resulting slopes (and the related trophic links in a mass–abundance scatter) are steeper for bacterivores, when compared with either fungivores or detritivores.

According to our previous studies (Mulder *et al.*, 2005b, 2008), at least one-fifth of the trophic links shows metazoans preying on other metazoans with a similar body mass up to one that is 4 orders of magnitude larger. The overlap in body-mass averages can be visualized in one undisturbed semi-natural grassland plotted along the three allometric dimensions of body mass, biomass and density, in Figure A1 (raw data in Mulder and Elser, 2009: site 245 worksheet C). This model (in different versions, i.e. as mass–abundance linear regression slope, also with reversed axes, and as size spectra) is the most suitable to assess changes in the "trait distribution" of differently sized species within real communities.

## C. Structure of Ecological Networks

Besides the detrital food web and the allometric scatterplot, food webs can be visualized in other ways. Biological networks have topologies defined by the relationships between the species and the physical limitations imposed on species interacting in space. The (trophic) structure of the food web can be plotted as in the upper right corner of Figure 9, where the same trophic links are shown as in the allometric scatterplot (Section IV.B). These diagrams are widely used for the so-called green food webs, such as in the case of mutualistic networks (e.g. Müller *et al.*, 1999; Novotny *et al.*, 2010), but can be used for other food webs as well. An allometric modification, focusing on the Eltonian distribution of the tritrophic components of each species (body mass, abundance and biomass) was applied twice in one pelagic system (Cohen *et al.*, 2003; Jonsson *et al.*, 2005; O'Gorman and Emmerson, 2010) and has much potential to visualize the dichotomy between the basic resources in "brown" food webs (Mulder, 2006) or the community responses to species loss (cf. De Visser *et al.*, 2011).

The true complexity of multitrophic interactions in an empirically observed food web can be represented as a "constellation diagram". The resulting diagram is shown in Figure 9 (bottom right) and enables the investigation of the structural effects on species persistence for a range of structural network models (Williams, 2008), such as variations in omnivory and trophicity, both properties that seem to be environmentally driven in terrestrial food webs (Denno and Fagan, 2003; Mulder *et al.*, 2006a). In many past network analysis, in fact, uniformity of resource was often taken as granted, a basic assumption that seems to be highly unlikely, given the prevalence of marked ontogenetic shifts in diets and trophic status (Rudolf and Lafferty, 2011; Woodward *et al.*, 2010b). Especially in the case of invertebrates, adult non-parasitic soil nematodes, micro-arthropods, and enchytraeids use different resources than those they exploit as larvae or juveniles.

**Table 4** Summary of the main computational possibilities for the food web models shown in Figure 9

| | Model structured as | Kind of architecture | Computing or visualizing | Main tools to assess |
|---|---|---|---|---|
| Detrital food web[a] *(only for soil systems)* | Structural model (static)/ cascade model (dynamic) | Directed graphs (flow) for functional groups | Model emphasizes dynamics: energy channels, dominant feeding relationships, loops, biomass and nutrient cycling | Supporting services and regulating services |
| Allometric scaling *(mass–abundance slopes and/or size spectra)* | Power law *(the exponent β is claimed to be close to –¾ resp. –⅔, or environmentally driven)* | Triangular web for taxa, trophic species and/or functional groups | Stability (departure from isometry), connectance, connectedness, prey overlap, biodiversity, trophic generality and vulnerability, polygonal relationships, distribution of trophic links and chain lengths, trophic heights, trophic transfer efficiency α, assimilation efficiency, resource: consumer ratios, trait distribution, stoichiometric imbalances | Supporting services, regulating services and provisioning services |
| Topological network | Static | Biopyramid for species, trophospecies and/or functional groups | Model emphasizes structure: biodiversity, size-delimited pathways, self-organization, trophic heights, nested hierarchy, parasitism | Provisioning services |
| Interaction network | Dynamic | Structural network for species, trophospecies and/or functional groups | Model visualizes both the dynamics and structure: connectance, complexity, biodiversity, metabolic rate, trophic transfer efficiency α, self-damping and mutualism, ontogenic status, population growth, distribution of trophic links and chain lengths, trophic generality and vulnerability, parasitism and cannibalism, risks, extinction, magnitude of feeding relationships | (In progress) |

More explanations in the text.
[a]See main text and references therein.

# V. AUTECOLOGY, BIOLOGICAL STOICHIOMETRY, AND ECOSYSTEM SERVICES

## A. Habitat–Response Relationships

There have been few studies describing the impact of grazing cattle on the numerical abundance and biodiversity of soil fauna, and there are even fewer published food webs comprising organisms with a body size less than 1 mm (Woodward *et al.*, 2005b). Effects of environmental pressure on the soil fauna related to increasing stocking intensity have been recognized within the micro-arthropods (Kay *et al.*, 1999; King *et al.*, 1976), where the numerical abundance of microbivores (Nanorchestidae, Tarsonemidae and Tydeidae) seemed to decline with livestock intensity: in reality, it is the entire nematofauna that reacts as a whole to different stocking levels (Mulder *et al.*, 2003a).

In 87 podzols (one *ST* spread over four *ET*s), the probability of occurrence of nematode genera seen as a cumulative function of independent predictors was investigated (Figure 10). Focusing on genera with an explained deviation higher than 60%, most nematode taxa decrease with increasing methane production by cattle, with especially among the fungal-feeding taxa. Since methane production is dependent on labile carbon pools and the decomposition of carbon pools in soils produces both $CH_4$ and $CO_2$ (Pendall *et al.*, 2004), additional methane production by cattle might not only affect the nematofauna, as shown in Figure 10, but might disturb further the entire carbon cycle process as well. With a total contribution of 15% methane by ruminants and 17% $N_2O$ by manure, livestock is a powerful and an increasing driver in agricultural changes worldwide (Power, 2010). At least two different processes run in parallel with increasing grazing intensity. On one hand, the bacterial biomass of all our sites on podzol decreases along the livestock gradient (Mulder *et al.*, 2003a, their Figures 6 and 7), whose effects result in a 50% decline in the abundance of bacterial-feeding nematodes. On the other hand, hyphal-feeding nematodes can be sensitive to intensive farming pressure, owing to the very low number of taxa and population density (Mulder *et al.*, 2003a, 2005a). Thus, not only did fungi and bacteria show contrasting trends with cattle and liming, but so did hyphal-feeding and bacterial-feeding nematodes. Also in other *ST*s, as in fluvisols, soil invertebrates reacted to compaction of the soil structure: for instance, preliminary results show that the total biomass of enchytraeids changes markedly according to the soil structure, being three times as high in fluvisols as in podzols (Didden *et al.*, unpublished results).

Other faunal patterns were recognized in agroecosystems after a change in land use. According to Postma-Blaauw *et al.* (2010), the conversion of

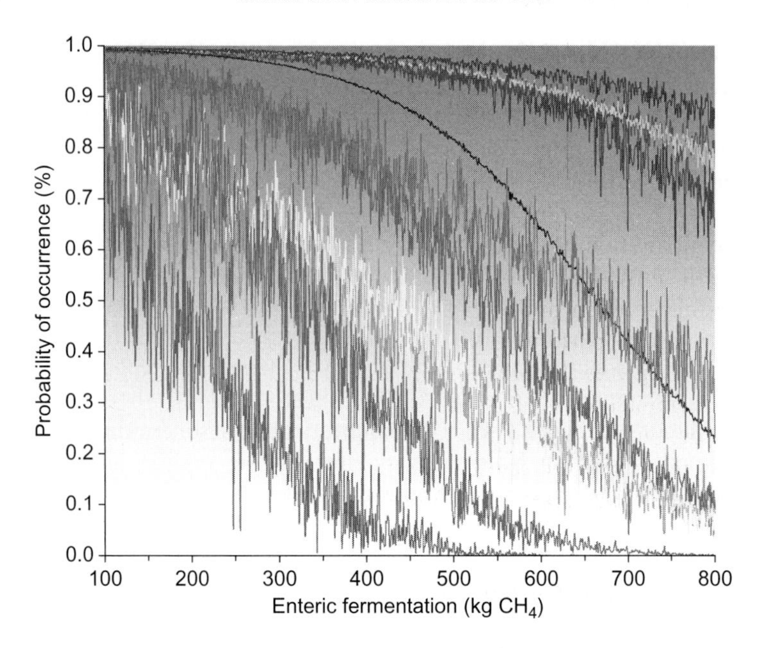


**Figure 10** Generalized Linear Model (Monte Carlo simulation) of the aboveground predictor $CH_4$ for a selection of nematodes. Methane emission by enteric fermentation according to the mechanistic model of Benchaar *et al.* (1998). Nematode scores on the vertical axis (0 = 100% absence, 1 = 100% presence; each genus has a different color). A pilot study on soil nematodes and livestock in Mulder *et al.* (2003a, their Figure 4) showed only two genera with increasing trends, namely *Chiloplacus* across the entire livestock gradient and the predatory *Thonus* under (semi)intensive management. Also there, the steepest trend in the occurrence was shown by Teratocephalidae, whose density score appears to be sensitive to the livestock density at the boundary between conventional and (semi)intensive farming.

grassland to arable field affects soil biota differentially: larger organisms (nematodes, arthropods, oligochaetes) were reduced relative to the smaller bacteria and protozoa, although they recovered after grassland re-establishment on formerly arable land. In this chapter, the protozoan communities were dominated by flagellates and although ciliates made up only a small percentage of protozoan abundance, they responded highly sensitive to land-use changes. In four experimental systems (long-term grassland, new arable field, long-term arable field and new grassland), Postma-Blaauw (2008) recognized that the ciliate abundances (1) increased in conventionally fertilized agroecosystems and (2) decreased after a conversion of grassland to arable land. It has been estimated that protozoa may account for ~70% of the soil animal respiration (Foissner, 1987) and for 14–66% and 20–40% of the soil C and N mineralization, respectively (Bloem *et al.*, 1994; Ekelund and Rønn, 1994; Griffiths, 1994). Food web model simulations have also suggested that protozoa are important contributors to N mineralization (De Ruiter *et al.*, 1993; Hunt *et al.*, 1987; Schröter *et al.*, 2003). Indirect contributions to nutrient cycling of protozoa may be even more important than their direct effects, since grazing stimulates microbial mineralization (Bonkowski *et al.*, 2000; Sohlenius, 1968).

The contribution of census area (in ha) and time (Julian Days) was not relevant in most sites (cf. Mulder *et al.*, 2003a, 2005a,b): a robust positive relationship between numerical abundance and territory is possibly one of the most general patterns in ecology and a close relationships between log density and log area across a set of 265 DSQN locations on podzols showed a significant intercept ($p < 0.0001$) for bacterial cells, nematodes and micro-arthropods, but no significance in the linear regression slope. It is the particular combination of soil acidity, SOM and FOM that seems to drive the trait distribution of body sizes in soil biota.

## B. Soil Acidity and Cations' Availability

Ecoenzymatic activities are the product of cellular metabolism regulated by the elemental concentrations in the environment, and are related to both resource availability and microbial growth (Sinsabaugh *et al.*, 2008, 2009). Microbial community composition is of key importance for driving the C:N:P stoichiometry of the total SOM (Cleveland and Liptzin, 2006; Manzoni *et al.*, 2010; Mulder *et al.*, 2006b), and terrestrial decomposition rates are influenced mainly by the chemical nature of carbon, the microclimate, the microbial pools (fungi, bacteria and protozoa), and by the soil nutrient balance (Beerling and Woodward, 2001; Moore *et al.*, 2004; Petersen and Luxton, 1982). Microbial pulses were observed in nutrient-rich farms with different redox states, suggesting high activities of glucose-responsive microbes (Mulder *et al.*, 2005b). Different substrate affinities during the microbial colonization of litter were modeled and observed (Moorhead and Sinsabaugh, 2006; Mulder *et al.*, 2006b): soil bacteria grown under different straws, for instance, show highly different utilization of carbon compounds, especially of carbohydrates (Mulder *et al.*, 2006b). Notwithstanding these temporal pulses, some ecoenzymatic activities show significant univariate regressions with soil pH (Sinsabaugh *et al.*, 2008, their Figure 2), suggesting that the potential for biological oxidation of detritus is driven by this key abiotic variable (Sinsabaugh *et al.*, 2008), as might be expected given the robust correlation between soil pH and the soil cation exchange capacity (Huston, 1993).

Soil pH reflects total exchangeable bases and the balance between soil nutrients (Mulder and Elser, 2009), and in addition to the opposing trends observed for bacteria and fungi in response to this variable the enzymatic potential during decomposition also reflects the chemical and physiological differences between the microbial pathways of the soil food web. Bacterial cells are composed of 80% easily decomposable material, in contrast to fungal remains, which are composed of 50% relatively more recalcitrant chitin (Gignoux *et al.*, 2001). Moreover, bacterial carbon-use efficiency is inversely related to the resource C: nutrient stoichiometry, whereas fungal carbon-use efficiency is directly related to it (Keiblinger *et al.*, 2010). Again, opposite trends within the soil microbial biomass help explain these dichotomies in the microbial pathways of the detrital soil food web (Wardle *et al.*, 2004).

In an authoritative and comprehensive review, Wardle (1998, p. 1632) recognized that the seasonal dynamics of soil microbial biomass, as expressed by their coefficient of variation, are correlated with pH in arable agroecosystems, with pH and organic carbon in grasslands, and with soil nitrogen in forests. Effects of pH on the structure and dynamics of aquatic food webs are now well known (e.g. Hildrew, 2009; Layer *et al.*, 2010, 2011) and comparable patterns might be expected in soil systems.

If we describe each of our soil communities with the average $\bar{m}$ (dry body mass) and $n$ (density) for each species within a given community on a log–log scale, the challenge of understanding below- and aboveground interrelationships can be illustrated in Figure 11. Here, each site's mass–abundance slope (a variable that summarizes the entire soil community in a way comparable to a diversity index, such as the Shannon-Wiener $H'$) is plotted as a function of each site's pH (cf. Layer *et al.*, 2010). [Soil pH is defined as the negative logarithm of the $[H^+]$ concentration: $\log(1/[H^+])$.] Pooling the *ET*s with a canopy of grasses, herbs and shrubs together ($n=110$ sites of Figure 11, all belonging to *ET*s #1–7), the dynamic interactions between SOM (SOM %, upper arrow), FOM (FOM %, vertical arrow) and decreasing $[H^+]$ concentration (hence, inversely related to pH, as the best independent sole predictor) accounts for more than 55% of the allometric variation in the mass–abundance slopes ($R^2$ from ordinary linear regression).

If we compare the highly diverse Scots Pine forests (low SOM, high FOM) with the conventionally managed arable fields (high SOM, low FOM, the only *ET* sampled in the Winter), we might make the erroneous assumption that it is mainly human disturbance that affects allometric scaling, as the mass–abundance slopes in the rejuvenated fields were more negative than in the undisturbed forests. However, the regression fitted across the allometric scaling of open-canopy *ET*s clearly reflects the nutrient:C ratios more strongly (Figure 11). Anthrosols show the lowest significance of the mass–abundance linear regression slope for all *ET*s ($R^2=0.27\pm0.16$). Still, for all

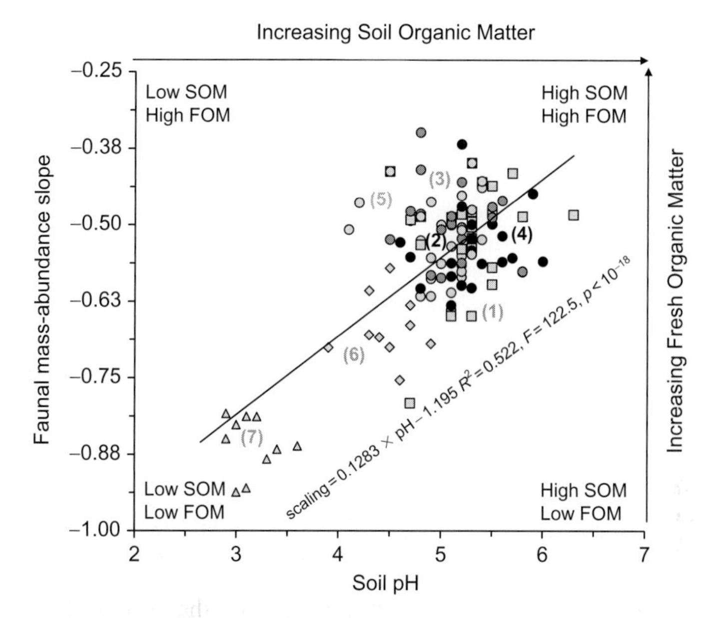


**Figure 11** The mass–abundance slope of the soil faunal assemblages of open canopies as predicted by $H^+$ cations (main predictor), soil organic matter (SOM) and fresh organic matter (FOM). The linear regression fitted through the allometric scaling of different ecosystems closely reflects the total exchangeable bases and the balance between soil nutrients,that is, the nutrient:C ratios (here, total soil N:C and P:C ratios). The Kruskal-Wallis ANOVA shows significant differences between nutrient: C in heathlands, abandoned and cultivated *ET*s ($\alpha = 0.05$). As the mass–abundance slopes from heathlands (*ET* #7) to abandoned (#6) or cultivated farms (#1–5) becomes increasingly less negative with decreasing soil acidity, it implies a corresponding higher increase in biomass of larger invertebrates at increasing pH. From mesic to acidic conditions, that is, from the right of the figure to its left, the increasing detrimental impact on large-bodied predators (often sit-and-wait strategists) accounts for steeper mass–abundance slopes. The *ET*s (all numbered as in Section III.C) did not add explanatory power to that provided by soil pH alone. The measurement of pH was performed in different solutions: for some *ET*s there was an adjustment, if necessary, to report the pH in KCl solution. Anthrosols and forests were excluded from this linear regression analysis. (For interpretation of the references to color in this figure legend, the reader is referred to the Web version of this chapter.)

the nine investigated *ET*s, the linearity tests of the allometric scaling hold for 95.2% of the soil food webs (Mulder *et al.*, 2006a, 2008, 2011; Reuman *et al.*, 2009). Our stoichiometrical model can accommodate this variation in the scaling exponent for fauna (i.e. Metazoa), fauna + fungi (i.e. Eukarya), or fauna + fungi + bacteria that is observed over a 15 orders of magnitude variation of body size (Mulder, 2010; Mulder and Elser, 2009).

# VI. SYNECOLOGY AND ECOLOGICAL STOICHIOMETRY

## A. Chemical Balance and Trophic Structure

Food web modeling provides novel ways to investigate to what extent (1) soil nutrient availability can enhance the multitrophic interactions in agroecosystems and (2) feedbacks to environmental pressures can enhance the predictability of ecosystem services. The differential sensitivity to the elemental resources of organisms with different body size helps to understand population dynamics and to increase the realism of our trait-based extrapolations, and it seems that it is the trophic level ("height") that matters for stoichiometric differences among ecosystems or within food web compartments (Elser *et al.*, 2000b; Statzner *et al.*, 2001; Sterner *et al.*, 2008; Sterner and Elser, 2002). The main implication for managed and abandoned soil systems is that elements such as C, N or P (Mulder and Elser, 2009), and micronutrients such as Zn and Cu (Mulder *et al.*, 2011), become important in:

(1) constraining the activity of individuals at a metabolic level,
(2) determining the numerical abundance at species level,
(3) defining the consumer–resource interactions at community level, and finally
(4) driving the soil food web organisation at ecosystem level.

All of the above are closely interlinked via the role that body size plays across these different organisational levels (Woodward *et al.*, 2010a). The total concentration and the availability of soil macro- and micronutrients seem to be reliable predictors of allometric scaling relationships (Mulder, 2010; Mulder and Elser, 2009). A good allometric predictability for soil biota is perhaps not surprising given the robust scaling occurring in the litter and aboveground between the leaf nitrogen concentrations and the leaf phosphorus concentrations for 9356 plant leaves (Reich *et al.*, 2010) and the well-known biological rates of metabolism that scaled with body mass (Brown *et al.*, 2004; Enquist *et al.*, 2009; Williams, 2008; Yodzis and Innes, 1992). Even in extreme environments, like the Atacama Desert in Chile, the whole-body elemental composition (especially P) within most arthropods decreases significantly with the species' body mass (González *et al.*, 2011), making a careful identification of realistic metabolic parameters for scaling rules and dynamic simulations possible.

Another example comes from our 35 well-graded sandy soils across The Netherlands sampled between 2002 and 2005 within the DSQN framework (Mulder, 2010; Mulder and Elser, 2009). Fifteen of these sites were organic farms sown with *Trifolium repens* and *T. pratense* (Mulder, 2010). Ten sites were abandoned grasslands, left unmanaged for 10 or more years, and 10 sites were heathlands (Section III.C). (These soil food webs are shown in Figure 11 as *ET*s #1, #6 and #7, respectively.) There were no significant differences in C concentration by *ET* (Kruskal-Wallis ANOVA, $p=0.15$), but N and P concentrations differed markedly (Kruskal-Wallis ANOVA, $p<0.0001$) being highest in organic farms and lowest in heathlands. High-P sites also had lower C:P and N:P ratios than low-P sites and supported invertebrate communities in which biomass increased with central $|m|$-bins in size spectra (values representing taxa belonging to the mesofauna which graze, browse, or ingest fungal remains). Low-P sites (heathlands) had higher C:P and N:P ratios and their faunal biomass decreased with increasing central $|m|$-bins in size spectra (Mulder *et al.*, 2009).

Of particular interest here is the P-limitation that causes a decrease in the numerical abundance of larger body-size classes of metazoans (Figure 12). The negative biomass spectrum slopes in heathlands might be ascribed to higher population densities of nematodes and microherbivorous mites, at the lower end of the range of body sizes. Positive biomass slopes of organic farms are associated with higher population densities of enchytraeids and springtails, at the upper end of the range of body sizes. Figure 12 shows how micro-arthropods and enchytraeids belonging to the second trophic level decline relative to the increasing abundance of soil nematodes, as P concentrations fall. Further, P-enriched organic farms are known to have much larger fungal biomasses than other *ET*s (Mulder and Elser, 2009).

Our allometric functions in Figure 12 integrate the observed differences between high small-to-large rates (i.e. microfauna-to-mesofauna and, to a lesser extent, resource-to-consumer ratios) due to the high numerical abundance of smaller metazoans in P-limited heathlands. Among the microbes, bacterial cells have the lowest average body mass and fungal mycelium the highest. Consequently, in all the possible trophic links from bacteria to each bacterivore, $m_C > m_R$, because the (individual) consumer (*C*) has an average body mass up to 8 orders of magnitude higher than its (single) bacterial cell resource (*R*), and the reverse ($m_C < m_R$) for all trophic links from hyphae to fungivore organisms (Mulder *et al.*, 2009). The environmental effects of ecological stoichiometry and soil quality (FOM and SOM) on the soil biota is apparently not coupled with a reduction in the resources available to larger metazoans, as shown in the resource-to-consumer ratios and the huge degree of omnivory (Figure A2). These resource-to-consumer ratios might reflect the rather high resilience of soil webs, in contrast to many marine, freshwater, or aboveground webs (Brose *et al.*, 2006).

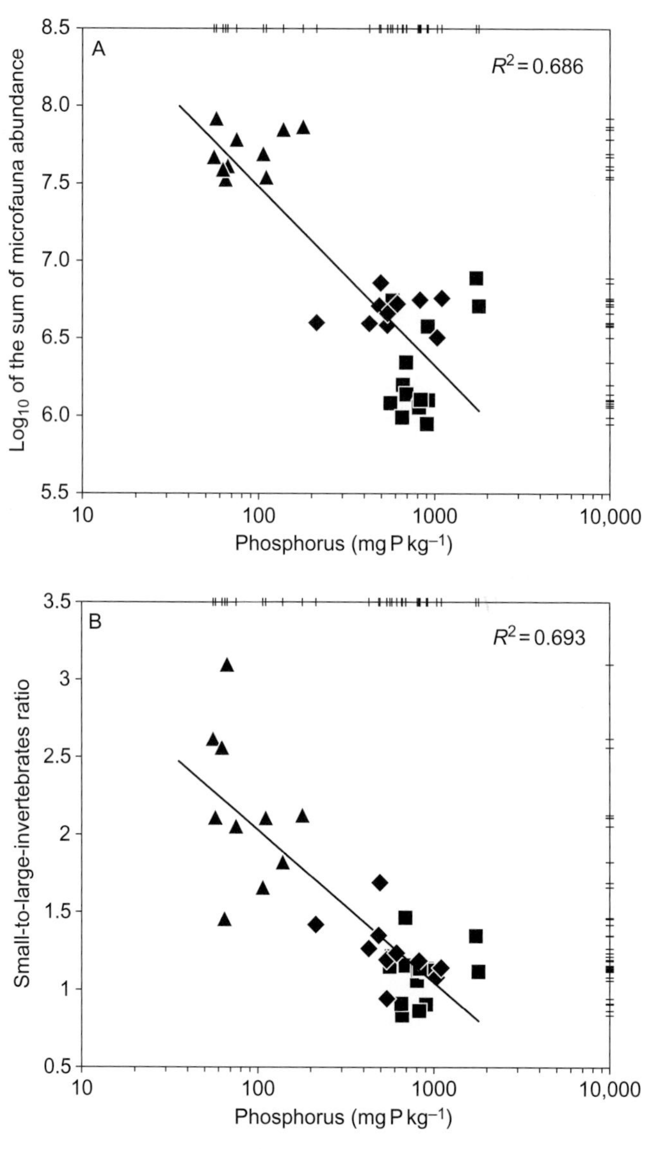


**Figure 12** Faunal body-size distributions as predicted by the total P concentration in the soil (log-scaled horizontal axis). Different vertical axes used; (A) log-transformed sum of the faunal abundances with body-mass averages lower than 1 $\mu$g (here: Nematoda); (B) log-transformed ratio of the aggregate abundance of all invertebrates with a body mass lower than 1 $\mu$g to the abundance of all invertebrates with a body mass higher than (or equal to) 1 $\mu$g. Changes in the nutritional quality of soils are clearly governing the composition of food webs and enable the coupling between vegetation (here: heathlands, triangles, abandoned grasslands, diamonds, and organic

## B. Elemental Availability Versus Prey Availability

Detritivory and omnivory are among the most commonly observed feeding modes in food webs, and can have particularly powerful influences on ecosystem processes in soil systems and in freshwater environments (Jepsen and Winemiller, 2002; Vanni and De Ruiter, 1996, respectively). Therefore, the multitrophic interactions within the soil food web might be expected to react in a comparable way to those of detrital-based aquatic systems (e.g. headwater streams), finally affecting the supporting ecosystem services provided by soil organisms with their different traits.

Stoichiometric constraints particularly affect the Net Primary Production (*NPP*) and the conversion efficiency from basal resources to the primary consumers (Andersen *et al.*, 2004; Woodward *et al.*, 2010a). Given the huge difference in the C:N ratios of woody stems and mycorrhizal tissues (from ca. 400:1 up to 10:1, respectively, see Johnson, 2010; Sterner and Elser, 2002, for details), the belowground invertebrates feeding on fungal tissues will get much more N than those feeding aboveground on plant tissues. Therefore, contrasting trends are expected along most *NPP* transects. Numerical abundance increases with $NPP^{3/4}$ for insects like soil ants (Kaspari, 2004) and the (physiologically optimal) body mass in multitrophic communities under predation risk decreases with *NPP* (Kozłowski, 1992). Both mass and abundance reflect environmental constraints that are acting as filters for biological traits (Hesse, 1924; Statzner *et al.*, 2001).

The synergy between biological and ecological stoichiometry and food webs provides a powerful way to show how tolerant a consumer of a given body size can be for a resource of poor elemental quality (Elser *et al.*, 2000a, b; Urabe *et al.*, 2010). Doi *et al.* (2010) showed the extent to what the stoichiometrical effects of a resource might modulate populations and communities:

$$\mathrm{TER}_{\mathrm{C:P}} = \left(\frac{\mathrm{GE}_{\mathrm{P}}^{\max}}{\mathrm{GE}_{\mathrm{C}}^{\max}}\right) \times \left(\frac{\mathrm{C}}{\mathrm{P}}\right)$$

where the threshold elemental ratio (TER) is the nutrient:carbon ratio of a resource below which the consumer's growth rate will be limited by the

farms, squares) and soil biota. Specifically, smaller invertebrates are favored by decreased P-availability as predicted by the stoichiometric theory (Mulder and Elser, 2009; Sterner and Elser, 2002), which predicts that organisms with higher phosphorus demands would suffer a competitive disadvantage due to poor stoichiometric food quality. Body size matters in the occupied territory (larger invertebrates are more mobile than smaller invertebrates), in the foraging strategy (mobile invertebrates can become selective), and, consequently, in the degree of omnivory of larger invertebrates (Mulder *et al.*, 2009).

resource's elemental content and $GE^{max}$ is the maximal growth efficiency for nutrient (here: P) and carbon (Doi *et al.*, 2010). If we assume the energetic requirements average per individual $\bar{E}$ to be proportional to $\bar{m}^{3/4}$ (Damuth, 1981), the energetic equivalence hypothesis (EEH) can be written as $n \propto \bar{E}/\bar{m}^{3/4}$. EEH does not assume that populations of every species absorb the same amount of energy from the environment (Brown and Gillooly, 2003; Reuman *et al.*, 2008) and $TER_{C:P}$ should not be affected by $\bar{m}$ (Doi *et al.*, 2010).

All growing organisms depend on a steady supply of P and (C/P) has been shown to be a robust independent parameter to explain allometric scaling either in the form of the mass–abundance slope $\beta$ or of the intercept $\alpha$ (Mulder *et al.*, 2006a, 2009, 2011), which have the same values as the coefficients $\alpha$ and $\beta$ in the allometrically general exponential equation $n = \alpha\bar{m}^{\beta}$. However, we can also see $\beta$ as constant (Brown *et al.*, 2004; but see White *et al.*, 2007). If we do, given the robustness of (C/P), we might assume that a large part of the stoichiometric signal is caught in the noise depicted by the departure from the model $\log(n) = -\beta \times \log(\bar{m}) + \alpha + \varepsilon_0$. Let us then define the departure $\varepsilon_0 = \varepsilon_1 + \varepsilon$ and rewrite the general 3/4 formula as:

$$\log(n) = -0.75 \times \log(\bar{m}) + \alpha + \varepsilon_1 + \varepsilon = -0.75 \times \log(\bar{m}) + \alpha + \left(\frac{C}{P}\right) + \varepsilon = -0.75 \times \log(\bar{m}) + TER_{C:P} \times \left(\frac{GE_C^{max}}{GE_P^{max}}\right) + \alpha + \varepsilon$$

making the cross-product between the quality of a resource (the soil for microbes, the microbes for consumers and all preys for predators) and the growth efficiency $GE^{max}$ (gross percentage of consumed biomass converted to body mass) an additional predictor for the subtle changes occurring in the size structure of soil food webs (Mulder and Elser, 2009; Figure 13).

New implications are that (1) the energetic interactions between EEH and TER have not been investigated yet besides an exploratory paper by Allen and Gillooly (2009, their Principle III) on the possibility to encompass EEH, and (2) the results are expected to contribute to the current discussion on the universality of allometric scaling: if the allometric model is really universal, soil abiotics could help to explain the observed deviations from the predicted values.

The importance of the detrital compartment in our simulation may be high because the largest fraction of nutrients is tied up in organic detritus, which itself acts as a buffer and increases the resilience of soil biota to external perturbations (DeAngelis *et al.*, 1989). N and P are considered to be the key limiting elements in the terrestrial biota, and living soil and litter (micro) organisms are especially competitive for P (Sterner and Elser, 2002). In temperate grasslands, N is typically seen as the limiting element in

**Figure 13** Topology of the multitrophic interactions in the soil of a abandoned, mature grassland. The food web analysis in NETWORK 3D (written by Rich Williams) spatially rearranges the original consumer–resource matrix with trophic level on the vertical $z$-axis and species' connectance on the $(x, y)$ plane in a cylindrical layout, so that highly connected nodes are placed closer to the $z$-axis. (The food web's connectance $c = l/S^2$ is here 21%.) All trophic links ($l$) were derived from literature survey (Mulder *et al.*, 2009: their online matrices) and all taxa ($S$) were evenly resolved at taxonomical level. Each node represents one "species" (mostly genera) sized according to the site-specific biomass. Orange, omnivore, light green, predator; dark green, detritivore; light blue, decomposer; dark blue, producers (fungal mycelium plotted towards the middle and bacterial population plotted near the edge). The soil microfauna is the most widespread across trophic levels, followed by the mesofauna. (For interpretation of the references to color in this figure legend, the reader is referred to the Web version of this chapter.)

meta-analysis depicting traits and production of invertebrates under N enrichment (González *et al.*, 2010, especially their Figure 4). However, in the soil food webs of comparable temperate grasslands, the empirical evidence of the actual importance of P-availability for the biotic demands of soil invertebrates is increasing (Mulder *et al.*, 2009, 2011; Mulder and Elser, 2009). As human impact worldwide impaired P-limited agroecosystems more strongly than N-limited agroecosystems (Wassen *et al.*, 2005), the consistently variation of the scaling relationships in soil biota modulated by pH and P-availability has broad implications because differently sized invertebrates mostly have different effects on ecosystem services (Peñuelas and Sardans, 2009).

# VII. TROPHIC INTERACTIONS

## A. Enhancing the Resolution and Quantification of Resource–Consumer Linkages

Techniques to study multitrophic interactions in terrestrial food webs have become increasingly powerful, as they have sought to overcome the limitations of more traditional approaches (e.g. gut content analysis, morphological examinations, field observations, and laboratory feeding trials). These new techniques include, besides the analysis of carbohydrases (used in this study, see Figures 9 and 13), the detection of prey pigments and proteins by chromatographic analysis, polyclonal antibodies and other DNA-based methodologies (reviewed by Harper *et al.*, 2005; Sheppard and Harwood, 2005; Symondson, 2002; Hagler, 2006; King *et al.*, 2008; Purdy *et al.*, 2010). Unfortunately, in the case of soil dwellers which tend to prefer deeper layers to litter quality (in contrast to litter dwellers and soil-surface dwellers), these biochemical and molecular techniques offer only a snapshot of what a soil invertebrate has recently consumed. Consequently, they do not show the full variety of the complete resource spectrum and, most importantly, they do not discriminate among resources in terms of assimilation and energetic contribution to the consumer's diet.

In the last 20 years, analyses of naturally occurring stable isotopes (SIAs) of carbon, nitrogen and, to a lesser extent, sulphur and phosphorus, have emerged as powerful techniques for answering pattern and process-related questions in animal ecology, and for disentangling the occurrence and strength of trophic relationships. The isotope composition of animal tissue reflects its diet during periods of growth and tissue turnover and thus SIA is effective at integrating long-term assimilation of nutrients (Fry, 2006). Carbon stable isotopes can trace the sources of organic carbon sustaining consumer communities, provided that the primary carbon sources differ in their isotopic signatures. In addition, nitrogen stable isotopes are generally reliable indicators of the trophic level (or, more correctly, "height") of organisms, due to the pronounced and relatively constant fractionation that occurs between consumers and resources (Post, 2002). C and N isotopic signatures can be used to identify the fractional contribution of a number of potential food sources to a consumer's diet by means of mixing models based on mass balance equations (Phillips, 2001), although SIA lacks the taxonomic resolution of gut contents analysis: ideally these two complementary approaches should be used together (Layer *et al.*, 2011).

Several isotopic studies have provided a considerable contribution to the analysis of terrestrial food webs (e.g. Albers *et al.*, 2006; Bokhorst *et al.*, 2007; Hyodo *et al.*, 2010; Pollierer *et al.*, 2009; Ponsard and Arditi, 2000; Schmidt *et al.*, 2004). Moreover, SIA has clarified some critical issues specific to decomposer communities, for example, species-specific fungi–detritivore interactions (e.g. Chahartaghi *et al.*, 2005; Semenina and Tiunov, 2010 and literature cited therein). However, although soil biota are highly diverse, they have a relatively homogeneous isotopic baseline represented by a mixture of humus, leaf litter, root litter and root exudates: a problem that is shared with many freshwater food webs (Layer *et al.*, 2011).

Unfortunately, a crucial limit in the generalized application of SIA to the scrutiny of resource–consumer interactions remains; fractionation in both C and N isotopes is rather variable and depends on resource heterogeneity in terms of elemental content, C:N ratios, isotopic signature of the resources, and assimilation efficiency (Caut *et al.*, 2009). These sources of uncertainty requires, for the use of SIA to be effective in food web research, the adoption of complementary approaches, based on controlled laboratory experiments or biochemical/molecular techniques (Boschker and Middelburg, 2002; Gannes *et al.*, 1997; Martínez Del Rio *et al.*, 2009; Wolf *et al.*, 2009).

More recent investigations have highlighted the potential of joint DNA-stable isotope analysis in clarifying the structure of aquatic food web studies (Carreon-Martinez and Heath, 2010 and literature cited therein); the possibility of adopting similar approaches to soil food webs is, as yet, unexplored and represents a stimulating research field for future efforts.

In contrast, fatty acids (FAs) analysis has long been used in the study of marine food webs and represents one of the most promising tools to trace trophic interactions and provide long-term dietary information in soil food webs (Stott *et al.*, 1997). Ruess *et al.* (2002) and Chamberlain *et al.* (2005) have successfully used FA biomarkers to determine the feeding strategies and the food sources of nematode and collembola consumers, and Pollierer *et al.* (2010) verified the trophic transfer of marker FAs from basal fungal resources to top predators. However, the efficacy of FAs as trophic tracers in belowground food webs has been demonstrated when used in combination with carbon stable isotope analysis, since newly synthesized FAs reflect the average C signature of food resources, while FAs with dietary routing reflect the signature of FAs of the specific resource. FAs carbon isotopic signatures have been used to identify bacterial substrates and the flux of carbon in microbial communities (Abraham *et al.*, 1998; Arao, 1999), but a growing number of investigations is corroborating the effectiveness of this method for terrestrial and soil food webs (Chamberlain *et al.*, 2004; Haubert *et al.*, 2006, 2009; Ruess *et al.*, 2005; Ruess and Chamberlain, 2010).

## B. Integrating Intraspecific Size Variation into Soil Food Webs

The trait-centred ecological theory currently under development is cast within a conceptual framework assuming a critical correspondence between a species and its body size (see the synthesis provided by Allen *et al.*, 2006). In other words, body size is typically assumed to be a species-specific trait. Inter-phenotypic size variation is a ubiquitous feature of animal populations and is expected to influence species abundance and dynamics strongly (Kendall and Fox, 2002). However, a myriad of factors may combine to produce wide variations in intraspecific sizes and of demographic rates (see Huston and DeAngelis, 1987, for a review of proposed mechanisms; and Chown and Gaston, 2010; Fujiwara *et al.*, 2004; Pfister and Stevens, 2002; Teder *et al.*, 2008). As an instructive example, data from Mercer *et al.* (2001) on individual body sizes of common aboveground arthropod adults from an oceanic island are presented in Figure A3 and briefly compared with our data from a Dutch pilot study on soil nematodes (Figure 14). In our agroecosystems we detected a comparable level of variation of the individual body mass in the dominant nematode groups, with coefficients of variation of the intraspecific body size between 18.6% and 160%, after taking all their life stages and both genders into account (Mulder and Vonk, 2011). This means that, although high (the intraspecific coefficient of variation in the nematofauna fluctuates between 10% and 151%), size variations in agroecosystems for juvenile, female and male nematodes seem to be much lower than those occurring in nature for litter-inhabiting arthropods adults.

Intraspecific body-size variation exerts at least two key effects at the community level. (1) It determines fluctuations in the size distribution of populations over time. These fluctuations can be synchronous, as determined by seasonal-related abiotic and biotic constraints on biological cycles, or asynchronous, as determined by species-specific factors affecting sex ratios,

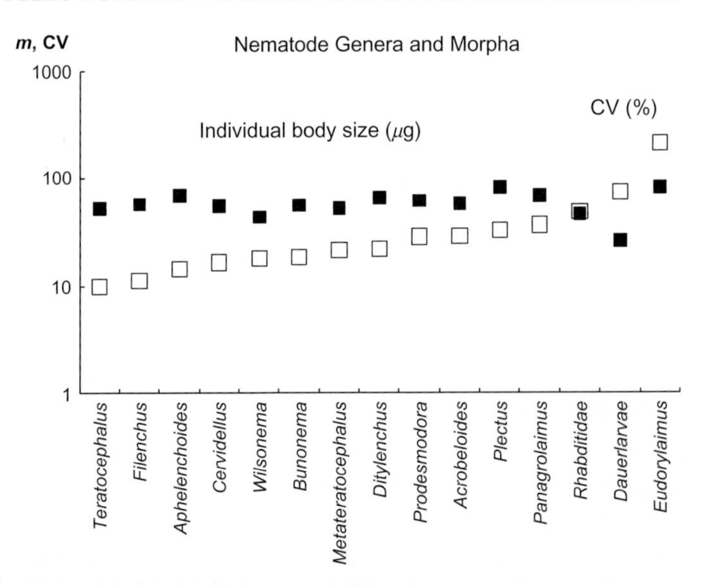


**Figure 14** Mean individual body sizes (μg dry weight; open square) and intraspecific CV (full square) of litterbag-grown nematodes. Resting stages and predators vary more, but to a lesser extent than the litter taxa of the meso- and macrofauna shown in Figure A3.

cannibalism, or intraspecific/intracohort competition. (2) It gives rise to an effect of inconstancy or even indeterminacy in prey–predator interactions due to size-specific differences in diet. Sexes, different age classes, morpha or individuals may specialize within a population at acquiring different resources, and an extensive literature on sexual dimorphisms (Slatkin, 1984), ontogenetic niche shifts (Polis, 1984; Werner and Gilliam, 1984), and resource polymorphisms (Smith and Skulason, 1996) demonstrates the high frequency and prominence of size-dependent intrapopulation variation in resource use (Bolnick *et al.*, 2002, 2003 and literature cited therein; Araujo *et al.*, 2007; Woodward *et al.*, 2010b). Current size-based food web investigations limit the resolution of the analysis at the scale of a species mean size overlooking its variance, disregarding what may represent both a strong limitation and a valuable complementary parameter for an effective use of body size as a tool for the interpretation of allometric scaling. Interestingly, intraspecific scaling of size-dependent processes with body size has also recently been shown to determine the slope of size spectra (Andersen *et al.*, 2009; Andersen and Beyer, 2006). No similar efforts have been performed for soil food webs; however, the dataset presented in this chapter has the potential to take on this issue and to verify it at an unprecedented scale.

# VIII. CAVEATS, CONCLUSIONS AND FUTURE DIRECTION

Scaling relationships help to disentangle the complexities of ecosystem functioning. An inverse allometric relationship between mass and abundance is well known and reflects how individual organisms within one species acquire and use resources as a function of their size. Major developments in the lineage originating from the metabolic theory of ecology (Brown *et al.*, 2004) using species composition, abundance, size, and mass have focused on energetic equivalence and trait distributions. True understanding of dynamic interplays within the soil biota is critical to recognize ecosystem stability and identify ecosystem services.

## A. Biodiversity Is the Ground Floor for Ecosystem Services

The disentangling of ecosystem processes has become an imperative goal in ecologygiven the intensity of changes to the biosphere (MA, 2005; Naeem *et al.*, 1995), although most investigations remain focused on primary producers, rather than the wider food web (Balvanera *et al.*, 2006). Classic examples of these studies are the model communities with controlled diversity levels in the Ecotron at Silwood Park (UK) and at the Cedar Creek Reserve in Minnesota (USA). However, primary producers are well known to be affected by both direct and indirect consequences of soil organisms in the belowground food web. Consequently, together with biological experiments, modeling of multitrophic systems has great potential for helping to understand many of the recently observed trends in the magnitude and direction of effects of human-induced changes on the soil biota and might assess the strong legacy of land-use history.

## B. What We Have Learned

Biodiversity might act according to the "Insurance Hypothesis" as a buffer to mitigate the effects of (anthropogenic) environmental perturbations upon ecosystem functioning (Loreau *et al.*, 2002; Yachi and Loreau, 1999): hence biodiversity is often seen as an ecosystem service in itself. Multimodality in the frequency distributions of traits like body size suggests that functional redundancy within the biota might provide a mechanism by which the insurance hypothesis may operate. According to differential species responses to environmental changes, a trait redundancy determines species life-history, material flow rates and energy conversion efficiencies, contributing to the ecosystem services, and, finally, maintaining stability.

## C. Towards a Universal Model?

The current mechanistic model (Figure 11) provided by allometric scaling, nutrient:C ratios and ecological stoichiometry held for all the open-canopies *ET*s. Since extremely shallow mass–abundance slopes have been reported in species-rich forested sites, where many larger (but rare) micro-arthropods dominate the primary consumer guild, it seems that the robustness of allometric scaling might decrease if too many monospecific genera occur (weakening the postulate of body mass providing the best independent sole predictor of abundance). And do rapid responses to increasing FOM (short-term pulses) in the occurrence of larger but rare micro-arthropods imply a lack of predictive power in their average body mass (in the case of forests, their average is based on the body mass of too few individuals)? Many approaches for approximating functional relationships exist, but the extent to which the synergy between individual-based biological stoichiometry and community-based ecological stoichiometry and the resource quality and the maximal growth efficiency or the intraspecific body-size variation (as described in Sections VI.B and VII.B, respectively) implement or generalize allometric scaling, could provide a "truly universal" remains to be seen. If this can be achieved, even partially (e.g. by contributing to a better fit to the data) it could represent a significant move towards the Metabolic Theory of Ecology proposed by Brown *et al.* (2004) and advocated by Allen and Gillooly (2009), amongst others.

## D. Elemental Content of Organisms

To define a comprehensive model for all terrestrial *ET*s, recent insights from parallel research conducted in aquatic ecology could help to provide powerful tools, especially when considering links between basal resources and primary consumers. For instance, bacterial biomass C:N ratio is, according to Cotner *et al.* (2010), relatively static, in contrast to the highly variable bacterial biomass C:P ratio, and it is possible that this might hold for protozoans too. Since soil total C:P is the best predictor for the allometric scaling for terrestrial detritivores (especially among the mesofauna, such as enchytraeids), the flux of P across trophic levels might provide a critical measure of overall ecosystem functioning. While many of these stoichiometric ratios have clear ecological significance, far more work is now needed to establish those that are important in the elucidation of the mechanisms behind the (provision of) ecosystem services.

## E. Data Paucity for Smaller Taxa

Plants with the greatest numerical abundances or biomass are often have the largest effects on ecosystem biogeochemistry (Grime, 1998), and this paradigm of biodiversity–ecosystem functioning research may well also hold for other kingdoms. If it does, seen the negative correlation between numerical abundance and body size, small organisms belonging to different kingdoms (such as bacteria, animal-like protozoa and nematodes) might have large effects on soil nutrient cycling. On the other hand, the inclusion of bacteria and protozoa in allometric models might introduce a bias due to the kind of aggregate data. If the average body mass is plotted in $\mu$g, $\log(\bar{m})=0$ represents the boundary between bacterial cells, protozoa and nematodes (microfauna) ($\log(\bar{m})<0$) and the mycelium, the mesofauna and the macrofauna ($\log(\bar{m})>0$); in that case, the effect of aggregation bias will be much stronger in the negative range of $\log(\bar{m})$, contributing to explain the different allometric scaling of single-celled organisms and larger multicellular organisms (Reiss *et al.*, 2010).

## F. Plea for a Missing Kingdom

The current knowledge on soil protozoa exists in The Netherlands only for some sites on fluvisols or cambisols, and biogeographical studies are necessary to investigate habitat–response relationships of protozoa. Frequency distributions of most testate amoebae (Rhizopoda) suggest a strong positive correlation with acid litter and oligo- up to mesotrophic conditions in podzols (Mulder and Janssen, 1999). But do other protozoa correlate with soil pH (directly, as bacteria, nematodes and testate amoebae do, or inversely, as fungi and micro-arthropods do)? And, if allometrically plotted, do protozoa increase the linearity of the scaling between $\log(\bar{m})$ and $\log(n)$ for prokaryotes and other eukaryotes?

## G. Conserving Ecosystem Services

Biodiversity and agricultural pressure are closely and negatively related, and a comprehensive survey of *all* the consequences of increasing farming pressure (resulting in changes in *ET*) for soil biodiversity and ecological services remains an important goal, particularly in Europe's heavily modified landscape. Although there seems to be a remarkable regularity over a size range of 15 orders of magnitude (from bacteria to earthworms), subtle changes in allometric scaling relationships show clear faunal responses to nutrient availability and strong effects of the soil biota on ecosystem services are emerging. The application of food web modeling can contribute, in an analogous way to the use of size spectra in commercial fisheries (Jennings, 2005; Petchey and Belgrano, 2010), to prevent ecosystem degradation or retrogression and to predict the rate of changes in soil processes and associated ecosystem services: this would effectively create a new way of applying theoretical ecology to address real-world problems.

**Box 1**

**Is ecology a set of contingent case studies or do universal laws apply?**

A paucity of data is increasingly claimed. However, we have never had such a plethora of information available as now. Borne (2008) states that "The more we know, the more we are driven to tweak or to revolutionize our models, thereby advancing our scientific understanding." Mahootian and Eastman (2009) described this historical shift in the conceptual process of investigation, from their theory-driven hypothetico-deductive framework, with all its logical implications like the classical "je pense donc je suis", later *Ego cogito, ergo sum* (Descartes, 1637, 1644, respectively), up to the hypothetico-inductive (balancing theory and empirical data at different levels) and observational-inductive frameworks. Internet access, intellectual networking, international research programmes and the vast amount of data have converged in the latter framework. Essentially, the traditional modes of empirical fieldwork and laboratory research and related descriptive assessment within the hypothetico-inductive framework were forced to change by surges in data increased volume, computing power and availability into an observational-inductive framework more closely focused on understanding causal relationships in the real world (Eastman *et al.*, 2005; Mahootian and Eastman, 2009; Sagarin and Pauchard, 2010). These frameworks reflect the strong dichotomy that arose between perception and calculation, even to the extent that mathematicians started to declare every argumentation not submitted to calculation as being potentially inaccurate or even invalid. Ecology progresses by testing fundamental conceptualizations of patterns and trends, which although they vary in complexity (Madin *et al.*, 2008), are essentially rooted within this Observational-Inductive framework. Although we may argue as to what the most appropriate kind of data-organisation might be, data-sharing always allows (and demands) greater insights. Databases published as supplementary files in the

(continued)

**Box 1 (continued)**

online section of peer-reviewed journals, in fact, are faced with another—still unmentioned—problem: the "publication bias" (Begg and Mazumdar, 1994; Rosenthal, 1979), also known as "file drawer effect" (Rosenthal, 1979). This tendency to report the results that are positive (i.e. "significant") differently from the results that are negative is widespread in pharmacological and biomedical worlds (Ioannidis, 2005; Moonesinghe *et al.*, 2007; Sutton *et al.*, 2000) and seems to occur in fundamental and applied sciences as well. Despite trait-based researches became very popular through studies on body-mass development (life-history experiments) in laboratory and in space and time (even as investigations at geological scales), debates regarding allometric scaling and biodiversity–ecosystem functioning possibly resemble this kind of bias. But we are also seeing, especially within the large-scale collaborative research programmes that span several European countries, a greater willingness and ability to exchange data openly, and this may continue to accelerate the rate of theoretical development within applied ecology in the near future.

## ACKNOWLEDGEMENTS

We thank Andrea Belgrano, Julia Blanchard, Josè Montoya and Guy Woodward for the organisation of the European Science Foundation SIZEMIC meeting in Barcelona in 2010 and Philippe Lemanceau for the coordination of the new EcoFINDERS, both projects which supported our work. We are grateful for helpful feedbacks to Ton Breure, Dick De Zwart, Jim Elser, Hans Helder, Mike Kaspari, Robert Middelveld, Shigeta Mori, Han Olff and Leo Posthuma. We also wish to thank Guy Woodward and an anonymous referee for constructive suggestions to improve further our work. All members of our program are credited with their contributions in our research, but we would particularly like to thank J. Bloem, J. Bogte, R. G. M. De Goede, H. A. Den Hollander, W. J. Dimmers, A. J. Schouten and N. J. M. Van Eekeren.

Here is no water but only rock
Rock and no water and the sandy road
T. S. Eliot, *The Waste Land*

# APPENDIX

## Overview of the used DSQN database

**Table A1** Main environmental measurements (with database entries)

| |
|---|
| Sampling date computed as Julian Days, that is, number of days that have elapsed each year since March, 1 (1069), Geographical coordinates (1050), Soil pH of oven-dried soil samples as measured in 1 M KCl (1001), SOM, soil organic matter as % dry weight (1001), Phosphate content after acetate-lactate (Al) buffer extraction including P in oxides on soil particles and in water-insoluble compounds (985), *lutum* fraction of all soil particles Ø < 2 μm (940), Phosphate content after water extraction at water-to-soil ratio 60:1 by volume (901), [Cr] (833), [Pb] (833), Total Soil Phosphorus (814), Soil pH in water (808), [Cu] (803), [Zn] (791), [Cd] (782), [Hg] (734), Total Soil Nitrogen (664), [Ni] (500), Total Soil Carbon [C] (463), [K] (375), Percentage cover grasslands (260), Percentage cover maize (242), [Mn] (236), Plant available phosphate after 0.01 M $CaCl_2$ extraction (226), Percentage cover croplands (205), [As] (196), [Fe] (194), Cation Exhange Capacity ($cmol^+$/kg dry weight) (194), Anthracene (194), Benzo(a)anthracene (194), Benzo(a)pyrene (194), Benzo(b)fluoranthene (194), Benzo(g,h,i)perylene (194), Benzo(k)fluoranthene (194), Chrysene (194), Fluoranthene (194), Indeno(1,2,3-cd)pyrene (194), Phenanthrene (194), Pyrene (194), Dibenz(a,h)anthracene (189), Average air temperatures for the latest 21 days before sampling (170), Average precipitation for the latest 21 days before sampling (170), Livestock units (167), *p,p′*-Dichlorodiphenyldichloroethylene (157), *p,p′*-Dichlorodiphenyltrichloroethane (138), Hexachlorobenzene (133), Lindane (108), Dieldrin (106), *o,p′*-Dichlorodiphenyltrichloroethane (93), *p,p′*-Dichlorodiphenyldichloroethane (81), Daily air temperatures (minima, maxima and averages) for the latest 28 days before sampling (79), Precipitation (minima, maxima and averages) for the latest 28 days before sampling (79), Fluorene (75), beta-Hexachlorocyclohexane (53), beta-Heptachlorepoxide (50) |

**Table A2** Bacterial and fungal trophospecies: monitored ecophysiological traits with entries

| |
|---|
| Bacterial growth rate by incorporation of $^{14}$C-leucine into proteins (1816), Bacterial growth rate by incorporation of $^{3}$H-thymidine into bacterial DNA (1816), Bacterial biomass μgC/g (1811), N mineralization (1750), $CO_2$ evolution (respiration) of soil microbiota (1694), $O_2$ consumption of soil microbiota (1676), Metabolic Quotient $qCO_2$ (1676), Numerical abundance of bacterial cells billions /g dry soil (1665), Thymidine to leucine ratio (1508), Average cell volume $\mu m^3$ (1470), Average cell shape (1470), Frequency of Dividing Cells % (1470), Thymidine (pmol):bacterial carbon (μgC h) ratio (1377), Leucine (pmol):bacterial carbon (μg C h) ratio (1377), Fungal hyphal length m/g soil (905), Fungal biomass μgC/g soil (905), Active fungi as percentage of hyphal length (880), Fungal C to bacterial C ratio (835), Unstained fungi (820), Potentially mineralizable N (816), Bacterial DNA banding patterns as analysed and quantified by image analysis (807), Bacterial DNA Shannon-Wiener Index (807), Bacterial DNA Evenness (807), Theoretical average response in BIOLOG (759), Colony Forming Units (759) |

**Table A3** All oligochaete taxa recorded in more than one location (with entries)

*Aporrectodea* (595), *Enchytraeus* (584), *Fridericia* (572), *Lumbricus* (519), *Henlea* (472), *Marionina* (437), Lumbricidae undiff (368), *Achaeta* (285), Lumbricidae with non-tanylobus prostomium (250), Juvenile enchytraeidae (212), Lumbricidae with tanylobous prostomium (204), *Cognettia* (124), *Enchytronia* (55), *Satchellius* (33), *Buchholzia* (32), *Eiseniella* (31), *Octolasion* (31), *Dendrobaena* (26), *Eisenia* (10), *Bryodrilus* (4), *Lumbricillus* (4), *Hemifridericia* (3), *Mesenchytraeus* (2)

**Table A4** All arthropod taxa recorded in more than one location (with entries)

*Eupodes* (574), *Lysigamasus* (574), *Lepidocyrtus* (571), *Isotoma* (563), *Pygmephorus* (514), *Scutacarus* (448), *Dendrolaelaps* (417), *Sminthurinus* (400), *Parisotoma* (357), *Alliphis* (348), Sminthuridae (331), *Sphaeridia* (319), Tydeidae (319), *Tarsonemus* (312), *Microtydeus* (308), *Proisotoma* (305), *Isotomiella* (303), *Tyrophagus* (301), *Arctoseius* (289), *Sminthurus* (268), *Isotomurus* (264), *Pergamasus* (253), *Hypoaspis* (246), *Cheiroseius* (241), *Mesaphorura* (227), Uropoda (218), *Ceratophysella* (210), *Friesea* (194), *Parasitus* (193), *Oppiella* (181), *Tectocepheus* (179), *Microppia* (168), *Hypogastrura* (160), *Pachylaelaps* (159), *Macrocheles* (158), Stigmaeidae (155), *Rhodacarellus* (149), *Medioppia* (145), *Nanorchestes* (143), *Folsomia* (126), *Brachystomella* (125), *Bdella* (121), *Onychiurus* (117), *Rhizoglyphus* (115), Mesostigmata (109), *Rhagidia* (108), *Histiostoma* (103), Oribatida (103), *Megalothorax* (100), Pachygnatidae (93), *Veigaia* (87), *Platynothrus* (86), Symphyla (80), *Pyemotes* (74), *Rhodacarus* (74), *Suctobelbella* (72), *Achipteria* (71), Astigmata (64), *Scheloribates* (62), Pauropoda (61), *Punctoribates* (59), *Eupelops* (58), Brachychthoniidae (55), *Schwiebea* (54), *Galumna* (50), *Nothrus* (49), *Hypochthonius* (46), *Sminthurides* (43), *Speleorchestes* (43), *Microtritia* (42), *Rhysotritia* (42), *Minunthozetes* (41), *Dendroseius* (38), *Ceratozetes* (37), Protura (37), *Stenaphorurella* (36), *Pseudachorutes* (35), *Liebstadia* (34), *Tydeus* (33), *Atropacarus* (31), *Lasioseius* (31), *Brachychthonius* (29), *Liochthonius* (29), *Banksinoma* (26), *Pantelozetes* (26), *Quadroppia* (26), *Asca* (25), *Phthiracarus* (24), *Steganacarus* (23), *Sellnickochthonius* (22), *Eniochthonius* (21), Ereynetidae (21), *Paratullbergia* (21), Penthalodidae (20), *Trichoribates* (19), *Arrhopalites* (16), *Zercon* (16), *Ceratoppia* (15), *Damaeobelba* (15), *Entomobrya* (15), *Tomocerus* (15), Trombidiidae (15), Erythraeidae (14), *Neojordensia* (14), *Adoristes* (13), *Epicriopsis* (13), *Siteroptes* (13), *Spatiodamaeus* (13), *Neotullbergia* (12), Oppiidae (11), Pachygnathidae (11), *Pseudisotoma* (11), *Trachytes* (11), *Amblyseius* (10), *Carabodes* (10), Diplura (10), *Dissorhina* (10), *Hemileius* (10), *Leioseius* (10), *Palaeacarus* (10), *Protodinychus* (10), *Chamobates* (9), Entomobryidae (9), Eriophyidae (9), *Odontocepheus* (9), *Uroseius* (9), *Dicyrtoma* (8), *Eugamasus* (8), *Gamasodes* (8), *Nanhermannia* (8), Anurida (7), *Diapterobates* (6), *Hermannia* (6), *Heteromurus* (6), *Labidostomma* (6), *Lauroppia* (6), *Neanura* (6), *Oribatula* (6), Rhodacaridae (6), Eupodidae (5), *Holoparasitus* (5), *Leptus* (5), *Parazercon* (5), *Pergalumna* (5), Prostigmata (5), *Protaphorura* (5), *Pseudoparasitus* (5), Belbidae (4), *Ctenobelba* (4), *Dinychus* (4), *Euzetes* (4), *Hermanniella* (4), Johnstonianidae (4), *Liacarus* (4), *Neelides* (4), *Orchesella* (4), *Trachyuropoda* (4), *Uroobovella* (4), *Adamaeus* (3), *Ameroseius* (3), *Antennoseius* (3), Anystidae (3), *Blattisocius* (3), *Damaeus* (3), *Epicrius* (3), *Gamasolaelaps* (3), *Geholaspis* (3), *Humerobates* (3), *Lipothrix* (3), *Peloptulus* (3), *Pseudolaelaps* (3), *Zercoseius* (3), *Berniniella* (2), *Camisia* (2), *Cilliba* (2), Cunaxidae (2), *Demodex* (2), *Erythraeus* (2), *Eulohmannia* (2), Eupalopsellidae (2), *Hypogeoppia* (2), *Ledermuelleria* (2), *Malaconothrus* (2), *Metabelba* (2), *Nenteria* (2), *Ophidiotrichus* (2), *Oribatella* (2), *Parasitidae* (2), *Phauloppia* (2), *Podocinum* (2), *Pseudosinella* (2), *Raphignathus* (2), *Scutovertex* (2), Uropodina (2), *Vertagopus* (2), *Vulgarogamasus* (2), *Willemia* (2)

**Table A5** All nematode taxa recorded in more than one location (with entries)

*Eucephalobus* (2559), *Tylenchorhynchus/Bitylenchus* (1432), *Pratylenchus* (1252), *Plectus* (1080), *Helicotylenchus* (1072), *Paratylenchus* (1031), *Aphelenchoides* (999), *Panagrolaimus* (983), Tylenchidae (976), *Acrobeloides* (944), Rhabditidae (800), *Anaplectus* (786), Dorylaimoidea (780), *Aporcelaimellus* (754), Dauerlarvae (697), Dolichodoridae (570), *Acrobeles* (538), *Meloidogyne* (537), *Filenchus* (522), *Prismatolaimus* (492), Neodiplogasteridae (456), Cephalobidae (387), *Aglenchus* (382), *Aphelenchus* (367), *Heterodera* (357), *Trichodorus* (344), *Teratocephalus* (312), *Malenchus* (310), *Alaimus* (297), *Cephalobus* (287), *Cervidellus* (265), *Eumonhystera* (244), *Metateratocephalus* (233), Thornenematidae (232), Qudsianematidae (220), *Tripyla* (199), *Eudorylaimus* (186), *Rhabditis* (185), *Diphtherophora* (171), *Ditylenchus* (139), Mononchidae (138), Monhysteridae (134), *Wilsonema* (130), *Clarkus* (127), *Psilenchus* (126), *Coslenchus* (122), Plectidae (114), *Tylolaimophorus* (113), *Paratrichodorus* (109), *Rotylenchus* (109), Dorylaimidae (103), *Longidorus* (102), *Pungentus* (102), *Tylenchus* (102), *Mesodorylaimus* (100), *Mylonchulus* (100), Hoplolaimidae (99), *Mononchus* (97), *Heterocephalobus* (92), *Bastiania* (91), *Prodorylaimus* (88), *Diploscapter* (87), Criconematidae (85), *Achromadora* (83), Diplogasteridae (82), *Amplimerlinius* (81), Panagrolaimidae (74), Trichodoridae (63), *Hemicycliophora* (61), *Dolichorhynchus* (59), *Paramphidelus* (56), Anguinidae (46), *Anatonchus* (43), *Seinura* (41), *Protorhabditis* (38), *Xiphinema* (38), *Chiloplacus* (37), *Bursilla* (34), *Cephalenchus* (34), *Epidorylaimus* (32), *Tylopharynx* (31), *Boleodorus* (25), *Bunonema* (21), *Chronogaster* (21), *Thonus* (21), *Tylencholaimus* (21), *Ecumenicus* (19), Pratylenchidae (19), Teratocephalidae (19), *Pristionchus* (18), *Rotylenchulus* (18), *Laimydorus* (17), *Merlinius* (16), *Cuticularia* (15), *Lelenchus* (15), *Scutylenchus* (15), *Discolaimus* (14), *Quinisulcius* (13), *Trophurus* (12), Alaimidae (11), Chromadoridae (11), *Labronema* (11), *Euteratocephalus* (10), *Monhystera* (10), *Rhabdolaimus* (10), *Ecphyadophora* (9), *Gracilacus* (9), *Cylindrolaimus* (8), Heteroderidae (8), *Mesorhabditis* (8), Myodonomidae (8), *Neopsilenchus* (8), *Acrobelophis* (7), *Nagelus* (7), Steinernematidae (7), *Thornia* (7), *Amphidelus* (6), Aphelenchoididae (6), *Domorganus* (6), *Drilocephalobus* (6), *Microdorylaimus* (6), *Pleurotylenchus* (6), *Prionchulus* (6), *Prodesmodora* (5), *Longidorella* (4), Nordiidae (4), *Pseudhalenchus* (4), *Theristus* (4), *Tylencholaimellus* (4), *Acrolobus* (3), *Coarctadera* (3), *Diplogaster* (3), *Geocenamus* (3), *Macrotrophurus* (3), *Microlaimus* (3), *Mononchoides* (3), *Tylocephalus* (3), Achromadoridae (2), Aphelenchidae (2), Aporcelaimidae (2), *Aporcelaimus* (2), Bastianiidae (2), *Butlerius* (2), Chromadorida (2), *Coomansus* (2), *Dorydorella* (2), Longidoridae (2), *Neothada* (2), *Nothotylenchus* (2), Paratylenchidae (2), *Pareudiplogaster* (2), *Pelodera* (2), Tripylidae (2)

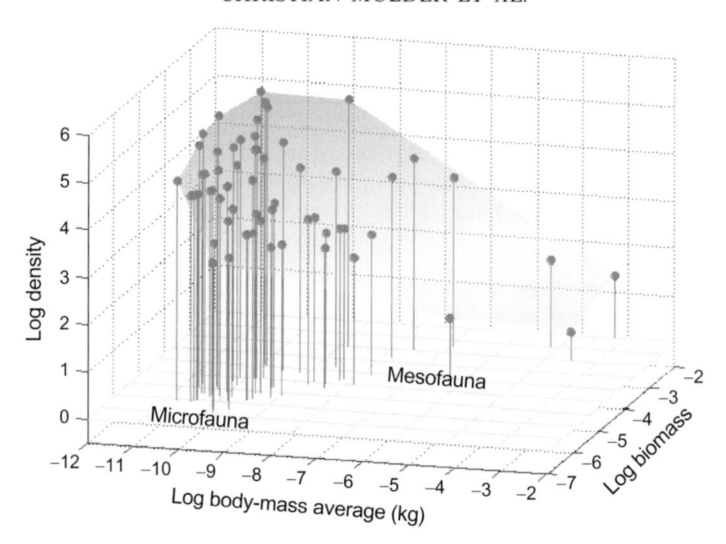


**Figure A1** 3D-scatterplot of one real community of soil invertebrates, showing the skewed distribution of smaller organisms along the body-mass average ($x$), biomass ($y$) and density ($z$) axes. The mass–abundance slope for the entire community (linear regression of all metazoans and trophospecies as unbinned log abundance on log mass) is $-0.7892$ (normal probability plot $p_{\mathrm{jb}}=0$; $p_{\mathrm{l}}<0.01$) and the size spectrum slope (size-binned biomass regression slope) is 0.35 (confidence interval 0.06–0.65) instead of the 0.21 regression slope expected from statistical theory: merging $\log(z)=\beta\times\log(x)+\alpha$ with $\log(y)=\log(x)+\log(z)$, we obtain $\log(y)=\log(x)+\beta\times\log(x)+\alpha=(1+\beta)\times\log(x)+\alpha$, thus these slopes should differ 1. This difference is due to the skewed distribution of real species (not trophic species) across a $\bar{m}$ gradient. Although the occurrence of few abundant and many rare species is already well known from macroecology, this surface response highlights these contrasting aspects of real soil biota.

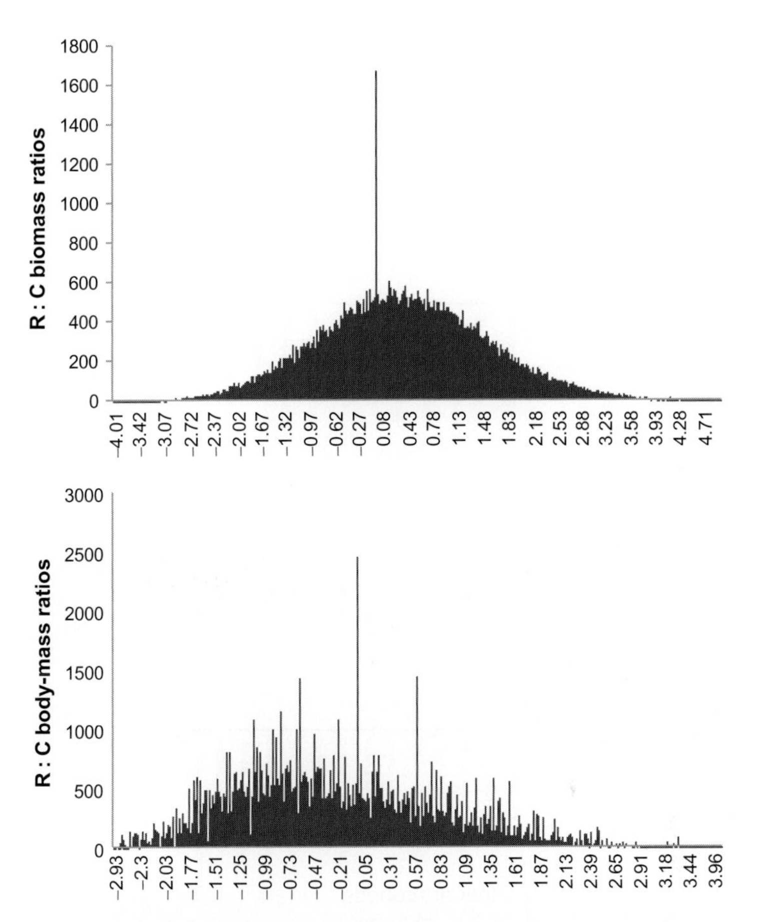


**Figure A2** Frequency distribution of the log-transformed resource ($R$) to consumer ($C$) biomass and body-mass ratios, showing the dominance of omnivory in our soil systems in contrast to the previous results of Brose *et al.* (2006); log-transformed $R$:$C$ ratios equal to 0 are cannibalistic links.

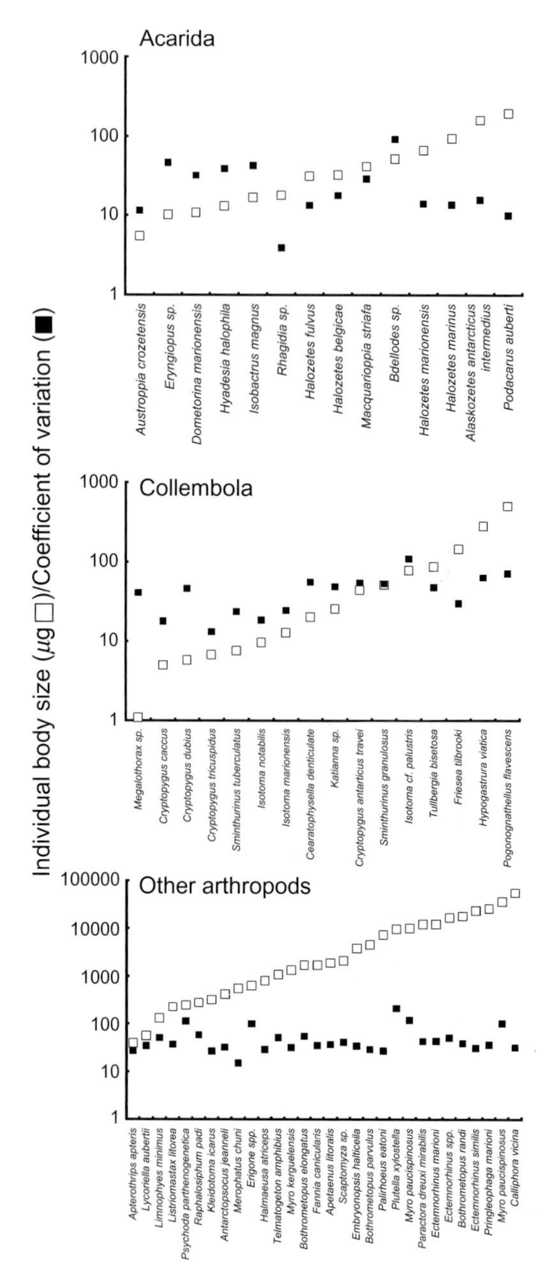


**Figure A3** Mean individual body sizes (adult fresh mass in $\mu$g; open square) and intraspecific coefficient of variation (full square) of dominant litter arthropods. For each group, taxa were ranked according to their increasing weight. Graphs based on Mercer *et al.* (2001) for Marion Island (data sampled between 1986 and 1999). Independently from the considered group, individual body masses varied considerably, with CVs in the range between 3.9% and 210%. Here, intraspecific ratios between the largest and the smallest collembolans fluctuate between 3 and 8, or even 10. Soil nematodes fluctuate much less, with a mean CV = 62% (Figure 14).

## REFERENCES

Abraham, W.R., Hesse, C., and Pelz, O. (1998). Ratios of carbon isotopes in microbial lipids as an indicator of substrate usage. *Appl. Environ. Microbiol.* **64**, 4202–4209.

Adl, S.M., Simpson, A.G., Farmer, M.A., Andersen, R.A., Anderson, O.R., Barta, J.R., Bowser, S.S., Brugerolle, G., Fensome, R.A., Fredericq, S., James, T.Y., Karpov, S., *et al.* (2005). The new higher level classification of Eukaryotes with emphasis on the taxonomy of Protists. *J. Eukaryot. Microbiol.* **52**, 399–451.

Albers, D., Schaefer, M., and Scheu, S. (2006). Incorporation of plant carbon into the soil animal food web of an arable system. *Ecology* **87**, 235–245.

Allen, A.P., and Gillooly, J.F. (2009). Towards an integration of ecological stoichiometry and the metabolic theory of ecology to better understand nutrient cycling. *Ecol. Lett.* **12**, 369–384.

Allen, C.R., Garmestani, A.S., Havlicek, T.D., Marquet, P.A., Peterson, G.D., Restrepo, C., Stow, C.A., and Weeks, B.E. (2006). Patterns in body mass distributions: Sifting among alternative hypotheses. *Ecol. Lett.* **9**, 630–643.

Andersen, K.H., and Beyer, J.E. (2006). Asymptotic size determines species abundance in the marine size spectrum. *Am. Nat.* **168**, 54–61.

Andersen, T., Elser, J., and Hessen, D. (2004). Stoichiometry and population dynamics. *Ecol. Lett.* **7**, 884–900.

Andersen, K.H., Beyer, J.E., and Lundberg, P. (2009). Trophic and individual efficiencies of size-structured communities. *Proc. R. Soc. Lond. B* **276**, 109–114.

Andrén, O. (1988). Ecology of arable land—An integrated project. *Ecol. Bull.* **39**, 131–133.

Arao, T. (1999). In situ detection of changes in soil bacterial and fungal activities by measuring $^{13}C$ incorporation into soil phospholipid fatty acids from $^{13}C$ acetate. *Soil Biol. Biochem.* **31**, 1015–1020.

Araujo, M.S., Bolnick, D.I., Machado, G., Giaretta, A.A., and Dos Reis, S.F. (2007). Using $\delta^{13}C$ stable isotopes to quantify individual-level diet variation. *Oecologia* **152**, 643–654.

Baer, S.G., Blair, J.M., Collins, S.L., and Knapp, A.K. (2004). Plant community responses to resource availability and heterogeneity during restoration. *Oecologia* **139**, 617–629.

Bakken, L.R., and Olsen, R.A. (1983). Buoyant densities and dry–matter contents of microorganisms: Conversion of a measured biovolume into biomass bacteria and fungi from soil. *Appl. Environ. Microbiol.* **45**, 1188–1195.

Balvanera, P., Pfisterer, A.B., Buchmann, N., He, J.-S., Nakashizuka, T., Raffaelli, D., and Schmid, B. (2006). Quantifying the evidence for biodiversity effects on ecosystem functioning and services. *Ecol. Lett.* **9**, 1146–1156.

Bardgett, R. (2005). *The Biology of Soil: A Community and Ecosystem Approach.* Oxford University Press, Oxford.

Beerling, D.J., and Woodward, F.I. (2001). *Vegetation and the Terrestrial Carbon Cycle: Modelling the First 400 Million Years.* Cambridge University Press, Cambridge, UK.

Begg, C.B., and Mazumdar, M. (1994). Operating characteristics of a rank correlation test for publication bias. *Biometrics* **50**, 1088–1101.

Benchaar, C., Rivest, J., Pomar, C., and Chiquette, J. (1998). Prediction of methane production from dairy cows using existing mechanistic models and regression equations. *J. Anim. Sci.* **76**, 617–627.

Bennett, E.M., Peterson, G.D., and Gordon, L.J. (2009). Understanding relationships among multiple ecosystem services. *Ecol. Lett.* **12**, 1–11.

Blanchard, J.L., Jennings, S., Law, R., Castle, M.D., McCloghrie, P., Rochet, M.-J., and Benoît, E. (2009). How does abundance scale with body size in coupled size-structured food webs? *J. Anim. Ecol.* **78**, 270–280.

Bloem, J., and Breure, A.M. (2003). Microbial indicators. In: *Bioindicators and Biomonitors: Principles, Concepts and Applications* (Ed. by B.A. Markert, A.M. Breure and H.G. Zechmeister), pp. 259–282. Elsevier Science, Oxford.

Bloem, J., Lebbink, G., Zwart, K.B., Bouwman, L.A., Burgers, S., Devos, J.A., and De Ruiter, P.C. (1994). Dynamics of microorganisms, microbivores and nitrogen mineralization in winter wheat fields under conventional and integrated management. *Agric. Ecosyst. Environ.* **51**, 129–143.

Bloem, J., Veninga, M., and Shepherd, J. (1995). Fully automatic determination of soil bacterium numbers, cell volumes and frequencies of dividing cells by confocal laser scanning microscopy and image analysis. *Appl. Environ. Microbiol.* **61**, 926–936.

Bobbink, R., Hornung, M., and Roelofs, J.G.M. (1998). The effects of air-borne nitrogen pollutants on species diversity in natural and semi-natural vegetation. *J. Ecol.* **86**, 717–738.

Bokhorst, S., Ronfort, C., Huiskes, A., Convey, P., and Aerts, R. (2007). Food choice of Antarctic soil arthropods clarified by stable isotope signatures. *Polar Biol.* **30**, 983–990.

Bolnick, D.I., Yang, L.H., Fordyce, J.A., Davis, J.M., and Svanback, R. (2002). Measuring individual-level resource specialization. *Ecology* **83**, 2936–2941.

Bolnick, D.I., Svanback, R., Fordyce, J.A., Yang, L.H., Davis, J.M., Hulsey, C.D., and Forister, M.L. (2003). The ecology of individuals: Incidence and implications of individual specialization. *Am. Nat.* **161**, 1–28.

Bonkowski, M. (2004). Soil protozoa and plant growth—The microbial loop in soil revisited. *New Phytol.* **162**, 617–631.

Bonkowski, M., Griffiths, B., and Scrimgeour, C. (2000). Substrate heterogeneity and microfauna in soil organic 'hotspots' as determinants of nitrogen capture and growth of ryegrass. *Appl. Soil Ecol.* **14**, 37–53.

Borne, K.D. (2008). Scientific data mining in astronomy. In: *Next Generation of Data Mining* (Ed. by H. Kargupta, J. Han, P.S. Yu, R. Motwani and V. Kumar), pp. 91–114. Chapman & HallCRC Data Mining and Knowledge Discovery Series. [arXiv:0911.0505v1].

Boschker, H.T.S., and Middelburg, J.J. (2002). Stable isotopes and biomarkers in microbial ecology. *FEMS Microbiol. Ecol.* **40**, 85–95.

Bouwman, L.A., and Zwart, K.B. (1994). The ecology of bacterivorous protozoans and nematodes in arable soil. *Agric. Ecosyst. Environ.* **51**, 145–160.

Brose, U., Jonsson, T., Berlow, E.L., Warren, P., Banasek-Richter, C., Bersier, L.-F., Blanchard, J.L., Brey, T., Carpenter, S.R., Cattin Blandenier, M.F., Cushing, L., Dawah, H.A., *et al.* (2006). Consumer–resource body-size relationships in natural food webs. *Ecology* **87**, 2411–2417.

Brown, J.H., and Gillooly, J.F. (2003). Ecological food webs: High-quality data facilitate theoretical unification. *Proc. Natl. Acad. Sci. USA* **100**, 1467–1468.

Brown, J.H., Gillooly, J.F., Allen, A.P., Savage, V.M., and West, G.B. (2004). Toward a metabolic theory of ecology. *Ecology* **85**, 1771–1789.

Brussaard, L. (1994). An appraisal of the Dutch Programme on Soil Ecology of Arable Farming Systems (1985–1992). *Agric. Ecosyst. Environ.* **51**, 1–6.

Brussaard, L., Behan-Pelletier, V.M., Bignell, D.E., Brown, V.K., Didden, W., Folgarait, P., Fragoso, C., Freckman, D.W., Gupta, V.V.S.R., Hattori, T.,

Hawksworth, D.L., Klopatek, C., *et al.* (1997). Biodiversity and ecosystem functioning in soil. *Ambio* **26**, 563–570.

Bulman, S.R., Kühn, S.F., Marshall, J.W., and Schnepf, E. (2001). A phylogenetic analysis of the SSU rRNA from members of the Plasmodiophorida and Phagomyxida. *Protist* **152**, 43–51.

Carpenter, S.R., Mooney, H.A., Agard, J., Capistrano, D., DeFries, R.S., Díaz, S., Dietz, T., Duraiappah, A.K., Oteng-Yeboah, A., Pereira, H.M., Perrings, C., Reid, W.V., *et al.* (2009). Science for managing ecosystem services: Beyond the Millennium Ecosystem Assessment. *Proc. Natl. Acad. Sci. USA* **106**, 1305–1312.

Carreon-Martinez, L., and Heath, D.D. (2010). Revolution in food web analysis and trophic ecology: Diet analysis by DNA and stable isotope analysis. *Mol. Ecol.* **19**, 25–27.

Caut, S., Angulo, E., and Courchamp, F. (2009). Variation in discrimination factors ($\Delta^{15}N$ and $\Delta\ ^{13}C$): The effect of diet isotopic values and applications for diet reconstruction. *J. Appl. Ecol.* **46**, 443–453.

Chahartaghi, M., Langel, R., Scheu, S., and Ruess, L. (2005). Feeding guilds in Collembola based on nitrogen stable isotope ratios. *Soil Biol. Biochem.* **37**, 1718–1725.

Chamberlain, P.M., Bull, I.D., Black, H.I.J., Ineson, P., and Evershed, R.P. (2004). Lipid content and carbon assimilation in Collembola: Implication for the use of compound-specific carbon isotope analysis in animal dietary studies. *Oecologia* **139**, 325–335.

Chamberlain, P.M., Bull, I.D., Black, H.I.J., Ineson, P., and Evershed, R.P. (2005). Fatty acid composition and change in Collembola fed differing diets: Identification of trophic biomarkers. *Soil Biol. Biochem.* **37**, 1608–1624.

Chown, S.L., and Gaston, K.J. (2010). Body size variation in insects: A macroecological perspective. *Biol. Rev.* **85**, 139–169.

Clarholm, M. (1985). Possible roles of roots, bacteria, protozoa, and fungi in supplying nitrogen to plants. In: *Ecological Interactions in Soil: Plants, Microbes, and Animals* (Ed. by A.H. Fitter, D. Atkinson, D.J. Read and M.B. Usher), pp. 355–365. Blackwell Scientific Publications, Oxford.

Clark, C.M., Cleland, E.E., Collins, S.L., Fargione, J.E., Gough, L., Gross, K.L., Pennings, S.C., Suding, K.N., and Grace, J.B. (2007). Environmental and plant community determinants of species loss following nitrogen enrichment. *Ecol. Lett.* **10**, 596–607.

Cleveland, C.C., and Liptzin, D. (2006). C:N:P stoichiometry in soil: Is there a "Redfield ratio" for the microbial biomass? *Biogeochemistry* **85**, 235–252.

Cohen, J.E., Łuczak, T., Newman, C.M., and Zhou, Z.-M. (1990). Stochastic structure and nonlinear dynamics of food webs: Qualitative stability in a Lotka–Volterra cascade model. *Proc. R. Soc. Lond. B* **240**, 607–627.

Cohen, J.E., Jonsson, T., and Carpenter, S.R. (2003). Ecological community description using the food web, species abundance, and body size. *Proc. Natl. Acad. Sci. USA* **100**, 1781–1786.

Costanza, R., and Pattern, B.C. (1995). Defining and predicting sustainability. *Ecol. Econ.* **15**, 193–196.

Costanza, R., D'Arge, R., De Groot, R., Farber, S., Grasso, M., Hannon, B., Limburg, K., Naeem, S., O'Neill, R.V., Paruelo, J., Raskin, R.G., Sutton, P., *et al.* (1997a). The value of the world's ecosystem services and natural capital. *Nature* **387**, 253–260.

Costanza, R., Cumberland, J., Daly, H., Goodland, R., and Norgaard, R. (1997b). *An Introduction to Ecological Economics*. St. Lucie Press/CRC, Boca Raton, FL.

Cotner, J.B., Hall, E.K., Scott, J.T., and Heldal, M. (2010). Freshwater bacteria are stoichiometrically flexible with a nutrient composition similar to seston. *Front. Microbiol.* **1**, a132.

Couteaux, M.M., and Palka, L. (1988). A direct counting method for soil ciliates. *Soil Biol. Biochem.* **20**, 7–10.

Cuddington, K., Byers, J.E., Wilson, W.G. and Hastings, A. (Eds.) (2007) *Ecosystem Engineers: Plant to Protists*. Academic Press, San Diego, CA, Theoretical Ecology Series.

Daily, G.C., and Matson, P.A. (2008). Ecosystem services: From theory to implementation. *Proc. Natl Acad. Sci. USA* **105**, 9455–9456.

Damuth, J. (1981). Population density and body size in mammals. *Nature* **290**, 699–700.

Darbyshire, J.F., Wheatley, R.E., Greaves, M.P., and Inkson, R.H.E. (1974). A rapid micromethod for estimating bacterial and protozoan populations in soil. *Rev. Écol. Biol. Sol.* **11**, 465–475.

De Bello, F., Lavorel, S., Díaz, S., Harrington, R., Cornelissen, J.H.C., Bardgett, R.D., Berg, M.P., Cipriotti, P., Feld, C.K., Hering, D., Martins da Silva, P., Potts, S.G., *et al.* (2010). Towards an assessment of multiple ecosystem processes and services via functional traits. *Biodiv. Cons.* **19**, 2873–2893.

De Deyn, G.B., Raaijmakers, C.E., and Van der Putten, W.H. (2004). Plant community development is affected by nutrients and soil biota. *J. Ecol.* **92**, 824–834.

De Ruiter, P.C., Moore, J.C., Zwart, K.B., Bouwman, L.A., Hassink, J., Bloem, J., De Vos, J.A., Marinissen, J.C.Y., Didden, W.A.M., Lebrink, G., and Brussaard, L. (1993). Simulation of nitrogen mineralization in the below-ground food webs of two winter wheat fields. *J. Appl. Ecol.* **30**, 95–106.

De Ruiter, P.C., Bloem, J., Bouwman, L.A., Didden, W.A.M., Lebbink, G., Marinissen, J.C.Y., De Vos, J.A., Vreeken-Buys, M.J., Zwart, K.B., and Brussaard, L. (1994). Simulation of dynamics in nitrogen mineralisation in the belowground food webs of two arable farming systems. *Agric. Ecosyst. Environ.* **51**, 199–208.

De Ruiter, P.C., Neutel, A.-M., and Moore, J.C. (1995). Energetics, patterns of interaction strenght, and stability in real ecosystems. *Science* **269**, 1257–1260.

De Visser, S.N., Freymann, B.P., and Olff, H. (2011). The Serengeti food web: Empirical quantification and analysis of topological changes under increasing human impact. *J. Anim. Ecol.* **80**, 484–494.

DeAngelis, D.L., Mulholland, P.J., Palumbo, A.V., Steinman, A.D., Huston, M.A., and Elwood, J.W. (1989). Nutrient dynamics and food-web stability. *Annu. Rev. Ecol. Syst.* **20**, 71–95.

Dekker, S.C., Scheu, S., Schröter, D., Setälä, H., Szanser, M., and Traas, T.P. (2005). Towards a new generation of dynamical decomposer food web model. In: *Dynamic Food Webs: Multispecies Assemblages, Ecosystem Development, and Environmental Change* (Ed. by P.C. De Ruiter, V. Wolters and J.C. Moore), pp. 258–266. Academic Press, San Diego, CA.

Denno, R.F., and Fagan, W.F. (2003). Might nitrogen limitation promote omnivory among carnivorous arthropods? *Ecology* **84**, 2522–2531.

Descartes, R. (1637). *Discours de la méthode pour bien conduire sa raison et chercher la verité dans les sciences*. Joannes Maire, Leiden.

Descartes, R. (1644). *Principia Philosophiae [1]. De principiis cognitionis humanae*. Ludovicum Elzevirium, Amstelodami(Elsevier: Amsterdam).

Díaz, S., Lavorel, S., McIntyre, S., Falczuk, V., Casanoves, F., Milchunas, D.G., Skarpe, C., Rusch, G., Sternberg, M., Noy-Meir, I., Landsberg, J., Zhang, W., *et al.* (2007). Plant trait responses to grazing—A global synthesis. *Global Change Biol.* **13**, 313–341.

Didden, W.A.M. (1993). Ecology of terrestrial Enchytraeidae. *Pedobiologia* **37**, 2–29.

Didden, W.A.M., Marinissen, J.C.Y., Vreeken-Buys, M.J., De Fluiter, R., Geurs, M., and Brussaard, L. (1994). Soil meso- and macrofauna in two agricultural systems: Factors affecting population dynamics and evaluation of their role in carbon and nitrogen dynamics. *Agric. Ecosyst. Environ.* **51**, 171–186.

Dighton, J. (2003). *Fungi in Ecosystem Processes*. Marcel Dekker, New York.

Doi, H., Cherif, M., Iwabuchi, T., Katano, I., Stegen, J.C., and Striebel, M. (2010). Integrating elements and energy through the metabolic dependencies of gross growth efficiency and the threshold elemental ratio. *Oikos* **119**, 752–765.

Eastman, T., Borne, K., Green, J., Grayzeck, E., McGuire, R., and Sawyer, D. (2005). eScience and archiving for space science. *Data Sci. J.* **4**, 67–76.

Ekelund, F. (1998). Enumeration and abundance of mycophagous protozoa in soil, with special emphasis on heterotrophic flagellates. *Soil Biol. Biochem.* **30**, 1343–1347.

Ekelund, F., and Rønn, R. (1994). Notes on protozoa in agricultural soil with emphasis on heterotrophic flagellates and naked amoebae and their ecology. *FEMS Microbiol. Rev.* **15**, 321–353.

Ekelund, F., Rønn, R., and Griffiths, B. (2001). Quantitative estimation of flagellate community structure and diversity in soil samples. *Protist* **152**, 301–314.

Elser, J.J., Fagan, W.F., Denno, R.F., Dobberfuhl, D.R., Folarin, A., Huberty, A., Interlandi, S., Kilham, S.S., McCauley, E., Schulz, K.L., Siemann, E.H., and Sterner, R.W. (2000a). Nutritional constraints in terrestrial and freshwater food webs. *Nature* **408**, 578–580.

Elser, J.J., Sterner, R.W., Gorokhova, E., Fagan, W.F., Markow, T.A., Cotner, J.B., Harrison, J.F., Hobbie, S.E., Odell, G.M., and Weider, L.W. (2000b). Biological stoichiometry from genes to ecosystems. *Ecol. Lett.* **3**, 540–550.

Emmerson, M.C., Montoya, J.M., and Woodward, G. (2005). Body size, interaction strength, and food web dynamics. In: *Dynamic Food Webs: Multispecies Assemblages, Ecosystem Development, and Environmental Change* (Ed. by P.C. De Ruiter, V. Wolters and J.C. Moore), pp. 166–178. Academic Press, San Diego, CA.

Enquist, B.J., West, G.B., and Brown, J.H. (2009). Extensions and evaluations of a general quantitative theory of forest structure and dynamics. *Proc. Natl. Acad. Sci. USA* **106**, 7046–7051.

ESBN, European Soil Bureau Network (2005). *Soil Atlas of Europe*. European Commission, Luxembourg, 128pp.

Feld, C.K., Birk, S., Bradley, D.C., Hering, D., Kail, J., Marzin, A., Melcher, A., Nemitz, D., Pedersen, M.L., Pletterbauer, F., Pont, D., Verdonschot, P.F.M., *et al.* (2011). From natural to degraded rivers and back again: A test of restoration ecology theory and practice. *Adv. Ecol. Res.* **44**, 119–210.

Finlay, B.J., Black, H.I., Brown, S., Clarke, K., Esteban, G., Hindle, R.M., Olmo, J.L., Rollett, A., and Vickerman, K. (2000). Estimating the growth potential of the soil protozoan community. *Protist* **151**, 69–80; *Corrigendum Protist* **151**, 367.

Fitter, A.H. (2005). Darkness visible: Reflections on underground ecology. *J. Ecol.* **93**, 231–243.

Fitter, A.H., Gilligan, C.A., Hollingworth, K., Kleczkowski, A., Twyman, R.M., and Pitchford, J.W., and the Members of the NERC SOIL BIODIVERSITY PROGRAMME (2005). Biodiversity and ecosystem function in soil. *Funct. Ecol.* **19**, 369–377.

Foissner, W. (1987). Soil protozoa: Fundamental problems, ecological significance, adaptations in ciliates and testaceans, bioindicators, and guide to the literature. *Prog. Protistol.* **2**, 69–212.

Foissner, W. (1997). Protozoa as bioindicators in agroecosystems, with emphasis on farming practices, biocides, and biodiversity. *Agric. Ecosyst. Environ.* **63**, 93–103.

Foissner, W. (2006). Biogeography and dispersal of micro-organisms: A review emphasizing protists. *Acta Protozool.* **45**, 111–136.

Foster, B.L., and Gross, K.L. (1998). Species richness in a successional grassland: Effects of nitrogen enrichment and plant litter. *Ecology* **79**, 2593–2602.

Frey, S.D., Gupta, V.V.S.R., Elliott, E.T., and Paustian, K. (2001). Protozoan grazing affects estimates of carbon utilization efficiency of the soil microbial community. *Soil Biol. Biochem.* **33**, 1759–1768.

Friberg, N., Bonada, N., Bradley, D.C., Dunbar, M.J., Edwards, F.K., Grey, J., Hayes, R.B., Hildrew, A.G., Lamouroux, N., Trimmer, M., and Woodward, G. (2011). Biomonitoring of human impacts in natural ecosystems: The good, the bad and the ugly. *Adv. Ecol. Res.* **44**, 1–68.

Fry, B. (2006). *Stable Isotope Ecology*. Springer Verlag, New York.

Fujiwara, M., Kendall, B.E., and Nisbet, R.M. (2004). Growth autocorrelation and animal size variation. *Ecol. Lett.* **7**, 106–113.

Gannes, L.Z., O'Brien, D., and Martínez Del Rio, C. (1997). Stable isotopes in animal ecology: Assumptions, caveats, and a call for laboratory experiments. *Ecology* **78**, 1271–1276.

Gignoux, J., House, J., Hall, D., Masse, D., Nacro, H.B., and Abbadie, I. (2001). Design and test of a generic cohort model of soil organic matter decomposition: The SOMKO model. *Global Ecol. Biogeogr.* **10**, 639–660.

Gillman, L.N., and Wright, S.D. (2010). Mega mistakes in meta-analyses: Devil in the detail. *Ecology* **91**, 2550–2552.

González, A.L., Kominoski, J.S., Danger, M., Ishida, S., Iwai, N., and Rubach, A. (2010). Can ecological stoichiometry help explain patterns of biological invasions? *Oikos* **119**, 779–790.

González, A.L., Fariña, J.M., Kay, A.D., Pinto, R., and Marquet, P.A. (2011). Exploring patterns and mechanisms of interspecific and intraspecific variation in body elemental composition of desert consumers. *Oikos* **120,** 10.1111/j.1600-0706.2010.19151.x.

Griffiths, B.S. (1994). Soil nutrient flow. In: *Soil Protozoa* (Ed. by J.F. Darbyshire), pp. 65–91. CAB International, Wallingford.

Grime, J.P. (1998). Benefits of plant diversity to ecosystems: Immediate, filter and founder effects. *J. Ecol.* **86**, 902–910.

Gurevitch, J., and Hedges, L.V. (2001). Meta-analysis: Combining the results of independent experiments. In: *Design and Analysis of Ecological Experiments* (Ed. by S.M. Scheiner and J. Gurevitch), pp. 347–369. Oxford University Press, Oxford.

Hagler, J.R. (2006). Development of an immunological technique for identifying multiple predator–prey interactions in a complex arthropod assemblage. *Ann. Appl. Biol.* **149**, 153–165.

Hansson, M., and Fogelfors, H. (1998). Management of permanent set-aside on arable land in Sweden. *J. Appl. Ecol.* **35**, 758–771.

Harper, G.L., King, R.A., Dodd, C.S., Harwood, J.D., Glen, D.M., Bruford, M.W., and Symondson, W.O.C. (2005). Rapid screening of invertebrate predators for multiple prey DNA targets. *Mol. Ecol.* **14**, 819–827.

Haubert, D., Häggblom, M.M., Langel, R., Scheu, S., and Ruess, L. (2006). Trophic shift of stable isotopes and fatty acids in Collembola on bacterial diets. *Soil Biol. Biochem.* **38**, 2004–2007.

Haubert, D., Birkhofer, K., Fließbach, A., Gehre, M., Scheu, S., and Ruess, L. (2009). Trophic structure and major trophic links in conventional versus organic farming systems as indicated by carbon stable isotope ratios of fatty acids. *Oikos* **118**, 1579–1589.

Hawksworth, D.L., Kirk, P.M., Sutton, B.C., and Pegler, D.N. (1995). *Ainsworth & Bisby's Dictionary of the Fungi.* 8th ed. International Mycological Institute, CAB International, Wallingford.

Hazeu, G.W., Schuiling, C., Dorland, G.J., Oldengarm, J., and Gijsbertse, H.A. (2010). Landelijk Grondgebruiksbestand Nederland versie 6 (LGN6). Vervaardiging, nauwkeurigheid en gebruik. Alterra Report 2012, Wageningen, The Netherlands.

Herbert, D. (1976). Stoichiometric aspects of microbial growth. In: *Continuous Culture 6: Application and New Fields* (Ed. by A.C.R. Dean, D.C. Ellwood, C.G. T. Evans and J. Melling), pp. 1–30. Ellis Horwood, Chichester, UK.

Hesse, R. (1924). *Tiergeographie auf Ökologischer Grundlage.* Fischer, Jena.

Hildrew, A.G. (2009). Sustained research on stream communities: A model system and the comparative approach. *Adv. Ecol. Res.* **41**, 175–312.

Hladyz, S., Åbjörnsson, K., Chauvet, E., Dobson, M., Elosegi, A., Ferreira, V., Fleituch, T., Gessner, M.O., Giller, P.S., Gulis, V., Hutton, S.A., Lacoursière, J.O., *et al.* (2011). Stream ecosystem functioning in an agricultural landscape: The importance of terrestrial-aquatic linkages. *Adv. Ecol. Res.* **44**, 211–276.

Högberg, M.N., Hogberg, P., and Myrold, D.D. (2007). Is microbial community composition in boreal forest soils determined by pH, C–to–N ratio, the trees, or all three? *Oecologia* **150**, 590–601.

Holmes, K.W., Roberts, D.A., Sweeney, S., Numata, I., Matricardi, E., Biggs, T.W., Batista, G., and Chadwick, O.A. (2004). Soil databases and the problem of establishing regional biogeochemical trends. *Global Change Biol.* **10**, 796–814.

Holterman, M., Van der Wurff, A., Van den Elsen, S., Van Megen, H., Bongers, T., Holovachov, O., Bakker, J., and Helder, J. (2006). Phylum-wide analysis of SSU rDNA reveals deep phylogenetic relationships among nematodes and accelerated evolution toward crown clades. *Mol. Biol. Evol.* **23**, 1792–1800.

Holterman, M., Rybarczyk, K., Van den Elsen, S., Van Megen, H., Mooyman, P., Santiago, R.P., Bongers, T., Bakker, J., and Helder, J. (2008). A ribosomal DNA-based framework for the detection and quantification of stress–sensitive nematode families in terrestrial habitats. *Mol. Ecol. Res.* **8**, 23–34.

Höper, H. (1999). Die Bedeutung abiotischer Bodeneigenschaften für bodenmikrobiologische Kennwerte. Ergebnisse aus der Bodendauerbeobachtung in Niedersachsen. *Mitteilungen der Deutschen Bodenkundlichen Gesellschaft* **89**, 253–256.

Huberty, L.E., Gross, K.L., and Miller, C.J. (1998). Effects of nitrogen addition on successional dynamics and species diversity in Michigan old-fields. *J. Ecol.* **86**, 794–803.

Hunt, H.W., Coleman, D.C., Ingham, E.R., Ingham, R.E., Elliott, E.T., Moore, J.C., Rose, S.L., Reid, C.P.P., and Morley, C.R. (1987). The detrital food web in a shortgrass prairie. *Biol. Fertil. Soils* **3**, 57–68.

Huston, M.A. (1993). Biological diversity, soils, and economics. *Science* **262**, 1676–1680.

Huston, M.A., and DeAngelis, D.L. (1987). Size bimodality in monospecific populations: A critical review of potential mechanisms. *Am. Nat.* **129**, 678–707.

Hyodo, F., Kohzu, A., and Tayasu, I. (2010). Linking aboveground and belowground food webs through carbon and nitrogen stable isotope analyses. *Ecol. Res.* **25**, 745–756.

Ioannidis, J.P.A. (2005). Why most published research findings are false. *PLoS Med.* **2**, 696–701.

Jagers op Akkerhuis, G.A.J.M., Dimmers, W.J., Van Vliet, P.C.J., Goedhart, G.F.P., Martakis, G.F.P., and De Goede, R.G.M. (2008). Evaluating the use of gel-based sub-sampling for assessing responses of terrestrial micro-arthropods (Collembola and Acari) to different slurry applications and organic matter contents. *Appl. Soil Ecol.* **38**, 239–248.

Jänsch, S., Römbke, J., and Didden, W. (2005). The use of enchytraeids in ecological soil classification and assessment concepts. *Ecotox. Environ. Saf.* **62**, 266–277.

Jeffery, S., Gardi, C., Jones, A., Montanarella, L., Marmo, L., Miko, L., Ritz, K., Peres, G., Römbke, J. and Van der Putten, W.H. (Eds.) (2010). *European Atlas of Soil Biodiversity*. European Commission, Publications Office of the European Union, Luxembourg.

Jennings, S. (2005). Size-based analyses of aquatic food webs. In: *Aquatic Food Webs: An Ecosystem Approach* (Ed. by A. Belgrano, U.M. Scharler, J. Dunne and R.E. Ulanowicz), pp. 86–97. Oxford University Press, Oxford.

Jepsen, D.B., and Winemiller, K.O. (2002). Structure of tropical river food webs revealed by stable isotope ratios. *Oikos* **96**, 46–55.

Johnson, N.C. (2010). Resource stoichiometry elucidates the structure and function of arbuscular mycorrhizas across scales. *New Phytol.* **185**, 631–647.

Jones, C.G., Lawton, J.H., and Shachak, M. (1994). Organisms as ecosystem engineers. *Oikos* **69**, 373–386.

Jones, D.L., Nguyen, C., and Finlay, R.D. (2009). Carbon flow in the rhizosphere: Carbon trading at the soil–root interface. *Plant Soil* **321**, 5–33.

Jonsson, T., Cohen, J.E., and Carpenter, S.R. (2005). Food webs, body size and species abundance in ecological community description. *Adv. Ecol. Res.* **36**, 1–84.

Kaspari, M. (2004). Using the metabolic theory of ecology to predict global patterns of abundance. *Ecology* **85**, 1800–1802.

Kaspari, M. (2005). Global energy gradients and size in colonial organisms: Worker mass and worker number in ant colonies. *Proc. Natl. Acad. Sci. USA* **102**, 5079–5083.

Kay, F.R., Sobhy, H.M., and Whitford, W.G. (1999). Soil microarthropods as indicators of exposure to environmental stress in Chihuahuan Desert Rangelands. *Biol. Fertil. Soils* **28**, 121–128.

Keiblinger, K.M., Hall, E.K., Wanek, W., Szukics, U., Hämmerle, I., Ellersdorfer, G., Böck, S., Strauss, J., Sterflinger, K., Richter, A., and Zechmeister-Boltenstern, S. (2010). The effect of resource quantity and resource stoichiometry on microbial carbon-use-efficiency. *FEMS Microbiol. Ecol.* **73**, 430–440.

Kendall, B.E., and Fox, G.A. (2002). Variation among individuals and reduced demographic stochasticity. *Cons. Biol.* **16**, 109–116.

Kerr, N.L. (1998). HARKing: Hypothesizing after the results are known. *Pers. Soc. Psychol.* **2**, 196–217.

King, K.L., Hutchinson, K.J., and Greenslade, P. (1976). The effects of sheep numbers on associations of Collembola in sown pastures. *J. Appl. Ecol.* **13**, 731–739.

King, R.A., Read, D.S., Traugott, M., and Symondson, W.O.C. (2008). Molecular analysis of predation: A review of best practice for DNA-based approaches. *Mol. Ecol.* **17**, 947–963.

Kozłowski, J. (1992). Optimal allocation of resources to growth and reproduction: Implications for age and size at maturity. *Trends Ecol. Evol.* **7**, 15–19.

Kristensen, H.L. (2001). High immobilization of $NH_4^+$ in Danish heath soil related to succession, soil and nutrients: Implications for critical loads of N. *Water Air Soil Pollut. Focus* **1**, 211–230.

Kühn, I., Durka, W., and Klotz, S. (2004). BiolFlor—A new plant-trait database as a tool for plant invasion ecology. *Diversity Distrib.* **10**, 363–365.

Laakso, J., and Setälä, H. (1999). Sensitivity of primary production to changes in the architecture of belowground food webs. *Oikos* **87**, 57–64.

LaBarbera, M. (1989). Analyzing body size as a factor in ecology and evolution. *Annu. Rev. Ecol. Syst.* **20**, 97–117.

Layer, K., Riede, J.O., Hildrew, A.G., and Woodward, G. (2010). Food web structure and stability in 20 streams across a wide pH gradient. *Adv. Ecol. Res.* **42**, 265–299.

Layer, K., Hildrew, A.G., Jenkins, G.B., Riede, J.O., Rossiter, S.J., Townsend, C.R., and Woodward, G. (2011). Long-term dynamics of a well-characterised food web: Four decades of acidification and recovery in the broadstone stream model system. *Adv. Ecol. Res.* **44**, 69–118.

Lebbink, G., Van Faassen, H.G., Van Ouwerkerk, C., and Brussaard, L. (1994). The Dutch programme on soil ecology of arable farming systems: Farm management, monitoring programme and general results. *Agric. Ecosyst. Environ.* **51**, 7–20.

Liiri, M., Setälä, H., Haimi, J., Pennanen, T., and Fritze, H. (2002). Relationship between soil microarthropod species diversity and plant growth does not change when the system is disturbed. *Oikos* **96**, 137–149.

Lindeman, R.L. (1942). The trophic-dynamic aspect of ecology. *Ecology* **23**, 399–418.

Loreau, M., Downing, A., Emmerson, M., Gonzalez, A., Hughes, J., Inchausti, P., Joshi, J., Norberg, J., and Sala, O. (2002). A new look at the relationship between diversity and stability. In: *Biodiversity and Ecosystem Functioning: Synthesis and Perspectives* (Ed. by M. Loreau, S. Naeem and P. Inchausti), pp. 79–91. Oxford University Press, Oxford.

Luftenegger, G., Petz, W., Foissner, W., and Adam, H. (1988). The efficiency of a direct counting method in estimating the numbers of microscopic soil organisms. *Pedobiologia* **31**, 95–101.

MA, Millennium Ecosystem Assessment (2005). *Ecosystems and Human Well Being.* Island Press, Washington, DC. [Encyclopedia of Earth, Retrieved December 10, 2010, http://www.eoearth.org/article/Millennium_Ecosystem_Assessment_Synthesis_Reports].

Madin, J., Bowers, S., Schildhauer, M., Krivov, S., Pennington, D., and Villa, F. (2007). An ontology for describing and synthesizing ecological observation data. *Ecol. Inform.* **2**, 279–296.

Madin, J.S., Bowers, S., Schildhauer, M.P., and Jones, M.B. (2008). Advancing ecological research with ontologies. *Trends Ecol. Evol.* **23**, 159–168.

Mahootian, F., and Eastman, T.E. (2009). Complementary frameworks of scientific inquiry: Hypothetico-deductive, hypothetico-inductive, and observational-inductive. *World Futures* **65**, 61–75.

Mäler, K.-G., Aniyar, S., and Jansson, Å. (2008). Accounting for ecosystem services as a way to understand the requirements for sustainable development. *Proc. Natl. Acad. Sci. USA* **105**, 9501–9506.

Manzoni, S., Trofymow, J.A., Jackson, R.B., and Porporato, A. (2010). Stoichiometric controls on carbon, nitrogen, and phosphorus dynamics in decomposing litter. *Ecol. Monogr.* **80**, 89–106.

Maron, J.L., Estes, J.A., Croll, D.A., Danner, E.M., Elmendorf, S.C., and Buckelew, S.L. (2006). An introduced predator alters Aleutian island plant communities by thwarting nutrient subsidies. *Ecol. Monogr.* **76**, 3–24.

Martínez Del Rio, C., Wolf, N., Carleton, S.A., and Gannes, L.Z. (2009). Isotopic ecology ten years after a call for more laboratory experiments. *Biol. Rev.* **84**, 91–111.

Mercer, R.D., Gabriel, A.G.A., Barendse, J., Marshall, D.J., and Chown, S.L. (2001). Invertebrate body sizes from Marion Island. *Antarct. Sci.* **13**, 135–143.

Moonesinghe, R., Khoury, M.J., and Janssens, A.C.J.W. (2007). Most published research findings are false—But a little replication goes a long way. *PLoS Med.* **4**, 218–221.

Moon-Van der Staay, S.Y., Tzeneva, V., Van der Staay, G.W.M., De Vos, W.M., Smidt, H., and Hackstein, J.H.P. (2006). Eukaryotic diversity in historical soil samples. *FEMS Microbiol. Ecol.* **57**, 420–428.

Moore, J.C. (1994). Impact of agricultural practices on soil food web structure: Theory and application. *Agric. Ecosyst. Environ.* **51**, 239–247.

Moore, J.C., Walter, D.E., and Hunt, H.W. (1988). Arthropod regulation of micro- and mesobiota in below-ground detrital food web. *Annu. Rev. Entomol.* **33**, 419–439.

Moore, J.C., Berlow, E.L., Coleman, D.C., De Ruiter, P.C., Dong, Q., Hastings, A., Johnson, N.C., McCann, K.S., Melville, K., Morin, P.J., Nadelhoffer, K., Rosemond, A.D., *et al.* (2004). Detritus, trophic dynamics and biodiversity. *Ecol. Lett.* **7**, 584–600.

Moorhead, D.L., and Sinsabaugh, R.L. (2006). A theoretical model of litter decay and microbial interaction. *Ecol. Monogr.* **76**, 151–174.

Mulder, C. (2006). Driving forces from soil invertebrates to ecosystem functioning: The allometric perspective. *Naturwissenschaften* **93**, 467–479.

Mulder, C. (2010). Soil fertility controls the size-specific distribution of eukaryotes. *Ann. NY Acad. Sci.* **1195**, E74–E81.

Mulder, C. (2011). World Wide Food Webs: Power to feed ecologists. *Ambio* **40**, 335–337.

Mulder, C., and Elser, J.J. (2009). Soil acidity, ecological stoichiometry and allometric scaling in grassland food webs. *Global Change Biol.* **15**, 2730–2738.

Mulder, C., and Janssen, C.R. (1999). Occurrence of pollen and spores in relation to present-day vegetation in a Dutch heathland area. *J. Veg. Sci.* **10**, 87–100.

Mulder, C., and Vonk, J.A. (2011). Nematode traits and environmental constraints in 200 soil systems: Scaling within the 60–6,000 $\mu$m body size range. *Ecology* (in press).

Mulder, C., De Zwart, D., Van Wijnen, H.J., Schouten, A.J., and Breure, A.M. (2003a). Observational and simulated evidence of ecological shifts within the soil nematode community of agroecosystems under conventional and organic farming. *Funct. Ecol.* **17**, 516–525.

Mulder, C., Breure, A.M., and Joosten, J.H.J. (2003b). Fungal functional diversity inferred along Ellenberg's abiotic gradients: Palynological evidence from different soil microbiota. *Grana* **42**, 55–64.

Mulder, C., Van Wijnen, H.J., and Van Wezel, A.P. (2005a). Numerical abundance and biodiversity of below-ground taxocenes along a pH gradient across the Netherlands. *J. Biogeogr.* **32**, 1775–1790.

Mulder, C., Cohen, J.E., Setälä, H., Bloem, J., and Breure, A.M. (2005b). Bacterial traits, organism mass, and numerical abundance in the detrital soil food web of Dutch agricultural grasslands. *Ecol. Lett.* **8**, 80–90.

Mulder, C., Den Hollander, H., Schouten, T., and Rutgers, M. (2006a). Allometry, biocomplexity, and web topology of hundred agro-environments in The Netherlands. *Ecol. Complex.* **3**, 219–230.

Mulder, C., Wouterse, M., Raubuch, M., Roelofs, W., and Rutgers, M. (2006b). Can transgenic maize affect soil microbial communities? *PLoS Comput. Biol.* **2**, 1165–1172.

Mulder, C., Baerselman, R., and Posthuma, L. (2007). Empirical maximum lifespan of earthworms is twice that of mice. *Age* **29**, 229–231.

Mulder, C., Den Hollander, H.A., and Hendriks, A.J. (2008). Aboveground herbivory shapes the biomass distribution and flux of soil invertebrates. *PLoS ONE* **3**, e3573.

Mulder, C., Den Hollander, H.A., Vonk, J.A., Rossberg, A.G., Jagers op Akkerhuis, G.A.J.M., and Yeates, G.W. (2009). Soil resource supply influences faunal size-specific distributions in natural food webs. *Naturwissenschaften* **96**, 813–826.

Mulder, C., Vonk, J.A., Den Hollander, H.A., Hendriks, A.J., and Breure, A.M. (2011). How allometric scaling relates to soil abiotics. *Oikos* **120**, 529–536.

Müller, C.B., Adriaanse, I.C.T., Belshaw, R., and Godfray, H.C.J. (1999). The structure of an aphid–parasitoid community. *J. Anim. Ecol.* **68**, 346–370.

Müller, P., and Döring, M. (2009). Isothermal DNA amplification facilitates the identification of a broad spectrum of bacteria, fungi and protozoa in *Eleutherococcus* sp. plant tissue cultures. *Plant Cell Tissue Organ Cult.* **98**, 35–45.

Naeem, S., Thomson, L.J., Lawler, S.P., Lawton, J.H., and Woodfin, R.M. (1995). Empirical evidence that declining species diversity may alter the performance of terrestrial ecosystems. *Phil. Trans. R. Soc. Lond. B* **347**, 249–262.

Neuhauser, S., and Kirchmair, M. (2009). *Ligniera junci*, a plasmodiophorid rediscovered in roots of *Juncus* in Austria. *Österreichische Zeitschrift für Pilzkunde* **18**, 151–157.

Neutel, A.M., Heesterbeek, J.A.P., and De Ruiter, P.C. (2002). Stability in real food webs: Weak links in long loops. *Science* **296**, 1120–1123.

Novotny, V., Miller, S.E., Baje, L., Balagawi, S., Basset, Y., Cizek, L., Craft, K.J., Dem, F., Drew, R.A.I., Huclr, J., Leps, J., Lewis, O.T., *et al.* (2010). Guild-specific patterns of species richness and host specialization in plant–herbivore food webs from a tropical forest. *J. Anim. Ecol.* **79**, 1193–1203.

O'Gorman, E., and Emmerson, M.C. (2010). Manipulating interaction strengths and the consequences for trivariate patterns in a marine food web. *Adv. Ecol. Res.* **42**, 301–419.

Oberholzer, H.-R., and Höper, H. (2000). Reference systems for the microbiological evaluation of soils. *Verband Deutscher Landwirtschaftlicher Untersuchungs- und Forschungsanstalten* **55**, 19–34.

Oberholzer, H.-R., Rek, J., Weisskopf, P., and Walther, U. (1999). Evaluation of soil quality by means of microbiological parameters related to the characteristics of individual arable sites. *Agribiol. Res.* **52**, 113–125.

Oostenbrink, M. (1960). Estimate nematode populations by some selected methods. In: *Nematology* (Ed. by J.N. Sasser and W.R. Jenkins), pp. 85–102. University of North Carolina Press, Chapel Hill, NC.

Patterson, D.J. (1996). *Free-Living Freshwater Protozoa*. Manson, London.

Payne, R.J., and Mitchell, E.A.D. (2009). How many is enough? Determining optimal count totals for ecological and palaeoecological studies of testate amoebae. *J. Paleolimnol.* **42**, 483–495.

Peltzer, D.A., Wardle, D.A., Allison, V.J., Baisden, W.T., Bardgett, R.D., Chadwick, O.A., Condron, L.M., Parfitt, R.L., Porder, S., Richardson, S.J., Turner, B.L., Vitousek, P.M., *et al.* (2010). Understanding ecosystem retrogression. *Ecol. Monogr.* **80**, 509–529.

Pendall, E., Bridgham, S., Hanson, P.J., Hungate, B., Kicklighter, D.W., Johnson, D.W., Law, B.E., Luo, Y., Megonigal, J.P., Olsrud, M., Ryan, M.G., and Wan, S. (2004). Below-ground process responses to elevated $CO_2$ and temperature: A discussion of observations, measurement methods, and models. *New Phytol.* **162**, 311–322.

Peñuelas, J., and Sardans, J. (2009). Elementary factors. *Nature* **460**, 803–804.

Petchey, O.L., and Belgrano, A. (2010). Body-size distributions and size-spectra: Universal indicators of ecological status? *Biol. Lett.* **6**, 434–437.

Petersen, H., and Luxton, M. (1982). A comparative analysis of soil fauna populations and their role in decomposition process. *Oikos* **39**, 288–388.

Petz, W., Foissner, W., and Adam, H. (1985). Culture, food selection and growth rate in the mycophagous ciliate *Grossglockneria acuta* Foissner, 1980: First evidence of autochthonous soil ciliates. *Soil Biol. Biochem.* **17**, 871–875.

Pfister, C.A., and Stevens, F.R. (2002). The genesis of size variability in plants and animals. *Ecology* **83**, 59–72.

Phillips, D.L. (2001). Mixing models in analyses of diet using multiple stable isotopes: A critique. *Oecologia* **127**, 166–170.

Pignatti, S. (1994). A complex approach to phytosociology. *Ann. Bot.* **52**, 65–80.

Polis, G.A. (1984). Age structure component of niche width and intraspecific resource partitioning: Can age groups function as ecological species? *Am. Nat.* **123**, 541–564.

Pollierer, M.M., Langel, R., Scheu, S., and Maraun, M. (2009). Compartmentalization of the soil animal food web as indicated by dual analysis of stable isotope ratios ($^{15}N/^{14}N$ and $^{13}C/^{12}C$). *Soil Biol. Biochem.* **41**, 1221–1226.

Pollierer, M.M., Scheu, S., and Haubert, D. (2010). Taking it to the next level: Trophic transfer of marker fatty acids from basal resource to predators. *Soil Biol. Biochem.* **42**, 919–925.

Ponge, J.F. (2000). Vertical distribution of Collembola (Hexapoda) and their food resources in organic horizons of beech forests. *Biol. Fertil. Soils* **32**, 508–522.

Ponsard, S., and Arditi, R. (2000). What can stable isotopes ($\delta^{15}N$ and $\delta\ ^{13}C$) tell about the food web of soil macro-invertebrates? *Ecology* **81**, 852–864.

Post, D.M. (2002). Using stable isotopes to estimate trophic position: Models, methods, and assumptions. *Ecology* **83**, 703–718.

Postma-Blaauw, M.B. (2008). *Soil Biodiversity and Nitrogen Cycling under Agricultural (De-)intensification.* PhD Thesis. Wageningen University, Wageningen, The Netherlands.

Postma-Blaauw, M.B., De Goede, R.G.M., Bloem, J., Faber, J.H., and Brussaard, L. (2010). Soil biota community structure and abundance under agricultural intensification and extensification. *Ecology* **91**, 460–473.

Power, A.G. (2010). Ecosystem services and agriculture: Tradeoffs and synergies. *Phil. Trans. R. Soc. Lond. B* **365**, 2959–2971.

Purdy, K.J., Hurd, P.J., Moya-Laraño, J., Trimmer, M., Oakley, B.B., and Woodward, G. (2010). Systems biology for ecology: From molecules to ecosystems. *Adv. Ecol. Res.* **43**, 87–149.

Queenborough, S.A., Cooke, I.R., and Schildhauer, M.P. (2010). Do we need an EcoBank? The ecology of data-sharing. *Bull. Br. Ecol. Soc.* **41**(3), 32–35.

Rayner, A.D.M. (1991). The challenge of the individualistic mycelium. *Mycologia* **83**, 48–71.

Reich, P.B., Oleksyn, J., Wright, I.J., Niklas, K.J., Hedin, L., and Elser, J.J. (2010). Evidence of a general 2/3-power law of scaling leaf nitrogen to phosphorus among major plant groups and biomes. *Proc. R. Soc. Lond. B* **277**, 877–883.

Reiss, J., Forster, J., Cássio, F., Pascoal, C., Stewart, R., and Hirst, A.G. (2010). When microscopic organisms inform general ecological theory. *Adv. Ecol. Res.* **43**, 45–85.

Reuman, D.C., Mulder, C., Raffaelli, D., and Cohen, J.E. (2008). Three allometric relations of population density to body mass: Theoretical integration and empirical tests in 149 food webs. *Ecol. Lett.* **11**, 1216–1228.

Reuman, D.C., Cohen, J.E., and Mulder, C. (2009). Human and environmental factors influence soil faunal abundance–mass allometry and structure. *Adv. Ecol. Res.* **41**, 45–85.

Ritz, K., and Trudgill, D.L. (1999). Utility of nematode community analysis as an integrated measure of the functional state of soils: Perspectives and challenges. *Plant Soil* **212**, 1–11.

Rooney, N., McCann, K.S., and Moore, J.C. (2008). A landscape theory for food web architecture. *Ecol. Lett.* **11**, 867–881.

Rosenthal, R. (1979). The "file-drawer problem" and tolerance for null results. *Psychol. Bull.* **86**, 638–641.

Rossberg, A.G., Brännström, Å., and Dieckmann, U. (2010). Food-web structure in low- and high-dimensional trophic niche spaces. *Interface* **7**, 1735–1743.

Rousk, J., Brookes, P.C., and Bååth, E. (2009). Contrasting soil pH effects on fungal and bacterial growth suggest functional redundancy in carbon mineralization. *Appl. Environ. Microbiol.* **75**, 1589–1596.

Rudolf, V.H.W., and Lafferty, K.D. (2011). Stage structure alters how complexity affects stability of ecological networks. *Ecol. Lett.* **14**, 75–79.

Ruess, L., and Chamberlain, P.M. (2010). The fat that matters: Soil food web analysis using fatty acids and their carbon stable isotope signature. *Soil Biol. Biochem.* **42**, 1898–1910.

Ruess, L., Häggblom, M.M., Zapata, E.J.G., and Dighton, J. (2002). Fatty acids of fungi and nematodes—Possible biomarkers in the soil food chain? *Soil Biol. Biochem.* **34**, 745–756.

Ruess, L., Tiunov, A., Haubert, D., Richnow, H.H., Haggblom, M.M., and Scheu, S. (2005). Carbon stable isotope fractionation and trophic transfer of fatty acids in fungal based soil food chains. *Soil Biol. Biochem.* **37**, 945–953.

Rutgers, M., Mulder, C., Schouten, A.J., Bloem, J., Bogte, J.J., Breure, A.M., Brussaard, L., De Goede, R.G.M., Faber, J.H., Jagers op Akkerhuis, G.A.J.M., Keidel, H., Korthals, G., *et al.* (2008). *Soil ecosystem profiling in the Netherlands with ten references for biological soil quality*. RIVM Report 607604009, Bilthoven, The Netherlands.

Rutgers, M., Schouten, A.J., Bloem, J., Van Eekeren, N., De Goede, R.G.M., Jagers op Akkerhuis, G.A.J.M., Van der Wal, A., Mulder, C., Brussaard, L., and Breure, A.M. (2009). Biological measurements in a nationwide soil monitoring network. *Eur. J. Soil Sci.* **60**, 820–832.

Sagarin, R., and Pauchard, A. (2010). Observational approaches in ecology open new ground in a changing world. *Front. Ecol. Environ.* **7**, 379–386.

Schaefer, M., and Schauermann, J. (1990). The soil fauna of beech forests: Comparison between a mull and a moder soil. *Pedobiologia* **34**, 299–314.

Schmidt, O., Curry, J.P., Dyckmans, J., Rota, E., and Scrimgeour, C.M. (2004). Dual stable isotope analysis ($\delta^{13}$C and $\delta^{15}$N) of soil invertebrates and their food sources. *Pedobiologia* **48**, 171–180.

Schneider, K., and Maraun, M. (2005). Feeding preferences among dark pigmented fungal taxa ("Dematiacea") indicate limited trophic niche differentiation of oribatid mites (Oribatida, Acari). *Pedobiologia* **49**, 61–67.

Schouten, A.J., Brussaard, L., De Ruiter, P.C., Siepel, H., and Van Straalen, N.M. (1997). Een indicatorsysteem voor life support functies van de bodem in relatie tot biodiversiteit. RIVM Report 712910005, Bilthoven, The Netherlands.

Schouten, A.J., Breure, A.M., Bloem, J., Didden, W., De Ruiter, P.C., and Siepel, H. (1999). Life support functies van de bodem: Operationalisering t.b.v. het biodiversiteitsbeleid. RIVM Report 607601003, Bilthoven, The Netherlands.

Schröter, D., Wolters, V., and De Ruiter, P.C. (2003). C and N mineralisation in the decomposer food webs of a European forest transect. *Oikos* **102**, 294–308.

Schulze, E.-D., Valentini, R., and Sanz, M.-J. (2002). The long way from Kyoto to Marrakesh: Implications of the Kyoto Protocol negotiations for global ecology. *Global Change Biol.* **8**, 505–518.

Semenina, E.E., and Tiunov, A.V. (2010). Isotopic fractionation by saprotrophic microfungi: Effects of species, temperature and the age of colonies. *Pedobiologia* **53**, 213–217.

Sheppard, S.K., and Harwood, J.D. (2005). Advances in molecular ecology: Tracking trophic links through predator–prey food-webs. *Funct. Ecol.* **19**, 751–762.

Siepel, H. (1994). Life–history tactics of soil microarthropods. *Biol. Fertil. Soils* **18**, 263–278.

Siepel, H., and Van de Bund, C.F. (1988). The influence of management practices on the microarthropod community of grassland. *Pedobiologia* **31**, 339–354.

Sinsabaugh, R.L., Lauber, C.L., Weintraub, M.N., Ahmed, B., Allison, S.D., Crenshaw, C., Contosta, A.R., Cusack, D., Frey, S., Gallo, M.E., Gartner, T.B., Hobbie, S.E., *et al.* (2008). Stoichiometry of soil enzyme activity at global scale. *Ecol. Lett.* **11**, 1252–1264.

Sinsabaugh, R.L., Hill, B.H., and Follstad Shah, J.J. (2009). Ecoenzymatic stoichiometry of microbial organic nutrient acquisition in soil and sediment. *Nature* **462**, 795–798.

Slatkin, M. (1984). Ecological causes of sexual dimorphism. *Evolution* **38**, 622–630.

Smith, T.B., and Skulason, S. (1996). Evolutionary significance of resource polymorphisms in fishes, amphibians, and birds. *Annu. Rev. Ecol. Syst.* **27**, 111–133.

Smith, M.L., Bruhn, J.N., and Anderson, J.B. (1992). The fungus *Armillaria bulbosa* is among the largest and oldest living organisms. *Nature* **356**, 428–431.

Sohlenius, B. (1968). Studies of the interactions between *Mesodiplogaster* sp., and other rhabditid nematodes and a protozoan. *Pedobiologia* **8**, 340–344.

Srivastava, D.S., and Vellend, M. (2005). Biodiversity–Ecosystem Function research: Is it relevant to conservation? *Annu. Rev. Ecol. Syst.* **36**, 267–294.

Statzner, B., Hildrew, A.G., and Resh, V.H. (2001). Species traits and environmental constraints: Entomological research and the history of ecological theory. *Annu. Rev. Entomol.* **46**, 291–316.

Sterner, R.W., and Elser, J.J. (2002). *Ecological Stoichiometry: The Biology of Elements from Molecules to the Biosphere*. Princeton University Press, Princeton, NJ.

Sterner, R.W., Anderson, T., Elser, J.J., Hessen, D.O., Hood, J.M., McCauley, E., and Urabe, J. (2008). Scale-dependent carbon: nitrogen: phosphorus seston stoichiometry in marine and freshwaters. *Limnol. Oceanogr.* **53**, 1169–1180.

Stevens, C.J., Dise, N.B., Mountford, J.O., and Gowing, D.J. (2004). Impact of nitrogen deposition on the species richness of grasslands. *Science* **303**, 1876–1879.

Stevens, C.J., Dise, N.B., Gowing, D.J.G., and Mountford, J.O. (2006). Loss of forb diversity in relation to nitrogen deposition in the UK: Regional trends and potential controls. *Global Change Biol.* **12**, 1823–1833.

Stott, A.W., Davies, E., Evershed, R.P., and Tuross, N. (1997). Monitoring the routing of dietary and biosythesised lipids through compound-specific stable isotope ($\delta^{13}C$) measurements at natural abundance. *Naturwissenschaften* **84**, 82–86.

Sutherland, W.J., Armstrong-Brown, S., Armsworth, P.R., Brereton, T., Brickland, J., Campbell, C.D., Chamberlain, D.E., Cooke, A.I., Dulvy, N.K., Dusic, N.R., Fitton, M., Frecleton, R.P., *et al.* (2006). The identification of 100 ecological questions of high policy relevance in the UK. *J. Appl. Ecol.* **43**, 617–627.

Sutton, A.J., Song, F., Gilbody, S.M., and Abrams, K.R. (2000). Modelling publication bias in meta-analysis: A review. *Stat. Methods Med. Res.* **9**, 421–445.

Symondson, W.O.C. (2002). Molecular identification of prey in predator diets. *Mol. Ecol.* **11**, 627–641.

Teder, T., Tammaru, T., and Esperk, T. (2008). Dependence of phenotypic variance in body size on environmental quality. *Am. Nat.* **172**, 223–232.

TEEB: The Economics of Ecosystems and Biodiversity (2010). *TEEB for Local and Regional Policy Makers.* UNEP.

Thibault, K.M., White, E.P., Hurlbert, A.H., and Ernest, S.K.M. (2011). Multimodality in the individual size distributions of bird communities. *Global Ecol. Biogeogr.* **20**, 145–153.

Tullgren, A. (1917). Ein sehr einfacher Ausleseapparat für terricole Tierformen. *Z. Angew. Entomol.* **4**, 149–150.

Ulanowicz, R.E. (1995). Ecosystem trophic foundations: Lindeman exonerata. In: *Complex Ecology: The Part-Whole Relation in Ecosystems* (Ed. by B. C. Patten and S.E. Jorgensen), pp. 549–560. Prentice-Hall, New York.

UNCED, United Nations Conference on Environment and Development (1992). Rio Declaration and Agenda 21. In: *The Earth Summit: The United Nations Conference on Environment and Development (UNCED), 1993* Johnson, S.P. (Ed.) (1992) Graham & Trotman/Martinus Nijhoff, London.

UNEP/CBD, United Nations Environment Programme / Convention on Biological Diversity (2000). *The Ecosystem Approach. UNEP/CBD/COP/5/23 Nairobi.* 15–26 May 2000, Decision V/6.

Urabe, J., Naeem, S., Raubenheimer, D., and Elser, J.J. (2010). The evolution of biological stoichiometry under global change. *Oikos* **119**, 737–740.

Urich, T., Lanzen, A., Qi, J., Huson, D.H., Schleper, C., and Schuster, S.C. (2008). Simultaneous assessment of soil microbial community structure and function through analysis of the meta–transcriptome. *PLoS ONE* **3**, e2527.

Van der Heijden, M.G.A., Verkade, S., and De Bruin, S. (2008). Mycorrhizal fungi reduce the negative effects of nitrogen enrichment on plant community structure in dune grassland. *Global Change Biol.* **14**, 2626–2635.

Van der Wal, A., Van Veen, J.A., Smant, W., Boschker, H.T.S., Bloem, J., Kardol, P., Van der Putten, W.H., and De Boer, W. (2006). Fungal biomass development in a chronosequence of land abandonment. *Soil Biol. Biochem.* **38**, 51–60.

Van der Wal, A., Geerts, R.H.E.M., Korevaar, H., Schouten, A.J., Jagers op Akkerhuis, G.A.J.M., Rutgers, M., and Mulder, C. (2009a). Dissimilar response of plant and soil biota communities to long-term nutrient addition in grasslands. *Biol. Fertil. Soils* **45**, 663–667.

Van der Wal, A., Bloem, J., Mulder, C., and De Boer, W. (2009b). Relative abundance and activity of melanized hyphae in different soil systems. *Soil Biol. Biochem.* **41**, 417–419.

Van Eekeren, N., De Boer, H., Hanegraaf, M., Bokhorst, J., Nierop, D., Bloem, J., Schouten, T., De Goede, R., and Brussaard, L. (2010). Ecosystem services in grassland associated with biotic and abiotic soil parameters. *Soil Biol. Biochem.* **42**, 1491–1504.

Van Veen, J.A., and Paul, E.A. (1979). Conversion of biovolume measurements of soil organisms, grown under various moisture tensions, to biomass and their nutrient content. *Appl. Environ. Microbiol.* **37**, 686–692.

Van Voorhies, W.A., Fuchs, J., and Thomas, S. (2005). The longevity of *Caenorhabditis elegans* in soil. *Biol. Lett.* **1**, 247–249.

Van Wezel, A.P., Van der Weijden, A.G.G., Van Wijnen, H.J., and Mulder, C. (2005). *Embedding soil quality in land-use planning*. RIVM Report 500025002, Bilthoven, The Netherlands.

Vanni, M.J., and De Ruiter, P.C. (1996). Detritus and nutrients in food webs. In: *Food Webs: Integration of Patterns and Dynamics* (Ed. by G.A. Polis and W. O. Winemiller), pp. 25–29. Chapman and Hall, New York.

Verbruggen, E., Röling, W.F.M., Gamper, H.A., Kowalchuk, G.A., Verhoef, H.A., and Van der Heijden, M.G.A. (2010). Positive effects of organic farming on belowground mutualists: Large-scale comparison of mycorrhizal fungal communities in agricultural soils. *New Phytol.* **186**, 968–979.

Vitousek, P.M., Mooney, H.A., Lubchenco, J., and Melillo, J.M. (1997). Human domination of Earth's ecosystems. *Science* **277**, 494–499.

Waide, R.B., Willig, M.R., Steiner, C.F., Mittelbach, G., Gough, L., Dodson, S.I., Juday, G.P., and Parmenter, R. (1999). The relationship between productivity and species richness. *Annu. Rev. Ecol. Syst.* **30**, 257–300.

Wall, D.H., Snelgrove, P.V.R., and Covich, A.P. (2001). Conservation priorities for soil and sediment invertebrates. In: *Conservation Biology: Research Priorities for the Next Decade* (Ed. by M.E. Soulé and G.H. Orians), pp. 99–123. Island Press, Society for Conservation Biology, Washington, DC.

Wardle, D.A. (1998). Controls of temporal variability of the soil microbial biomass: A global-scale synthesis. *Soil Biol. Biochem.* **30**, 1627–1637.

Wardle, D.A. (1999). How soil food webs make plants grow. *Trends Ecol. Evol.* **14**, 418–420.

Wardle, D.A. (2002). *Communities and Ecosystems: Linking the Aboveground and Belowground Components*. Princeton University Press, Princeton, NJ.

Wardle, D.A., and Bardgett, R.D. (2004). Human-induced changes in large herbivorous mammal density: The consequences for decomposers. *Front. Ecol. Environ.* **2**, 145–153.

Wardle, D.A., Bardgett, R.D., Klironomos, J.N., Setälä, H., Van der Putten, W.H., and Wall, D.H. (2004). Ecological linkages between aboveground and belowground biota. *Science* **304**, 1629–1633.

Wassen, M.J., Olde Venterink, H., Lapshina, E.D., and Tanneberger, F. (2005). Endangered plants persist under phosphorus limitation. *Nature* **437**, 547–550.

WCED, World Commission on Environment and Development (1987). *Our Common Future*. Oxford University Press, New York.

Werner, E.E., and Gilliam, J.F. (1984). The ontogenetic niche and species interactions in size structured populations. *Annu. Rev. Ecol. Syst.* **15**, 393–425.

White, C.R., Cassey, P., and Blackburn, T.M. (2007). Allometric exponents do not support a universal metabolic allometry. *Ecology* **88**, 315–323.

Whitlock, M.C., McPeek, M.A., Rausher, M.D., Rieseberg, L., and Moore, A.J. (2010). Data archiving. *Am. Nat.* **175**, 1–2.

Whittaker, R.H. (1975). *Communities and Ecosystems*. Macmillan, New York.

Whittaker, R.J. (2010a). Meta-analyses and mega-mistakes: Calling time on meta-analysis of the species richness–productivity relationship. *Ecology* **91**, 2522–2533.

Whittaker, R.J. (2010b). In the dragon's den: A response to the meta-analysis forum contributions. *Ecology* **91**, 2568–2571.

Williams, R.J. (2008). Effects of network and dynamical model structure on species persistence in large model food webs. *Theor. Ecol.* **1**, 141–151.

Wolf, N., Carleton, S.A., and Martínez Del Rio, C. (2009). Ten years of experimental animal isotopic ecology. *Funct. Ecol.* **23**, 17–26.

Woodward, G., Ebenman, B., Emmerson, M., Montoya, J.M., Olesen, J.M., Valido, A., and Warren, P.H. (2005a). Body size in ecological networks. *Trends Ecol. Evol.* **20**, 402–409.

Woodward, G., Ebenman, B., Emmerson, M.C., Montoya, J.M., Olesen, J.M., Valido, A., and Warren, P.H. (2005b). Body-size determinants of the structure and dynamics of ecological networks: Scaling from the individual to the ecosystem. In: *Dynamic Food Webs: Multispecies Assemblages, Ecosystem Development, and Environmental Change* (Ed. by P.C. De Ruiter, V. Wolters and J.C. Moore), pp. 179–197. Academic Press, San Diego, CA.

Woodward, G., Benstead, J.P., Beveridge, O.S., Blanchard, J., Brey, T., Brown, L., Cross, W.F., Friberg, N., Ings, T.C., Jacob, U., Jennings, S., Ledger, M.E., *et al.* (2010a). Ecological networks in a changing climate. *Adv. Ecol. Res.* **42**, 72–138.

Woodward, G., Blanchard, J., Edwards, F.K., Jones, J.I., Figueroa, D., Warren, P. H., and Petchey, O.L. (2010b). Individual-based food webs: Species identity, body size and sampling effects. *Adv. Ecol. Res.* **43**, 211–266.

Yachi, S., and Loreau, M. (1999). Biodiversity and ecosystem productivity in a fluctuating environment: The insurance hypothesis. *Proc. Natl. Acad. Sci. USA.* **96**, 1463–1468.

Yeates, G.W. (2010). Nematodes in ecological webs. *Encyclopedia of Life Sciences.* 10.1002/9780470015902.a0021913. Wiley, Chichester.

Yeates, G.W., and Foissner, W. (1995). Testate amoebae as predators of nematodes. *Biol. Fertil. Soils* **20**, 1–7.

Yeates, G.W., Bongers, T., De Goede, R.G.M., Freckmann, D.W., and Georgieva, S.S. (1993). Feeding habits in nematode families and genera. An outline for soil ecologists. *J. Nematol.* **25**, 315–331.

Yodzis, P. (1982). The compartmentation of real and assembled ecosystems. *Am. Nat.* **120**, 551–570.

Yodzis, P., and Innes, S. (1992). Body size and consumer-resource dynamics. *Am. Nat.* **139**, 1151–1175.

Yoon, I., Williams, R.J., Levine, E., Yoon, S., Dunne, J.A., and Martinez, N.D. (2004). Webs on the Web (WoW): 3D visualization of ecological networks on the WWW for collaborative research and education. In: *Proceedings of the IS&T/SPIE Symposium on Electronic Imaging, Visualization and Data Analysis Section*, pp. 124–132.

Yoon, I., Williams, R.J., Yoon, S., Dunne, J.A., and Martinez, N.D. (2005). Interactive 3D visualization of highly connected ecological networks on the WWW. In: *ACM Symposium on Applied Computing (SAC 2005), Multimedia and Visualization Section*, pp. 1207–1217.

Zwart, K.B., Kuikman, P.J., and Van Veen, J.A. (1994). Rhizosphere protozoa: Their significance in nutrient dynamics. In: *Soil Protozoa* (Ed. by J.F. Darbyshire), pp. 93–121. CAB International, Wallingford.

# Index

# Advances in Ecological Research Volume 1–44

# Cumulative List of Titles